T3-BWY-615

Cholinesterases

Structure, Function, Mechanism, Genetics, and Cell Biology

Cholinesterases

Structure, Function, Mechanism, Genetics, and Cell Biology

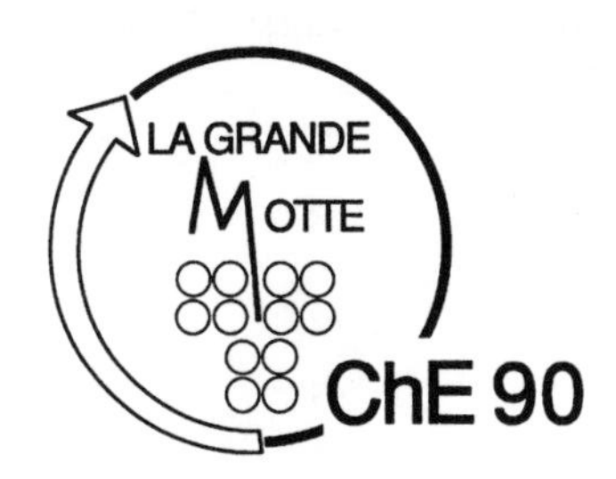

EDITED BY

Jean Massoulié
Centre National de la Recherche Scientifique

Francis Bacou
Institut National de la Recherche Agronomique

Eric Barnard
Medical Research Council

Arnaud Chatonnet
Institut National de la Recherche Agronomique

Bhupendra P. Doctor
Walter Reed Army Institute of Research

Daniel M. Quinn
University of Iowa

Proceedings of the Third International Meeting on Cholinesterases, La Grande-Motte, France, May 12–16, 1990

American Chemical Society, Washington, DC 1991

Library of Congress Cataloging-in-Publication Data

International Meeting on Cholinesterases (3rd: 1990: La Grande Motte, France)

Cholinesterases: structure, function, mechanism, genetics, and cell biology / Jean Massoulié, editor ... [et al.].

p. cm.—(Conference proceedings series)

Includes index.

"Proceedings of the Third International Meeting on Cholinesterases, La Grande-Motte, France, May 12–16, 1990."

ISBN 0–8412–2008–5: $89.95

1. Cholinesterases—Congresses. 2. Cholinesterase genes—Congresses. 3. Cholinesterase inhibitors—Congresses.

I. Massoulié, Jean, 1938– . II. Title. III. Series: Conference proceedings series (American Chemical Society)

QP609.C4I58 1990
599'.019'253—dc20

91–12371
CIP

The paper used in this publication meets the minimum requirements of American National Standard for Information Sciences—Permanence of Paper for Printed Library Materials, ANSI Z39.48–1984. ♾

PRINTED IN THE UNITED STATES OF AMERICA

ACS Conference Proceedings Series

M. Joan Comstock, *Series Editor*

ACKNOWLEDGEMENTS
REMERCIEMENTS

Support from the following Associations and Organisms is acknowledged with gratitude:

Les Associations et Organismes suivants sont particulièrement remerciés pour leur soutien à ce Congrès:

- Association Française contre les Myopathies
- Centre National d'Etudes Spatiales
- Centre National de la Recherche Scientifique
- Institut National de la Recherche Agronomique
- Institut National de la Santé et de la Recherche Médicale
- International Society for Neurochemistry
- International Union of Biochemistry
- Région Languedoc-Roussillon

And the following Companies:

Ainsi que les Sociétés:

- AB Draco
- FIDIA s.p.a.
- Kontron Instruments S.A.
- OSI
- Procida/Roussel Uclaf
- Sandoz

and the Municipality of La Grande-Motte
ainsi que la Municipalité de La Grande-Motte

CENTRE NATIONAL DE LA RECHERCHE SCIENTIFIQUE

Contents

CELLULAR BIOLOGY OF CHOLINESTERASES

GENE STRUCTURE AND EXPRESSION OF CHOLINESTERASES

CATALYTIC MECHANISM OF CHOLINESTERASES: STRUCTURE–FUNCTION RELATIONSHIPS OF ANTICHOLINESTERASE AGENTS, NERVE AGENTS, AND PESTICIDES

PHARMACOLOGICAL UTILIZATION OF ANTICHOLINESTERASE AGENTS. NEUROPATHOLOGY OF CHOLINERGIC SYSTEMS

NONCHOLINERGIC ROLES OF CHOLINESTERASES

APPENDIX

INDEX

Preface

"ACETYLCHOLINESTERASE NEVER CEASED TO AMAZE, EXCITE, OR CHARM US, with its wide ramifications, unexpected roles, ... strange forms, and complex inhibition," according to M. Brzin, E. A. Barnard, and D. Sket.[1] "We doubt that any other enzyme has so large an international following." These words described perfectly the atmosphere at the meeting on which this book is based. In fact, the charm of acetylcholinesterase (AChE) is becoming more powerful as we learn more about it. This meeting witnessed important advances resulting from new concepts and methodologies, such as monoclonal antibodies and molecular genetics. We will cite just a few examples.

The catalytic mechanism and molecular structure of cholinesterases (ChEs) have been investigated by different methods. It is impossible to identify the elements of a classic Asp–His–Ser triad by comparison with well-investigated proteases, such as trypsin, because there is no homology between the sequences of ChEs and these enzymes except for the immediate environment of the serine residue. Analyses of the effect of D_2O–H_2O on the kinetics of ChEs cast a serious doubt on the very existence of a triad mechanism.

In any case, modifications engineered by site-directed mutagenesis have already revealed the importance of a specific histidine residue for the catalytic mechanism of ChEs. However, a monoclonal antibody that blocks activity without blocking the reaction of organophosphorus inhibitors with the active site serine will certainly help to clarify the catalytic mechanism. In other words, monoclonal antibodies capable of modifying enzyme activity toward substrates and inhibitors will help to clarify the catalytic mechanisms and to identify protein sequences responsible for specificity of ChEs toward their substrates and inhibitors. The analysis of natural mutants, such as AChE from pesticide-resistant insect strains and human butyrylcholinesterase (BuChE) variants, has already provided some interesting hints.

Comparisons of ChE sequences will also suggest parts of the molecules that may be important for the specificity of the enzymes. Chemical approaches, using affinity ligands that can be made photoactive, for example, will point to some of the residues involved in the active and peripheral sites. Finally, it will be possible to confirm such identifications by site-directed mutagenesis and by constructing chimeric molecules by combining parts of different cholinesterases, for example, AChE and BuChE. The final elucidation of the catalytic mechanism (existence of a triad?) and of the relationship between the peripheral and catalytic sites will require a combination of

[1]Brzin, M.; Barnard, E. A.; Sket, D. *Cholinesterases: Fundamental and Applied Aspects*, Walter de Gruyter, Berlin, Germany, 1984.

these methods with structural analyses, notably X-ray crystallography. Crystals of a solubilized, lytic form of dimeric AChE from the genus *Torpedo* have already yielded diffraction data.

One of the most intriguing aspects of ChEs is the multiplicity of their molecular forms. The origin and extent of this polymorphism are now at least partly understood: Several catalytic subunits may be produced by alternative splicing, modified posttranslationally, and combined with other structural elements. The cloning of the collagenic subunit, which characterizes the structure of the asymmetric forms of vertebrate ChEs, was reported at the meeting. It will thus become possible to investigate the interactions that associate the catalytic and collagenic subunits and to ask whether these subunits are unique to collagen-tailed molecular forms of ChEs.

Some complex asymmetric forms incorporate two kinds of globular subunits, associated with the collagenlike tail (such as AChE from *Torpedo*, which contains a noncatalytic 100-kDa subunit) and the hybrid AChE–BuChE asymmetric forms from avian and mammalian embryos. The existence of such molecules suggests highly conserved quaternary interactions.

Studies of the biosynthesis of ChEs have raised several interesting problems. For example, although the enzyme is first synthesized as an inactive precursor, the nature of the maturation process through which it acquires activity is still mysterious. The major part of newly synthesized AChE, at least in murine and avian cell cultures, is degraded intracellularly, either as inactive protein or as active monomers or dimers, and the reason for this is not understood. In fact, it was reported at the meeting that normal tissues, in vivo, may contain a very important fraction of inactive AChE.

Another question concerns the nature of monomeric and dimeric forms of AChE that possess a hydrophobic domain apparently distinct from the glycolipid anchors of some AChE dimers and from the hydrophobic subunit of membrane-bound tetramers. Transfection experiments suggest that such forms may be produced from subunits that generate asymmetric forms, soluble tetramers, and probably membrane-bound tetramers, and that do not contain a hydrophobic peptide. The origin and nature of their hydrophobic domain are therefore puzzling.

The expression of ChEs is of major interest. The influence of various conditions, both in culture and in vivo, on their synthesis and on the distributions of their molecular forms has been investigated. It is remarkable that various modes of exercise elicit distinct patterns of molecular forms in muscles; in particular, the expression of G_4 appears to be specifically regulated, a result suggesting that this form may play an important physiological role in neuromuscular junctions, together with asymmetric forms.

An anomalous expression of both AChE and BuChE, sometimes accompanied by gene amplification, has been reported in various types of carcinomas. The relationship between this expression and oncogenicity is intriguing. On the other hand, AChE, as a part of cholinergic function, may constitute a prime indicator of various pathological disorders, among which is the senile dementia of Alzheimer's type. A very interesting report demonstrated that a particular monoclonal antibody directed against rabbit AChE could penetrate the central nervous system and was able to induce a permanent loss of cholinergic neurons. This type of antibody may be a powerful tool for the exploration of central cholinergic systems.

The possibility that ChEs may fulfill other functions than inactivation of acetylcholine at cholinergic synapses has been put on firm ground by studies of their expression in embryos, of their influence on the activity of neurons in situ, and of the growth of neurites in culture. Additional catalytic functions have been controversial, whereas the arylacylamidase activity is based on the same catalytic site as the cholinesterasic activity.

The possibility that ChEs may possess noncatalytic functions has been suggested because of their homology with noncatalytic proteins, including an adhesion protein from *Drosophila*. ChE-like proteins constitute an ancient family representatives of which are found in the slime mold.

On the whole, this volume contains a wealth of new information and opens a number of avenues for future research. The combination of experiments on the structure, pharmacology, catalytic mechanisms, cell biology, and molecular genetics that was described at the meeting attests to the excitement and vigor that pervade ChE research. The outlook for future major advances is optimistic.

JEAN MASSOULIÉ
Centre National de la Recherche Scientifique
Paris, Cedex 05, France

FRANCIS BACOU
Institut National de la Recherche Agronomique
34060 Montpellier Cedex 1, France

ERIC BARNARD
Medical Research Council
Molecular Neurobiology Unit
Cambridge, England CB2 2QH

ARNAUD CHATONNET
Institut National de la Recherche Agronomique
34060 Montpellier Cedex 1, France

BHUPENDRA P. DOCTOR
Walter Reed Army Institute of Research
Washington, DC 20307–5100

DANIEL M. QUINN
University of Iowa
Iowa City, IA 52242

November 6, 1990

These informal photographs were taken during a banquet at the 3rd International Meeting on Cholinesterases at La Grande-Motte, France, May 12–16, 1990.

Jean Massoulie'

Francis Bacou

left to right: Daniel Quinn, Andrea Quinn, Bhupendra Doctor, and Ellen Doctor

Eric Barnard

No photograph of Arnaud Chatonnet was available.

POLYMORPHISM AND STRUCTURE OF CHOLINESTERASES

THE STRUCTURE AND SIGNIFICANCE OF MULTIPLE CHOLINESTERASE FORMS

Jean Massoulié*, Suzanne Bon, Eric Krejci, Françoise Coussen, Nathalie Duval, Jean-Marc Chatel, Alain Anselmet, Claire Legay, and François-Marie Vallette, Laboratoire de Neurobiologie, E.N.S., 46, rue d'Ulm, 75230 Paris Cedex, France
Jacques Grassi, L.E.R.I., C.E.N. Saclay, 91191 Gif sur Yvette Cedex, France

Cholinesterases are highly polymorphic, as shown by a variety of analytical methods. Our group has been interested in this polymorphism for many years, and we attempted to unravel its complexity by designing simple analytical procedures, which allow an unambiguous classification of cholinesterase molecules (Massoulié and Bon, 1982; Massoulié and Toutant, 1988).

In this article, we first describe the stepwise characterization of these molecular forms. We then discuss the diversity of their subunits and the mechanisms which generate them. Finally, we show that cholinesterases may be considered as prototypes of a superfamily of proteins and suggest that they may exert structural as well as catalytic functions.

Classification of cholinesterase molecular forms: necessity of sequential operational criteria

1) Asymmetric, or collagen-tailed forms

Asymmetric forms are characterized by the presence of a collagenous element. Their main characteristics are a large Stokes radius, a specific sensitivity to collagenase, and a capacity to aggregate in low salt conditions in the presence of polyanions. The asymmetric forms consist of either one, two or three catalytic tetramers linked to the subunits of the triple-helical collagenic tail. These forms are noted respectively as A_4, A_8 and A_{12} and sediment around 9-10 S, 12-14 S and 16-21 S, depending on the species. They may be included in the extracellular matrix (basal lamina), through ionic interactions.

2) Amphiphilic and non-amphiphilic globular forms

Globular (G) forms do not possess a collagenic tail, and as a first approximation correspond to monomers (G_1), dimers (G_2) and tetramers (G_4) of the catalytic subunits.

Globular forms may be operationally subdivided into amphiphilic and non-amphiphilic forms: the amphiphilic forms are defined by their capacity to bind micelles of non denaturing detergents. These interactions are demonstrated by sedimentation shifts, increase in Stokes radius, or changes in electrophoretic migration under non denaturing conditions (Massoulié et al., 1988).

It has to be emphasized that the distinction between amphiphilic and non amphiphilic molecules does not coincide with fractionation based on solubility in aqueous and detergent buffers. Amphiphilic forms may frequently be solubilized, at least partially, without detergents, e.g. in the mammalian CNS (Grassi et al., 1982), in chicken or rabbit muscles (Bon et al., 1991). An amphiphilic G_1 form has even been found to be secreted in cultures of T_{28} murine cells (Lazar et al., 1984).

3) Hydrophobic domains of amphiphilic forms

The capacity of amphiphilic forms to associate with detergent micelles implies the presence of a hydrophobic domain. Two types of amphiphilic domains have been identified. The membrane-bound G_4 form of AChE from mammalian CNS contains a 20 kDa hydrophobic subunit (Gennari et al., 1987; Inestrosa et al., 1987). The organiza-

0–8412–2008–5/91/0002$06.00/0

tion of this molecule appears very similar to that of asymmetric forms, so that it may be called "hydrophobic-tailed".

A number of amphiphilic AChE dimers, notably from *Drosophila*, from *Torpedo* electric organs and from mammalian erythrocytes have been shown to possess a glycophosphatidylinositol anchor, linked through an ethanolamine to the C-terminal residue of the catalytic subunit (reviewed in Silman and Futerman, 1988).

We recently analyzed the amphiphilic character of globular AChE and BuChE forms in *Torpedo* tissues (Bon et al., 1988 a, b). We distinguish two types of amphiphilic G_2 forms. Type I corresponds to glycolipid-anchored dimers of AChE, occurring not only in electric organs, but also in nerves, muscles, erythrocytes, etc...

Type II dimers of AChE occur in soluble form in the plasma and are also clearly amphiphilic. In contrast with type I dimers, they are insensitive to specific phospholipases (PI-PLC) and never aggregate in the absence of detergent, but only show a limited shift in sedimentation. BuChE dimers and monomers from *Torpedo* heart resemble type II AChE dimers, as well as G_2 and G_1 amphiphilic AChE obtained from chicken and rabbit muscles, cultures of T_{28} cells, a murine neural cell line (Bon et al., 1991). We suggest that these molecules possess a novel kind of hydrophobic domain.

4) Glycosylation and electrophoretic variants

The glycolipid-anchored G_2 forms of AChE obtained from *Torpedo* tissues differ in their electrophoretic mobility in non-denaturing conditions (Bon et al., 1988b). An analysis of their interactions with lectins suggests that these differences are due to glycosylation, which therefore seems to impose a tissue-specific label on otherwise identical molecules. We also found a difference in the lectin binding characteristics of bovine G_2 AChE from lymphocyte and erythrocyte membranes (Méflah et al., 1984). In this case, the two enzymes also differ in their K_m, suggesting that the carbohydrate chains may modulate the binding of substrate to the active site.

Biosynthesis of AChE molecular forms

Molecular genetics studies of *Torpedo* AChE showed that a single gene produces all the molecular forms (Sikorav et al., 1987; Maulet et al., 1990). Alternative splicing generates multiple transcripts (Sikorav et al., 1987, 1988; Schumacher et al., 1988). Variation in the 5' untranslated region suggests the existence of two origins of replication (Sikorav et al., 1987). Alternative splicing of the 3' part of the coding sequence generates several C-terminal peptides.

The precursors of the catalytic subunits thus consist of a signal peptide, a common catalytic domain (535 aminoacids) and divergent C-terminal peptides, generated by alternative splicing. The C-terminal peptide (40 aminoacids) is maintained in the mature A forms. In the case of the glycolipid-anchored form, most of the C-terminal peptide (29 aminoacids) is cleaved and exchanged for the ethanolamine-glycophosphatidylinositol, leaving only Ala-Cys from the divergent peptide (Gibney et al., 1988).

Analyses of cDNA clones and S1 protection experiments suggested the existence of a third, minor, type of mRNA, diverging downstream from the same position of the A and G coding sequences (Sikorav et al., 1988).

The C-terminal peptides of the subunits engaged in A forms of *Torpedo* AChE and in the soluble G_4 form of human BuChE are clearly homologous. In addition, the quaternary structures of the collagen-tailed A forms and of the amphiphilic "hydrophobic-tailed" G_4 form appear very similar. These observations suggest strongly that all these molecules incorporate the same type of catalytic subunits. It is intriguing that the same A subunits, although they do not possess a hydrophobic domain, also produce amphiphilic monomers and dimers (N. Duval et al., in preparation).

The possible correspondence between precursors of *Torpedo* AChE produced through alternative splicing and mature cholinesterase forms is illustrated in Figure 1.

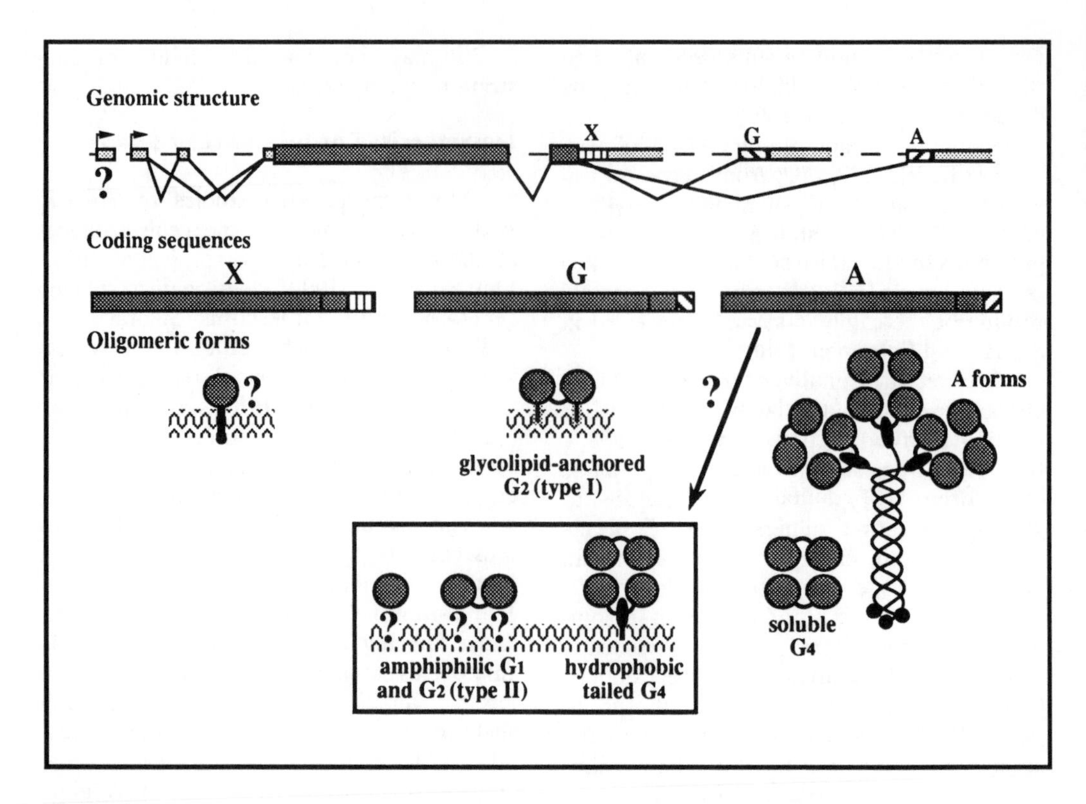

Figure 1

Alternative splicing generates several AChE catalytic subunits, associated into multiple oligomeric forms

In *Torpedo*, we observed three types of cDNAs and coding sequences. In the X sequence, the divergent 3' region is contiguous with the last common exon on the genome. The putative subunit encoded by this sequence has not been identified but may possess a transmembrane peptide. In the oligomers, the catalytic subunits are symbolized by shaded circles. They are linked by disulfide bonds as dimers, or to non catalytic subunits. The collagenic subunits of the A forms are represented with globular domains at both extremities (black ovals), as suggested by the predicted primary sequence, deduced from tryptic peptides and cDNA sequence. The amphiphilic monomers (G_1) and dimers of type II (G_2) are most likely derived from the A precursor, as well as the "hydrophobic-tailed" tetramers. The similarity of the latter form with collagen-tailed forms suggests that part of the hydrophobic 20 kDa subunit may present homology with a globular domain, through which the collagenic subunits associate with the catalytic subunits.

Associated non-catalytic elements: cloning of a collagenic subunit

We recently cloned and sequenced a collagenic subunit of the A forms of *Torpedo* AChE (Krejci et al., 1991). We identified the collagenic subunits from purified A forms by their sensitivity to collagenase, and determined partial aminoacid sequences of several tryptic peptides. One of these sequences was used for PCR amplification of a cDNA fragment, which then served as a probe to isolate cDNA clones. These clones predict a coding sequence of 471 aminoacids. This includes a signal peptide, a proline-rich domain, a collagenic domain of about 190 aminoacids, with the characteristic repetition of Gly every three residues, and a C-terminal, Cys-containing domain. This domain is entirely maintained in the mature subunit, as indicated by the sequenced peptides.

Comparison of several cDNA clones indicates the presence of variations, due to alternative splicing, in particular in the C-terminal domain. The coexistence of multiple variants in the mature A forms is consistent with the observed peptidic sequences.

It is of course tempting to imagine that the domain of the collagenic subunit which interacts with the catalytic subunits in A forms is identical or similar to the corresponding domain in the 20 kDa subunits of "hydrophobic-tailed" G_4 forms.

The cholinesterase-like domain in other proteins

The common domain of cholinesterases may be considered as the prototype of a new family of protein motifs, involved in both catalytic and structural functions (E. Krejci, P. Vincens, A. Chatonnet, and J. Massoulié, in preparation).

A number of proteins have now been found to present clear homology with cholinesterases. This includes esterases, in which the active serine is present, and proteins which lack this serine and are not known to possess any catalytic activity. Two *Drosophila* proteins, glutactin (Olson et al., 1990) and neurotactin (De la Escalera et al., 1990) belong to the latter group. Glutactin is a basal lamina component, and neurotactin is a cell surface component. Evidence is growing that these proteins are involved in cellular interactions (Barthalay et al., 1990).

The cholinesterase-like domain may have been used in evolution to fulfil both catalytic and adhesion functions. It appeared very early, as shown by the presence of two esterases of this family in the slime mould, *Dictyostelium discoideum*, which diverged from animal lineages more than a thousand million years ago.

It seems possible that cholinesterases themselves are thus involved in adhesion phenomena. A subset of AChE from *Electrophorus* and *Torpedo* electric organs carries the Elec-39/HNK-1 glycanic epitope (Bon et al., 1987). This suggests that these molecules are engaged in structural functions, perhaps related to the formation and/or stability of the cholinergic synapse, since the epitope has been considered a hallmark of adhesive proteins, and seems to be responsible for such interactions (Keilhauer et al., 1985).

Conclusion

A considerable variety of molecular species of cholinesterases is generated from a single gene, by alternative splicing of the transcripts and post-translational modifications of the precursors. These modifications include replacement of a C-terminal peptide with a glycolipid anchor, formation of homo- and hetero-oligomers with non structural subunits and tissue-specific glycosylation. These processes are probably designed mainly to produce molecules which may be secreted or anchored at strategic positions in cholinergic synapses, i.e. in plasma membranes and in the extracellular matrix. Glycosylation may modulate catalytic activity and participate in protein-protein interactions.

In addition to the hydrolysis of acetylcholine, cholinesterases, like other proteins possessing a cholinesterase-like domain, may present specific adhesive or recognition functions, possibly explaining their expression in non cholinergic contexts.

Acknowledgements

We thank Lisa Oliver for critical reading of this manuscript. This research was supported by the Centre National de la Recherche Scientifique, the Direction des Recherches et Etudes techniques, the Muscular Dystrophy Association of America and the Association Française contre les Myopathies.

References

Barthalay, Y., Hipeau-Jacquotte, R., De la Escalera, S., Jiménez, F. and Piovant, M., *EMBO J.* **1990,** *9* 3603-3609.

Bon, S., Méflah, K., Musset, F., Grassi, J. and Massoulié, J., *J. Neurochem.* **1987,** *49,* 1720-1731.

Bon, S., Toutant, J.P., Méflah, K. and Massoulié, J., *J. Neurochem.* **1988a,** *51,* 776-784.

Bon, S., Toutant, J.P., Méflah, K. and Massoulié, J., *J. Neurochem.* **1988b,** *51,* 786-794.

Bon, S., Rosenberry, T.L. and Massoulié, J., *Cell. and Mol. Neurobiol.* **1991,** *11,* 157-172.

De la Escalera, S., Bockamp, E.-O., Moya, F., Piovant, M. and Jiménez, F., *EMBO J.* **1990,** *9,* 3593-3601.

Gennari, K., Brunner, J. and Brodbeck, U., *J. Neurochem.* **1987,** *49,* 12-18.

Gibney, G., MacPhee-Quigley, K., Thompson, B., Vedvick, T., Low, M., Taylor, S.S. and Taylor, P., *J. Biol. Chem.* **1988,** *263,* 1140-1145.

Grassi, J., Vigny, M. and Massoulié, J., *J. Neurochem.* **1982,** *38,* 457-469.

Inestrosa, N.C., Roberts, W.L., Marshall, T.L. and Rosenberry, T.L., *J. Biol. Chem.* **1987,** *262,* 4441-4444.

Keilhauer, G., Faissner, A. and Schachner, M., *Nature* **1985,** *316,* 728-730.

Krejci, E., Coussen, F., Duval, N., Chatel, J.-M., Legay, C., Puype, M., Vandekerckhove, J., Cartaud, J., Bon, S. and Massoulié, J., *EMBO J.* **1990** (in press).

Lazar, M., Salmeron, E., Vigny, M. and Massoulié, J., *J. Biol. Chem.* **1984,** *259,* 3703-3713.

Massoulié, J. and Bon, S., *Annu. Rev. Neurosci.* **1982,** *5,* 57-102.

Massoulié, J. and Toutant, J.P., *Handbook of Exp. Pharm.* **1988,** *86,* 67-224.

Massoulié, J., Toutant, J.P. and Bon, S., Post-translational modifications of proteins by lipids (Brodbeck, U. and Bordier, C., eds.), Springer Verlag, 1988 pp. 132-142.

Maulet, Y., Camp, S., Gibney, G., Rachinsky, T.L., Ekström, T.J. and Taylor, P., *Neuron* **1990,** *4,* 289-301.

Méflah, K., Bernard, S. and Massoulié, J., *Biochimie* **1984,** *66,* 59-69.

Olson, P.F., Fessler, L.I., Nelson, R.E., Sterne, R.E., Campbell, A.G. and Fessler, J.H., *EMBO J.* **1990,** *9,* 1219-1227.

Schumacher, M., Maulet, Y., Camp, S. and Taylor, P., *J. Biol. Chem.* **1988,** *35,* 18879-18987.

Sikorav, J.L., Krejci, E. and Massoulié, J., *EMBO J.* **1987,** *6,* 1865-1873.

Sikorav, J.L., Duval, N., Anselmet, A., Bon, S., Krejci, E., Legay, C., Osterlund, M., Reimund, B. and Massoulié, J., *EMBO J.* **1988,** *7,* 2983-2993.

Silman, I. and Futerman, A.H., *Eur. J. Biochem.* **1987,** *170,* 11-22.

Structural Studies on Acetylcholinesterase from *Torpedo californica*

Joel Sussman, Michal Harel, Felix Frolow, Department of Structural Chemistry, The Weizmann Institute of Science, Rehovot 76100, Israel
Christian Oefner, Central Research Unit, F. Hoffmann-La Roche Co. Ltd., CH-4002 Basel, Switzerland
Lili Toker and Israel Silman*, Department of Neurobiology, The Weizmann Institute of Science, Rehovot 76100, Israel

The dimeric form of acetylcholinesterase from electric organ tissue of *Torpedo californica*, purified by affinity chromatography subsequent to solubilization with phosphatidylinositol-specific phospholipase C of bacterial origin, was crystallized from concentrated ammonium sulfate. Trigonal crystals were obtained which diffracted out to 2.8 Å. They belong to space group $P3_121$ or its enantiomorph, $P3_221$, with unit cell dimensions a=b=110.9 Å, c=136.9 Å. A putative heavy atom derivative was obtained by soaking the crystals in uranyl nitrate. The difference Patterson map for the uranyl derivative, calculated at 3.5 Å resolution, is interpretable in terms of 2 uranyl sites located in general positions in the asymmetric unit of the unit cell.

The principal biological role of acetylcholinesterase (AChE) is the termination of impulse transmission at cholinergic synapses by rapid hydrolysis of the neurotransmitter, acetylcholine (Barnard, 1974). AChE is, accordingly, characterized by a remarkably high specific activity, especially for a serine hydrolase (Quinn, 1987), functioning at a rate approaching that of a diffusion-controlled reaction (Bazelyansky *et al.*, 1986). The powerful acute toxicity of organophosphate and carbamate poisons is attributed primarily to their action as potent covalent inhibitors of AChE (Koelle, 1963). Elucidation of the three-dimensional structure of AChE is thus of fundamental interest in understanding its remarkable catalytic efficacy, and of great importance in the development of therapeutic approaches to organophosphorus poisoning. It may also be of value in the design of anticholinesterases for management of Alzheimer's disease (Becker and Giacobini, 1988).

The electric organs of the electric fish, *Torpedo* and *Electrophorus*, provide rich sources of AChE. These forms are structurally and functionally homologous to the corresponding forms in vertebrate nerve and muscle (Massoulié and Bon, 1982); highly purified preparations from these sources have, therefore, provided much of our knowledge concerning both the structure and function of AChE (Silman and Futerman, 1987; Quinn, 1987). AChE from *Torpedo californica* has been cloned (Schumacher *et al.*, 1986), and both its primary sequence and arrangement of disulfide bridges are known (MacPhee-Quigley *et al.*, 1986). In *Torpedo* electric organ, one of the principal forms of AChE is a disulphide-linked catalytic subunit dimer belonging to a recently described class of membrane proteins which are linked to the plasma membrane via covalently attached phosphatidylinositol (PI) (Low *et al.*, 1986). In such proteins, the PI is attached at the carboxyl-terminus of the polypeptide chain through an intervening oligosaccharide sequence which is added post-translationally (Ferguson and Williams, 1988). PI-anchored proteins can often be solubilized by a PI-specific phospholipase C of bacterial origin, which detaches the

0–8412–2008–5/91/0007$06.00/0

diglyceride moiety of the PI (Low *et* al., 1986), and this is also the case for the *Torpedo* AChE dimer (Futerman *et al.*, 1983).

The 11 S tetrameric form of AChE from *Electrophorus electricus* was first crystallized by Leuzinger and Baker (1967) and preliminary characterization of crystals obtained from this form was subsequently reported (Chothia and Leuzinger, 1975; Schrag *et al.*, 1988). Sequence information concerning *Electrophorus* AChE is not yet available. Furthermore, the 11 S tetramer possesses proteolytic 'nick' sites due to its mode of preparation (for a discussion, see Anglister and Silman, 1980), and the best crystal form reported by Schrag *et al.* (1988) diffracts anisotropically to between 3.5 and 6 Å, depending on direction. Therefore, we selected the dimer from *Torpedo* electric organ as a more suitable candidate for crystallization and for subsequent structural studies. For this purpose we adopted a novel and mild purification procedure, involving solubilization with PIPLC followed by affinity chromatography, to yield a highly purified water-soluble preparation of the *Torpedo* dimer which differed from the native enzyme only in the removal of the diglyceride moiety of the membrane anchor (Futerman *et al.*, 1985; Sussman *et al.*, 1988). In an earlier publication we reported that crystals of this *Torpedo* AChE preparation could be obtained from PEG200. These crystals diffracted well and were used to obtain preliminary crystallographic data (Sussman *et al.*, 1988).

Crystallization and Crystal Characterization

Recently, a trigonal crystal form was obtained using standard vapor-diffusion techniques in hanging drops at room temperature, with concentrated ammonium sulfate at pH 7.0 as the precipitant. Crystals routinely grow as flat pyramids up to ~1 mm in their longest and ~0.5 mm in their shortest dimension, over a period of several weeks after nucleation (Fig. 1). The crystals diffract to ~2.8 Å resolution, are stable for about 2 days in the X-ray beam at room temperature, and survive indefinitely at 90 K upon shock cooling according to Hope *et al.* (1987). They were ascribed to either space group $P3_121$ or $P3_112$ with cell dimensions a=b=110.9 Å and c=136.9 Å, as determined by X-ray studies at low temperature (90 K), using a Rigaku AFC5-R rotating-anode diffractometer.

In order to distinguish unambiguously between the space groups, $P3_121$ and $P3_112$, with their Laue symmetry P3m1 and P31m, respectively, one needs to examine symmetry related reflections, which has been done from zonal hk0 and hk1 precession photographs. The hk0 zone, showing hexagonal symmetry, does not allow one to distinguish between a twofold axis and a mirror plane, which are located between the axes of the primitive cell in $P3_121$ or $P3_112$, respectively. A twofold axis, present in the space group $P3_121$, will vanish in the hk1 zone, whereas a mirror plane, present in the space group $P3_112$, will remain, a fact which allows one to distinguish between the two space groups. The hk1 zone clearly indicates the loss of symmetry between the axes of the primitive cell, indicating that the symmetry element in the hk0 zone is a twofold axis. This confirms the space group as $P3_121$ or its enantiomorph, $P3_221$.

Data Collection and Processing

Data were collected on a Siemens/Xentronics area detector installed on a Rigaku-RU300 rotating anode generator. Data frames of 0.25° were collected, with exposure times of ~2 min, depending on the quality of the crystals (Fig. 2). The data were processed with the Xengen (Howard *et al.*, 1987) and XDS (Kabsch, 1988) software packages, and subsequent crystallographic calculations were performed using the CCP4 computing package, obtained from the Daresbury Laboratory, U.K.

A data set to 3.2 Å resolution was collected for a native AChE crystal with an overall R_{sym} of 0.106. Of ca. 20 compounds soaked into the AChE crystals, in order to produce heavy atom derivatives, one, namely uranyl nitrate, yielded a good derivative, which diffracted as well as the native crystal. A difference Patterson map, calculated at 3.5 Å resolution, is interpretable in terms of 2 uranyl sites located in general positions in the asymmetric unit of the unit cell. These sites have been confirmed by the anomalous Patterson map, as well as by both the automatic Patterson solution routine and the direct methods of SHELX-86. Figures 3a and 3b, respectively, show the Harker section at w=1/3 of the difference and anomalous Patterson maps. Heavy atom parameters were refined by using centric data to 3.5 Å. The overall R factor obtained is 43.6%. Phasing to 3.5 Å resolution, taking the isomorphous and anomalous contribution of the uranyl sites, gives a figure-of-merit of 0.71.

A search for additional heavy atom derivatives is in progress, in order to produce an interpretable MIR electron density map.

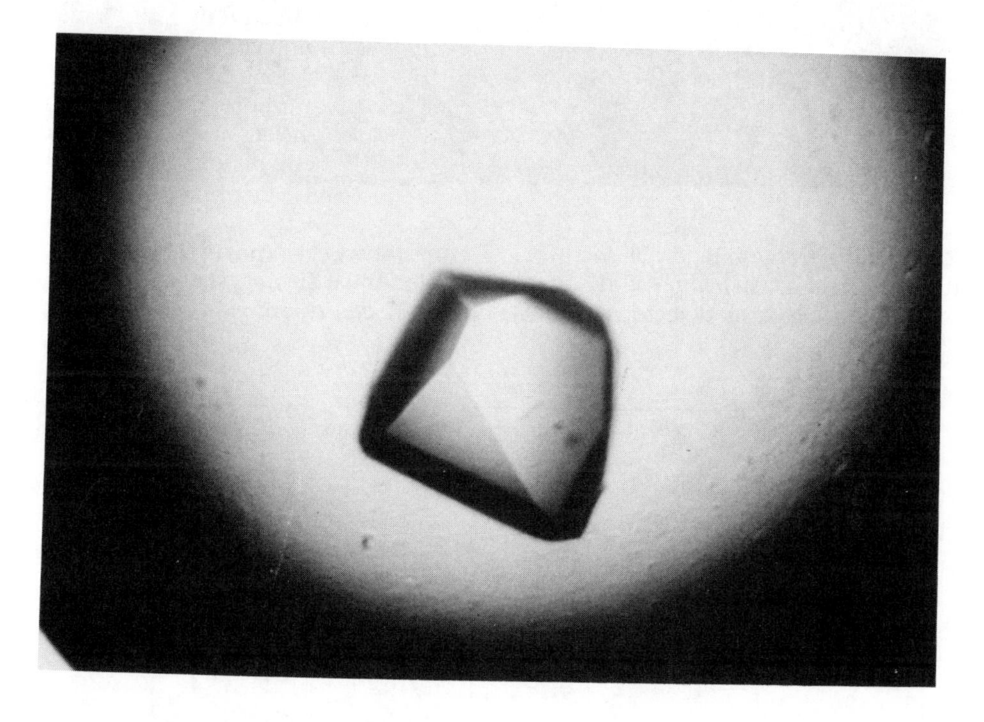

Fig. 1. Crystal of AChE from *Torpedo californica* obtained by precipitation from concentrated ammonium sulfate. The crystal shown is ~1 mm in its longest dimension.

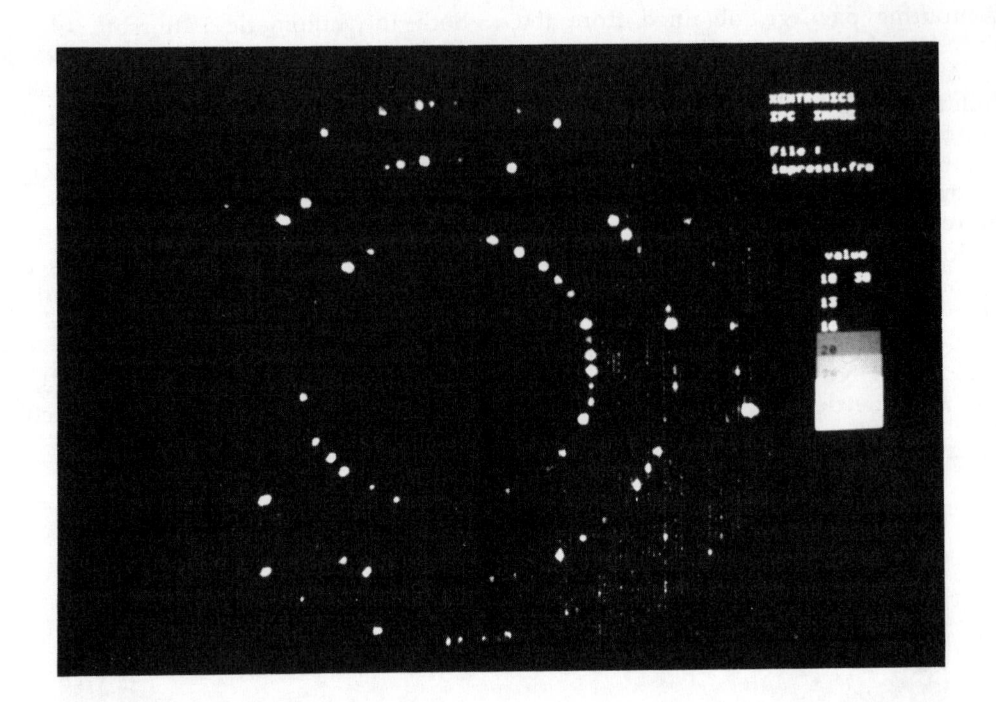

Fig. 2. A single frame image obtained for a native crystal of *Torpedo* AChE on a Siemens/Xentronics area detector at room temperature. Exposure time, 120 sec; oscillation, 0.25°; crystal-to-detector distance, 12 cm; 2θ=10°.

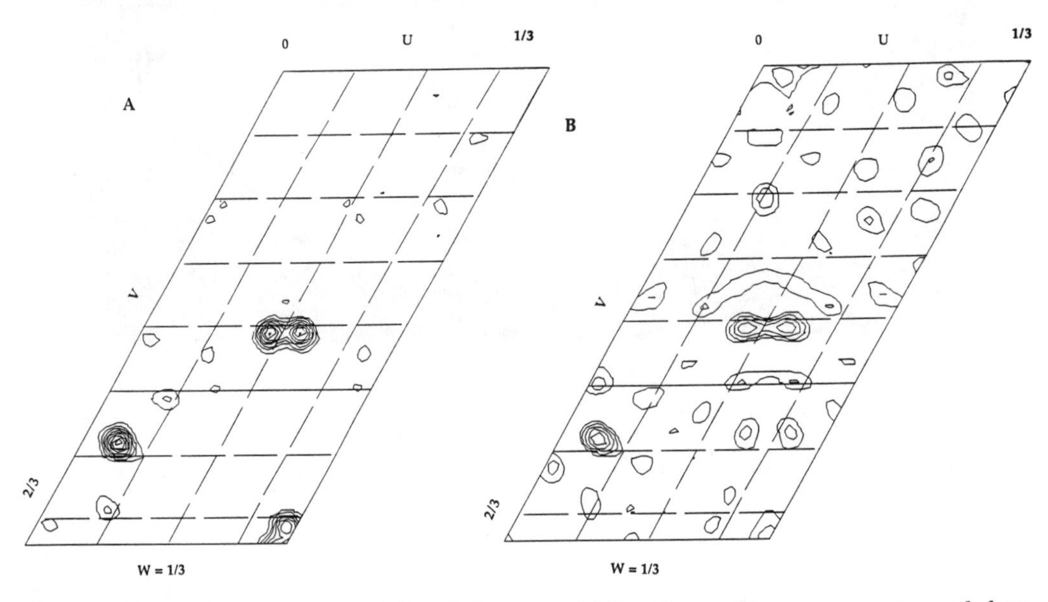

Fig. 3. Harker section at w=1/3 of the difference Patterson map (A), and the anomalous Patterson map (B), calculated at resolutions of 3.5 and 5 Å, respectively, for the $UO_2(NO_3)_2$ derivative of *Torpedo* AChE. The sections are contoured from 1σ, in equal intervals of 1σ, where σ is the standard deviation from the average density of the map.

Acknowledgments

This research was supported by Contract No. DAMD17-89-C-9063 from the United States Army Medical Research and Development Command, and by grants from the Charles Revson Foundation and from the Minerva Foundation.

References

Anglister, L., Silman, I., *J. Mol. Biol.*, **1978,** *125,* 293-311.

Barnard, E.A., In *The Peripheral Nervous System,* Hubbard, J.I., Ed., Plenum, New York, NY, 1974, pp. 201-224.

Bazelyansky, M., Robey, E.C., Kirsch, J.F., *Biochemistry,* **1986,** *25,* 125-130.

Becker, R.E., Giacobini, E., *Drug. Dev. Res.,* **1988,** *12,* 163-195.

Chothia, C., Leuzinger, W., *J. Mol. Biol.,* **1975,** *97,* 55-60.

Ferguson, M.A.J., Williams, A.F., *Ann. Rev. Biochem.,* **1988,** *57,* 285-320.

Futerman, A.H., Low, M.G., Silman, I., *Neurosci. Lett.,* **1983,** *40,* 85-89.

Futerman, A.H., Low, M.G., Ackermann, K.E., Sherman, W.R., Silman, I., *Biochem. Biophys. Res. Comm.,* **1985,** *129,* 312-317.

Hope, H., Frolow, F., Sussman, J.L., *Rigaku J.,* **1987,** *4,* 3-10.

Howard, A.J., Gilliland, G.L., Finzel, B.C., Poulos, T.L., Ohlendorf, D. H., Salemme, R.F., *J. Appl. Cryst.,* **1987,** *20,* 383-387.

Kabsch, W., *J. Appl. Cryst.,* **1988,** *21,* 916-924.

Koelle, G.B., *Handbuch der Experimentellen Pharmakologie,* Cholinesterases and Anticholinesterase Agents, Springer-Verlag, Berlin, 1963, Vol. XV.

Leuzinger, W., Baker, A.L., *Proc. Natl. Acad. Sci. USA,* **1967,** *57,* 446-451.

Low, M.G., Ferguson, M.A.J., Futerman, A.H., Silman, I., *Trends Biochem. Sci.,* **1986,** *11,* 211-214.

MacPhee-Quigley, K., Vedvick, T.S., Taylor, P., Taylor, S., *J. Biol. Chem.,* **1986,** *261,* 13565-13570.

Massoulié, J., Bon, S., *Ann. Rev. Neurosci.,* **1982,** *5,* 57-106.

Quinn, D.M., *Chem. Revs.,* **1987,** *87,* 955-979.

Schrag, J., Schmid, M.F., Morgan, D.G., Phillips, G.N., Jr., Chiu, W., Tang, L., *J. Biol. Chem.,* **1988,** *263,* 9795-9800.

Schumacher, M., Camp, S., Maulet, Y., Newton, M., MacPhee-Quigley, K., Taylor, S., Friedman, T., Taylor, P., *Nature,* **1986,** *319,* 407-409.

Silman, I., Futerman, A.H., *Eur. J. Biochem.,* **1987,** *170,* 11-22.

Sussman, J.L., Harel, M., Frolow, M., Varon, L., Toker, L., Futerman, A.H., Silman, I., *J. Mol. Biol.,* **1988,** *203,* 821-823.

Amphiphilic G_1 and G_2 Forms of Acetylcholinesterase: Sensitivity or Resistance to Phosphatidylinositol-specific Phospholipase C

Jean-Pierre Toutant, Physiologie animale, INRA, place Viala, 34060 Montpellier
Nicole R. Murray, Jennifer A. Krall, Michael K. Richards, Terrone L. Rosenberry, Department of Pharmacology, Case Western Reserve University, 44106 Cleveland, Ohio.

Globular forms of acetylcholinesterase (G_4, G_2 and G_1 forms) may be associated with plasma membranes or secreted outside the cell (review in Massoulié and Toutant, 1988). Membrane-bound forms are often referred to as **amphiphilic** because they contain a large hydrophilic domain, containing the active site of the enzyme and a small hydrophobic portion interacting with the membrane. These two domains are easily separated by mild proteolyic treatments. Secreted forms are usually **hydrophilic**.

The structure of devices mediating the membrane attachment of G_2 forms of AChE in certain tissues has recently been elucidated. In most amphiphilic G_2 forms studied (those of "class I", see Bon *et al.*, 1988a,b), the catalytic polypeptide possesses a glycolipid domain covalently bound to the C-terminal amino acid. The terminal lipid is responsible for membrane insertion (see Ferguson and Williams, 1988 and Low, 1989 for a review of glycolipid-anchored proteins). The membrane association of other G_2 forms (those of "class II", see Bon *et al.*, 1988a,b) might involve a sequence of hydrophobic amino acids and this might be also the case for G_1 forms in rabbit and chicken muscles (Bon *et al.*, 1990).

In the present study, the sensitivity or resistance of G_2 and G_1 forms to phosphatidylinositol-specific phospholipase C (EC 3.1.4.10) was used as indirect but useful information on the structure of the hydrophobic domain (see Rosenberry *et al.*, 1989). The amphiphilic dimers were extracted from *Drosophila* heads and mammalian erythrocyte membranes. We also report the properties of amphiphilic G_2 and G_1 forms of AChE produced in cultures of human erythroleukemia K562 cells.

Identification of Amphiphilic and Hydrophilic Forms

In the present paper the **amphiphilic** G_1 and G_2 forms are noted G_1a and G_2a (as in Bon *et al.*, 1988a,b). They can be converted to their **hydrophilic** counterparts noted G_1h and G_2h (instead of the previous G_1na and G_2na, Bon *et al.*, 1988a,b).

The interaction between the hydrophobic domain of amphiphilic forms and nondenaturing detergents in solution results in the formation of one detergent micelle on each hydrophobic "tail" (Rosenberry and Scoggin, 1984). This binding confers particular physical properties to amphiphilic molecules. For example, the migration of amphiphilic forms in nondenaturing electrophoreses (in the presence of neutral detergents such as Triton X100) is strikingly slower than their hydrophilic counterparts (Arpagaus and Toutant, 1985; Toutant, 1986). Fig. 1 shows this for AChE from *Drosophila* heads. In lane a, the major band corresponds to the G_2a form and three minor bands correspond to G_1a, G_2h and G_1h forms in the order of increasing mobilities (see Toutant *et al.*, 1988). A further characterization of the different components is shown in lane b where the disulfide bond in G_2 forms was reduced by dithiothreitol: the only bands remaining after reduction are G_1a and G_1h.

Sensitivity of G_2a Forms to PI-PLC

In most cases, AChE G_2a forms are anchored to the membrane via a glycosylphosphatidylinositol (Futerman *et al.*, 1985a; Rosenberry *et al.*, 1986; Roberts *et al.*, 1987). They correspond to the class I of *Torpedo* amphiphilic dimers (see Bon *et al.*, 1988b). In these forms the hydrophobic portion is confined to the terminal lipid moiety of the glycolipid domain which can be cleaved by bacterial PI-PLC (Low and Finean, 1977). Thus PI-PLC converts the amphiphilic AChE to a hydrophilic component. Such a conversion is nicely detected in the nondenaturing electrophoresis described above. Figure 2 shows the effects of PI-PLC on the migration of AChE G_2a forms extracted from *Drosophila* heads (A) or bovine erythrocyte membranes (B). In both cases we observed an almost complete conversion of G_2a to G_2h.

Resistance of G2a Forms to PI-PLC

Certain G2a forms of AChE may be resistant to PI-PLC digestion because they are associated to the membrane by another device than a glycolipid anchor, but resistance to PI-PLC also occurs in G2a forms that do possess a glycolipid anchor as in human erythrocyte (E^{hu}) AChE (Futerman *et al.*, 1985b; Roberts *et al.*, 1987). In this case, the resistance results from the acylation of the inositol ring in the glycolipid domain (Roberts *et al.*, 1988a, 1988b). We tested whether a deacylation with alkaline hydroxylamine renders E^{hu} G2a AChE sensitive to a further treatment with PI-PLC (Toutant *et al.*, 1989). We obtained a good conversion of G2a to G2h only in the case where the incubation with hydroxylamine preceded the PI-PLC treatment. The treatments had no effect if performed in the reverse order (Toutant *et al.*, 1989). We attributed the partial conversion obtained to a partial deacylation (the use of alkaline pH is limited by the inactivation of AChE). AChE in mouse erythrocyte membranes is also a G2a form that resists conversion to G2h with PI-PLC (Futerman *et al.*, 1985b and Fig. 3, lanes a and b). A deacylation with alkaline hydroxylamine (lane c) was sufficient to render the enzyme partially susceptible to PI-PLC (lane d). Thus an additional acylation of the glycolipid anchor in mouse erythrocyte AChE most likely explains the resistance to a direct digestion with PI-PLC.

AChE Molecular Forms in Human K562 Cells

The human erythroleukemia cell line K562 is considered as a model of erythrocyte differentiation. In particular, K562 cells synthesize glycolipid-anchored G2 forms of AChE comparable to those of erythrocytes. In two sublines of K562 cells, we found that G2a forms of AChE were either directly sensitive to PI-PLC digestion (subline K562-243) or resistant (K562-48). In the latter, a prior treatment with alkaline hydroxylamine was necessary to render the G2a forms sensitive (Toutant *et al.*, 1990a). This suggests that AChE in the two sublines differs in the acylation of the glycolipid domain.

In nondenaturing gels of AChE extracted from the two sublines, we observed a minor component migrating in front of the G2a form. This band was identified as a G1a form (see Fig. 4, lane a, for the subline K562-243). G1a forms were resistant to digestion with PI-PLC even in the subline 243 where G2a forms were directly sensitive (Fig. 4, lane c). In addition the electrophoretic migration of G1a forms was unaffected by a treatment with neuraminidase whereas G2a forms were retarded (lanes b and d). The slower migration of G2a forms after neuraminidase treatment is due to the release of terminal sialic acids from their glycoconjugates. The lack of effect of neuraminidase on G1a forms suggests that they might be intracellular precursors of G2a forms. In this hypothesis the resistance of G1a forms to PI-PLC might result either from the presence of the C-terminal peptidic hydrophobic extension predicted by the cDNAs of all glycolipid-anchored proteins (Ferguson and Williams, 1988) or to the presence of an acylated form of glycolipid anchors that should be deacylated with further maturation in the subline 243. Since alkaline hydroxylamine treatment inhibited G1a form, it was not yet possible to decide between these two possibilities (Toutant *et al.*, 1990a).

In a parallel investigation of the Decay Accelerating Factor (DAF) in the two sublines K562-48 and -243, we observed that DAF was PI-PLC-resistant in subline 48 and PI-PLC-sensitive in the subline 243. The PI-PLC-resistant form of DAF became sensitive after a treatment with alkaline hydroxylamine (Walter *et al.*, 1988). This indicates that the type of glycolipid synthesized is cell-specific (see discussion in Toutant *et al.*, 1990b).

Other Structural Features of Glycolipid Anchors of AChE.

Alkaline hydroxylamine

We observed that treatment of G2a forms of AChE from *Drosophila* heads or bovine erythrocyte membranes **by alkaline hydroxylamine alone** did not affect their migration in nondenaturing electrophoresis (Toutant *et al.*, 1990b). This means that the detergent micelle bound to each hydrophobic "tail" remained attached to the rest of the molecule by a hydroxylamine-resistant bond. In bovine and human AChE glycolipid anchors, it is known that there is an ether bond on the C1 of glycerol (O-alkyl chain) and an ester bond (acyl chain) on the C2 (Roberts *et al.*, 1988b,c). Therefore treatment by alkaline hydroxylamine alone allows to distinguish between anchors possessing a **diacylglycerol** (conversion into hydrophilic forms by alkaline hydroxylamine) and an **alkylacylglycerol** (no conversion). AChEs tested thus far (in human, bovine and mouse erythrocytes and in *Drosophila* heads) are all of the second category.

Phospholipases A1

Incubations in alkaline hydroxylamine (pH 10.7 to 11) as described above often result in inactivation of AChE. This precludes the use of this treatment when the AChE activity available is low. We searched for milder treatments that would bring the same information (Alkylacyl versus diacyl species in AChE anchors). Phospholipases A1 (for example from *Rhizopus arrhizus*) hydrolyze **ester** bonds on C1 **and** C2. As previously shown by El Tamer *et al.* (1984), the cleavage occurs first on C2 and then on C1. In the conditions used the cleavage of C1 acyl chains with *Rhizopus* lipase is often incomplete. When applied to cholinesterases, treatment with *Rhizopus* lipase can differentiate between molecules with a diacylglycerol from those possessing an alkylacylglycerol.

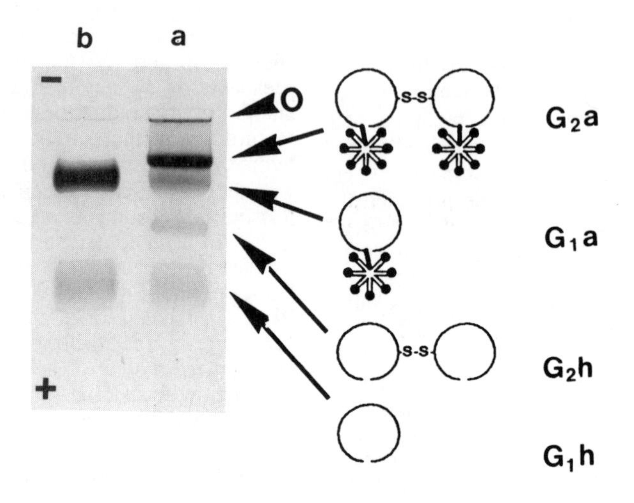

Fig.1. Nondenaturing electrophoresis of AChE components in a detergent extract of *Drosophila* heads and their schematic representation.
a: control sample. b: sample reduced by DTT
G_2a and G_1a: amphiphilic dimer and monomer
G_2h and G_1h: hydrophilic dimer and monomer
O: origin of migration
(adapted from Toutant *et al.*, 1990a).

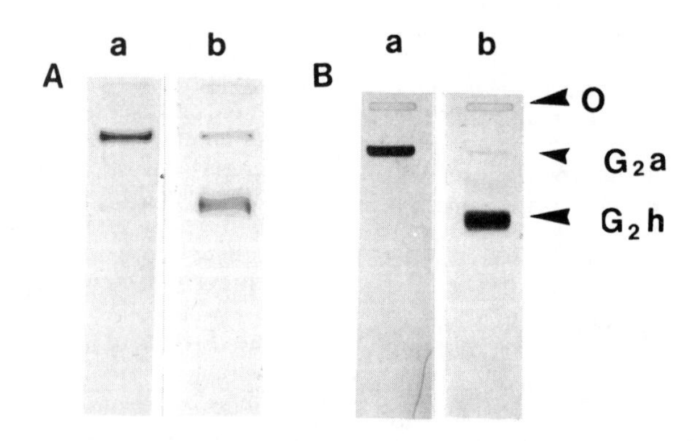

Fig.2. Effect of PI-PLC from *Bacillus thuringiensis* on AChE G_2a forms from *Drosophila* heads (A) and bovine erythrocyte membranes (B).
a: control samples. b: PI-PLC-treated samples (conversion to G_2h).
(Adapted from Toutant *et al.*, 1990b).

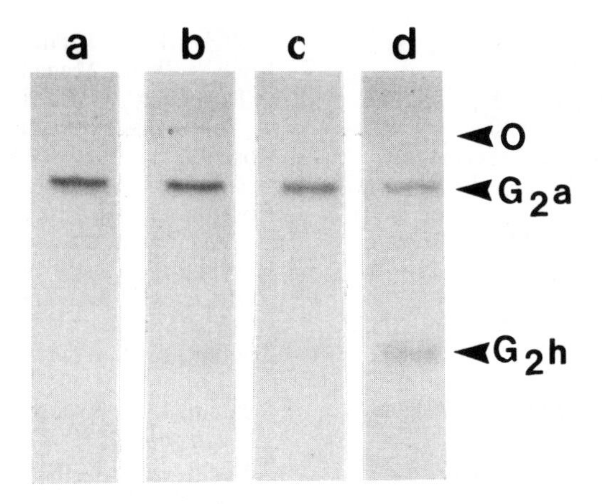

Fig.3. Effect of alkaline hydroxylamine on AChE G_2a form of mouse erythrocyte membranes.
a: control sample. b: sample treated with PI-PLC. c: treated with alkaline hydroxylamine alone. d: treated sequentially with hydroxylamine and PI-PLC.
(Adapted from Toutant *et al.*,1990b).

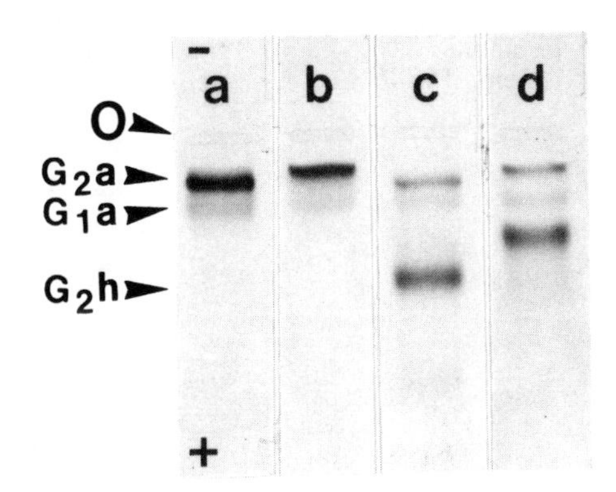

Fig.4. Molecular forms of AChE synthesized in cultures of the subline 243 of K562 cells.
a: control sample: G_2a and G_1a forms
b: treatment with neuraminidase (sialidase)
c: treatment with PI-PLC
d: treatment with PI-PLC **and** neuraminidase
Note the decrease of G_2a (and G2h) migration after neuraminidase in lanes b and d and the resistance of G_1a forms to both PI-PLC and neuraminidase treatments (adapted from Toutant *et al.*, 1990a).

Acknowledgements

This work was supported by grants from the "Institut National de la Recherche Agronomique" and the "Association Française contre les Myopathies" (to J.P.T.), from the National Institutes of Health and the Muscular Dystrophy Association (to T.L.R). N.R.M. was a recipient of an Eva L. Pancoast award, decerned by Case Western Reserve University and of an OECD fellowship.

References

Arpagaus, M., Toutant, J.-P., *Neurochem. Int.*, **1985,** *7*, 793-804.
Bon, S., Toutant, J.-P., Méflah, K., Massoulié, J., *J. Neurochem.*, **1988a,** *51*, 776-785.
Bon, S., Toutant, J.-P., Méflah, K., Massoulié, J., *J. Neurochem.*, **1988b,** *51*, 786-794.
Bon, S., Rosenberry, T.L., Massoulié, J., *Cell. Molec. Neurobiol.*, **1990**, in the press.
El Tamer, A., Record, M., Fauvel, J., Chap,

H., Douste-Blazy, L., *Biochim. Biophys. Acta*, **1984**, *793*, 213-220.
Ferguson, M.A.J., Williams, A.F., *Ann. Rev. Biochem.*, **1988**, *57*, 285-320.
Futerman, A.H., Low, M.G., Ackermann, K.E., Sherman, W.R., Silman, I., *Biophys. Res. Commun.*, **1985a**, *129*, 312-317.
Futerman, A.H., Low, M.G., Michaelson, D.M., Silman, I., *J. Neurochem.*, **1985b**, *45*, 1487-1494.
Inestrosa, N.C, Roberts, W.L., Marshall, T., Rosenberry, T.L., *J. Biol. Chem.*, **1987**, *262*, 4441-4444.
Low, M.G., *Biochim. Biophys. Acta*, **1989**, *988*, 427-454.
Low, M.G., Finean, J.B., *FEBS Lett.*, **1977**, *82*, 143-146.
Massoulié, J., Toutant, J.-P., *Handbook of Expl. Pharmacol.*, **1988**, *86*, 167-224.
Roberts, W.L., Kim, B.H., Rosenberry, T.L., *Proc. Natl. Acad. Sci. USA*, **1987**, *84*, 7817-7821.
Roberts, W.L., Myher, J.J., Kuksis, A., Low, M.G., Rosenberry, T.L., *J. Biol. Chem.*, **1988a**, *263*, 18766-18775.
Roberts, W.L., Santikarn, S., Reinhold, V.N., Rosenberry, T.L., *J. Biol. Chem.*, **1988b**, *263*, 18776-18784.
Roberts, W.L., Myher, J.J., Kuksis, A., Rosenberry, T.L., *Biochem. Biophys. Res. Commun.*, **1988c**, *150*, 271-277.
Rosenberry, T.L., Scoggin, D.M., *J. Biol. Chem.*, **1984**, *259*, 5643-5652.
Rosenberry, T.L., Roberts, W.L., Haas, R., *Fed. Proc.*, **1986**, *45*, 2970-2975.
Rosenberry, T.L., Toutant, J.P., Haas, R., Roberts, W.L., *Meth. Cell Biol.*, **1989**, *32*, 231-255.
Toutant, J.-P., *Neurochem. Int.*, **1986**, *9*, 111-119.
Toutant, J.-P., Arpagaus, M., Fournier, D., *J. Neurochem.* **1988**, *50*, 209-218.
Toutant, J.-P., Roberts, W.L., Murray, N.R., Rosenberry, T.L., *Eur. J. Biochem.*, **1989**, 180, 503-508.
Toutant, J.-P., Richards, M.K., Krall, J.A., Rosenberry, T.L., *Eur. J. Biochem.*, **1990a**, 187, 31-38.
Toutant, J.-P., Krall, J.A., Richards, M.K., Rosenberry, T.L., *Cell. Molec. Neurobiol.*, **1990b**, in the press.
Walter, E.I., Toutant, J.-P., Rosenberry, T.L., Tykocinski, M.L., Medof, M.E., *Parasitology Meeting of the NIH*, **1988**, Bethesda.

Bovine Brain Acetylcholinesterase Sequence Involved in Intersubunit Disulfide Linkages

Terrone L. Rosenberry and **William L. Roberts**, Department of Pharmacology, Case Western Reserve University, Cleveland, OH 44106
Bhupendra P. Doctor, Division of Biochemistry, Walter Reed Army Institute of Research, Washington, D.C. 20307

The A_{12} and G_2 forms of torpedo acetylcholinesterase (AChE) differ only in their C-terminal peptide sequences and reflect alternative 3' exon splicing of the AChE mRNA. In this report, the cysteine involved in intersubunit disulfide linkages in bovine brain AChE was selectively reduced and radioalkylated with iodoacetamide and the major radiolabeled tryptic fragment was isolated. Its amino acid sequence was determined to be CSDL, identical to the C-terminus of fetal bovine serum AChE and closely homologous to the C-terminal sequence of torpedo A_{12} AChE. We conclude that the mammalian brain G_4 AChEs utilize the same exon splicing pattern as the A_{12} AChEs and that factors other than the primary sequence regulate assembly with alternative noncatalytic subunits.

Molecular forms of acetylcholinesterase (AChE) show tissue-specific variations in their structure and assembly (see Massoulie and Bon, 1982). These variations involve differences not only in the size of the oligomeric assemblies but also in the subunits which attach the enzyme to membranes. Three classes of membrane-bound AChEs have been identified as diagrammed in Fig. 1. A_{12} AChE, found in eel electric organ, is composed of twelve catalytic subunits. These subunits are arranged in three tetrameric groups in which two subunits are linked by a direct intersubunit disulfide bond while the other two are disulfide-linked directly to 35 kDa noncatalytic, collagen-like subunits (Rosenberry et al., 1982). Similar A_{12} AChE structures appear to represent the predominant AChE species in mammalian neuromuscular junctions (Massoulie and Bon, 1982) and are localized in the junctional basement membrane matrix. G_2 AChE, present in mammalian erythrocytes, insect central nervous system, and torpedo electric organ, is localized in membranes by a glycoinositol phospholipid covalently linked to the C-terminal amino acid of the catalytic subunits (Futerman et al., 1985; Roberts et al., 1988). No evidence of additional noncatalytic subunits in membrane-bound G_2 AChEs has been found. The third class of membrane-bound AChE is represented by G_4 AChE in the mammalian central nervous system. These tetramers, like those in A_{12} AChE, involve two subunits linked by a direct intersubunit disulfide bond while the other two are disulfide-linked to the membrane-binding, noncatalytic subunit. In this case the noncatalytic subunit is an apparent 20 kDa hydrophobic structure (Inestrosa et al., 1987).

Molecular cloning studies have so far failed to find evidence of more than one AChE gene in torpedo (Maulet et al., 1990) or other species. Analysis of the gene structure of torpedo AChE, however, has revealed that the A_{12} and G_2 catalytic subunits reflect alternative mRNA splicing of small exons at the 3' end of the coding region and consequently express different C-terminal peptide sequences (Sikorav et al., 1989; Maulet et al., 1990). Since torpedo does not express a significant amount of membrane-bound G_4 AChE, the question has remained open as to whether brain G_4 AChEs that are disulfide-linked to 20 kDa hydrophobic subunits (Fig. 1) reflect yet a third alternative exon encoding a distinct C-terminal peptide. To address this question, we have obtained peptide sequences from bovine brain and erythrocyte AChEs and compared them to the recently obtained complete amino acid sequence of bovine fetal serum AChE (Doctor et al., 1990).

0–8412–2008–5/91/0017$06.00/0

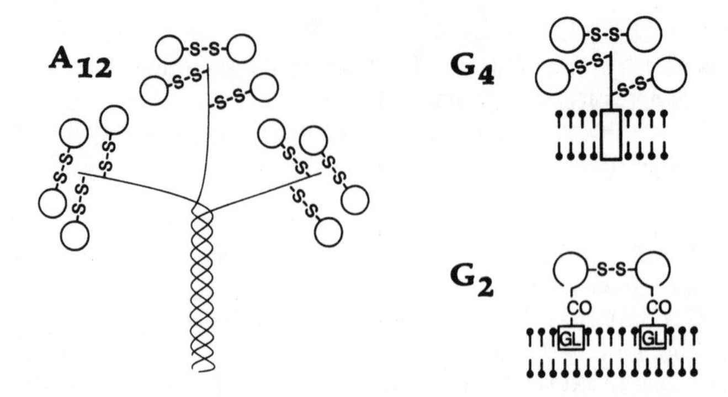

Fig. 1. Schematic models of three classes of membrane-bound AChE forms. In each model, *open circles* designate catalytic subunits of about 70 kDa and intersubunit disulfide bonds involving these subunits are shown as *S-S*. G_4 and G_2 AChEs interact directly with phospholipid membranes (*closed circles* with *tails*). The glycoinositol phospholipid anchor in G_2 AChE is represented by *GL*.

Amino acid sequences of intact bovine AChEs.

Bovine AChEs from fetal serum (De La Hoz et al., 1986), brain (Inestrosa et al., 1987), and erythrocytes (Roberts et al., 1987) were purified by affinity chromatography and subjected to gas phase sequencing. The N-terminal sequences of residues 1-38 of the brain and erythrocyte AChEs were identical to that shown for the fetal serum AChE in Fig. 2 with the exception of residue 16, where a glutamate residue in the fetal serum enzyme was replaced by an arginine in the brain and erythrocyte AChEs (Roberts et al., 1990). The basis for this difference is unclear, but it probably arises from an allelic variation. Alternative sequences that suggest allelic variants at residues 478 and 480 in fetal bovine serum AChE have been found (B. P. Doctor, unpublished observations).

Radioalkylation of sulfhydryl groups involved in intersubunit disulfide linkages.

All AChE and butyrylcholinesterase sequences obtained to date contain three conserved intrasubunit disulfide bonds within the catalytic subunits (Lockridge et al., 1987). In addition to these six Cys residues, torpedo AChE catalytic subunits contain two additional Cys residues. One of these, Cys^{572} near the C-terminus of torpedo A_{12} AChE, has been identified previously as a site of intersubunit crosslinking (MacPhee-Quigley et al., 1986). In the alternative C-terminal peptide of torpedo G_2 AChE, Cys^{537} at the mature C-terminus is the likely site of intersubunit disulfide formation (Gibney et al., 1988). In contrast to the torpedo AChEs, the sequence of fetal bovine serum AChE (Fig. 2) shows only one Cys in addition to the six involved in intrasubunit disulfides, Cys^{580} homologous to Cys^{572} in the torpedo enzyme. If this Cys residue could be selectively reduced and radioalkylated, we anticipated that a tryptic peptide could be isolated to provide a C-terminal catalytic subunit sequence that would yield important information about exon splicing in brain AChE.

Both eel A_{12} and human G_2 AChEs have no free sulfhydryl groups available to radiolabeled alkylating agents prior to treatment with disulfide reducing agents (Rosenberry, 1975; Rosenberry and Scoggin, 1984). Furthermore, intersubunit disulfide linkages are reduced more readily than intrasubunit disulfides in both enzymes. When these AChEs are selectively reduced with dithiothreitol in the presence of the active site inhibitor edrophonium and in the absence of

EGPEDPELLVMVR GGE*LR GLR LMAPR

GPVSAFLGIPFAEPPVGPR R FLPPEPK R

PWPGVLNATAFQSVCYQYVDTLYPGFEGTEMWNPNR

ELSEDCLYLNVWTPYPR PSSPTPVLVWIYGGGFYSGASSLDVYDGR

FLVQAEGTVLVSMNYR VGAFGFLALPGSR EAPGNVGLLDQR

LALQSVQENVAAFGGDPTSVTLFGESAGAASVGMHLLSPPSR

GLFHR AVLQSGAPNGPwATVGVGEAr R R ATLLAR

LVGCPPGGAGGnDTELVACLr AR PAQDLVDHEWR

VLPQEHVFR FSFVPVVDGDFLSDTPEALINAGDFVGLQVLVGVVK

DEGSYFLVYGAPGFSK DNESLISR AQFLAGVR

VGVPQASDLAAEAVVLHYTDWLHPEDPAR WR

EALSDVVGDhNVVCPVAQLAGr LAAQGAR VYAYIFEHR

ASTLSWPLWMGVPHGYEIEFIFGLPLEPSLNYTIEER TFAQR LMR

YWANFAR TGDPNDPR APK APQWPPYTAGAQQYVSLNLR

PLGVPQASR AQACAFWNR FLPK LLN*ATDTLDEAER

QWK AEFHR WSSYMVHWK NQFDHYSK QDR CSDL

Fig. 2. Homology of bovine brain AChE to fetal bovine serum AChE (Roberts et al., 1990). The complete amino acid sequence of fetal bovine serum AChE reported by Doctor et al. (1990) is broken into the tryptic fragments predicted by cleavage at Arg and Lys residues. Both intact enzyme and tryptic peptides obtained from bovine brain AChE were subjected to gas phase sequencing, and the residues identical to those in fetal bovine enzyme are indicated by the underline. For those cycles in which no residue could be assigned, a lower case letter is shown. The asterisks indicate residues which differed from the serum AChE: Glu^{16} was changed to Arg^{16} and Asn^{541} was changed to Ser^{541} in the brain AChE. The underlined sequences correspond, respectively, to residues 1-38, 225-245, 248-253, 254-274, 365-380, 396-417, 535-549, 552-559, and 580-583.

denaturants, 1.0-1.5 sulfhydryl groups per catalytic subunit can be radioalkylated. Bovine brain AChE behaves in a similar fashion. Selective reduction and radioalkylation of the brain enzyme dissociated AChE oligomers to radiolabeled 70 kDa catalytic subunits and 20 kDa subunits nearly as efficiently as complete reduction in the presence of SDS (Roberts et al., 1990). The selectively reduced enzyme incorporated 1.3 $\pm$ 0.1 mols of [^{14}C]iodoacetamide per mol of catalytic subunit. Thus it appears that selectively reduced intersubunit disulfides were the predominant sites of radioalkylation.

Isolation and sequencing of tryptic fragments that contain Cys residues derived from intersubunit disulfides

Selectively reduced and radioalkylated bovine brain AChE was run on SDS-PAGE and electroblotted on nitrocellulose. The blotted 70 kDa catalytic subunit band was digested with trypsin, and tryptic fragments released from the blot were fractionated by narrow bore HPLC as shown in Fig. 3. The absorbance profile at 215 nm revealed a number of peaks. Peptides in major peaks at 56, 81 and 111 min were subjected to gas phase sequencing, and the observed sequences were identical, respectively, to those of the tryptic fragments from residues 248-253 and 225-245 and the first part of the tryptic fragment from residues 365-393 as indicated in Fig. 2. The peak at 41 min yielded a sequence identical to the tryptic fragment from residues 555-559. The most prominent radioalkylated tryptic fragment eluted at 32 min, and its observed sequence was C^*SDL, where C^* was the radiolabeled cysteine residue. This corresponds precisely to the C-terminal tetrapeptide in fetal bovine serum AChE (Fig. 2). Minor radiolabeled tryptic fragments eluting at 75 min and 86 min corresponded to residues 254-274 and 396-417 in the fetal bovine serum AChE. These fragments indicated slight reduction and radioalkylation of intrasubunit disulfides within the catalytic subunits. In addition to tryptic fragments indicated in Fig. 3, two radiomethylated tryptic fragments obtained from brain AChE labeled with $[^{14}C]$HCHO and $NaCNBH_3$ (Roberts et al. 1990) yielded sequences corresponding to residues 535-549 and 552-558. This sequence was identical to that in Fig. 2 except for Asn^{541}, a potential site for N-linked glycosylation in the serum enzyme, which yielded only Ser in the brain enzyme.

A model of AChE tetramer assembly with noncatalytic subunits

The extensive identity in the primary amino sequences of bovine brain and erythrocyte AChEs with that previously published for fetal bovine serum AChE (Doctor et al., 1990) is an important finding. This sequence identity provides strong evidence that the torpedo AChE gene structure applies also to mammalian AChEs. In torpedo, the complete AChE catalytic subunit sequences are encoded by three exons. The first two exons correspond to amino acid residues 1-535 and are common to both A_{12} and G_2 AChEs while alternative third exons encode different C-terminal sequences for residues 536-575 in the A_{12} AChE and residues 536-565 in the G_2 AChE (Maulet et al., 1990). The overall homology of fetal bovine serum G_4 AChE to torpedo A_{12} is 61%, but homology is even higher at the C-terminus: 30 of 40 residues corresponding to the torpedo third exon and 12 of 13 residues corresponding to the C-terminal end of the second exon are identical. Thus it seems clear that the exon which encodes the C-terminus of fetal bovine serum AChE is the bovine homolog of the torpedo exon that encodes the A_{12} AChE C-terminus.

The predominant radiolabeled tryptic fragment obtained following selective reduction and radioalkylation of bovine brain AChE was CSDL. This observation clearly confirmed Cys^{580} as a major site of intersubunit disulfide bonding in bovine brain G_4 AChE catalytic subunits. Furthermore, this sequence is precisely the same as that observed at the C-terminus of fetal bovine serum AChE. Thus we propose the scheme in Fig. 4 in which the mammalian brain membrane-bound G_4 AChEs utilize the same catalytic subunit exon splicing pattern as that of soluble G_4 AChEs and basement membrane-bound A_{12} AChEs, and that factors other than the primary sequence of the AChE catalytic subunits dictate assembly with either the collagen-like noncatalytic subunits or the 20 kDa hydrophobic subunit.

While the scheme in Fig. 4 is the simplest one to accomodate our data, other proposals involving brain G_4 AChE catalytic subunits that are heterogeneous in their C-terminal sequences are possible. For example, the catalytic subunits that are directly linked by a single disulfide may involve the C-terminal sequence in Fig. 2 while the catalytic subunits that are linked to the noncatalytic 20 kDa subunit may involve an

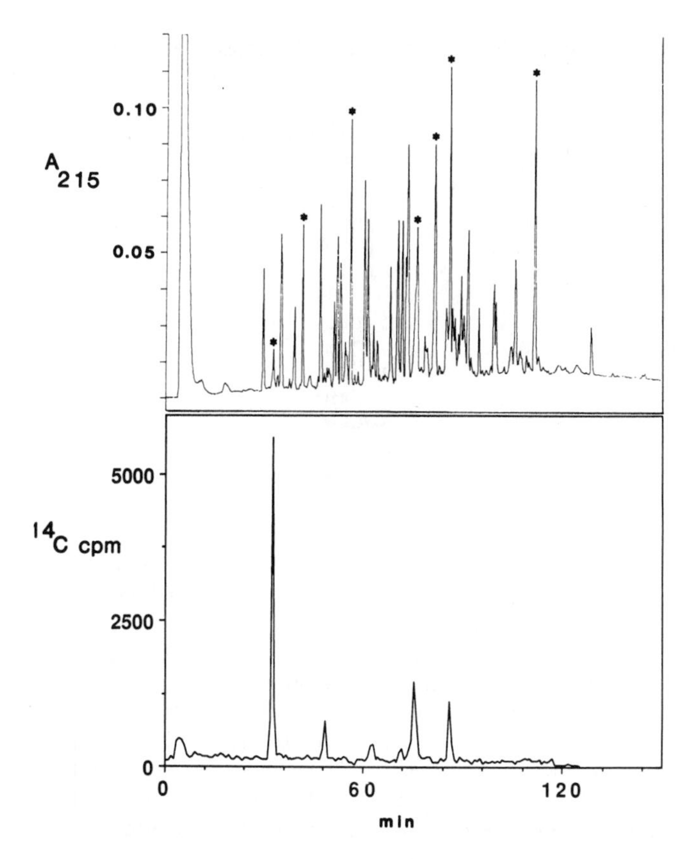

Fig. 3. Fractionation of a tryptic digest of electroblotted 70 kDa catalytic subunit from bovine brain AChE on narrow bore HPLC (Roberts et al., 1990). AChE selective reduction and radioalkylation, electroblotting, trypsin digestion and HPLC procedures are outlined elsewhere (Roberts et al., 1990). The absorbance profile at 215 nm (top) was obtained from the HPLC UV detector, and the radioactivity of corresponding fractions (bottom) was determined by scintillation counting. The recovery of cpm in peak fractions at 32 min (46% of the cpm loaded on the HPLC) corresponded to 82 pmol of radioalkylated Cys. Peptides in fractions corresponding to peaks indicated by the asterisks in the top panel were subjected to gas phase sequencing.

alternative C-terminal sequence. No evidence for such catalytic subunit heterogeneity has been reported for any AChE, but an apparent A_{12} species in chick muscle has been reported in which both AChE and butyrylcholinesterase catalytic subunits coassemble (Tsim et al., 1988). The nonreduced tryptic fragment corresponding to the C-terminus of torpedo A_{12} catalytic subunit disulfide-linked to itself has been identified (MacPhee-Quigley et al., 1986), but no fragments from any AChE in which a catalytic subunit peptide is disulfide-linked to a noncatalytic subunit peptide have been reported. To pursue further this possibility, we plan either to confirm the C-terminal CSDL sequence or to identify the bovine brain AChE catalytic subunit sulfhydryl group(s) involved in intersubunit disulfide linkages to the 20 kDa subunit in the 160 kDa dimer that is a major band of SDS-PAGE gels prior to disulfide reduction (Inestrosa et al., 1987).

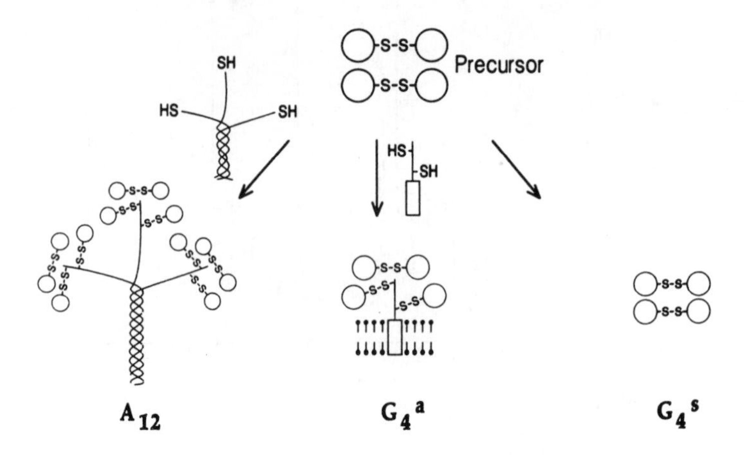

Fig. 4. Proposed scheme for the assembly of A_{12} AChE with collagen-like noncatalytic subunits, brain G_4 AChE with a 20 kDa noncatalytic subunit, and secreted G_4 AChE for which no evidence of noncatalytic subunits has been obtained (Roberts et al., 1990). The primary sequence of the catalytic subunits in these three AChE species is presumed to be identical and to result in the formation of a tetrameric precursor composed of a dimer of disulfide-linked catalytic subunit dimers. The distribution of the three AChEs in a given cell is suggested to be determined by factors that regulate their post-translational assembly. Such factors could include the extent of cosynthesis of the noncatalytic subunits from distinct genes or the presence of specific enzymes that catalyze insertion and disulfide bond formation with distinct noncatalytic subunits.

Acknowledgements: We express our appreciation for the excellent technical assistance provided by Jennifer Krall and John Foster. This work was supported by National Institutes of Health Grant NS16577 and by grants from the Muscular Dystrophy Association.

REFERENCES

De La Hoz, D., Doctor, B. P., Ralston, J. S., Rush, R. S., and Wolfe, A. D., *Life Sci.*, **1986**, *39*, 195-199.

Doctor, B. P., Chapman, T. C., Christner, C. E., Deal, C. D., De La Hoz, D. M., Gentry, M. K., Ogert, R. A., Rush, R. S., Smyth, K. K., and Wolfe, A. D., *FEBS Lett.*, **1990**, in press.

Futerman, A. H., Fiorini, R.-M., Roth, E., Low, M. G., and Silman, I., *Biochem. J.*, **1985**, *226*, 369-377.

Gibney, G., MacPhee-Quigley, K., Thompson, B., Vedvick, T., Low, M. G., Taylor, S. S. and Taylor, P., *J. Biol. Chem.*, **1988**, *263*, 1140-1145.

Inestrosa, N. C., Roberts, W. L., Marshall, T. L., and Rosenberry, T. L., *J. Biol. Chem.*, **1987**, *262*, 4441-4444.

Lockridge, O. L., Adkins, S., and La Du, B. N., *J. Biol. Chem.*, **1987**, *262*, 12945-12952.

MacPhee-Quigley, K., Vedvick, T. S., Taylor, P., and Taylor, S. S., *J. Biol. Chem.*, **1986**, *261*, 13565-13570.

Massoulie, J. and Bon, S., *Annu. Rev. Neurosci.*, **1982**, *5*, 57-106.

Maulet, Y., Camp, S., Gibney, G., Rachinsky, T.L., Ekstrom, T.J., and Taylor, P., *Neuron*, **1990**, *4*, 289-301.

Roberts, W. L., Kim, B. H., and Rosenberry,

T. L., *Proc. Natl. Acad. Sci. USA*, **1987**, *84*, 8917-7821.

Roberts, W. L., Santikarn, S., Reinhold, V. N., and Rosenberry, T. L., *J. Biol. Chem.*, **1988**, *263*, 18776-18784.

Roberts, W. L., Doctor, B. P., and Rosenberry, T. L., **1990**, submitted.

Rosenberry, T. L., *Adv. Enzymol.*, **1975**, *43*, 103-218.

Rosenberry, T. L., Barnett, P., and Mays, C., *Meths. Enzymol.* **1982**, *82*, 325-339.

Rosenberry, T. L., and Scoggin, D. M., *J. Biol. Chem.*, **1984**, *259*, 5643-5652.

Sikorav, J.-L., Duval, N., Anselmet, A., Bon, S., Krejci, E., Legay, C., Osterlund, M., Reimund, B., and Massoulié, J., *EMBO J.*, **1988**, *7*, 2983-2993.

Tsim, K. W. K., Randall, W. R., and Barnard, E. A., *Proc. Natl. Acad. Sci. USA*, **1988**, *85*, 1262-1266.

Structure and Function of Cholinesterase from Amphioxus

*Leo Pezzementi, Michael Sanders, Todd Jenkins, and Dan Holliman, Division of Science and Mathematics, Birmingham-Southern College, 900 Arkadelphia Road, Birmingham, Alabama, 35254, U.S.A. (205)226-4882
Jean-Pierre Toutant, Physiologie Animale, INRA, 2 place Viala, F-34060 Montpellier Cedex 01, France
Ronald J. Bradley, Department of Psychiatry, School of Medicine, University of Alabama at Birmingham, University Station, Birmingham, Alabama 35294, U.S.A.

It has become customary to classify the Cholinesterase (ChE) of vertebrates as either acetylcholinesterase (AChE) or butyrylcholinesterase (BuChE) on the basis of substrate preference and inhibition, as well as diagnostic inhibitor specificity (Silver, 1974). Recent information from molecular biology has substantiated this division (Massoulié & Toutant, 1988; Chatonnet & Lockridge, 1989). Higher vertebrates are generally thought to have both types of ChE. However, BuChE has not been reported in the lamprey (Pezzementi, 1987; 1989), and in Torpedo the enzyme closely resembles AChE (Toutant et al., 1985) In the phylogenetically related deuterostome invertebrates, the amphioxus Branchiostoma lanceolatum (Flood, 1974), the sea squirt Ciona intestinalis (Meedel & Whittaker, 1979; Meedel, 1980), and the sea urchin Strongylocentrotus purpuratis (Ozaki, 1976) the ChE present has been classified as AChE, although not all criteria have been met, and these classifications should be considered tentative. These observations have led to the proposal that the two closely related enzymes resulted from a gene duplication event early in the evolution of the vertebrates, with subsequent structural and functional divergence (Toutant et al., 1985; Chatonnet & Lockridge, 1989). Recently, reports have come to light describing in teleosts a ChE that does not fit easily into either category; but, instead, appears to have properties intermediate to AChE and BuChE (Lundin, 1966; Brodbeck et al, 1973; Stieger et al., 1989). These data have confounded the phylogenetic origins of the ChE.

Additionally, it is generally assumed that vertebrates possess parallel sets of globular and asymmetric forms of AChE and BuChE, while deuterostome invertebrates have only globular forms of ChE; asymmetric enzyme has not been reported in the two organisms studied: Ciona (Meedel & Whittaker, 1979; Meedel, 1980) and Strongylocentrotus (Ozaki, 1976). These observations suggest that the gene duplication event occurred subsequent to the appearance of globular and asymmetric forms in the ancestral enzyme; perhaps early in vertebrate evolution (Toutant, 1989).

As part of our investigation of the evolution of ChE, we have begun a characterization of the ChE of the invertebrate chordate amphioxus Branchiostoma floridae. We have found that the ChE activity of amphioxus is due primarily to a globular esterase, apparently G_2, which exists in both hydrophilic and amphiphilic forms, with the latter possessing a glycoinositol phospholipid tail. Small amounts of asymmetric A_{12} ChE were also observed. Finally, we determined that the kinetic and pharmacologic characteristics of the ChE are intermediate to those of AChE and BuChE.

Materials and Methods

Specimens of amphioxus were collected in Mobile Bay at the Dauphin Island Sea Lab. and kept in aerated sea water or frozen at -70°C. Because of their small size and novel muscle innervation, involving central motor endplates (Flood, 1966), which precludes separation of muscle from the central nervous system, whole specimens were homogenized.

For analysis of molecular forms of ChE, the sequential extraction technique of Younkin et al. (1982) was used. Tissue was extracted twice in low ionic strength buffer (LIS buffer; 10mM $NaHPO_4$, pH7, 1% Triton X-100, 1mM EDTA, 1ug/ml pepstatin, 1mM iodoacetamide.), and then reextracted in high ionic strength buffer containing the above and 1M NaCl (HIS).

For analysis of the amphiphilicity of the ChE, a modification of the sequential extraction technique of Toutant (1986) was used. Tissue was extracted

0-8412-2008-5/91/0024$06.00/0

in low-salt buffer (LS buffer; 10mM $NaHPO_4$, pH7, 1mM EDTA, 1ug/ml pepstatin, 1mM iodoacetamide and then reextracted in low-salt buffer containing 1% Triton X-100 (DS buffer).

Velocity sedimentation was performed on 5-20% linear sucrose gradients (Pezzementi et al., 1987).

ChE was assayed by the method of Ellman et al. (1961). Gradient fractions were assayed by the microtiter assay of Doctor et al. (1987).

Nondenaturing polyacrylamide electrophoresis, and digestion of ChE by phosphatidylinositol-specific phospholipase C (PtdIns-specific PLC; purified from B. thurigiensis by Dr. Martin Low) were performed as described previously (Toutant et al., 1989).

Results and Discussion

The molecular forms of ChE present in amphioxus were characterized by sequential extraction (Younkin et al., 1982) and velocity sedimentation on sucrose gradients. Specimens were homogenized first in LIS buffer, which preferentially solubilizes globular ChE, and then in HIS buffer, which selectively extracts asymmetric enzyme. We found that 89% of the activity was extracted into LIS buffer, and another 7% of the activity was solubilized by HIS buffer; 4% was not extracted (Table 1).

Table 1. Sequential extraction of ChE

Fraction	Buffer	AChE Activity nmol/mg-min	% Total
S1	LIS	9.25 ± 0.70	78
S2	LIS	1.33 ± 0.47	11
S3	HIS	0.87 ± 0.27	7
H4	HIS	0.44 ± 0.16	4

Units of ChE activity are nanomoles of acetylthiocholine hydrolyzed per minute and milligram (wet weight) of tissue. Means and SE of 7 experiments are given.

When ChE from S1 was analyzed by velocity sedimentation, it sedimented at 5.6S (SE=0.1; N=13; Fig. 1). Since S1 contains primarily globular ChE, we concluded that this peak represents a globular form, probably G_2 (see below). Analysis of the S3 fraction on sucrose gradients revealed two peaks of activity, a prominent peak at 5.5S (SE=0.1, N=12; Fig. 1), and a small peak at 17S (SE=0.4, N=11; Fig. 1), representing less than 0.1% of the total ChE activity. The prominent peak at 5.5S probably represents contaminating G_2. On the other hand, by virtue of its sedimentation coefficient and its relative absence from S1, the 17S peak probably corresponds to A_{12}. To confirm this identification, we attempted to perform collagenase digestion of the putative asymmetric enzyme. However, we were unable to extract the 17S form in the EDTA-deficient buffer normally used for the experiment (Bon & Massoulié, 1978), suggesting that the enzyme is a Class II asymmetric ChE, requiring high ionic strength and chelating agents for extraction (Rodrìquez-Borrajo et al., 1982). The 5.5S peak was not shifted by collagenase digestion (not shown).

Subsequently, we confirmed that the 17S peak represents A_{12} ChE by precipitating it at low ionic strength (Bon et al., 1978). The S3 fraction from sequential extraction was dialyzed in LIS buffer, and aliquots of the supernatant and the precipitate that formed were analyzed on sucrose gradients. The 17S peak was seen only in the low ionic strength precipitate, consistent with it being A_{12} (Fig. 2). The 5.5S form was found in both fractions, probably because of its relatively high concentration in S3.

Thus, we have demonstrated the presence of an asymmetric ChE in an invertebrate, implying that asymmetric enzyme evolved prior to the emergence of the vertebrates, perhaps in an ancestral chordate. The lamellar muscle of amphioxus is differentiated into fast, intermediate, and slow fiber types (Flood, 1968; Guthrie & Banks, 1970). In the vertebrates, the asymmetric forms of ChE are preferentially localized to the endplates of fast-twitch muscle (Barnard et al., 1984). It is tempting to speculate that the A_{12} ChE of amphioxus is localized to the central motor endplates of the fast muscle fibers, which stain intensely for ChE activity (Flood, 1974), while the globular enzyme is distributed on the surface of the muscles, the nervous system, and the notochord, accounting for the diffuse staining pattern seen in these organs. The small amounts of A_{12} that were observed can then, in part, be accounted for by the small surface area of the central motor endplates of the fast muscle. It is

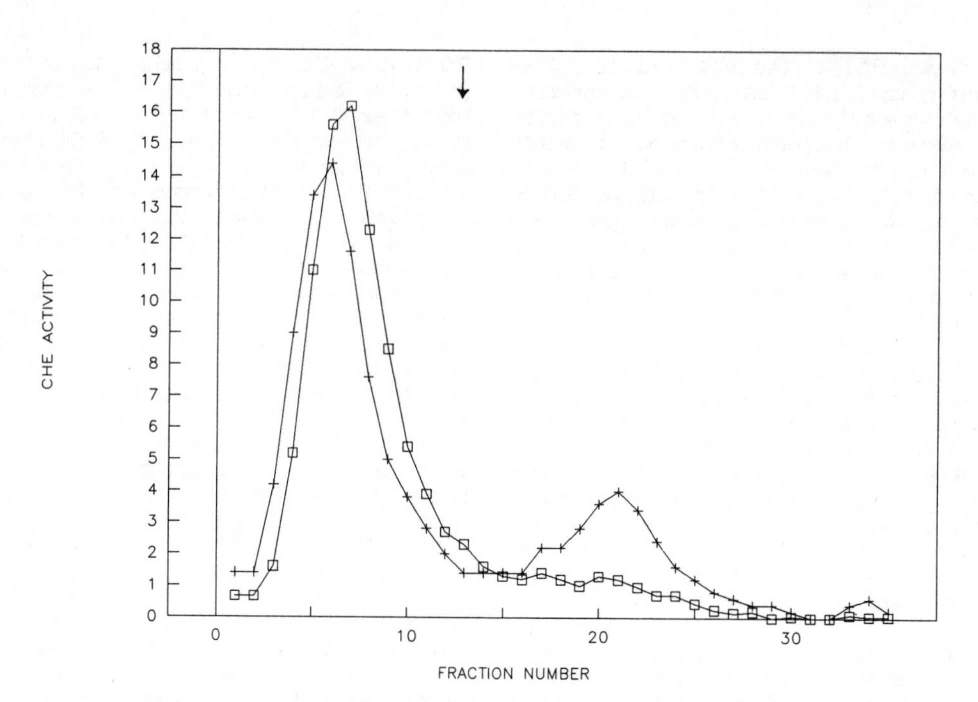

Figure 1. Sedimentation profiles of ChE from amphioxus. Molecular forms of ChE in S1 (□) and S3 (+) were separated on sucrose gradients and assayed for ChE activity. The arrow indicates the position of catalase (11.3S).

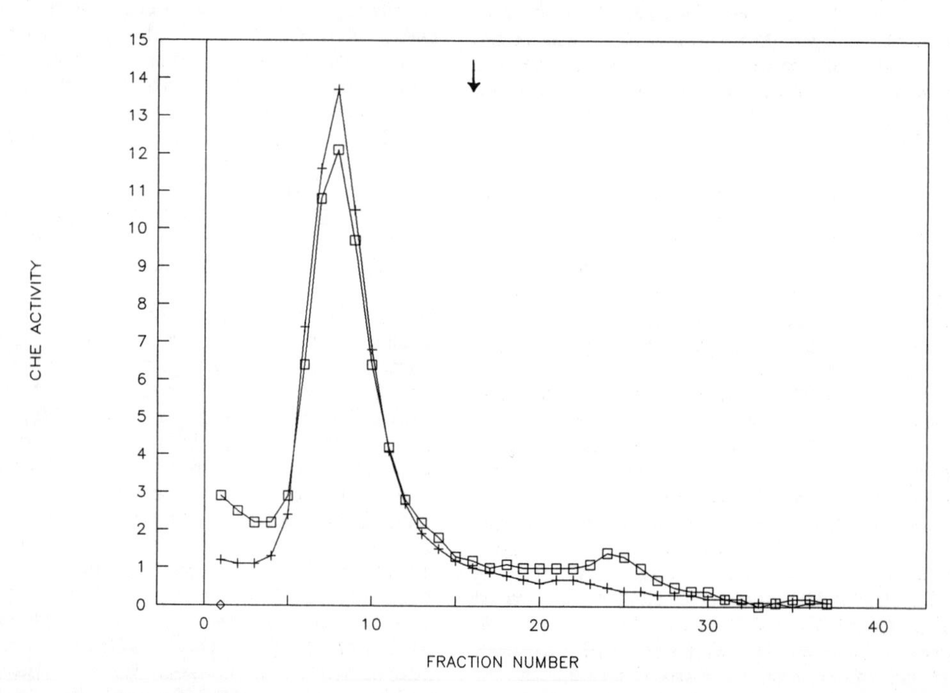

Figure 2. Precipitation of asymmetric ChE at low ionic strength. Aliquots of S3 were dialyzed against LIS buffer. Samples of supernatant (+) and resolubilized precipitate (□) were analyzed on sucrose gradients. The arrow indicates the position of catalase (11.3S).

worth noting that the muscle of Ciona is not differentiated into fiber subtypes.

To investigate the amphiphilic characteristics of the globular ChE, we used an alternate form of sequential extraction (Toutant, 1986), which separates low-salt soluble (LSS) from detergent-soluble (DS) enzyme, in conjunction with velocity sedimentation on sucrose gradients in the presence and absence of detergent. The globular enzymes in both fractions sedimented at 6.7S (LSS: SE=0.1, N=12; DS: SE=0.1, N=6; Fig. 3) in the absence of detergent, while in the presence of Triton X-100, the sedimentation coefficients of the peaks were shifted to considerably lower values (LSS: X=5.8, SE=0.1, N=12; DS: X=5.0, SE=0.1, N=6), indicating interactions with detergent. Thus, by hydrodynamic criteria, both fractions contain amphiphilic ChE. These data also confirm that the ChE is G_2, since the sedimentation coefficient obtained in the absence of detergent is characteristic of G_2, and shoulders that could correspond to G_1 and G_4 are seen in the gradients.

The amphiphilic nature of the G_2 ChE suggested to us that it might have a glycoinositol phospholipid tail. We investigated this possibility by nondenaturing polyacrylamide gel electrophoresis, which is also able to separate forms of ChE that can not be resolved on sucrose gradients (Toutant, 1986). Amphiphilic and hydrophilic forms are found in both the LSS and DS extracts (Fig 4, lanes a & b). The amphiphilic ChE in both fractions could be converted into hydrophilic forms by digestion with PtdIns-specific PLC (Fig 4, lanes c-f), indicating that the amphiphilic ChE of amphioxus has a glycoinositol phospholipid anchor (Rosenberry et al., 1989). The hydrophilic forms are probably due to varying degrees of degradation of the amphiphilic ChE by different endogenous proteases and phospholipases present in the LSS and DS extracts. Thus, ChE from deuterostome vertebrates and invertebrates, as well as protostomes is modified by a glycoinositol phospholipid (Massoulié & Toutant, 1988).

To determine the nature of the ChE of amphioxus, we performed a series of kinetic and pharmacologic experiments on direct HIS extracts, in which >99% of the activity is due to G_2. The catalytic properties of the esterase from amphioxus appear to be intermediate to AChE and BuChE. Although the ChE activity could be completely inhibited by BW284C51, an inhibitor of AChE, higher concentrations than those typically required in our laboratory were necessary. Also, the enzyme was inhibited by relatively low concentrations of ethopropazine, an inhibitor of BuChE. In contrast, iso-OMPA, another BuChE inhibitor, had no effect (Fig. 5). Similar monophasic inhibition curves were obtained when butyrylthiocholine, instead of acetylthiocholine, was used as substrate (not shown). These data indicate that the ChE activity is due to a single type of enzyme present in the extracts.

Further evidence for the intermediate nature of the ChE is found in the results of kinetic studies (Table 2). The butyrylthiocholine/acetylthiocholine V_{max} ratio is high, but still within the range of AChE. However, the enzyme has an eight-fold lower K_m for butyrylthiocholine than it does for acetylthiocholine. Finally, the esterase exhibits substrate inhibition by both butyrylthiocholine and acetylthiocholine, but butyrylthiocholine is the more potent inhibitor.

Table 2. Kinetics of ChE from amphioxus

Substrate	Km [uM]	Ki [mM]	Vmax nmol/mg-min
AsCh (A)	330	28	9.12
BusCh (B)	41	1.3	1.01
Ratio B/A	0.12	0.05	0.11

Extracts were assayed in triplicate for enzyme activity with acetylthiocholine (AsCh) and butyrylthiocholine (BusCh).

The kinetic and pharmacologic characteristics of the ChE from amphioxus are compared to those of vertebrates in Table 3. To minimize procedural differences among laboratories, we have used K_m and V_{max} ratios for hydrolysis of butyrylthiocholine and acetylthiocholine by ChE, and pIC_{50} ratios for inhibition of ChE by ethopropazine and BW284C51. Inspection of the data reveals that there are three classes of ChE. In general, AChE exhibits substrate inhibition, has a low V_{max} ratio, a K_m ratio of $\geq$1, and a pIC_{50} ratio of ~0.5. In contrast, BuChE does not show inhibition by excess substrate, has V_{max} and pIC_{50} ratios of $\geq$1, and a K_m ratio of <1. The third class of ChE is represented by the

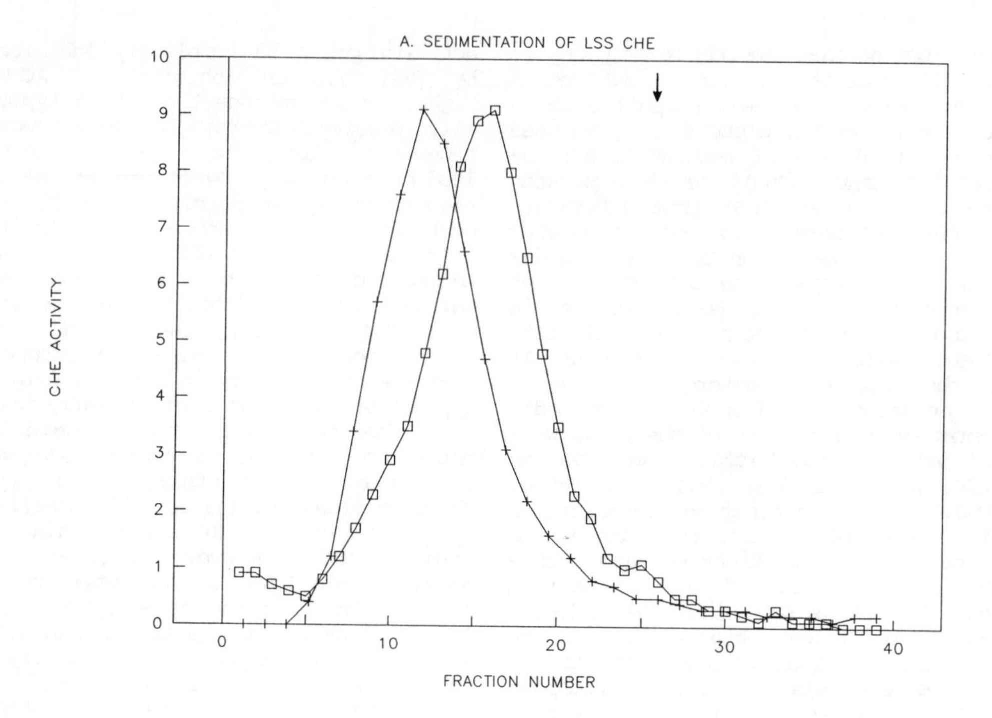

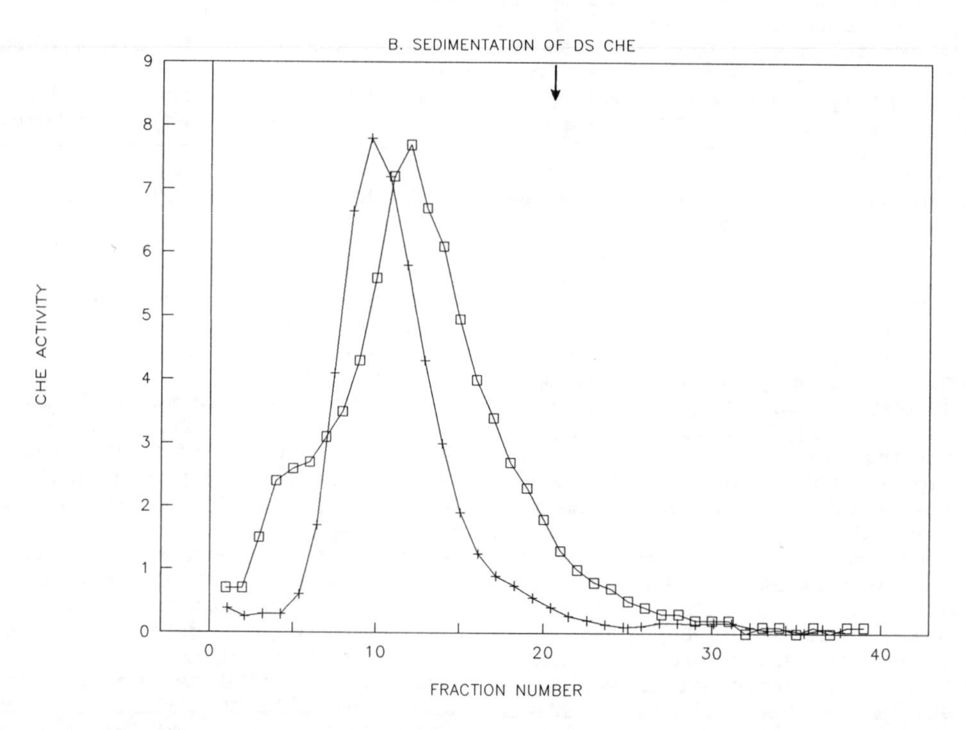

Figure 3. Sedimentation profiles of LSS and DS ChE from amphioxus. LSS and DS extracts were analyzed on sucrose gradients in the presence (+) and absence (□) of Triton X-100. Arrows indicate the position of catalase (11.3S).

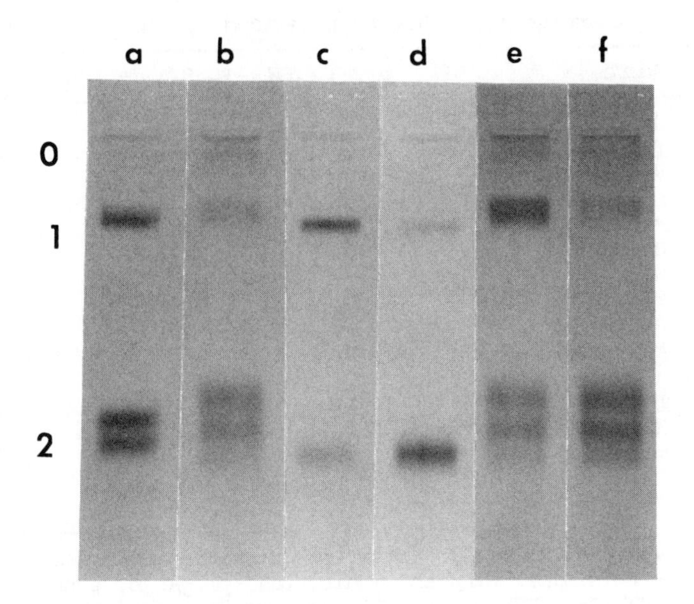

Figure 4. Characterization of globular ChE from amphioxus by nondenaturing electrophoresis. Samples of LSS and DS ChE were subjected to nondenaturing electrophoresis in the presence of Triton X-100. Lanes: a, DS ChE; b, LSS ChE; c, DS ChE diluted (1:3); d, DS ChE diluted (1:3) and digested with PtdIns-specific PLC; e, LSS ChE; f, LSS ChE digested with PtdIns-specific PLC. 0, origin of migration; 1, amphiphilic ChE; 2, hydrophilic ChE.

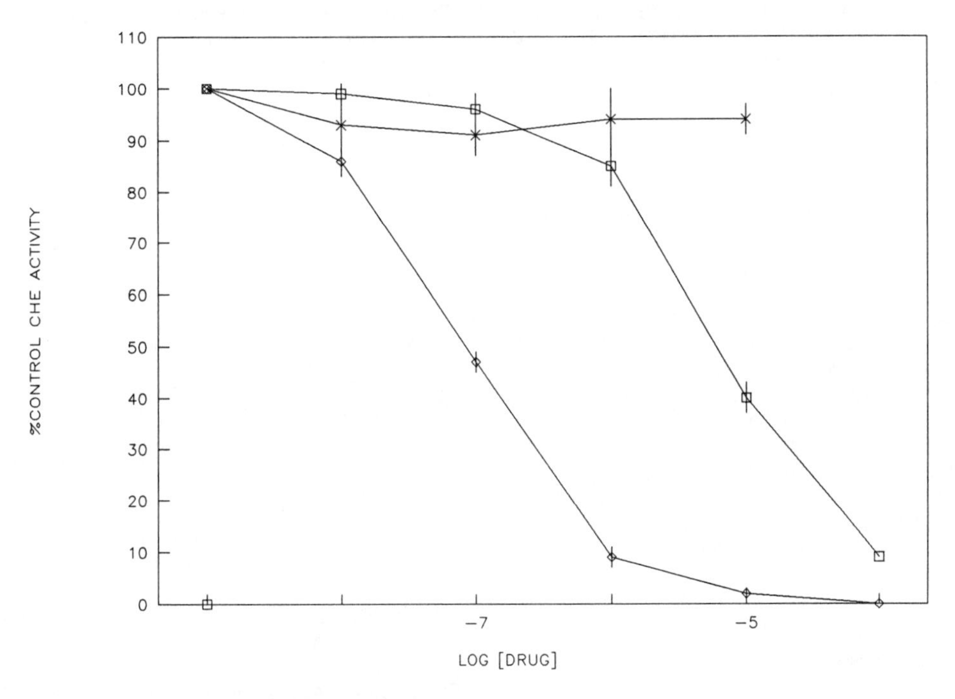

Figure 5. Pharmacologic analysis of ChE from amphioxus. HIS extracts were incubated with BW284C51 (<>), ethopropazine (□), or iso-OMPA (X) for twenty minutes prior to being assayed for ChE activity. Means and SE are given.

Table 3. Comparison of ChE from various species of chordates

Enzyme	Source	SI	pIC_{50} E/BW	K_m B/A	V_{max} B/A
AChE	Eel (1)	+	?	16.4	0.006
	Torpedo (2)	+	0.51	?	0
	Flounder (3)	+	0.55	?	0.005
	Lamprey (4)	+	0.53	2.55	0.01
	Hu. RBC (5)	+	0.50	1.25	0.01
	Chicken (6)	+	0.75*	0.33	0.004
ChE	Flounder (3)	+	0.65	0.15	0.22
	Plaice (7)	+	?	0.20	0.40
	Amphioxus	+	0.72	0.12	0.11
BuChE	_Torpedo_ (2)	-	1.0	0.13	0.30
	Hu. Serum (8)	-	1.4	0.31	2.2
	Eq. Serum (9)	-	?	0.50	1.6
	Pig (10)	-	?	0.70	0.94
	Chicken* (6)	-	1.7	1	0.89

Michaelis constants (K_m), relative maximal velocities (V_{max}), substrate inhibition (SI) and inhibition (pIC_{50}) by ethopropazine (E) and BW284C51 (BW) of the ChE were determined with acetylthiocholine (A) and butyrylthiocholine as substrates except where indicated by an (*). In these studies acetylcholine and butyrylcholine were used as substrates and BW62C47 was used instead of BW284C51. The data in this table were obtained from the following sources: (1) Rosenberry (1975); (2) Toutant et al. (1985); (3) Stieger et al. (1989); (4) Pezzementi et al. (1987; 1989); (5) Gnagey et al. (1987), Brimijoin & Hammond (1988); (6) Rotundo (1984), Blaber & Cuthbert (1962), Myers (1953); (7) Lundin (1968), Brodbeck et al. (1973); (8) Heilbronn (1959), Brimijoin & Hammond (1988); (9) Lee & Harpst (1971), Main et al. (1977); (10) Tucci & Seifter (1969).

ChE of amphioxus and the teleosts, which exhibits substrate inhibition, but has a K_m ratio below those of AChE and BuChE, and V_{max} and pIC_{50} ratios intermediate to the other two esterases. The presence of this intermediate form of ChE in the primitive chordate amphioxus suggests that it may resemble the ancestral cholinesterase. AChE from chicken and BuChE from _Torpedo_ are exceptions to the above classification scheme by their resemblance to the intermediate form of ChE; perhaps, because they have retained some ancestral features.

Acknowledgements

This research was supported by an AREA grant (NS-24943) from the National Institutes of Health to L.P.

References

Barnard, E.A., Barnard, P.J., Jarvis, J., Jedrzejczyk, J., Lai, J., Pizzey, J.A., Randall, W.R. _In Cholinesterases: Fundamental and Applied Aspects_; Brzin, M., et al., Eds.; de Gruyter: Berlin, **1984**, pp. 49-71.

Blaber, L.C., Cuthbert, A.W. _Biochem. Pharm._ **1962**, _11_, 113-124.

Bon, S., Massoulié, J. _Eur. J. Biochem._ **1978**, _89_, 89-94.

Bon, S., Cartaud, J., Massoulié, _J. Eur. J. Biochem._ **1978**, _85_, 1-14.

Brimijoin, S., Hammond, P. _J. Neurochem._ **1988**, _51_, 1227-1231.

Brodbeck, U., Gentinetta, R., Lundin, S.J. _Acta Chem. Scand._ **1973**, _27_, 561-572.

Chatonnet, A., Lockridge, O. _Biochem. J._ **1989**, _260_, 625-634.

Doctor, B.P., Toker, L., Roth, E., Silman, I. _Analyt. Biochem._ **1987**,

166, 399-403.
Ellman, G.L., Courtney, K.D., Andres, V., and Featherstone, R.M. Biochem. Pharmacol. **1961**, 7, 88-95.
Flood, P. J. Comp. Neurol. **1966**, 126, 181-217.
Flood, P. J. Comp. Neurol. **1974**, 157, 407-419.
Flood, P. Z. Zellforsch. **1968**, 84, 389-416.
Gnagey, A. L., Forte, M., Rosenberry, T.L. J. Biol. Chem. **1987**, 262, 13290-13298.
Guthrie, D.M., Banks, J.R. J. Exp. Biol. **1970**, 52, 401-417.
Heilbronn, E. Acta Chem. Scand. **1959**, 13, 1547-1560.
Lee, J.C., Harpst, J.A. Arch. Biochem. Biophys. **1971**, 145, 55-63.
Lundin, S.J. Acta Chem. Scand. **1968**, 22, 2183-2190.
Main, A.R., McKnelly, S.C., Burgess-Miller, S.K. Biochem. J. **1977**, 167, 367-376.
Massoulié, J., Toutant, J.-P.In Handbook of Pharmacology; Whittaker, V.P., Ed.; Springer: Heidelberg, **1988**, Vol. 86; pp. 167-224.
Meedel, T.H., Whittaker, J.R. J. Exp. Zool. **1979**, 210, 1-10.
Meedel, T.H. Biochim. Biophys. Acta **1980**, 625, 360-369.
Myers, D.K. Biochem. J. **1953**, 55, 67-78.
Ozaki, H. Dev. Growth Differ. **1976**, 18, 245-257.
Pezzementi, L., Reinheimer, E.J., Pezzementi, M.L. J. Neurochem. **1987**, 48, 1753-1760.
Pezzementi, L., Nickson, H., Dunn, R., Bradley, R. Comp. Physiol. Biochem. **1989**, 92B, 385-387.
Rodrìguez-Borrajo, C., Barat, A., Ramìrez, G. Neurochem. Int. **1982**, 4, 563-568.
Rosenberry, T.L. Adv. Enzymol. Relat. Areas Mol. Biol. **1975**, 43, 103-218.
Rosenberry, T.L., Toutant, J.-P., Haas, R., Roberts, W.L. Methods Cell Biol. **1989**, 32, 231-255.
Rotundo, R.L. J. Biol. Chem. **1984**, 259, 13186-13194.
Silver, A., The Biology of Cholinesterase, North Holland: Amsterdam, **1974**.
Stieger, S., Gentinetta, R., Brodbeck, U. Eur. J. Biochem. **1989**, 181, 633-642.
Toutant, J.-P. Neurochem. Int. **1986**, 9, 111-119.
Toutant, J.-P. Prog. Neurobiol. **1989**, 32, 423-446.
Toutant, J.-P., Roberts, W.L., Murray, N.R., Rosenberry, T.L. Eur. J. Biochem. **1989**, 180, 503-508.
Toutant, J.-P., Massoulié, J., Bon, S. J. Neurochem. **1985**, 44, 580-592.
Tucci, A.F., Seifter, S. J. Biol. Chem. **1969**, 24, 841-849.
Younkin, S.J., Rosenstein C., Collins, P.L., Rosenberry, T.L. J. Biol. Chem. **1982**, 257, 13630-13637.

Studies on the Tetrameric Amphiphilic Acetylcholinesterase from Bovine Caudate Nucleus: Hydrophobic Domains, Membrane Anchoring, and State of Glycosylation

Harald Heider, Pascal Meyer, Jian Liao, Susi Stieger and Urs Brodbeck*,
Institute of Biochemistry and Molecular Biology, University of Bern, Bühlstrasse 28, 3012 Bern, Switzerland

Mammalian brain acetylcholinesterase (AChE) mainly exists as tetrameric globular enzyme (G_4 form) of which approximately 80% are amphiphilic AChE extractable only with detergent containing buffer (for reviews see Brodbeck, 1986; Rakonczay, 1986). Amphiphilic brain AChE is membrane bound through a structural subunit of about 20 kD which appears to be linked to one pair of the four catalytic subunits by disulfide bridges (Gennari et al., 1987; Inestrosa et al., 1987). This situation is reminiscent of the attachment of the asymmetric A_{12}-forms of AChE to collagen of basal laminae (for reviews see Massoulié and Toutant, 1988). Similar to A_{12} AChE, the second pair of subunits of mammalian brain AChE is also disulfide linked, and the tetrameric assembly is either hold together by strong non-covalent hydrophobic interactions or by additional disulfide bonds as proposed by Fuentes et al. (1988). The assembly of these four catalytic subunits together with the attached hydrophobic anchor migrates on SDS-PAGE with an apparent molecular mass of approximately 300 kD, while the monomer has an apparent molecular mass of about 68 kD.

On the other hand, G_2 AChE from erythrocytes and a number of other sources consists of two disulfide linked catalytic subunits which are membrane bound through a glycosylphosphatidylinositol (GPI) moiety covalently attached to the C-terminus of each subunit (for reviews see Silman and Futermann, 1987). Immunological data suggest high homology between G_2 AChE from erythrocytes and the G_4 form from brain.

To date, full length sequence information is available for the following mammalian acetylcholinesterases: human and murine brain AChE as well as fetal bovine serum (fbs) AChE (sequences presented at the 3rd International Meeting on Cholinesterases in La Grande Motte, 1990). Our group could identify around 130 amino acids of bovine brain AChE as well as bovine erythrocyte AChE (Heider et al., 1990). Chhajlani et al. (1989) published the N-terminal sequences of five tryptic peptides of human erythrocyte AChE. The mammalian sequences show a high degree of homology to other vertebrate cholinesterases such as Torpedo AChE (Schumacher at al., 1986; Sikorav et al., 1987), or human butyrylcholinesterase (BChE) (Lockridge et al., 1987; Prody et al., 1987).

While the complete chemical structure of the GPI-membrane anchor of dimeric AChE from human eythrocytes has been worked out (Roberts et al., 1987, 1988 a,b), relatively little is known about the anchor of the brain enzyme. By labelling with the photoactivatable reagent 3-(trifluoro-methyl)-3-(*m*[^{125}I]-iodophenyl) diazirine ([^{125}I]TID), a non-catalytic structural subunit of about 20 kD molecular mass was identified in human brain AChE (Gennari et al., 1987) and in bovine brain (Inestrosa et al., 1987). The [^{125}I]TID-label could in part be removed by treatment with hydroxylamine indicating that the [^{125}I]TID-labelled structural subunit is not only composed of hydrophobic amino acids but also contains other hydrophobic constituents (Inestrosa et al., 1987; Brodbeck and Stieger, 1988).

Acetylcholinesterases are ecto enzymes and have varying amounts of carbohydrates attached to the core protein. The glycoprotein nature of bovine brain AChE was described by Grossmann and Liefländer (1979), that of bovine erythrocyte by Ruess et al., (1976), and of human erythrocytes by Ott et al. (1975). Sequence analysis of the known primary structures of fbs AChE (Doctor et al., 1990) gives evidence for five asparagine linked carbohydrates. The present communication gives further information on the primary structure of the catalytic subunits as well as on the nature of the hydrophobic anchor and the state of glycosylation of bovine brain AChE.

Results and Discussion

Amphiphilic G_4 AChE was isolated from bovine caudate nucleus according to the procedure described for the human brain enzyme by Sørensen et al. (1982). The preparation was labelled with ^{3}H-diisopropyl fluorophosphate (DFP), precipitated with acetone, and the pellet redisolved in 6 M guanidine hydrochloride with subsequent subjection to SDS-PAGE in non-reducing conditions. Protein stain as well as autoradiogra-

0–8412–2008–5/91/0032$06.00/0

phy revealed that the major part of the enzyme migrated to a position in the gel corresponding to about 150 kD indicating that non-reduced denatured brain AChE predominantly exists as a dimer. Brain enzyme applied to a SDS-gel without prior treatment with 6 M guanidine hydrochloride migrated under non-reducing conditions to a position in the gel corresponding to about 300 kD molecular mass. Prolonged denaturation in boiling SDS led to the formation of increasing amounts of dimers and - to a certain extent - monomers. The appearance of monomeric AChE could be due to preassembled dimers which are not yet interconnected by disulfide bonds or due to the action of a copurified protease.

From our results, we conclude that the amphiphilic G_4-form of brain AChE is composed of pairs of disulfide linked dimers (one dimer being formed directly by inter-chain disulfide bonds, the other connected via disulfide bonds to the hydrophobic 20 kD structure). This intersubunit disulfide bond pattern is in agreement with the 7 cysteine residues found in fbs-AChE. The corresponding intra-chain disulfide bonding pattern conserved in AChE and BChE shows 3 intrachain disulfide loops leaving one SH group free for one inter-chain disulfide bond. Our results further show that pairs of dimers assemble to the G_4 form by strong non-covalent hydrophobic interactions which are only overcome by either boiling the enzyme in SDS prior to electrophoresis, or dissolving it in high concentrations of guanidine hydrochloride. The subunit assembly of brain AChE is thus analogous to that found in the asymmetric A_{12} AChE.

To further characterize these interactions and to possibly identify hydrophobic intersubunit contact areas, bovine brain AChE was labelled by [^{125}I]TID essentially as described by Stieger et al. (1984). About 80% of all label incorporated into bovine brain AChE was recovered in the hydrophobic membrane anchor, and about 20% of the label was distributed over the four catalytic subunits. The labelled enzyme was cleaved by cyanogen bromide, and the resulting four radioactive peptides were resolved by SDS-PAGE in Tricine buffer (Schägger and von Jagow, 1987). N-terminal sequence analysis of three of these peptides revealed three stretches alignable to the sequence of fbs-AChE (Fig. 1). The two peptides with the highest molecular masses (31 and 34 kD) had identical N-terminal sequences originating at position 212 (numbering according to fbs-AChE). At position 223, we identified isoleucine and histidine in equal amounts whereas in fbs-AChE, Doctor et al. (1990) detected only histidine at this position. The third peptide with an apparent molecular mass of 11 kD has an N-terminal sequence which is homologous to the sequence of fbs-AChE starting at residue 478.

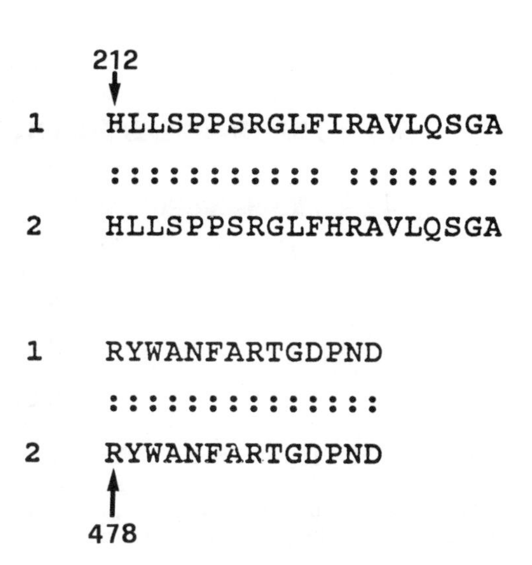

Fig. 1. N-terminal sequences of ^{125}I-TID-labelled CNBr peptides of the catalytic subunits from bovine brain AChE (1) aligned to the corresponding sequence stretches of fetal bovine serum AChE (2). The numbers at the beginning of the sequence stretches are referring to the numbering of fetal bovine serum AChE.

From this result, it is concluded that the hydrophobic contact areas between the two pairs of dimers are located C-terminally of the active site serine. The fourth peptide represented the anchor which migrated on the Tricine-gel with an apparent molecular mass of 18 kD. N-terminal sequencing of this peptide failed.

In non-reducing conditions, the CNBr cleavage pattern of the [^{125}I]TID-labelled enzyme was significantly changed. The band corresponding to the anchor became larger with an apparent molecular mass of 23 kD. The band at 11 kD disappeared, and the two 31 and 34 kD bands shifted to molecular masses of about 42 and 45 kD. Our group (Gennari et al., 1987) and Inestrosa et al. (1987) have shown earlier that conversion of amphiphilic brain AChE to the hydrophilic enzyme by proteinase K treatment not only detached the anchor without grossly reducing the molecular mass of the catalytic subunits but also led to the monomerization of the tetramer on SDS-gels in non-reducing conditions. These data and the CNBr cleavage pattern of the unreduced enzyme give strong evidence that the membrane anchor is disulfide bonded to the C-terminal CNBr peptide which is generated by the cleavage after methionine 558 (numbering according to fbs-AChE) as shown in Fig. 2. The intersubunit disulfide bridges most probably are also mediated via cysteines located within this C-terminal peptide in agreement with the findings

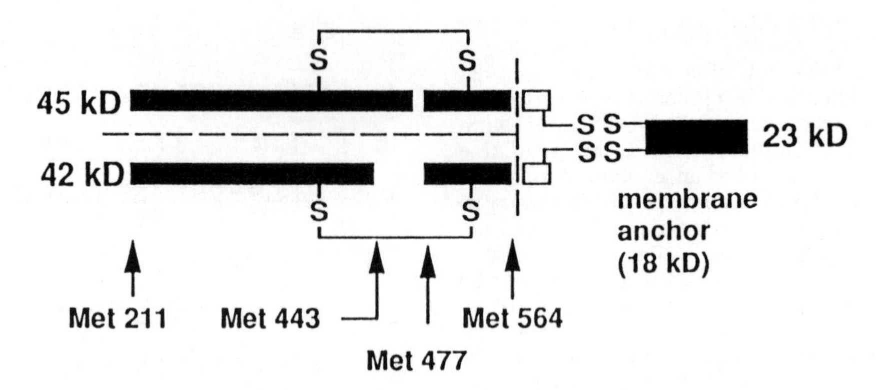

Fig. 2. ^{125}I-TID-labelled CNBr peptides of bovine brain AChE (black bars) with their corresponding disulfide bridges. Dashed lines separate the three radioactive CNBr peptides detectable upon SDS-PAGE of unreduced ^{125}I-TID-labelled brain AChE.

of McPhee-Quigley et al. (1986) who could show that in the asymmetric AChE of Torpedo two monomers are interconnected via disulfide bonds between the cysteines at positions 572. Most recently, Rosenberry and coworkers (presented at La Grande Motte Meeting 1990) showed that the cysteine involved in intersubunit disulfide bonding of bovine brain AChE is located in the amino acid pattern Cys-Ser-Asp-Leu which is completely homologous to the C-terminus of fbs-AChE indicating that the interchain disulfide bond is located at the last cysteine residue in the amino acid sequence, a result which is in complete agreement with our findings.

From the results of Inestrosa et al. (1987) as well as from our own observations (Brodbeck and Stieger, 1988), it appears that the anchor of brain AChE not only contained hydrophobic amino acids but might also be glycosylated, acylated, alkylated, or contain another hydrophobic residue such as a prenyl one. N-linked glycosylation could not be detected as treatment of the TID-labelled enzyme by N-glycosidase F did not result in a change in migration behaviour of the labelled anchor peptide on SDS-PAGE. On the other hand, treatment of [^{125}I]TID-labelled brain AChE with 1 M hydroxylamine at pH 7.4 released 20% of the label from the anchor without decreasing the amount of label in the catalytic subunits. Since the amount of label released at pH 7.4 was the same as the amount released in alkaline conditions, it is concluded that presumably there exists a fatty acid in thiol-linkage to the hydrophobic anchor. This could be further substantiated by treatment of the TID-labelled enzyme with sodium methylate which, by transesterification, yields methyl esters of fatty acids. Analysis of the reaction product by TLC revealed a radioactive spot which migrated with an R_f-value closely similar to standard fatty acid methyl ester. Density gradient centrifugation of hydroxylamine treated [^{125}I]TID-labelled brain AChE in the absence and presence of detergent revealed no major difference in the sedimentation patterns as compared to the untreated enzyme. This result showed that the removal of a putative fatty acid did not render the enzyme hydrophilic and had no major influence on the amphiphilic nature of bovine brain AChE.

Treatment of SDS-denatured bovine brain AChE with N-glycosidase F resulted in a new band which migrated on SDS-gel with a molecular mass of about 9 kD less than the untreated one. In contrast, the molecular mass of the erythrocyte enzyme was decreased from 77 to 58 kD upon N-glycosidase F digestion of the denatured enzyme. N-acetylgalactosamine could neither be detected in bovine brain nor in bovine erythrocyte AChE indicating that both enzyme forms contain no O-linked carbohydrates.

A monoclonal antibody MAB 2G8 (subclass IgG2a) raised against AChE from electric organs of Torpedo nacline timilei (Xin et al., 1989) showed interesting properties towards AChE from bovine brain and bovine erythrocytes. This antibody crossreacted with AChE from bovine and human brain whereas no significant reaction was found between MAB 2G8 and AChE from bovine and human eythrocytes. Treatment of the brain enzyme with N-glycosidase F abolished binding of MAB 2G8 suggesting that the epitope, or part of it, consists of N-linked carbohydrates. The remarkable difference in reactivity of MAB 2G8 towards AChE from brain and erythrocytes as well as the differences observed after treatment with N-glycosidase F suggest different glycosylation patterns in dimeric erythrocyte and in tetrameric brain AChE.

The data presented in this paper support the model shown in Fig.3.

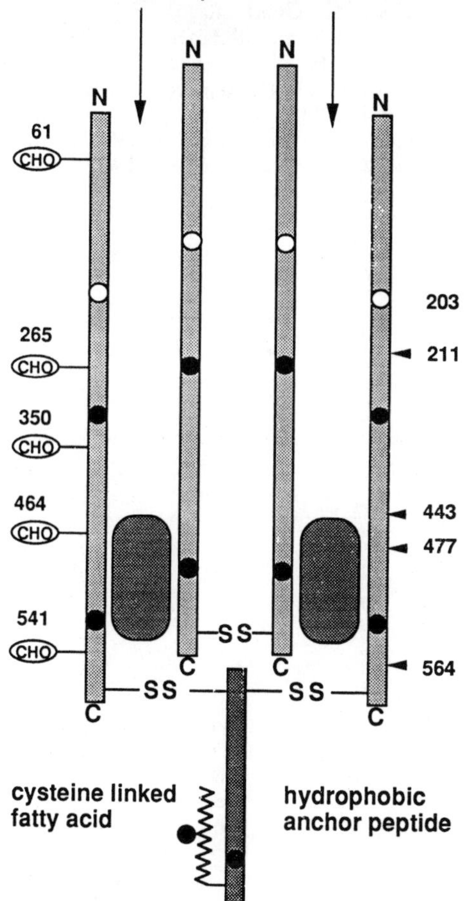

Fig. 3. Model of the putative assembly of bovine brain AChE. Numbers with arrows at right refer to CNBr cleavage sites which lead to ^{125}I-TID-labelled peptides (numbering according to fetal bovine serum AChE (Doctor et al., 1990). Open circles represent the active site serine residues of the subunits (position 203). The black circles stand for covalently attached ^{125}I-TID-label (orientation within the corresponding CNBr peptides is arbitrary). CHO symbolizes the potential sites of glycosylation. In this schematic representation, CNBr cleavage sites and sites of glycosylation are depicted on one subunit for reasons of clarity only, although they exist on all four identical catalytic subunits.

Acknowledgements

We thank Dr. J. Schaller for amino acid sequencing. This work was supported by the Swiss National Science Foundation grant No. 3.070-0.87 and by funds of the Hochschulstiftung der Universität Bern, the Sandoz Stiftung, and the Roche Research Foundation.

References

Brodbeck, U. In *Progress in Protein-Lipid Interactions*, Watts, A., DePont, J.J.H.H.M., Eds., Elsevier, Amsterdam, **1986**, *2*, 303-338.

Brodbeck, U., Stieger, S., *Neurochem. Internat. Suppl.*, *1*, **1988**, *13*, 32.

Chhajlani, V., Derr, D., Earles, B., Schmell, E., August, T., *FEBS Lett.*, **1989**, *247*, 279-282.

Doctor, B.P., Gentry, M.K., Wu, S.-J., De La Hoz, D.M., **1990**, *this volume*.

Fuentes, M.-E., Rosenberry, T.L., Inestrosa, N.C., *Biochem. J.*, **1988**, *256*, 1047-1050.

Gennari, K., Brunner, J., Brodbeck, U., *J. Neurochem.*, **1987**, *49*, 12-18.

Grossmann, H., Liefländer, M., *Z. Naturforsch.*, **1979**, *34c*, 721-725.

Hall, L.M., Spierer, P., *EMBO J.*, **1986**, *5*, 2949-2954.

Heider, H., Litynski, P., Stieger, S., Brodbeck, U., *Cell. Mol. Neurobiol.*, **1990**, in press.

Inestrosa, N.C., Roberts, W.L., Marshall, T.L., Rosenberry, T.L., *J. Biol. Chem.*, **1987**, *262*, 4441-4444.

Lockridge, O., Bartels, C.F., Vaughan, T.A., Wong, C.K., Norton, S.E., Johnson, L.L., *J. Biol. Chem.*, **1987**, *262*, 549-557.

Massoulié, J., Toutant, J.-P., In *Handb. Exp. Pharm. The Cholinergic Synapse*, Whittaker, V.P., Ed., Springer-Verlag, Berlin, **1988**, *86*, 167-224.

McPhee-Quigley, K., Vedvick, T.S., Taylor, P., Taylor, S.S., *J. Biol. Chem.*, **1986**, 261, 13565-13570.

Ott, P., Jenny, B., Brodbeck, U., *Eur. J. Biochem.*, **1975**, *57*, 469-480.

Prody, C.A., Zevin-Sonkin, D., Gnatt, A., Goldberg, O., Soreq, H., *Proc. Natl. Acad. Sci.*, **1987**, *84*, 3555-3559.

Rakonczay, Z., In *Neuromethods*, Boulton, A.A., Baker, G.B., Yu, P.H., Eds., Human Press Inc., Clifton, N.J., **1986**, *5*, 319-360.

Roberts, W.L., Kim, B.H., Rosenberry, T., *Proc. Natl. Acad. Sci.*, **1987**, *84*, 7817-7821.

Roberts, W.L., Myher, J.J., Kuksis, A., Low, M.G., Rosenberry, T.L., *J. Biol.Chem.*, **1988a**, *263*, 18766-18775.

Roberts, W.L., Santikarn, S., Reinhold, V.N., Rosenberry, T.L., *J. Biol. Chem.*, **1988b**, *263*, 18776-18784.

Ruess, K.P., Weinert, M., Liefländer, M., *Hoppe-Seyler's Z. Physiol. Chem.*, **1976**, *357*, 783-793.

Schägger, H., von Jagow, G., *Anal. Biochem.*, **1987**, *166*, 368-379.
Schumacher, M., Camp, S., Maulet, Y., Newton, M., McPhee-Quigley, K., Taylor, S.S., Friedmann, T., Taylor, P., *Nature*, **1986**, *319*, 407-409.
Sikorav, J.-L., Krejci, E., Massoulié, J., *EMBO J.*, **1987**, *6*, 1865-1873.
Silman, I., Futermann, T., *Eur. J. Biochem.*, **1987**, *170*, 11-22.
Smyth, K.K., De La Hoz, D.M., Christner, C.E., Rush, R.S., De La Hoz, F., Doctor, B.P., *FASEB J.*, **1988**, *2*, A 1745.
Sørensen, K., Gentinetta, R., Brodbeck, U., *J. Neurochem.*, **1982**, *39*, 1050-1060.
Stieger, S., Brodbeck, U., Reber, B., Brunner, J., *FEBS Lett.*, **1984**, *168*, 231-234.
Xin, Y.B., Yan, G.Z., Xu, P.F., Li, F.Z., Zhang, H., Wang, J.B., Sun, M.C., *Chinese J. Pharmacol. and Toxicol.*, **1989**, *3*, 68-72.

Fetal Bovine Serum Acetylcholinesterase: Structure-Function Correlation

Bhupendra P. Doctor*, Mary K. Gentry, Shuenn-Jue Wu, Y. Ashani and Denise M. De La Hoz, Division of Biochemistry, Walter Reed Army Institute of Research, Washington DC 20307-5100 USA

The complete amino acid sequence of the catalytic subunit of fetal bovine serum acetylcholinesterase (FBS AChE) has been elucidated. Monoclonal antibodies (mAb) which affect the catalytic function of this enzyme have been employed to map the topography of the functional regions of the molecule.

Primary structures of the catalytic subunit of acetylcholinesterase (EC 3.1.1.7, AChE) isolated from *Torpedo californica* (Schumacher et al., 1986), *Torpedo mamorata* (Sikorav et al., 1987) *Drosophila melanogaster* (Hall & Spierer, 1986) fetal bovine serum (Doctor et al., 1990), mouse (Rachinsky et al., 1990), and butyrylcholinesterase (EC 3.1.1.8, BChE) isolated from human serum (Lockridge et al., 1987) and fetal human brain (Prody et al., 1987) have recently been reported. Partial sequences of AChEs isolated from human erythrocytes (Chhajlani et al., 1984) and human brain (Soreq et al., 1989) have also been reported. Pharmacological studies and enzyme kinetic studies with these enzymes, including multiple molecular forms, have demonstrated that they possess similar catalytic properties. The similarities and any differences in their catalytic functions should be reflected in the corresponding similarities and differences in the structural domains of their catalytic subunit(s). The esteratic regions of several cholinesterases share a high degree of amino acid sequence homology (MacPhee-Quigley et al., 1985) while some other regions are markedly dissimilar (Doctor et al., 1990).

In this report we present the comparison of the amino acid sequence of FBS AChE with other cholinesterases in various regions of the molecule which may be involved in catalytic function and a glimpse of the topography of some regions of the molecule. Using monoclonal antibodies, we have determined some topographical characteristics of the enzyme.

Fig. 1 describes the complete amino acid sequence of the catalytic subunit of AChE isolated from fetal bovine serum. The FBS AChE sequence shares an identity of 89% with mouse AChE (Rachinsky et al., 1990) 61% with *T. californica* AChE (Schumacher et al., 1986) and 36% with Drosophila AChE (Hall & Spierer,1986). It also has a relatively high degree of identity (52%) with butyrylcholinesterase from human serum (Lockridge et al., 1987) and human fetal tissues (Prody et al., 1987), 33% with rabbit and 35% with rat liver carboxyesterase (Long et al., 1983), 29% with Drosophila esterase-6 (Oakeshott et al., 1987), 32% with bovine thyroglobulin (Mercken et al., 1985) and 32% with rat lysophospholipase (Han et al., 1987).

There are seven cysteine residues in FBS AChE, all of which are located at the same position in all AChEs, except for Torpedo AChE in which there are four fewer amino acid residues in the small S-S loop (cys-257 to cys-272). There are five presumed N-linked glycosylation sites.

Since multiple molecular forms of AChE isolated from various species possess similar catalytic properties, the active site serine-containing region (ser-203) might be expected to have structural similarities. This was shown to be true by MacPhee-Quigley et al., 1985, who compared the amino acid sequence of Torpedo AChE with human serum BChE in this region. Fig. 2 shows the comparison of amino acid sequences of ChEs in several regions of the molecule. As shown in Fig. 2A there is identity among 16 amino acids of bovine, mouse and human AChEs and matching of 12 of 16 with Torpedo AChE, 11 of 16 with Drosophila AChE, and 15 of 16 with human serum BChE. The sequence PXXXTXFGESAGXXSV is common to all six cholinesterases. The folding pattern of the active site domain may be pivotal in conferring on these molecules the specificity of catalytic function. Using polyclonal and monoclonal antibodies we have recently shown that this region is located in a pocket-like conformation (Ogert et al., 1990).

Figs. 2B and 2C present sequences in two other regions of the molecule which may be potential candidates for catalytic site(s). The amino acid sequences of FBS AChE in the regions of asp-95 and his-447 are compared with the sequences of other ChEs in the same regions. These two amino acids, in addition to ser-203, are suggested to be involved in a catalytic triad mechanism. The sequence in this region is highly conserved. Based on results of the binding of

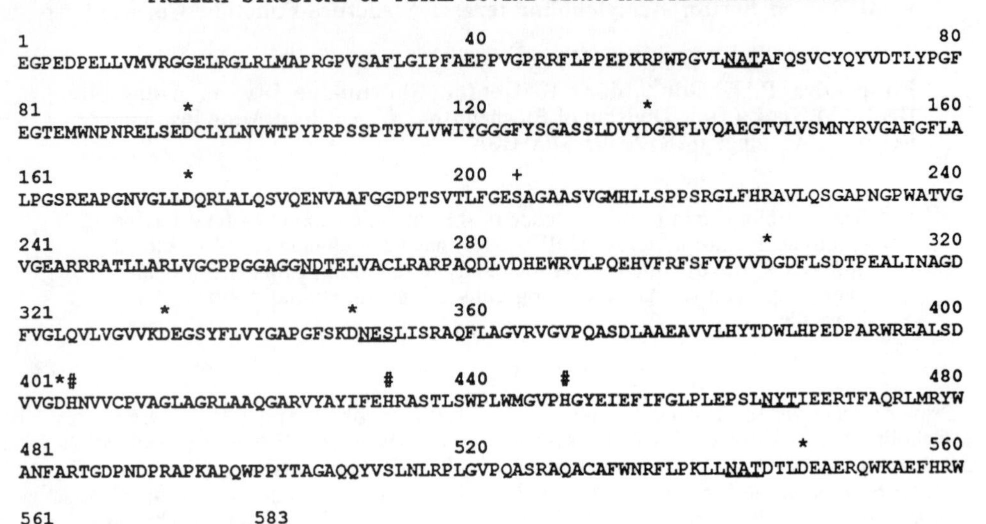

Fig. 1. Amino acid sequence of fetal bovine serum acetylcholinesterase. Active site serine is marked with (+). Homologous asp residues with (*) and homologous his residues with (#). Apparent N-linked glycosylation sites are underlined. S--S loops are shown by solid lines. (Adopted from Doctor et al. 1990).

```
                                *
           FBS AChE   PTSVTLFGESAGAASV
         HUMAN AChE   PTSVTLFGESAGAASV
         MOUSE AChE   PMSVTLFGESAGAASV
2A.    TORPEDO AChE   PKTVTIFGESAGGASV
          DROS AChE   PEWMTLFGESAGSSSV
         HUMAN BChE   PKSVTLFGESAGAASV
          CONSENSUS   P---T-FGESAG--SV
```

```
                                 *
           FBS AChE   WNPNRELSEDCLYLNVWTP
         MOUSE AChE   WNPNRELSEDCLYLNVWTP
       TORPEDO AChE   WNPNREMSEDCLYLNIWVP
2B.       DROS AChE   WNPNTNVSEDCLYINVWAP
      HM. SR. BChE   WNPNTDLSEDCLYLNVWIP
          CONSENSUS   WNPN---SEDCLY-N-W-P
```

```
                                          *
       FBS AChE   HRASTLSWPLWMGVPHGYEIEFIFG
     HM.BR.AChE   HRASTLSW--WMGVPHGYEIEFTFG
     MOUSE AChE   HRASTLTWPLWMGVPHGYEIEGIGG
      TORP AChE   HRASNLVWPEWMGVIHGYEIEYFFG
2C.   DROS AChE   HRTSTSLWGEWMGVLHGDEIEYFFG
     HM SR BChE   HRSSKLPWPEWMGVMHGYEIEFVFG
      CONSENSUS   HR-S---W--WMGV-HG-EIE---G
```

Fig. 2. Comparison of amino acid sequences in various regions of FBS AChE with other cholinesterases. Sources of sequences are as noted in the text. Amino acids are numbered according to FBS AChE. Fig. 2A. Sequences in the region of ser-203 (*). Fig. 2B. Sequences in the asp-95 region (*). Fig. 2C. Sequences in the his-447 region (*).

inhibitory monoclonal antibody AE-2 to this region of FBS AChE and the sequence homology in the region surrounding asp-95, it is likely that this amino acid may be involved in the catalytic triad (Doctor et al., 1989). Fig. 2C shows the amino acid sequence following his-432 and around his-447, both of which are conserved in all ChEs. The high degree of sequence homology surrounding his-447, compared to his-432, makes it more likely to be a candidate in catalytic triad (Doctor et al., 1989). Since his-447 and asp-95 are located in disulfide loops (MacPhee-Quigley et al., 1986) and asp-95 is located next to cys-96, which may affect its net charge, we feel that his-447 and asp-95 are likely candidates for involvement with ser-203 in the charge relay system.

Finally, the comparison of amino acid sequences of FBS AChE with mouse, Torpedo, Drosophila AChEs and human serum BChE in the region of a small disulfide loop (amino acids 254-275, Fig. 3) shows that FBS AChE, mouse and Drosophila AChE have four extra amino acids in the loop. The presence of these additional amino acids in the loop in mammalian AChEs would result in a different conformation in this region.

In general, mammalian AChEs appear to have structures more in common with each other than with Torpedo AChE or human serum BChE.

We have prepared monoclonal antibodies directed against various forms (natural or modified) of AChEs with the objective of using them to map not only the topography of the entire surface region of the enzyme but also those regions of the molecule which may be involved in its catalytic function. One series of mAbs were directed against the 25-amino acid synthetic peptide corresponding to the amino acid sequence surrounding the active site serine of *Torpedo californica* AChE (Ogert et al., 1990). These monoclonal antibodies were designed to determine whether this region is located on the surface of the molecule or is located in a pocket of restricted access. If this region is located on the surface of the enzyme, the antibodies would be able to bind to the enzyme in its native conformation. If, on the other hand, this region is located in a pocket-like conformation as in the serine proteases, the mAb would bind to the denatured enzyme but not to the native conformation. Results obtained from these studies showed that one group of these mAbs recognized the denatured form of the enzyme but not the native form. This group of antibodies showed similar specificity for FBS AChE and Torpedo AChE. The epitope for one of this group of mAbs consists of a truncated region within the peptide which contains the active-site serine. Based on these results it appears that in native form the active site serine containing regions of both Torpedo AChE and FBS AChE are located in a pocket and thus are not accessible for binding to antibodies generated against a peptide corresponding to the amino acid sequence of this region. The elucidation of the detailed structure of the AChE catalytic site pocket will await x-ray diffraction data on single crystals. However, these results provide a first glimpse of the catalytic site in terms of suspected topography.

Obviously, the most useful mAbs are the ones that influence the catalytic function of AChE. One inhibitory mAb, 25Bl, generated against DFP-inhibited FBS AChE has been characterized as to its anti-ChE properties (Ashani et al., 1990). It was shown that this mAb binds to a conformational epitope in the region of the peripheral anionic site near the active site of the enzyme and nearly completely inhibits its catalytic activity. Anti-idiotypic antibodies to this inhibitory antibody would have an internal image of the peripheral anionic site of FBS AChE.

Using this inhibitory antibody we have generated antibodies in rabbits serum which presumably contains a sub population of antibodies whose combining sites resemble the conformation of the region of FBS AChE to which the mAb 25B1 binds (Wu et al., 1990). The IgG enriched fraction containing anti-idiotype Ab was obtained from rabbit serum by ammonium sulphate precipitation (0-50% saturation) and chromatography on Protein A Sepharose. The extent of binding of 25B1 to the anti-idiotype and to normal rabbit immunoglobulin was determined by ELISA. Binding of the Fab fragment of mAb 25Bl to the anti-idiotype and to normal rabbit serum immunoglobulin was also determined. As shown in Fig 4 both forms of the 25Bl are equally recognized by anti-idiotypic antibody. Thus, the epitope for the anti-idiotype is located in the Fab region.

As mentioned above, mAb 25Bl almost completely inhibits the catalytic activity of FBS AChE (>99%), presumably by binding to the enzyme in the region of the peripheral anionic site. When a fixed amount of anti-idiotype was preincubated with varying amounts of mAb 25Bl, and the complex was assayed for catalytic activity, the results show (Fig. 5) that mAb 25Bl when complexed with anti-idiotype appears to lose its ability to inhibit the catalytic activity of

FBS AChE	C?PGGAGGNDTELVAC
MOUSE AChE	CPPGGAGGNDTELIAC
TORPEDO AChE	CNLNS----DEELIHC
DROS AChE	CNASMLKTNPAHVMSC
HUMAN BChE	CSRENETEII----KC
CONSENSUS	C--------------C

Fig. 3. Comparison of amino acid sequences in small S-S loop region of various cholinesterases.

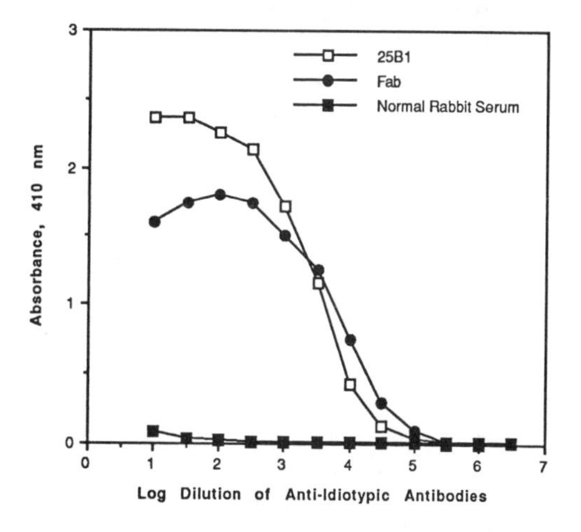

Fig. 4. Binding of purified rabbit antiidotypic antibodies to Fab fragments of mAb 25B1 and mAb 25B1 using indirect enzyme linked immuno-absorbent assay (ELISA).

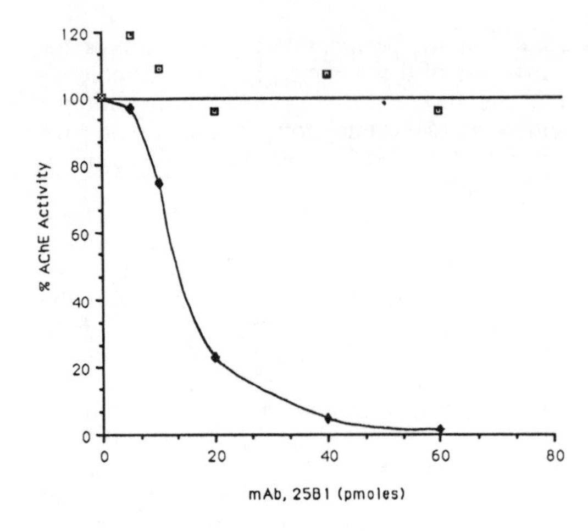

Fig. 5. Prevention of inhibition of FBS-AChE activity by mAb 25B1 in the presence of anti-idiotypic antibodies. A constant amount of purified anti-idiotypic (—□—) or normal serum (—◆—) antibodies were added to each tube containing increasing amount of mAb 25B1. The mixtures were incubated overnight at 4°C. Five units of FBS-AChE were added to each tube and incubated for 4 hours at room temperature. The enzyme activity in each tube was determined by the Ellman assay.

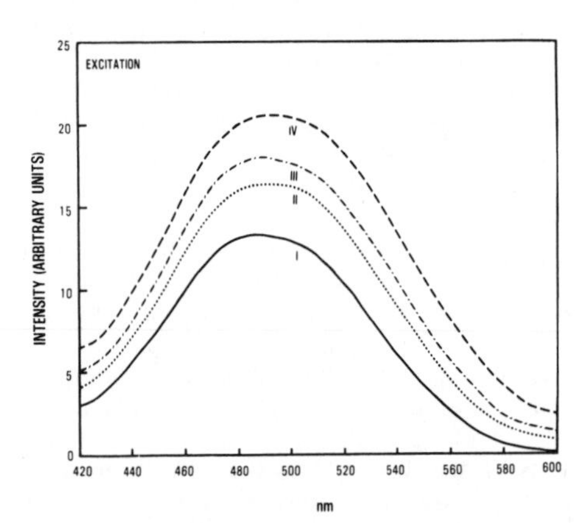

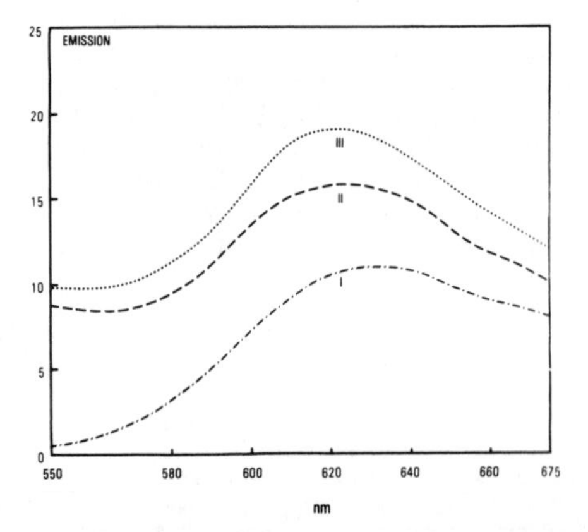

Fig. 6A. Binding of anti-idiotypic antibodies to propidium. The binding was dose dependent when tested with three concentrations (II-8.7 μM, III-17.3 μM, IV-35 μM) of anti-idiotypic antibodies. When purified normal rabbit serum was used as control, both the excitation and emission spectra superimposed with those of propidium alone (I).

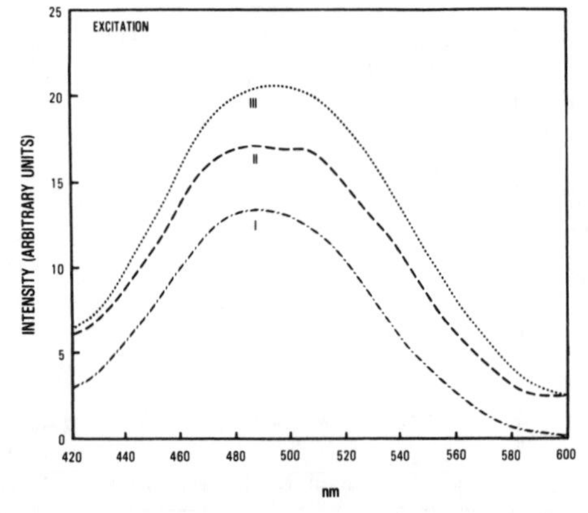

Fig. 6B. The dissociation of propidium from Ag:Ab complex. With the addition of 3.7 μM mAb 25B1 (II) to the mixture of 35 μM anti-idiotypic antibodies and 2.4 μM propidium (III), partial dissociation of propidium from its conjugate with anti-idiotypic antibodies was observed in both excitation and emission spectra compared with propidium alone (I).

FBS AChE. The Ag:Ab complex under the present experimental conditions was soluble, as evidenced by high speed centrifugation.

Further evidence for generation of an internal image of the peripheral anionic site of FBS AChE on the rabbit anti-idiotype was obtained by showing the binding of propidium, a peripheral anionic site ligand (Taylor & Lappi, 1975) to the anti-idiotype fraction. As seen in Fig. 6A both excitation and emission fluorescence spectral maxima were increased in a concentration dependent manner when an increasing amount of anti-idiotype was added to a constant amount of propidium. When mAb 25B1 was added to the anti-idiotype : propidium complex, the fluorescence maxima decreased, indicating that mAb 25B1 has a higher binding constant for the anti-idiotype than propidium (Fig. 6B).

In summary, the complete amino acid sequence of a mammalian AChE, FBS AChE is described. Asp-95, ser-203 and his-447 appear to be potential candidates to participate in a charge relay mechanism for the catalysis of acetylcholine hydrolysis. Using specific monoclonal antibodies, it has been shown that the active site serine containing region of the molecule is located in a pocket-like conformation. Rabbit anti-idiotypic antibodies were generated against an inhibitory antibody that was shown to bind to a peripheral anionic site of FBS AChE. These antibodies were characterized to demonstrate that they may contain an internal image of the peripheral anionic site of the enzyme.

References

Ashani, Y., Gentry, M.K., Doctor, B.P., *Biochemistry*, **1990**, *29*, 2456-2463.

Chhajlani, V., Derr, D., Earles, B., Schmell, E., August, T., *FEBS Lett.*, **1989**, *247*, 279-282.

Doctor, B.P., Chapman, T.C., Christner, C.E., Deal, C.D., De Le Hoz, D.M., Gentry, M.K., Ogert, R.A., Rush, R.S., Smyth, K.K., Wolfe, A.D., *FEBS Lett.*, **1990**, *266*, 133-137.

Doctor, B.P., Smyth, K., Gentry, M., Ashani, Y., Christner, C., De La Hoz, D., Ogert, R., Smith, S. In *Progress in Clinical and Biological Research*; Rein, R.; Golombek, A., Eds.; Alan R. Liss, Inc., New York, NY, **1989**, Vol. 289; pp. 305-316.

Hall, L., Spierer, P., *EMBO J.*, **1986**, *5*, 2949-2954.

Han, J.H., Stratowa, C., Rutter, W.J., *Biochemistry*, **1987**, *26*, 1617-1625.

Lockridge, O., Bartels, C., Vaughn, T., Wong, C., Norton, S., Johnson, L., *J. Biol. Chem.*, **1987**, *262*, 549-557.

Long, R.M., Satoh, H., Martin, B.M., Kimura, S., Gonzalez, F. J., Pohl, L.R., *Biochem. Biophys. Res. Comm.*, **1988**, *156*, 866-873.

MacPhee-Quigley, K., Taylor, P., Taylor, S., *J. Biol. Chem.*, **1985**, *260*, 12183-12189.

MacPhee-Quigley, K., Vedvick, T.S., Taylor, S., *J. Biol. Chem.*, **1986**, *261*, 13565-13570.

Mercken, L., Simons, M.-J., Swillens, S., Massaer, M., Vassart, G., *Nature*, **1985**, *316*, 647-651.

Oakeshott, J.G., Collet, C., Phillis, R.W., Nielsen, K.M., Russell, R.J., Chambers, G.K., Ross, V., Richmond, R.C., *Proc. Natl. Acad. Sci. USA*, **1987**, *84*, 3359-3363.

Ogert, R., Richardson, E., Gentry, M.K., Abramson, S., Alving, C., Taylor, P., Doctor, B.P., *J. Neurochem.*, **1990**, *55*, 756-763.

Prody, C., Zevin-Sonkin, D., Genatt, A., Goldberg, 0., Soreq, H., *Proc. Natl. Acad. Sci. USA*, **1987**, *84*, 3555-3559.

Rachinsky, T.L., Camp, S., Taylor, P., *FASEB J.*, **1990**, *4*, 1183.

Schumacher, M., Camp, S., Maulet, Y., Newton, M., MacPhee-Quigley, K., Taylor, S., Friedman, T., Taylor, P., *Nature*, **1986**, *319*, 407-409.

Sikorav, J.L., Krejci, E., Massoulie, J., *EMBO J.*, **1987**, *6*, 1865-1873.

Soreq, H., Prody, C. In *Progress in Clinical and Biological Research*; Rein, R.; Golombek, A., Eds.; Alan R. Liss, Inc., New York, NY, 1989, Vol. 289; pp. 347-359.

Taylor, P., Lappi, S., *Biochem.*, **1975**, *14*, 1989-1997

Wu, S-J., Gentry, M.K., Ashani, Y., R.K., Doctor, B.P., *FASEB J.*, **1990**, *4*, 1184.

Molecular Heterogeneity of Human Plasma Cholinesterase

Patrick Masson, Centre de Recherches du Service de Santé des Armées, Unité de Biochimie, BP87, 38702, La Tronche cedex, France

Molecular heterogeneity of butyrylcholinesterase refers to the existence of multiple molecular forms. The main forms of the plasma BChE are size isomers G_1, G_2, G_4 (some of them contain cuts in their polypeptide backbone) and a covalent conjugate G_1–ALB. The inconstant C_5 isoenzyme is a non covalent hybrid of G_4 with a 60kD component that remains to be identified.

The existence of a non specific cholinesterase (butyrylcholinesterase, BChE ; EC .3.1.1.8) in plasma was recognized at the beginning of the thirties . This enzyme has been extensively investigated (for reviews see : Whittaker, 1986 ; Chatonnet, 1989).

The plasma BChE is a 24 % carbohydrate sialoglycoprotein (exclusively ?) synthesized in the liver (Khoury, 1987).Though its synthesis is governed by a single gene (Arpagaus, 1990) located on the q arm of chromosome 3 (Sparkes, 1984 ; Zakut, 1989), this enzyme shows a complex genetic and molecular polymorphism.

This communication deals with the latter which appears to be mainly the result of posttranslational modifications of the gene products. Our purpose is to summarize the structural organization of the multiple molecular forms of the enzyme.

Genetic Polymorphism

The genetic polymorphism of human plasma BChE has initially been identified by an exaggerated response to succinylcholine, a short acting myorelaxant (Whittaker, 1986).

Variants are allelozymes of the BChE gene, previously named CHE_1 or E_1 gene. The genotyping methods using the polymerase chain reaction amplification have already allowed identification of point mutations responsible for several variants (McGuire, 1989 ; La Du, 1990) and a frameshift mutation leading to a "silent" truncated variant (Nogueira, 1990).

Molecular polymorphism

In plasma, the products of the BChE gene exist in multiple molecular forms. Four main forms can be recognized by gel electrophoresis ; they are designated in order of decreasing mobility as C_1, C_2, C_3 and C_4. The C_4 form represents about 90 % of the activity. In serum, an additional form of intermediate mobility between C_3 and C_4 is observed. This form is progressively formed at the expense of C_4 upon action of trypsin, plasmin and related proteases (Saeed, 1971). Another nomenclature allows designation of the five serum forms : ChE_1 for C1, ChE_2 for C_2, ChE_3 for C_3, ChE_4 for the protease–generated form and ChE_5 for C_4. Besides the above–mentioned molecular forms, a slow–migrating variant called C_5 because it migrates behind C_4 has been described (Harris, 1962). C_5 has no counterpart in the ChE nomenclature system. The use of the two nomenclature systems is confusing. Confusion may also arise from BChE uncommon electrophoretic patterns of normal subjects (Van Ross, 1966 ; Yoshida, 1969 ; Delbrück, 1979) and of patients with malignant diseases (Gallango, 1969 ; Ogita, 1975) where bands slower than C_5, noted C_6 to C_{10} have been observed. These slow–migrating forms are different from other slow–migrating components called "storage bands" S_1 and S_2 (Harris, 1962) whose intensity increases at the expense of C_4 during storage of serum at low temperature. But other more complicated patterns can be seen : e. g . splitting in two of all the bands ; increasing the number of bands up to 12 in 1–D gel (Juul, 1968) and up to 22 in 2–D gel (Mascall, 1977) ; fast migrating bands faster than C_1 or bands of mobility intermediate between C_3 and ChE_4. A possible clinical significance of these forms cannot be ruled out but most of them are undoubtedly artifacts depending on the method of analysis e.g. buffer–salt induced conformers. Therefore, a nomenclature based on the molecular structure as for AChE (Massoulié, 1988 ; Chatonnet, 1989) would be more convenient than numbering the forms according to their electrophoretic mobility. For that, molecular characterization of the forms is needed. Attention will be focused on C_2, C_4 and C_5.

Size isomers.

C_1, C_3 and C_4 represent globular (G) monomeric, dimeric and tetrameric forms of the enzyme (Muensch, 1976 ;Masson, 1979) ; they will be noted G_1, G_2 and G_4 according to the number of subunits. The subunit is a 85kDa monomer of 574 aminoacids with 9 asparagine–linked carbohydrate chains and it possesses one active center (Lockridge, 1987a). The G_4 form is a 340kDa homotetramer arranged as a dimer of dimers (Lockridge, 1979). It appears to be an asymmetric enzyme in which the two dimers are associated in a quasi linear fashion (Masson, 1979, 1984). The two subunits in each dimer are linked by a single disulfide bridge located at four aminoacids from the C–terminus (Lockridge, 1987b). The 3 size isomers are partially interconvertible (La Motta, 1965 ; Masson, 1979). Thus, G_1 and G_2 may be precursors or / and degradation products of G_4. Some related forms may also result from a differential attachment or cleavage of carbohydrate units.

Microheterogeneity of G_4

Though G_4 migrates on non denaturing gels as a single species there is structural and kinetic evidence that it is not homogeneous.

G_4 is a highly stable enzyme (Masson, 1986). The four subunits are held together by strong noncovalent bonds and the assembly is stabilized by the intersubunit

0–8412–2008–5/91/0042$06.00/0

disulfides (Lockridge, 1979). However, former works showed that a part of C_4 can dissociate spontaneously into G_1 and G_2 (La Motta, 1965 ; Masson, 1979). Moreover, drastic physical treatment of plasma or purified G_4 by hydrostatic pressure (3.5 kbar) or ultrasounds results in a progressive loss of BChE activity and generates monomer and dimer (in preparation), but dissociation of G_4 is never complete without adding denaturing agent. This suggests that a part of tetramer population is fragilised. It has been shown that treatment of G_4 by trypsin cleaves off a C−terminal peptide (< 5kDa) that contains the interchain disulfide (Lockridge, 1982, 1987b). The loss of the bridge lowers the stability of G_4 and leads to its dissociation into active monomeric, dimeric and trimeric species noted G'_1, G'_2 and G'_3 in this paper. These forms have mobility and molecular properties similar to the naturally−occuring C_1 and C_3 forms and to ChE_4 which is generated upon action of clotting enzymes, respectively. Direct evidence for disulfide−depleted G_4 in highly purified preparations of BChE tetramer is given by the presence of monomer band on SDS−gels in non reduring conditions. Internal cuts in the polypeptide backbone of BChE have been found (Lockridge, 1986) ; they may be due also to limited action of peptidases.

Besides proteolysis, other reactions alter the G_4 form. After denaturation of highly purified $[^3H]$ DFP labeled BChE tetramer, a minor (~10%) non reducible labeled dimer is always present on SDS gels in the presence of a reducing agent (Lockridge, 1979 ; Chatonnet, 1986). Analogous patterns have also been noted for horse serum BChE (Teng, 1976) and for electric eel , *Torpedo* and human erythrocyte AChE (Rosenberry, 1977, 1984 ; Lee, 1982). There is no explanation for this covalent dimer. It could be the product of crosslinking of some monomeric subunits, but to date neither the group involved in the bonding (aminoacid side chain or sugar) nor the mechanism of formation of this species are known.

The kinetics of BChE− catalyzed reactions displays several complexities that could reflect microheterogeneity : a). substrate hydrolysis deviates from michaelian kinetics (Kalow, 1964 ; Christian, 1968 ; Eriksson, 1979) ; b). irreversible inhibition does not follow first−order kinetics (Main, 1969) ; c). kinetics presents a non linear temperature dependence with a break at 18−21°C in Arrhenius plots (Ferro, 1987). Many attempts have been made to elucidate the mechanism of these deviations, but to date no conclusive answer has been given. In addition, though there are 4 active sites per tetramer (Lockridge, 1978 ; Eriksson, 1979), lower active site numbers have been reported, e.g. 2 (Muensch, 1976). These discrepancies may be explained either by the presence of large amounts of inactive BChE in plasma or by enzyme inactivation during purification.

The last indication for microheterogeneity comes from thermal inactivation. Irreversible thermo−inactivation of BChE follows a complex kinetics (Masson, 1988 ;Payne, 1989). Moreover, Eyring plots for inactivation of different preparations of tetramer at pH 7 showed significant differences between preparations. Results could not be interpreted in terms of thermally−induced quaternary structure changes. Indeed, as shown by transverse urea−gradient electrophoresis, at 60°C the tetrameric structure is maintained up to 7 M urea (Masson, 1986a). These complexities may reflect the existence of intermediates differing both in thermostability and molecular catalytic activity or structural microheterogeneity. Therefore, the kinetic oddities and differences in behavior within BChE preparations could be interpreted in terms of molecular microheterogeneity i.e. coexistence of populations of native, protease−cleaved, and crosslinked dimer−containing tetramers in varying proportions. We do not know whether this microheterogeneity results from epigenetic modifications of G_2 and G_4 and / or side reactions during purification and storage.

The C_2 form : an albumin conjugate G_1−ALB .

The nature of C_2 has just been elucidated (Masson, 1989). It appeared that C_2 did not belong to the size isomer family. In particular, Ferguson plot analysis of gel electrophoresis patterns allowed estimation of its apparent molecular mass to be 150 kDa, but showed that its charge / mass ratio was different from that of G_1, G_2 and G_4 (Masson, 1979). Moreover,its isoelectric point (Masson, 1979) as well as its mobility in alkaline buffers (Harris, 1962 ; Masson, 1980) were found to be slightly greater than those of the size isomers. Reduction of C_2 in mild conditions released active G_1. Therefore, we concluded that C_2 was a dimer formed by the association of G_1 with a 65 kDa protein through a solvent−exposed disulfide and that the small subunit was responsible for its specific electrical properties.

Purification of C_2 was achieved by affinity chromatography. The 65 kDa subunit could not be labeled by $[^3H]$ DFP and its apparent molecular mass did not change after treatment by endoglycosidase F. C_2 and G_1 were subjected to to electrophoresis on hydrophobic gels designed for affinity electrophoresis of hydrophobic proteins. It was found that C_2 (but not G_1)binds strongly to the immobilized ligand and that its apparent dissociation constant was similar to that of human serum albumin. Demonstration of the identity of the 65 kDa subunit with albumin was completed by specific adsorption of C_2 on an immunoadsorbent for albumin.Therefore C_2 is refered as to G_1−ALB The.disulfide bond is very likely between free thiols : albumin cys−34 and BChE cys−571. The albumin subunit exerts steric hindrance on the BChE subunit and partially masks its active center. In addition, albumin protects desialylated C_2 against self−association. Albumin probably acts by interacting with a hydrophobic area that promotes self−association of sialic acid−depleted BChE. Analysis of hepatocellular carcinoma cell culture fluids indicates that the conjugation very likely occurs in liver cells.

The C_5 isoenzyme : a noncovalent association G_4X

The presence of C_5 in plasma increases the BChE activity up to 30% . Its incidence fluctuates from 0 to 29% among populations and does not show a clear geographical or ethnic variation (Steegmüller, 1975). This variant is not an allelozyme of the BChE gene ; it appears to be a hybrid enzyme composed of BChE subunits associated with an unknown protein (Scott, 1974 ; Muensch, 1978). The genetics of C_5 is complex and its production is controlled by a gene previously named E_2 and now designated CHE2. Though C_5 is an autosomal dominant trait, inheritance of CHE2 is non mendelian in certain cases. (Robson, 1966 ; Ashton, 1966 ; Eiberg, 1989) In addition, this expression is variable : e.g., C_5^+ individuals may transitory turn as C_5^- (Masson, 1990).

The occurence of C_5 raised the question of the multiplicity of BChE genes. Chromosomal mapping of CHE2 has been attempted. First genetic linkage analysis suggested that CHE2 could be assigned to the q arm of chromosome 16 (Lovrien, 1978). Refined in situ hybridization of chromosomes with a BChE cDNA probe mapped a site on chromosome 16 (16 q 21) that could correspond to a second BChE gene (Zakut, 1989). However, a recent study using all published linkage analysis data (Marazita, 1989) failed to assign CHE2 to a particular chromosome but allowed exclusion of substantial portions of the genome. In particular, data are not in favor of a linkage between CHE2 and haptoglobin (16 q 22). Another recent genetic linkage analysis showed a tight linkage between the γ-cristallin gene cluster and CHE2 (Eiberg, 1989) indicating that the CHE2 gene may be assigned to chromosome 2, region q33–35.

The pedigree analysis of a family carrying both the C_5 variant and a silent gene showed that homozygous individuals for the silent gene have no BChE in plasma and are of C_5^- phenotype (Goto, 1988). This data supported the hypothesis that only one gene encodes all BChE subunits, including those of C_5. Isolation and characterization of genomic clones covering the entire sequence of BChE established that the human enzyme is coded by one gene present in a single copy (Arpagaus, 1990). Nevertheless, to determine whether C_5^+ individuals possess a second BChE gene, we isolated white blood cell DNA of 3 unrelated blood donors of C_5 phenotype. Southern blot hybridization of genomic DNA restriction fragments with probes corresponding to each of the 4 exons of the BChE gene provided evidence that individuals carrying C_5 do not have a second BChE gene (Masson, 1990). Moreover, amplification of genomic DNA by the polymerase chain reaction never shown any band heterogeneity that could support the existence of a second gene (McGuire, 1989 ; Nogueira, 1990). Therefore, it may be assumed that CHE2 directs the production of the non BChE component of C_5. Our next goal was to get insight into the molecular structure of C_5.

After several unsuccessful attempts due to the instability of C_5 we succeded in purifying it from plasma of C_5^+ donors, (cf Masson's poster presentation in this volume).

The apparent molecular mass of C_5 is 400kDa. This value is 60kDa higher than that of G_4 : this indicates C_5 cannot originate from epigenetic modification of G_4 by a circulating neuraminidase like enzyme (Ogita, 1975 ; Sakoyama, 1977). As for G_1–ALB, Ferguson plot analysis indicated that C_5 has a charge / mass ratio different from that of G_1, G_2 and G_4. Preparations of C_5 were unstable and progressively released G_4–, G_2– and G_1– like forms. Trypsin treatment also converted C_5 to G_4– like species. Lastly, [^{3}H]DFP labeling of C_5 followed by SDS electrophoresis showed that BChE subunits of C_5 are of G_1 type. Thus, C_5 appears to be formed by the association of G_4 with a subunit (X) of 60kDa.

Reducing agents had no effect on C_5 as well as pressure (3.5kbar) which is known to disrupt hydrophobic bonds and to dissociate oligomeric proteins. On the other hand, high salt concentrations (NaCl 1.5–4M)dissociated C_5 giving rise to G_4. This indicated that X is noncovalently linked to G_4 and that ionic interactions are dominating. The isoelectric point of C_5 was estimated to be 3.95 (3.50 for G_4). Though endoglycosidase F removed the sugar content of BChE , partially deglycosylated C_4 and C_5 exist in different molecular states : removal of sialic acid units by neuraminidase induced self–association of the two enzymes ; but treatment by a mixture of neuraminidase, galactosidase and fucosidase did not allow formation of polydisperse aggregates of C_5 whereas aggregates of C_4 occured. This may reflect small glycosylation differences between G_4 and C_5 BChE subunits or may be due to the presence of X in C_5.

So far we failed to identify X. However, we have rejected several hypotheses. a) First, we considered that C_5 could be a C_2 aggregate : as probed by immunoadsorption experiments and immunodetection after Western blotting, X is not albumin. b) Since the presence of C_5 was reported in blood of myeloma patients (Gallango, 1969), we considered the possibility that X could be an Ig fragment : antibodies to κ and λ Ig chains failed to detect such a fragment. c) C_5 could be a BChE / AChE hybrid similar to hybrid tetramers reported by Tsim, (1988). After gel rod electrophoresis of C_5 plasma, gels were incubated for 30min in buffered solutions of Bambuterol (10^{-8}–to 10^{-4}M), a specific inhibitor of BChE (Tunek, 1988) ; densitometric analysis of activity stained gels showed parallel inhibition curves for G_4 and C_5. d) We found that C_5 binds to heparin and this was used for its purification. Since we do not know the tissue origin of C_5, this property suggested that X could be a tail component. Unfortunately, C_5 was found to be insensitive to collagenase. Nevertheless, X could be another basement membrane protein. Since the fibronectin gene was mapped on a chromosome 2 site : q 32.3–ter (Koch, 1982) close to the region assigned by Eiberg to CHE2, it was tempting to postulate that X could be the heparin–binding fragment of fibronectin. Unfortunately, C_5 was not recognized by polyclonal antibodies to human fibronectin. e) Lastly, we found that C_5 has a hydrophobic character : it binds long–chain fatty acid derivatives. Moreover, it binds heparin at low ionic strenght. It has been claimed that some forms of BChE are complexed with lipoproteins or lipid molecules (Kutty, 1980). Thus, purified C_5 was treated by phospholipase A_2, C, D and PIPLC. The insensitivity of C_5 to these enzymes indicated that X is not a phospholipid–containing protein.

Conclusion

Many questions remain to be solved regarding the molecular polymorphism of BChE. Slow–migrating components have not yet been characterized. Some of them appear to be associations with plasma proteins, e. g. β–lipoproteins, α_2–macroglobulin (Masson, unpublished). Other slow–migrating forms could be amphiphilic forms fitted with a hydrophobic anchor. Such forms have never been described in human tissues but presence of amphiphilic BChE have been demonstrated in *Torpedo* heart (Bon, 1988). Alternative exons coding for hydrophobic C–terminus have been demonstrated for *Torpedo* AChE (Maulet, 1990) but to date there is no evidence for alternative splicing of the human BChE gene (Arpagaus, 1990).

The tissue distribution of the different molecular forms of BChE has not been systematically investigated and there is little information on the tissue origin and on the mechanisms of formation of plasma isoenzymes. Since there is no conclusive evidence for a definite physiological function of BChE (Chatonnet, 1989), the significance of the molecular polymorphism of this enzyme is obscure.

References

Arpagaus, M., Kott, M., Vatsis, K.P., Bartels, C.F., La Du, B.N., Lockridge, O., *Biochemistry*, **1990**, *29*, 124–131

Ashton, G.C., Simpson, N.E., Am. J. Hum. Genet., **1966**, *18*, 438–447

Bon, S., Toutant, J.P., Meflah,K., Massoulié, J., *J. Neurochem.*, **1988**, *51*, 786–794.

Chatonnet, A., Masson, P., *Biochimie*, **1986**, *68* , 657–667.

Chatonnet, A., Lockridge, O.,*Biochem. J.*, **1989**,*260*, 625–634.

Christian, T.S., and Beasley, J.G., *J. Pharmacol. Sci.*, **1968**, *6*, 1025– 1027.

Delbrück, A. , Henkel, E., *Eur. J. Biochem.*, **1979**, *99*, 65–69.

Eiberg, H., Nielsen, L.S., Klausen, J., Dahlén, M., Kristensen, M., Bisgaard, M.L., Moller, N., Mohr, J., *Clin. Genet.*,**1989**, *35*, 313–321.

Eriksson, H., Augustinsson, K.B., *Biochim.Biophys. Acta*, **1979**, *567*, 161–173.

Ferro, A.,Masson, P., *Biochim Bio phys .Acta*, **1987**, *916*, 193–199.

Gallango, M.L., Arends,T., *Human genetik*, **1969**, *7*, 104–108.

Goto, H., Hoshino, T., Fukui, T., Nakahara, R.,Koizumi, H.,Kumasaka, K., Sudo, K., Uchiyama, K.,Ikawa, S., *Rinsho Byori*, **1988**, *36*,694– 698.

Harris, H., Hopkinson, D.A., Robson, E.B., *Nature*,**1962**, *196*, 296–1298.

Juul, P., *Clin. Chim. Acta*, **1968**, *19*, 205–213.

Kalow, W., *Can.J.Physiol.Pharmacol.*, **1964** *42*, 161–168.

Khoury, G.F., Brill, J., Walts, L., Busuttil, R.W., *Anesthesiol.*,**1987**, *67*, 273–274.

Koch, G.A., Schoen, R.C.,Klebe, R.J., Shows, T.B., *Exp. Cell. Res.*, **1982**, *141*, 293–302.

Kutty, K.M., *Clin. Biochem.*, **1980**, *13*, 239–243.

La Du, B.N., Bartels, C.F.,Nogueira, C.P., Hajra, A., Lightstone, H., Van der Spek, A., Lockridge, O., *Clin. Biochem.*, **1990**, in press.

La Motta, R.V., McComb, R.B., Wetstone, H.J., *Can. J. Physiol. Pharmacol.*, **1965**, *43*, 313–318.

Lee, S.L., Taylor, P., *J.Biol.Chem.*, **1982**, *57*, 12292–12301.

Lockridge, O., La Du, B.N., *J. Biol. Chem.*, **1978**, *253*, 361–366.

Lockridge, O. Eckerson, H.W., La Du, B.N., *J. Biol. Chem.*, **1979**, *254*, 8324–8330.

Lockridge, O., La Du, B.N., *J. Biol. Chem.*, **1982**, *257*, 12012–12018.

Lockridge, O., La Du,; B.N.,*Biochem. Genet.*, **1986**, *24*, 485–498.

Lockridge, O., Bartels, C.F., Vaughan, T.A., Wong,C.K., Norton, S.E., Johnson, L.L., *J. Biol.Chem.*, **1987a**, *262*, 549–557.

Lockridge, O., Adkins, S., La Du, B.N., *J. Biol. Chem.*, **1987b**, *262*, 12945–12952.

Lovrien, E.W., Magenis, R.E., Rivas, M.L., Lamvik, N., Rowe, S., Wood, J., Hemmerling,J., *Cytogenet. Cell. Genet.*, **1978**, *22*, 324–326.

Main, A.R., *J. Biol. Chem.* , **1969**, *244*, 829–840.

Marazita, M.L., Keats, B.J.B., Spence, M.A., Sparkes, R.S.,Field, L.L., Sparkes, M.C., Crist, M.,*Hum. Genet.*, **1989**, *83*, 139–144.

Mascall, G.C., Evans, R.T., *J.Chroma togr.*, **1977**, *143*, 77–82.

Masson, P., *Biochim. Biophy.Acta*, **1979**, *578*, 493–504.

Masson, P., Anguille, J., *J. Chroma togr.*, **1980**, *192*, 402–407.

Masson, P., Marnot, B., Lombard, J Y., Morelis, P., *Biochimie*, **1984**, *66*, 235–249.

Masson, P., Goasdoué, J.L., *Biochim. Biophys. Acta*, **1986**, *869*, 304–313.

Masson, P., Laurentie, M., *Biochim. Biophys. Acta*, **1988**, *957*, 111–121.

Masson, P., *Biochim. Biophys. Acact*, **1989**, *988*, 258–266.

Masson, P., Chatonnet, A. Lockridge, O., *FEBS Lett.*, **1990**, *262*,115– 118.

Massoulié, J., Toutant, J.P., in *The Cholinergic Synapse* ;Whittaker, V.P. Ed ; Springer–Verlag Berlin, **1988** ; pp. 167–224.

Maulet, Y., Camp, S., Gibney, G., Rachinsky, T.L., Ekström, T.J., Taylor, P., *Neuron.*, **1990**, *4*, 289–301.

McGuire, M.C., Nogueira, C.P.,Bar– tels, C.F., Lighstone, H.,Hajra, A., Van der Spek, A.F.L., Lockridge, O., La Du, B.N., *Proc. Natl. Acad. Sci.* USA, **1989**, *86*, 953–957.

Muensch, H., Goedde,H–W.,Yoshida, A.,*Eur. J. Biochem*, **1976**, *70*,217–223.

Muensch, H., Yoshida, A., Altland, K., Jensen, W., Goedde, H.W., *Am. J. Hum. Genet.*, **1978**, *30*, 302–307.

Nogueira, C.P., McGuire, M.C., Graeser, C., Bartels, C.F., Arpagaus, M., Van der Spek, A.F.L., Lightstone, H., Lockridge, O., La Du, B.N., *Human Genet*, **1990**, *46*, in press.

Ogita, Z.I., In *IsozymeII, Physiological function*, Markert, C.L., Ed, Academic Press , **1975**, pp, 289–314.

Payne, C.S., Saeed, M., Wolfe, A. D., *Biochim. Biophys. Acta*, **1989**, *999*, 46–51.

Robson, E.B., Harris, H., *Ann. Hum. Genet.*, **1966**, *29*, 403–408.

Rosenberry, T.L., Richardson, J.M., *Biochemistry*, **1977**, *16*, 3550–3558.

Rosenberry, T.L. Scoggin, D.M.,*J. Biol. Chem.*, **1984**, *259*, 5643–5652.

Saeed, S.H., Chadwick, G.R., Mill, P.J., *Biochim Biophys Acta*, **1971**, *229*, 186–192.

Scott, E.M., Powers, R.F., *Am.J. Hum. Genet.*, **1974**, *26*, 189–194.

Sparkes, R.S., Field, L.L., Sparkes, M.C., Crist, M., Spence, M.A., James, K., Garry, P.J., *Hum.Herd.*, **1984**, *34*, 96–100.

Steegmüller, H., *Humangenetik*, **1975**, *26*, 167–175.
Teng, T.L., Harpst, J.LC., Zinn, A., Carlson, D.M., *Arch. Biochem. Biophys.*, **1976**,*176*, 71–81.
Tsim, K.W.K., Randall, W.R., Barnard, E.A., *Proc.Nath. Acad. Sci.* USA,**1988**, *85*, 1262–1266.
Tunek, A., Svensson, L.A., *Drug. Metal. Dispos.*, **1988**, *16*, 759–764.
Van Ross, G., Druet, R., *Nature*, **1966**, *212*, 543–544.
Whittaker, M.,*Cholinesterase*,Karger, Basel, **1986**.pp. 1–132
Yoshida, A., Motulsky, A.G., *Am. J. Hum. Genet.*, **1969**, *21*, 486–498.
Zakut, H., Zamir, R., Sindel, L., and Soreq, H., *Human Reprod.*, **1989**, *4*, 941–946.

Biochemical Characteristics of Cholinesterases from Bar Tissues

Mohamed Yassine, Francis Bacou, INRA Physiologie animale, Place Viala, 34060 Montpellier Cedex 1
Hélène Alami-Durante, INRA Hydrobiologie, St-Pée sur Nivelle, 64310 Ascain, France.

Bar (*Dicentrarchus labrax*) tissues contain two types of cholinesterases (ChE) that differ in molecular form and in substrate specificity. One type of ChE is a true acetylcholinesterase (AChE). The second type of enzyme hydrolyses acetylcholine (ACh), butyrylcholine (BCh), propionylcholine and benzoylcholine. It corresponds to the atypical ChE already found in surgeonfish (Leibel, 1988) and flounder muscle (Stieger et al., 1989). The distribution of both enzymes is tissue-dependent. In muscle, most of the high ChE activity is related to the atypical enzyme, and to true AChE. Liver contains, with a much lower activity, both type of ChEs. In the Central Nervous System (CNS), a high level of AChE only is detected. The atypical enzyme from muscle and liver is inhibited by Eserine (pI50=9), Iso OMPA (pI50=7), Bambuterol (pI50=6), and much less by BW 284 C51 (pI50=4.5) with either ACh or BCh as substrates. The Km and Vmax of this enzyme varies with the length of the substrate acyl chain; it is inhibited at high substrate concentration. AChE, as measured with ACh as substrate in CNS, shows inhibition characteristics close to those found in vertebrate tissues: Eserine (pI50=7.5), Iso OMPA (pI50=4.8), Bambuterol (pI50=2.5) and BW 284 C51 (pI50=7.2). The enzyme is inhibited at high substrate concentration.

In liver and muscle, the atypical ChE occurs only as a form sedimenting under our experimental conditions at 4.5S in the presence of Triton X-100. AChE occurs in multiple molecular forms, the percentage of which varying according to tissues. In muscle, globular forms occur as G4 (10.5S) and as a lighter form sedimenting at 5.5S; collagen-tailed asymmetric forms are sedimenting at 17.5S (A12), 13.5S (A8) and 8S (A4). In CNS, AChE occurs mainly as a tetramer sedimenting at 10.5S; a globular form (5.5S), and asymmetric forms A12 (17.5S) and A8 (13.5S) are also detected.

Our data show that the atypical ChE presents intermediate characteritics of AChE and butyrylcholinesterase which suggest that this enzyme originates from a common polymorphic ChE. These results also show that the classification of ChEs as either "true" (E.C. 3.1.1.7.) or "pseudo" (E.C.3.1.1.8.) is inadequate to the atypical enzyme of muscle sea-fishes.

References

Leibel, W.S., *J. Exp. Zool.*, **1988**, *247*, 198-208.
Stieger, S., Gentinetta R., Brodbeck U., *Eur. J. Biochem.*, **1989**, *181*, 633-642.

0–8412–2008–5/91/0047$06.00/0

Molecular Species Analysis of the Glycosylphosphatidylinositol Anchor of *Torpedo marmorata* Acetylcholinesterase

P. Bütikofer, F.A. Kuypers and C. Shackleton, Children's Hospital Oakland Research Institute, 747-52nd Street, Oakland, CA 94609,USA
U. Brodbeck and S. Stieger*, Institute of Biochemistry and Molecular Biology, University of Bern, Bühlstrasse 28, 3012 Bern, Switzerland

We report on the application of high performance liquid chromatography (HPLC) separation with UV detection and direct, on-line, structural analysis by mass spectrometry of the phosphatidylinositol (PI) moiety of the glycosylphosphatidylinositol (GPI) anchor of acetylcholinesterase (AChE) and of the major phospholipid classes of membranes of the electric organ of *Torpedo marmorata.*

Amphiphilic dimeric AChE from the electric organ of *Torpedo marmorata* was purified by affinity chromatography. Subsequently, the protein was digested with PI-specific phospholipase C from B. cereus in order to release the diradylglycerol moiety from the membrane anchoring domain of AChE. Individual phospholipid classes from membrane tissue of the electric organ of *Torpedo marmorata* were resolved from total lipid extracts by thin-layer chromatography (TLC) and subsequently treated with non-specific phospholipase C. The diradylglycerols of membrane phospholipids and of the PI moiety of the GPI anchor of AChE were then derivatized into their diradylglycerobenzoates and separated by TLC into subclasses (alk-1-enylacyl, alkylacyl, and diacyl types). The molecular species within each subclass were resolved by HPLC with an octadecyl reversed-phase column in acetonitrile/ isopropanol (80/20 v/v). Individual peaks were quantitated at the picomole level by measuring absorbance at 230 nm. After post-column addition of methanol/0.2 M ammonium acetate (50/50 v/v), peaks were introduced through the thermospray interface into a VG Masslab 30-250 quadrupole mass spectrometer. The mass values of the sodium adducts of the molecular ions $(M+22)^+$ permitted easy deduction of the overall fatty acyl composition while analysis of the fragments gave information on the position of the fatty acyl groups in the individual HPLC peaks.

We found that the PI moiety of the GPI anchor of *Torpedo marmorata* AChE consisted exclusively of diacyl molecular species. This is in contrast to AChE from human or bovine red blood cells, where only alkylacyl species were reported. Over 80% of the molecular species contained 16:0, 18:0, and 18:1 fatty acyl chains in the sn-1 and sn-2 positions. Only 15% of the molecular species of the PI moiety of the GPI anchor contained polyunsaturated fatty acyl chains as compared to more than 65% of the species of the PI from the total membrane of the electric organ. This suggests that the assembly of the GPI anchor of AChE is a highly specific process involving the attachment of very defined molecular species of PI to the glycan moiety.

The major phospholipid classes of the electric organ, phosphatidylcholine and -ethanolamine, contained substantial amounts of the alkylacyl (46% and 28%, respectively) and alk-1-enylacyl (5% and 46%, respectively) subclasses. More than 95% of the molecular species of both phosphatidylserine and PI were of the diacyl subclass.

0–8412–2008–5/91/0048$06.00/0

Immunocytochemical Localization of Phosphatidylinositol-Anchored Acetylcholinesterase in Excitable Membranes of *Torpedo ocellata*

Jerry Eichler, Israel Silman, Department of Neurobiology, The Weizmann Institute of Science, Rehovot 76100, Israel
Mary K. Gentry, Division of Biochemistry, Walter Reed Army Institute of Research, Washington, DC 20307, USA
Lili Anglister*, Department of Anatomy and Embryology, The Hebrew University-Hadassah Medical School, P.O. Box 1172, Jerusalem 91010, Israel

In *Torpedo* electric organ much of the acetylcholinesterase is a 'globular' dimer (G_2), anchored to the plasma membrane via covalently attached phosphatidylinositol and solubilized by a bacterial phosphatidylinositol-specific phospholipase C (PIPLC) (Futerman *et al.*, 1985). This suggested that selective solubilization with PIPLC, coupled with immunocytochemistry, might be used to localize G_2 acetylcholinesterase in excitable tissues of *Torpedo*. Cryostat sections of electric organ, electromotor nerve, electric lobe and back muscle from *Torpedo ocellata* were labelled, using three different antibody preparations to *Torpedo* acetylcholinesterase, followed by a fluorescent second antibody, before and after exposure to the phospholipase. Sites of innervation on electrocytes and myofibers were labelled selectively, as were motor and electromotor nerves. In all these cases labelling was substantially diminished by prior exposure to the phospholipase. The results support our previous assignment, based on biochemical evidence, for a neuronal and synaptic localization of the G_2 acetylcholinesterase in *Torpedo* (Futerman *et al.*, 1987). Electric lobe acetylcholinesterase appears insensitive to the phospholipase treatment and lacks certain epitopes present in both electric organ and electromotor nerve enzyme. This suggests that substantial processing of the G_2 form occurs concomitantly with its movement from the electric lobe into the electromotor nerve.

References

Futerman, A. H., Low, M. G., Michaelson, D. M., Silman, I., *J. Neurochem.*, **1985,** *451*, 487-494.

Futerman, A. H., Raviv, D., Michaelson, D. M., Silman, I., *Mol. Brain Res.*, **1987,** *2*, 105-112.

0–8412–2008–5/91/0049$06.00/0

Effects of Exposure to PIPL-C on AChE Activity of Schistosoma Mansoni

Bertha Espinoza, Israel Silman[1], Ruth Arnon, and Rebecca Tarrab-Hazdai.

Departments of Chemical Immunology and [1]Neurobiology, The Weizmann Institute of Science, Rehovot 76100, Israel.

We have previously shown that two ectoenzymes, acetylcholinesterase (AChE) and alkaline phosphatase (APase), are released, by exposure to PIPL-C of bacterial origin, from the cell surface and from particulate fractions of the parasite S. mansoni. By such exposure to PILP-C it is possible to remove large amounts of AChE from the surface of intact, viable schistosomula in culture. A function for this surface/released AChE is less easily understood. Concomitantly, with the release of surface AChE a rapid increase is observed in overall levels of AChE in the parasite. Exposure to PIPL-C in the presence of the protein synthesis inhibitor, cyclohexamide, totally blocks the increase in AChE activity. This suggests that de novo synthesis of the enzyme occurs with its release by PIPL-C.

Since the breakdown products of phosphoinositides produced by cleavage with phospholipase C may serve as second messengers, we investigated the effect of diacylglycerols (DAGs) on the synthesis of AChE in S. mansoni.

Three different DAGs were tested as possible inducers of AChE activity in the parasite. Two of them, 1-oleoyl-2-acetyl-sn-glycerol (C18:1) and 1,2 dioctanoyl-sn-glycerol (C8:0), are known to activate protein kinase C; the third, dimyristin (C14:0), has been reported to be present in the glycolipid membrane-anchoring domain of the variant surface glycoprotein of Trypanosoma brucei.

Our results demonstrate that at final concentrations of 25 μg, both 1-oleoyl-2-acetyl-sn-glycerol and dimyristin are able to increase AChE activity by 35-40%. A higher concentration of 1,2-dioctanoyl-sn-glycerol (70 μg) was needed to produce a similar effect. Furthermore, addition of phorbol 12-myristate 13-acetate, together with the Ca^{+2} ionophore, A23187, produced a similar increase in AChE activity. Our results suggest that GPI membrane-anchor breakdown products may serve as putative second messengers in the parasite S. mansoni.

references

Goldlust, A., Arnon, R., Silman, I. and Tarrab-Hazdai, R. Journal of Neuroscience Research, 1985, 15, 569-581.

Espinoza, B., Tarrab-Hazdai, R., Silman, I. and Arnon, R. Mol. Biochem. Parasitol., 1988, 29, 171-179.

Antibodies Against the Cross Reacting Determinant of Glycosylphosphatidylinositol Anchored Acetylcholinesterase

Karin Jäger*, Susi Stieger and Urs Brodbeck, Institute of Biochemistry and Molecular Biology, University of Bern, Bühlstrasse 28, 3012 Bern, Switzerland

Dimeric acetylcholinesterase is attached to the lipid bilayer through a glycosylphosphatidylinositol anchor. A similar anchoring principle has been identified in many proteins. Analysis of different glycosylphosphatidylinositol anchors showed that they share common structural features including linkage to the C-terminus of the polypeptide via ethanolamine phosphate and the presence of three mannose residues connected to the phosphatidylinositol by glucosamine. Additional molecules such as galactose or ethanolamine may be linked to the carbohydrate backbone or an additional fatty acid to inositol.

The glycosylphosphatidylinositol anchored proteins can be released from the membrane by phosphatidylinositol-specific phospholipases. In case of phosphatidylinositol-specific phospholipase C, a cyclic phosphate is created at the inositol residue, and an antigenic epitope is revealed which is cryptic in the native amphiphilic form. This epitope is called cross reacting determinant since it is present in most glycosylphosphatidylinositol bearing proteins after digestion with phosphatidylinositol-specific phospholipase C.

Polyclonal antibodies were raised against the cross reacting determinant of acetylcholinesterase (Jäger et al., 1990). The antibodies recognized only the hydrophilic forms of bovine erythrocyte and Torpedo acetylcholinesterase, and of variant surface glycoproteins from trypanosomes after treatment with phosphatidylinositol-specific phospholipase C but not the corresponding amphiphilic proteins. Competition experiments on western blots showed that the binding of our antibodies to hydrophilic acetylcholinesterase from Torpedo was inhibited by constituents of the anchor. The presence of myo-inositol-1,2-cyclic phosphate completely abolished binding, while glucosamine only reduced the binding of the antibodies. Thus, both substances interfered with the recognition of the antibodies on western blots. Reductive methylation of glucosamine in the anchor also led to markedly reduced binding of the antibodies. After digestion with phosphatidylinositol-specific phospholipase C, N-methylated acetylcholinesterase from Torpedo was hardly recognized by the antibodies, whereas the modified enzyme from bovine erythrocytes did not react at all with the antibodies. This indicated that methylation of the amine groups alters an important site in the antigenic epitope. From these results, we conclude that myo-inositol-1,2-cyclic phosphate and glucosamine, especially the free amine group of this residue, significantly contribute to the epitope recognized by our antibodies.

Reference

Jäger, K., Meyer, P., Stieger, S., Brodbeck, U., *Biochim. Biophys. Acta*, **1990**, in press.

Drosophila Acetylcholinesterase in Wild Type Flies and in Thermosensitive or Nonconditional Mutants of the Ace Locus

Jean-Pierre Toutant, Physiologie animale, INRA, 2 place Viala, Montpellier Cedex
Martine Arpagaus, Pharmacology, University of Michigan, Ann Arbor 48109-0626

AChE in *Drosophila*, as well as in other insects, is predominantly a disulfide-linked dimer of globular, amphiphilic subunits (G_2a form, Toutant *et al.*, 1988; Toutant, 1989). Each catalytic subunit possesses a C-terminal glycosyl phosphatidylinositol that mediates the attachment to the plasma membrane (Gnagey *et al.*, 1987; Fournier *et al.*, 1988). This domain is cleaved by phosphatidylinositol-specific phospholipase C from *Bacillus thuringiensis* (PI-PLC) and by an anchor-specific phospholipase D from rabbit serum (PLD).

A deacylation of G_2a AChE from *Drosophila* with alkaline hydroxylamine does not result in the production of any hydrophilic dimer (G_2h) indicating that at least one hydrophobic chain in the anchor is linked to the glycerol backbone by a NH_2OH-resistant bond. As suggested for mammalian AChEs, it is likely that *Drosophila* AChE glycolipid anchor contains an **alkylacylglycerol**.

Mild treatments of G_2a AChE with pronase or proteinase K cleave off the C-terminal cysteine of the catalytic peptide to which the glycolipid anchor is covalently attached. These treatments gave rise to G_2h but not to G_1h (hydrophilic monomer). This indicates that the interchain disulfide bridge in the dimer does not involve the terminal Cys residue in position 615 but rather the other free Cys at position 328 (Hall and Spierer, 1986; Fournier *et al.*, 1989).

Several mutations of the *Ace* locus have been isolated by J. Hall *et al.* (1980). Homozygous flies Ace^{J29}/Ace^{J29} are cold-sensitive (die when they are raised at 18ºC but survive at 25-29ºC). Ace^{J40}/Ace^{J40} animals are heat-sensitive and survive only at 18ºC. Other mutations of the *Ace* locus are nonconditionally lethal (J^{19}, J^{50}, lm^{38}). We raised flies bearing different combinations of the above mutations at permissive temperatures. In all cases we observed a striking reduction in the level of AChE activity per fly, but the structure of the enzyme was unchanged. In particular, the major molecular form was an amphiphilic dimer, converted into a hydrophilic G_2 by treatment with PI-PLC or with serum PLD. Thus, neither the type of membrane association nor the assembly of monomer into dimers were affected.

Amplification primers (31 and 32-mers) were designed from DNA sequences chosen within exon 4 in the *Ace* locus (Hall and Spierer, 1986; Fournier *et al.*, 1989). The 2 primers delineated a 550bp-long segment containing the sequence coding for the active site of the enzyme. This region was amplified from genomic DNAs of the different mutants by the Polymerase Chain Reaction. 40 cycles of 1.5 min denaturation at 94ºC, 1.5 min hybridization at 50ºC and 3 min elongation at 72ºC were used. A band of 550bp was obtained for each mutant, cloned in M13mp18 and M13mp19 and sequenced.

Fournier, D.; Bergé, J.-B.; Bordier, C. *J. Neurochem.* **1988,** *50,* 1158-1163.

Fournier, D.; Karch, F.; Bride, J.-M.; Hall, L. M.C.; Bergé, J.-B.; Spierer, P. *J. Mol. Biol.* **1989,** *210,* 15-22.

Gnagey, A.L.; Forte, M.; Rosenberry, T.L. *J. Biol. Chem.* **1987,** *262,* 13290-13298.

Hall, L.M.C.; Spierer, P. *EMBO J.* **1986,** *5,* 2949-2954.

Hall, J.C.; Alahiotis, S.N.; Strumpf, D.A.; White, K. *Genetics* **1980,** *96,* 939-965.

Toutant, J.-P.; Arpagaus, M.; Fournier, D. *J. Neurochem.* **1988,** *50,* 209-218.

Toutant, J.-P. *Prog. Neurobiol.* **1989,** *32,* 423-446.

Comparison of Methods for Detecting Acetylcholinesterase in Drosophila Melanogaster

Erno Zador, Department of Genterics, University of Glasqow, Glasgow G11 5js, Scotland, U

The method of Karnovsky and Roots (1964, K&R) developed for mammalian tissues gives less sensitive detection of a electromorphs of Drosophila acetyl-chloinesterase (AChE) than its modified version by Morton and Singh (1980, M&S). The bands 1, 3 and 4 on Fig. 1 probably correspond to the amphiphilic dimer, the hydrophilic dimer and monomer AChE, respectively, and the 1st band might be a double one containing both the amphiphilic dimer and monomer (J. P. Toutant pers. com.).

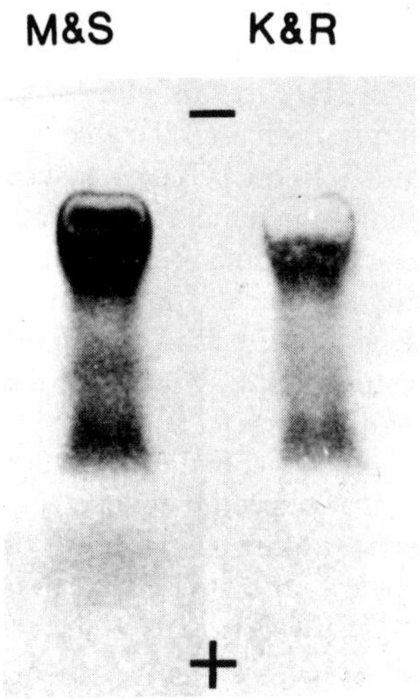

Figure 1

Each of the five (six) AChE bands detected by M&S (Fig. 1) is in the fly brain, while in embryos and larvae only bands 1, 3 and 4 are from the central nervous system (CNS, Zador 1989).

Deletions of the AChE gene (87 EL-5) eliminate all of the AChE bands (Zador et al., 1986). Embryos homozygous for DF (3r)Ace, ^{HD1} a small deficiency of the distal part of the AChE gene, show only the non-CNS-specific bands (fig. 2C). The pattern of the extract of wild type (Oregon R) embryos (Fig. 2A) was not altered by coincubation with the extract or Ace^{HD1}/Ace^{HD1} embryos (Fig. 2B) suggesting that bands 2 and 5 are not degraded forms of bands 1, 3 or 4.

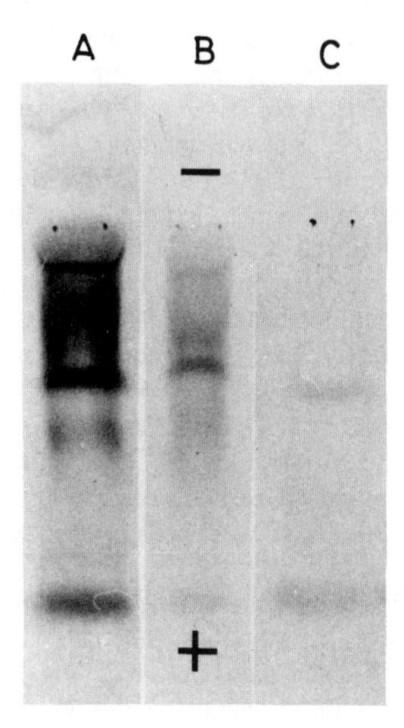

Figure 2

References

Karnovsky, M.J., Roots L., *J. Histochem. Cytochem.*, **1964**, *12*, 219-222

Morton, R.A., Singh, R. W., *Biochem. Genet.*, **1980**, *18*, 439-454

Zador, E., *Mol. Gen. Genet.*, **1989**, *218*, 487-490

Zador, E., Gausz, J., Maroy, P., *Mol. Gen. Genet.*, **1986**,*204*, 469-472

0–8412–2008–5/91/0053$06.00/0

Inhibitory Monoclonal Antibodies Raised against Acetylcholinesterase from *Torpedo nacline timilei*, and Crossreaction with Mammalian Acetylcholinesterases

Jian Liao* and Urs Brodbeck, Institute of Biochemistry and Molecular Biology, University of Bern, Bühlstrasse 28, 3012 Bern, Switzerland
Yan-Bin Xin and Man-Chi Sun, Institute of Pharmacology and Toxicology, Beijing 100850, China

Acetylcholinesterase (AChE, EC 3.1.1.7) from *Torpedo nacline timilei* was purified by affinity chromatography to a specific activity of 12'500IU/mg of protein. Monoclonal antibodies (McAbs) were raised by hybridoma technique. BALB/c mice were immunized by intraperitoneal injection of 200 μg of AChE with an equal volume of Freund's incomplete adjuvant. Anti-AChE McAbs were purified by ammonium sulfate precipitation and by chromatography on DE-52 cellulose and protein A-Sepharose CL-4B. They were pure as judged by polyacrylamide gel electrophoresis. Antibody titers ranged from 1:5'000 to 1:300'000. Three out of eleven anti-AChE McAbs (i.e. 3F3, 2G8, and 1H11) significantly decreased enzyme activity when acetylcholine (ACh) was used as substrate while, with indophenyl acetate (IPA), only 3F3 caused a significant inhibition. This differential inhibition showed that McAbs 2G8 and 1H11 impeded charged substrates from reaching the active site, presumably by binding to the sites near the active center. On the other hand, McAb 3F3 most likely bound to the active site of the enzyme.

The crossreactivity of McAb 1H11 with detergent-soluable mammalian AChE (DS-AChE) from rat brain, bovine brain and erythrocytes, human brain and erythrocytes, was studied using the enzyme antigen immunoassay (EAIA) developed for AChE. It could be shown that 1H11 crossreacted with mammalian brain forms suggesting that the epitope to which 1H11 bound, had been highly conserved during evolution. Binding of McAb 1H11 to *Torpedo marmorata* DS-AChE strongly decreased enzyme activity when acetylthiocholine (ACTh) was used as substrate, but almost no inhibition was found in the case of mammalian forms. This result suggested that binding site for McAb 1H11 is in the vicinity of the active site in Torpedo AChE, while in mammalian AChE, it is remote thereof. The remarkable difference in reacting of McAb 1H11 towards mammalian brain and erythrocyte AChE suggests that a specific epitope is shared in mammalian brain forms which is not present in AChE from erythrocytes. Further studies excluded that McAb 1H11 binds to the anionic site of AChE since edrophonium-chloride (Tensilon), an inhibitor directly binding to the anionic site, showed no influence on the binding of McAb 1H11 to AChE.

0–8412–2008–5/91/0054$06.00/0

Anti-Idiotypic Antibodies Exhibiting an Acetylcholinesterase Abzyme Activity

Laurent Joron, Alain Friboulet*, Marie-Hélène Rémy, Gianfranco Pancino, Alberto Roseto and Daniel Thomas.

Génie Enzymatique et Cellulaire, URA 523 du CNRS

Université de Technologie de Compiègne, 60206 Compiègne Cedex France.

Anti-idiotypic antibodies (anti-IdAbs) can be raised against the binding sites of other antibodies. According to the idiotypic-network theory, these anti-IdAbs may mimic the interactions of the starting antigen (receptor or enzyme binding sites) with their specific ligands. Polyclonal anti-IdAbs have been elicited to AE-2, an inhibitory mAb raised against an anionic site of human erythrocyte acetylcholinesterase (AChE) (Fambrough et al., 1982).

Rabbits were immunized with purified AE-2. Binding of mAb AE-2 to AChE was inhibited by immune serum, but not by a non immune serum. Anti-IdAbs were isolated by sequential adsorption on protein A-Sepharose then AE-2-Sepharose columns. Purified anti-IdAbs were adsorbed on anti-rabbit IgG, and AChE activity was assayed. The curves indicate linearity between adsorbed anti-AE-2 level and AChE activity . When a normal rabbit serum was subjected to the same purification steps and then to the same test, no rabbit IgG binding, nor AChE activity could be detected. Thus, AChE activity cannot be explained by the presence of contaminating enzyme, but could be due to the formation of a specific complex between anti-IdAbs and AChE. This hypothesis was ruled out by sedimentation velocity analysis.

Hydrolysis of acetylthiocholine by anti-IdAbs follows Michaelis-Menten kinetics, in contrast to classical AChEs which exhibit usual excess substrate inhibition. Moreover, the measured high K_M value (65 mM) unambiguously indicates that the activity cannot be due to a classical AChE. However, from inhibition studies with active site-specific inhibitors (DFP, BW), and from substrate specificity experiments, it can be concluded that the anti-IdAbs exhibit a specific AChE activity.

The described study using "internal imagery" could be a general method to study enzyme molecules by creating similar three-dimensional structures with different aminoacid sequences.

Reference

Fambrough, D.M., Engel, A.G., Rosenberry, T.L., *Proc. Natl. Acad. Sci. U.S.A*, **1982**, *79*, 1078-1082.

A comparative Raman spectroscopic study of cholinesterases

Dimitrina Aslanian*, Pal Grof, Laboratoire de Physique des Solides, Université P. & M. Curie, 75005 Paris, France.
Patrick Masson, Unité de Biochimie, CRSSA, 38702 La Tronche Cedex, France.
Palmer Taylor, Dept of Pharmacology, University of California at San Diego, La Jolla, CA 92093, USA.
Suzanne Bon, Jean-Marc Chatel, Jean Massoulié, Laboratoire de Neurobiologie, ENS, 75005, Paris, France.

Classical Raman spectroscopy makes it possible to evaluate the contribution of the various types of secondary structure in the conformation of a purified protein. It also provides indications regarding the environment of specific amino acid residues. We have used this method to study the following cholinesterases: lytic tetrameric formes(G4) obtained by triptic digestion of asymmetric AChE from *T. californica* and *Electrophorus*, a PI-PLC-treated dimeric form(G2) of AChE from *T. marmorata* and the soluble form(C4) of BuChE from human plasma. Fig. a and b shows the Raman spectrum of *Electrophorus* (G4) AChE. Deconvolution of the spectrum in the 1500-1800 cm^{-1} region (Fig. b) permits : 1/ to improve the resolution of the aromatic amino acid vibrations which overlap the amide I band and had to be subtracted (doted line in Fig. b); 2/ to improve the resolution of the amide I band.

Table 1

Enzyme	Structure (%) α	β	T	U
AChE(G4) *T. californica*	49	23	11	16
AChE (G4) *Electrophorus*	49	25	14	10
AChE (G2) *T. marmorata*	51	23	14	11
BuChE (C4) human serum	47	26	16	10

α: total helix; β: total sheet; T-turns; U-undefined.

Each deconvoluted peak is, however, a sum of two or three secondary structure elements. The secondary structures of the enzymes were quantitatively estimated (Table 1) according to R. W. Williams and A. K. Dunker (J. Mol. Biol., 1981, 152 783).

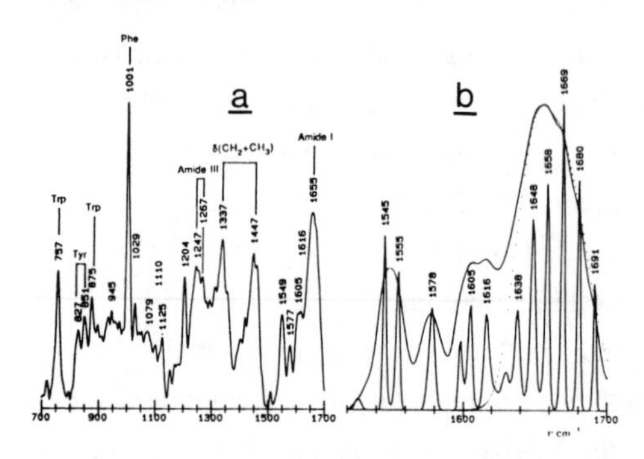

The main result is that the Raman spectra of these enzymes are very similar. The major peaks are due to the vibrations of amide I group (1640-1700 cm^{-1}), to the Phe-, Tyr- and Trp-residues (755, 827/851, 875, 1001, 1204, 1549, 1577, 1605 and 1616 cm^{-1}) as well as to the vibrations of CH2- and CH3-groups (1337 and 1447 cm^{-1}) (Fig. a). There are, however, interesting differences between the Raman spectra, concerning the vibrations of the Tyr doublet (827/851 cm^{-1}), the region of skeletal (C-C) and (C-N) stretching modes (800-1150 cm^{-1}) and, finally, the methylene bending vibrations at 1337 cm^{-1}.

0–8412–2008–5/91/0056$06.00/0

An Inactive Form of Acetylcholinesterase with Intact Binding Site

R.Raba*, KEMOTEX AE, Estonian Acad. Sci., 17 Akadeemia tee, Tallinn 200026, Estonia, U.S.S.R.
K. Tônismägi, A. Aaviksaar, Institute of Chemical Physics and Biophysics, Estonian Acad. Sci., P.O.B. 670, Tallinn 200026, Estonia, U.S.S.R.

An inactive form of cobra venom acetylcholinesterase (AChE) was separated from native enzyme preparation by immunoaffinity chromatography on an antibody clone. The inactivated form (i-AChE) was present in all enzyme preparations as well as in the crude venom and its content increased spontaneously from 2-3% in fresh enzyme preparations up to 15-20% in aged purified AChE samples. The i-AChE had the same molecular weight and electrophoretical properties as the native one, and it coeluted with the native AChE on an affinity column with the N,N,N-trimethyl-m-phenylendiamine ligand. By SDS-electrophoresis it was demonstrated that an intramolecular split of enzyme polypeptide chain had occurred in i-AChE dividing native 70kD molecule into 30kD and 40kD parts. The parts were bound together with disulphide bridges.

The ligand binding properties of i-AChE were studied by fluorescence measurements using N-methylacridinium. The i-AChE showed similar binding properties compared to the native enzyme. The binding constants K_d estimated from Scatchard plots were 1.05μM and 0.99μM for the native enzyme and i-AChE, respectively.

It is possible, that occurrence of a specific intramolecular cleavage site is common for all acetylcholinesterases. It has been shown for electric organ AChE that a similar spontaneous subunit split occurs (Morrod, Marshall, Clark, 1975).

The antibody clone obtained is a valuable tool in AChE purification. The i-AChE can be used in "pure binding" studies of AChE substrates which are hydrolysed by native enzyme.

References

Morrod, P.J., Marshall, A.G., Clark, D.G., *Biochem. Biophys. Res. Commun.*, **1975**, *63*(1), 335-342.

0–8412–2008–5/91/0057$06.00/0

Partial Sequence of Cobra Venom Acetylcholinesterase – AChE and Lysophospholipase Have Similar Sequences

Christoph Weise, Hans-Jürgen Kreienkamp, Ferdinand Hucho*, Institute of Biochemistry, Free University of Berlin, Thielallee 63, 1000 Berlin 33, Germany, Phone (49)-30-838 55 45
Raivo Raba, Aavo Aaviksaar, Institute of Chemical Physics of the Academy of Sciences of the Estonian SSR, PO Box 670, Tallinn 200 026, USSR

The venom of a variety of elapid snakes has been known as a rich source of acetylcholinesterase (AChE). The enzyme has been purified from several snake venoms including the venom of the cobra Naja Naja oxiana (Raba et al., 1979). It is isolated from the venom as a monomer and consists of a single polypeptide chain with a molecular weight of 67 kDa (Raba et al., 1982).

About 30 % of the primary structure of the acetylcholinesterase (AChE) from Cobra venom were determined by protein chemical methods. The sequence of the tryptic peptide around the active site serine labeled by diisopropylfluorophosphate (DFP) was found to be TVTLFGESAGAASVGM which is similar to other cholinesterases. The part of the structure determined shows 76 % identity with AChE from Torpedo and 42 % identity with the enzyme from Drosophila (Weise et al., 1990).

A surprisingly large sequence identity (42 % in the sequence determined) was found with lysophospholipase from rat (E.C. 3.1.1.5.) as determined by Han et al., 1987. The sequence immediately flanking the active site serine is invariably FGESAG in both enzymes.

This prompted us to investigate if AChE exhibits hydrolytic activity towards lysophospholipid substrates. AChE from Torpedo as well as from Cobra venom has no detectable lysophospholipase activity towards lysophosphatidylcholine (lyso-PC) and lysophosphatidylethanolamine (lyso-PE) under standard test conditions (pH 7.2, with and without addition of 10 µM Ca^{2+}). From this lack of overlapping enzymatic activity we conclude that even the very pronounced sequence similarity around the active serine residue is not sufficient to confer substrate specifity upon the enzymes.

References

Han, J., Stratowa, C., Rutter, W., *Biochemistry*, **1987**, 26, 1617-1625

Raba, R., Aaviksaar, A., Raba, M., Siigur,J., *Eur.J.Biochem.*, **1979**, 96,151-158

Raba, R., Aaviksaar, A. *Eur.J.Biochem.*, **1982**, 127, 507-512

Weise, C., Kreienkamp, H.-J., Raba, R., Aaviksaar, A., Hucho, F. *J.Prot.Chem.*, **1990**, 9, 53-57

0-8412-2008-5/91/0058$06.00/0

Multiple Forms of Non-Specific Cholinesterases in Human Serum and Cerebrospinal Fluid

D.S.Navaratnam, A.McIlhinney & J.D.Priddle
University Department of Pharmacology, South Parks Rd, Oxford. OX1 3QT U.K.

Non-specific cholinesterase, "butyryl-cholinesterase" (BChE) - E.C. 3.1.1.8 - is present in human CSF and serum. It has been postulated that the enzyme in CSF is identical to that in serum and may be derived from a passive transfer of serum enzyme across the blood-brain barrier (Atack *et al.*, 1987; Rao *et al.*, 1989). Using polyacrylamide iso-electric focussing we demonstrate differences in the iso-electric points of BChE from serum and CSF. These differences are shown, at least in part, to be due to a greater sialation of the serum enzyme.

Methods. Vertical iso-electric focussing was a modification of the method of Giulian *et al.* (1984). A mixture made of acrylamide (5%C, 2.6%T),10% glycerol and 7% Ampholine pH 3.5 - 9.5 (LKB) was cast in a Hoefer SE250 apparatus containing a GelBond PAG (FMC Co.) support. The gel was "pre-run" for 10 min. at 200V, 300V and 400V each. Degassed NaOH and acetic acid (10mM) served as catholyte and anolyte respectively. 15 mU of enzyme was concentrated to approximately 20μl in a 10,000 NMWL ultrafilter (Millipore), and applied to the cathodic end. After focussing for 3.5 h. at 1000V, BChE activity was detected as described by Chubb & Smith (1975). Desialation of the sample was achieved by Neuraminidase treatment (Ott *et al.*,1975).

Results. BChE was present in two regions; that in the acidic region (pH 3.5-4.5) was present in serum and CSF, while that in the more basic region (pH 6.0-8.5) was present only in CSF (Fig. 1 *A,B*). Treatment of both enzymes with Neuraminidase resulted in a shift of enzyme from the acidic to the more basic region (Fig. 1 *C,D*). These differences may be accounted for by secretion from brain tissue into CSF of forms less sialated than the serum enzyme, or by the passively transferred serum enzyme being desialated in CSF (Svensmark, 1961); desialated enzyme in serum being absent due to its removal by the asialoglycoprotein receptor.

References

Atack,J.R. *et al.* (1987) *J. Neurochem.* 48, 1845-1850.
Chubb, I.W. and Smith, A.D. (1975) *Proc. R. Soc. Lond.* 191, 245-261
Giulian, G.G. *et al* (1984) *Anal. Biochem.* 142, 421-436
Ott, P.*et al* (1975) *Eur. J. Biochem.* 57, 469-480
Rao, R.V. *et al* (1989) *Clin. Chim. Acta* 183:2, 135-145
Svensmark, O. (1961) *Acta physiol. scand.* 52, 267-275

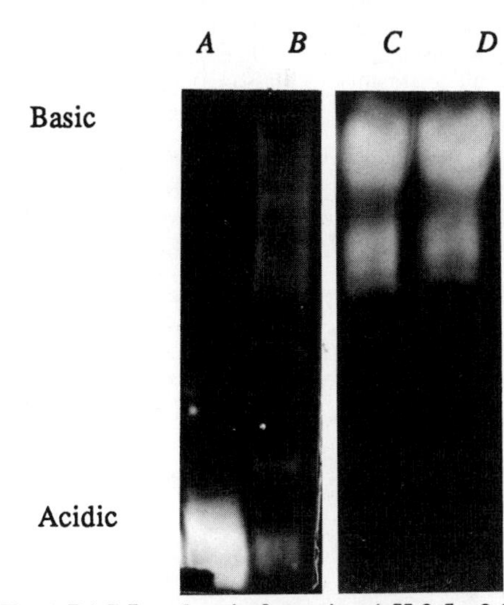

Fig. 1 PAG Iso-electric focussing (pH 3.5 - 9.0) of Human CSF and Serum. BChE activity staining. *Tracks A + B, Serum and CSF; Tracks C + D, Neuraminidase treated Serum and CSF.*

Molecular characterization of the C_5 human plasma cholinesterase variant

Patrick Masson, Marie–Thérèse Froment, Jean–Christophe Audras, Frederique Renault,
Centre de recherches du Service de Santé des Armées, Unité de Biochimie, BP 87, 38702 La Tronche, France.

The C_5 component of human plasma cholinesterase is a butyrylcholinesterase isoenzyme which migrates slower than the major form C_4 (or G_4) of the enzyme on nondenaturing gels.

C_5 is present in 8–10% of Caucasians (Whittaker, 1986), but its incidence in populations is heterogeneous.This isoenzyme has long been considered as the product of a gene designated as E_2 or CHE2. This gene, likely located on chromosome 2 ,q33–q35 region. (Eiberg, 1989), is not allelic to the BChE gene (previously named E_1 or CHE1), assigned to 3q21 which directs the common enzyme. It has been shown that there is only one gene for human BChE (Arpagaus, 1990) and that individuals carrying C_5 have no other BChE gene (Masson, 1990). Therefore, as already suggested (Scott, 1974), C_5 has to be considered as a hybrid of BChE and a non BChE protein (X) whose synthesis is directed by CHE2.

In order to determine the oligomeric structure of C_5 and to characterize X, we purified C_5 from plasma of regular blood donors of C_5 phenotype. Due to variability of the expression of C_5 and instability of the isoenzyme purification was very difficult. Nevertheless, purification was achieved by several affinity chromatography steps using procainamide, heparin, Cibacron Blue and phenyltrimethylammonium as ligands.

The molecular weight of C_5 was determined to be 400kDa by non denaturing polyacrylamide gradient gel electrophoresis and disc–electrophoresis in gels of varying acrylamide concentration). Ferguson plots showed that C_5 does not belong to the size isomer serie G_1, G_2 and G_4. However, it contains G_4 because G_4 is released from C_5 by high concentration salt treatment (NaCl 1.5–4 M) and because trypsin and proteases convert C_5 into G_4–, G_2– and G_1– like forms. [^{3}H]DFP labeling followed by SDS electrophoresis showed that the BChE tetramer of C_5 is analogous to the common G_4, i.e., a dimer of dimers of BChE subunits. Therefore, X appears to be a 60kDa protein.

Reducing compounds do not dissociate C_5 as well as high pressure (3.5kbar) which is known to disrupt hydrophobic bonds. Therefore, due to the dissociating action of NaCl, salt bridges appear to be the dominating forces which stabilize the noncovalent association of G_4 and X.

The isoelectric point of C_5 (3.95) is 0.5 unit higher than that of G_4, indicating that X may contain positively charged groups. The fact that C_5 binds to heparin in low ionic strenght buffers supports this hypothesis. In addition, X has a hydrophobic character : it binds octadecyl derivatives. However, C_5 is insensitive to phospholipases A_2, C and D from various sources and to phosphatidyl–inositol specific phospholipase C. Thus, C_5 is not a phospholipid–containing protein.

We failed to identify X but we rejected several hypotheses concerning C_5 structure. We showed that X is not : a). acetylcholinesterase (hybrid AChE / BChE) ; b). albumin (C_2 aggregate hypothesis, Masson 1989) ; c). an immunoglobulin fragment ; d). a collagenous tail fragment ; e). a heparin–binding fibronectin fragment (fibronectin gene was mapped on chromosome 2, q32.5–ter).

References

Arpagaus, M. Kott, M., Vatsis, K.P., Bartels, C.F., La Du, B.N., Lockridge, O., *Biochemistry*, **1990**, *29*, 124–131

Eiberg, H., Nielsen, L.S., Klausen, J., Dahlén, M., Kristensen, M., Bisgaard, M.L., Moller, N., Mohr, J., *Clin. Genet.*, **1989**, *35*, 313–321.

Masson, P., *Biochim. Biophys. Acta*, **1989**,*988*, 258–266.

Masson, P., Chatonnet, A., Lockridge, O., *FEBS Lett.*, **1990**, *262*, 115–118.

Scott, E.M., Powers, R.F., *Am. J. Hum. Genet.*, **1974**, *26*, 189–194.

Whittaker, M., *Cholinesterase*, Karger, Basel, **1986**, pp. 1–132.

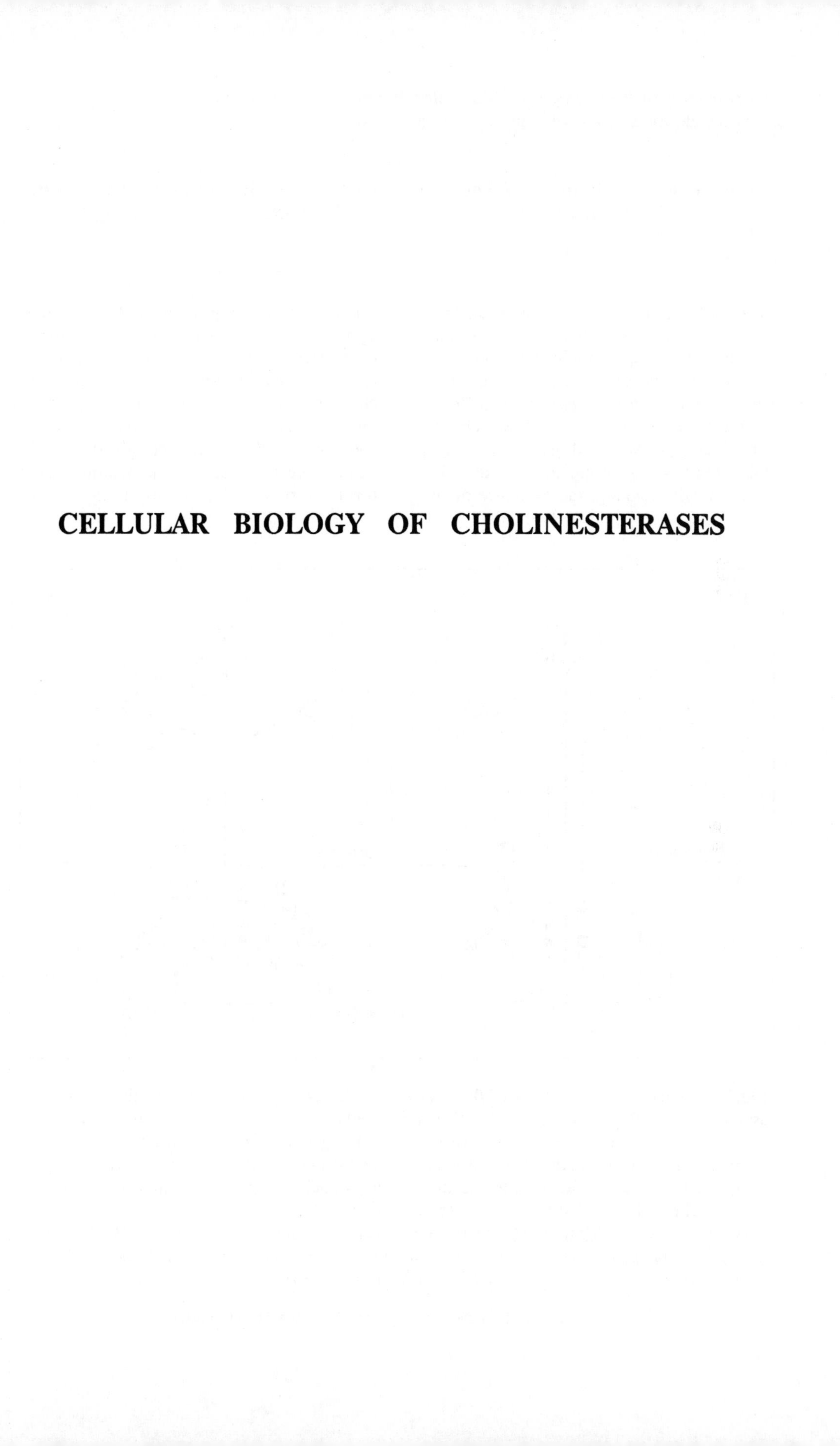

CELLULAR BIOLOGY OF CHOLINESTERASES

Properties of Two Types of Molecular Forms of Asymmetric Acetylcholinesterase in Chick Skeletal Muscle

Galo Ramírez*, Jordi Pérez-Tur, Milagros Ramos, Xavier Busquets, Ana Barat,
Centro de Biología Molecular (CSIC-UAM), Universidad Autónoma, Cantoblanco, 28049 Madrid, Spain.

Our earlier discovery that a combination of high ionic strength and chelating agents could solubilize asymmetric molecular forms (A-forms) of acetylcholinesterase (AChE) from central nervous system tissues, in different vertebrate species (Barat et al, 1980; Gómez-Barriocanal et al, 1981; Rodríguez-Borrajo et al, 1982) led us to propose, at the 1983 Cholinesterase Meeting, the existence of two classes or types of asymmetric AChE species. Type I (i.e., the conventional A-forms already characterized in peripheral tissues) could be readily extracted by high salt concentrations (1M NaCl used in routine experiments), while type II required the presence of divalent cation chelators, in addition to high salt, for solubilization (Fig.1). These two populations of A-forms were found to co-exist, in different

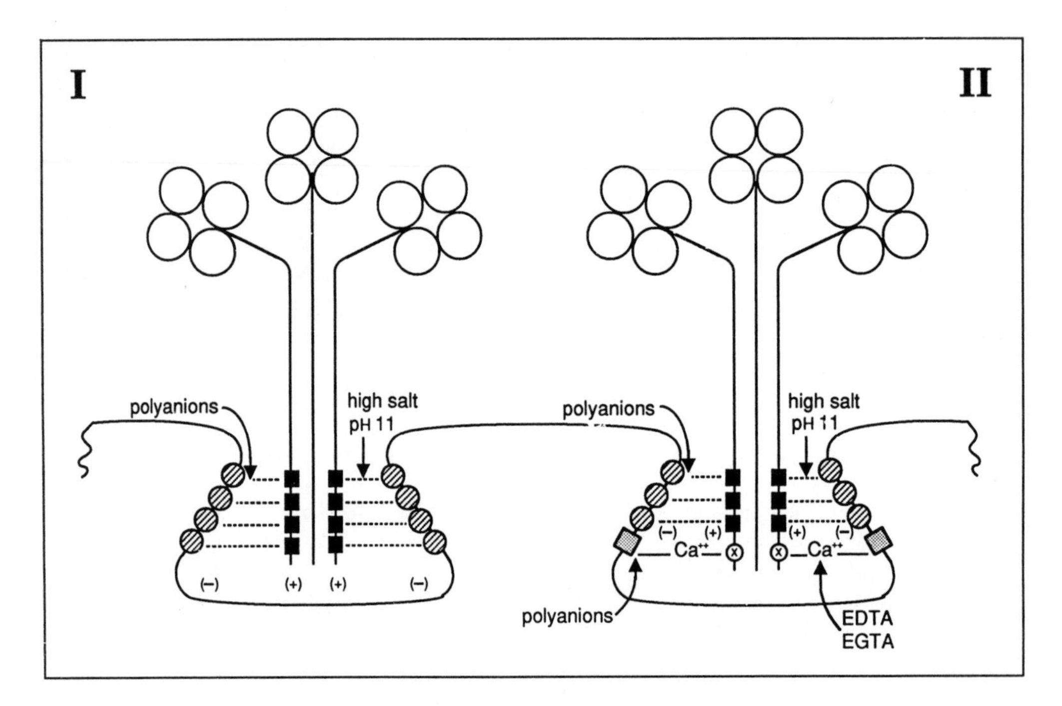

Fig.1. Schematic representation of the possible mechanisms of attachment of the A^{I} and A^{II} asymmetric AChE types to the muscle ECM, as inferred from solubilization experiments. High-salt-sensitive electrostatic interactions between the ECM negative charges (polyanions?) and some hypothetical positively charged domains in the tail are common to both types. In the case of type II, additional Ca^{++} bridges (sensitive to chelating agents) would reinforce the electrostatic anchorage system. The tail end of these bridges (**x**) could either be an alternative domain of the tail polypeptide (variable tail hypothesis), or a linker molecule provided by the ECM (unique tail/variable target hypothesis). The site of action of polyanions would explain their ability to extract both types of A-forms even in the absence of EDTA (Barat et al, 1986).

relative proportions, in nearly all cholinergic tissues and systems examined, from Torpedo electric organ to mammalian brain (Barat et al, 1984; Ramírez et al, 1984). The characteristic action of chelating agents (EDTA, EGTA) and the reversible aggregation of type II (but not type I) A-forms by an excess of Ca^{++} furthermore suggested a role for Ca^{++} bridges in the attachment of the tail of the class II enzyme to the extracellular matrix (ECM) (Gómez-Barriocanal et al, 1981; Barat et al, 1984; Ramírez et al, 1984).

Trying to find some physiological basis for this new manifestation of the already complex polymorphism of AChE, we looked for differential properties of the two A-form types in muscle. For instance, in the rat diaphragm, type II A-forms were found to be mainly extracellular and concentrated in end-plate areas. On the other hand, chick fast- and slow-twitch muscles contained similar proportions of the two A-form classes. Also relevant to function could be the fact that, upon denervation, chick hindlimb muscle A^{I} species disappeared earlier than the their type II counterparts (Barat et al, 1984). This apparent difference in neural regulation of the two A-form types has been confirmed by Fadić and Inestrosa (1989).

Consistent Patterns of Development of the Two Asymmetric AChE Types in Chick Hindlimb Muscle

Looking for further differences in the regulation of the two A-form fractions in the chick hindlimb muscle we have recently examined the developmental profiles of both tailed species, in vivo and in cell culture, and their recovery upon irreversible inhibition of muscle AChE in vivo with diisopropylfluorophosphate (DFP)(Busquets et al, 1990). During normal development, A^{I} appears and accumulates earlier than A^{II} so that a A^{II}/A^{I} ratio of 1.0 is not reached until about hatching time. The potential interest of this finding is suggested by the precocious onset of locomotion in this avian species. Something similar happens in muscle cell culture, where A^{I} develops earlier and to a greater extent than A^{II}. Finally, our studies on the recovery of A-forms upon treatment of the chick peroneal muscle with DFP, in vivo, show again an earlier reappearance and faster recovery of A^{I} as compared with A^{II}. The cell culture results would suggest that intrinsic muscle regulatory phenomena also play a part in the consistent precedence of A^{I} over A^{II} in all experimental situations involving de novo synthesis of the two A-form classes.

A Systematic Exploration of the Origins of A-form Diversity

In looking for an explanation as to the existence of two variants of tailed AChE species, several possibilities could be considered. Differences in the molecular make-up could of course easily explain the dissimilar behavior of the two asymmetric enzyme forms, alternative tail polypeptides being a rather more likely possibility than different catalytic subunits. However, purified A^{I} and A^{II} enzyme forms show identical electrophoretic patterns and similar reactivity towards a number of antibodies available to us (either catalytic subunit- or tail-specific; our unpublished results). Furthermore, type II A-forms no longer aggregate in the presence of Ca^{++}. Differential post-translational processing and interconversion could also explain the generation of the two tailed forms, although pulse-chase experiments have so far failed to support the latter possibility. The anchorage specificity could alternatively depend on the target molecule recognized by the tail in the ECM (or elsewhere), or upon the existence of linker molecules associated to the tail, as previously suggested (Gómez-Barriocanal et al, 1981). In the following paragraphs we will describe our recent work on the mechanisms of attachment of the tail to the ECM in chick muscle.

Interaction of Glycosaminoglycans and other Polyanions with Asymmetric and Globular Molecular Forms of AChE

The possible association of tailed AChE forms to sulfated glycosaminoglycans (GAGs) in the synaptic basal lamina of the electric organ was first suggested by Massoulié and coworkers (Bon et al, 1978). Torres and Inestrosa (1983) later found out that heparin could solubilize A-forms from rat muscle in the absence of salt.

Our first experiments with heparin revealed that this sulfated GAG solubilized both types of A-forms with the same efficiency, under appropriate conditions (Barat et al, 1986). We have also demonstrated that, although crude A-form extracts behave differently, purified A^{I} and A^{II} forms are both quantitatively bound to heparin-agarose columns (Ramírez et al, 1990). The solubilizing effect of heparin on A-forms, and the affinity of the tail for immobilized heparin (Brandan and Inestrosa, 1984; Brandan et al, 1985) led Inestrosa and coworkers to suggest a role for heparan sulfate proteoglycans in the anchorage of A-forms to the ECM.

We have extended our studies on heparin-affinity to the globular forms of AChE (G-forms). At low salt concentration, G-forms (both low-salt-soluble and detergent-soluble) are also quantitatively bound to the column. Actually, about one third of the G-forms remain attached to the column at 0.15 M NaCl (Ramírez et al, 1990). So, if heparin-affinity is taken as indicative of association to the ECM, both A- and G-forms should be considered as potentially synaptic in localization and function (Jedrejczyk et al, 1984; Nicolet et al, 1987).

However, heparin-affinity could also be construed as a non-specific anionic interaction phenomenon. Actually, both A- and G-forms are also efficiently bound, for instance, to immobilized dextran sulfate columns. Our experiments with different polyanions show that polysulfates are generally good displacers of AChE forms (A and G) from heparin-agarose columns. Among them, heparin, dextran sulfate and polyvinyl sulfate are 100% effective. Surprisingly, heparan sulfate is a poor displacer, even worse than chondroitin sulfates. Reduction (de-N-sulfated heparin), or total absence of sulfate groups (hyaluronate, dextran) result in negligible displacement of AChE from the heparin-agarose column. Polycarboxylates (poly-aspartate) and polyphosphates (including polynucleotides) show moderate to low efficiency in this type of experiments (Ramírez et al, 1990; Pérez-Tur et al, submitted).

Solubilization of A-forms from Chick Muscle by Different Polyanions

When the same polyanions are tested for their ability to solubilize A-forms directly from chick muscle, under controlled extraction conditions, the results are, however, quite different. Thus, while heparin and dextran sulfate are still very efficient (32 and 31% of total AChE solubilized as A-forms, respectively), polyvinyl sulfate releases only 17%; on the other hand, heparan sulfate (23%), chondroitin sulfate B (15%) and de-N-sulfated heparin (10%) perform much better than it would be expected from the results observed in heparin-agarose columns. Even more surprisingly, polyaspartate is now the most active of the group, with a controlled solubilization efficiency of over 35% (Pérez-Tur et al, submitted). These two sets of experiments with polyanions support then the involvement of negatively charged molecules in the attachment of AChE to the synaptic basal lamina and extrasynaptic ECM without offering any direct evidence for an exclusive role of sulfated GAGs (including heparan sulfate) in the anchorage mechanism(s) (Fig.1).

Release of AChE from Chick Muscle by Glycosaminoglycanases

Looking for direct proof of the implication of GAGs in securing the AChE molecular forms to the ECM we tested several heparinases (including heparitinase) and chondroitinases for their ability to release A and G enzyme species. In contrast with the results of Brandan et al (1985), heparinases were unable to release AChE from chick muscle, while chondroitinase ABC detached both A-forms (40% of 1M NaCl control) and G-forms (150% of control) from a chick hindlimb muscle preparation previously extracted with detergent (Pérez-Tur et al, submitted). This would appear to confirm the early hypothesis of the involvement of chondroitin/dermatan sulfates in the attachment of the tailed enzyme to the synaptic basal lamina (Bon et al, 1978), although the observed effects could also depend to some degree on the disorganization of the ECM following the massive digestion of this family of GAGs.

This work was supported by grants from the Dirección General de Investigación Científica y Técnica (PB87-0244), the Fundación Ramón Areces and the Laboratorios Dr. Esteve, S.A.

References

Barat, A., Escudero, E., Gómez-Barriocanal, J., Ramírez,G., Biochem. Biophys. Res. Commun.,**1980**, 96, 1421-1426.

Barat, A., Gómez-Barriocanal, J., Ramírez, G., Neurochem. Int., **1984**, 6, 403-412.

Barat, A., Escudero, E., Ramírez, G., FEBS Letters, **1986**, 195, 209-214.

Bon, S., Cartaud, J., Massoulié, J., Eur. J. Biochem., **1978**, 85, 1-14.

Brandan, E., Inestrosa, N.C., Biochem. J., **1984**, 221, 415-422.

Brandan, E., Maldonado, M., Garrido, J., Inestrosa, N.C. J. Cell Biol., **1985**, 101, 985-992.

Busquets, X., Pérez-Tur, J., Rosario. P., Ramírez, G., Cell. Mol. Neurobiol., **1990**, in press.

Fadić, R., Inestrosa, N.C., J. Neurosci. Res., **1989**, 22, 449-455.

Gómez-Barriocanal, J., Barat, A., Escudero, E., Rodríguez-Borrajo, C., Ramírez, G., J. Neurochem., **1981**, 37, 1239-1249.

Jedrejczyk, J., Silman, I., Lai, J., Barnard, E.A., Neurosci. Lett., **1984**, 46, 283-289.

Nicolet, M., García, L., Dreyfus, P.A., Verdière-Sahuqué, M., Pinçon-Raymond, M., Rieger, F., Neurochem. Int., **1987**, 11, 189-198.

Pérez-Tur, J., Ramos, M., Barat, A., Ramírez, G., submitted.

Ramírez, G., Gómez-Barriocanal, J., Barat,A.,Rodríguez-Borrajo, C., In Cholinesterases; Brzin, M., Sket, D., Barnard, E.A., Eds.; Walter de Gruyter, Berlin-New York, **1984**; pp 115-128.

Ramírez, G., Barat, A., Fernández, H., J. Neurochem., **1990**, 54, 1761-1768.

Rodríguez-Borrajo, C., Barat, A., Ramírez, G., Neurochem. Int., **1982**, 4, 563-568.

Torres, J.C., Inestrosa, N.C., FEBS Letters, **1983**, 154, 265-268.

Intrinsic Regulation and Neural Modulation of Acetylcholinesterase in Skeletal Muscles

J. Sketelj and M. Brzin*, Institute of Pathophysiology, School of Medicine, 61105 Ljubljana, Yugoslavia

Development of motor endplates after early postnatal denervation of rat soleus muscles and the changes of AChE molecular forms pattern in slow and fast muscles after experimental disturbances of neural input were studied in order to elucidate the importance of neural influences vs. intrinsic muscle mechanisms in nerve-muscle interactions. The postsynaptic-like sarcolemma differentiation including both AChE and acetylcholine receptor accumulation and growth of sarcolemmal folds continued after early postnatal denervation and even expanded to much greater length than that of normal motor endplates at the same age. Exaggerated expansion of the sarcolemmal postsynaptic-like differentiation after denervation may be due to release from a neural inhibitory influence on accumulation of synaptic components in immediate vicinity of the motor endplates. Procedures which disrupted normal stimulation pattern of rat muscles (tenotomy, denervation, reinnervation of regenerating muscles) did not change the typical features of the patterns or distribution of AChE molecular forms in slow and fast muscles. The results support the hypothesis that intrinsic differences in AChE regulation largely determine the specific AChE patterns in slow and fast rat muscles.

Many properties of skeletal muscles are determined and regulated by interactions of muscles with their motor nerves (Vrbova et al., 1978). Neural influences may be local, affecting the muscle fibre underneath the nerve ending, e.g. development of the motor endplate. On the other hand, the nerve may exert its influence also along the whole length of muscle fiber as in regulation of activity and pattern of molecular forms of acetylcholinesterase (AChE) in muscles (Sketelj and Brzin, 1980; Lai et al., 1986). It is well documented that within both above mentioned examples of nerve-muscle interaction the muscle itself, i.e. in the absence of motor nerves, is able to express some phenotypic features that are normally regulated by the motor nerve. Aneural muscles developing in the embryo or regenerating after injury or growing in culture form postsynaptic-like sarcolemmal specializations without contact with the nerve (Silberstein et al., 1982; McMahan and Slater, 1984; Sketelj et al., 1988; Sohal, 1988), and produce the asymmetric AChE molecular forms which are normally concentrated in the motor endplates (for review see Toutant and Massoulie, 1987). Therefore, neural modulation of intrinsic muscle abilities seems to underlie these regulatory processes in innervated muscles.

Two experimental approaches were chosen to throw more light on this relationship. First, the fate of the motor endplates was studied after early postnatal denervation of rat muscles thus creating a situation in which neural influences were eliminated in the middle of motor endplate development. Second, neural influences to slow soleus (SOL) and fast extensor digitorum longus (EDL) muscles of adult rats were modified to study the role of nerve impulse pattern vs. intrinsic muscle mechanisms in determining specific expression of AChE molecular forms and their distribution in these two muscles.

Methods

Wistar strain of albino rats was used in experiments. Ether or pentobarbitone sodium (50 mg/kg i.p.) were used during surgery or muscle isolation.

0–8412–2008–5/91/0066$06.00/0

1. Surgical procedures

a) Denervation: was performed in two day old animals and in adult rats (180 to 200 g) by high axotomy and ligation of the sciatic nerve.
b) Tenotomy: distal SOL tendon was dissected free and about 5 mm of the tendon was excised.
c) Regeneration: nerve intact SOL and EDL muscle regenerates were prepared after severing all blood vessels to the muscles (Carlson et al., 1981). In addition, bupivacaine was injected into muscles to make degeneration more complete, together with phospholine to irreversibly inhibit the preexistent AChE.

2. Histochemical procedures

Double staining for AChE (Koelle's acetylthiocholine procedure) and acetylcholine receptor (AChR - rhodamine labeled α-bungarotoxin binding) was performed in frozen muscle sections essentially as described elsewhere (Do Thi et al., 1984). A modification of acetylthiocholine method was used for electron microscopic visualization of AChE in motor endplates (Brzin and Pucihar, 1976).

3. Analysis of AChE molecular forms

AChE was extracted from muscles in high salt-Triton X-100 medium with antiproteases and analysed by velocity sedimentation in linear 5-20% sucrose gradients as described earlier (Sketelj et al., 1987).

Results and discussion

A. Development of subsynaptic-like sarcolemmal specializations in places of former neuromuscular junctions after early postnatal denervation

Denervated and contralateral innervated SOL muscles were isolated 14-16 days after unilateral sciatic nerve section in two day old rats. Longitudinal frozen sections were cut from muscles and doubly stained for AChE and AChR so that both synaptic components can be visualized in the same section (Fig. 1). In normal SOL muscles at this age, both AChE and AChR are concentrated in the motor endplates (Fig. 1 a, b). However, the appearance of the motor endplates is still fairly immature: they are small (about 10-15 um in diameter), rounded and nonsegmented,

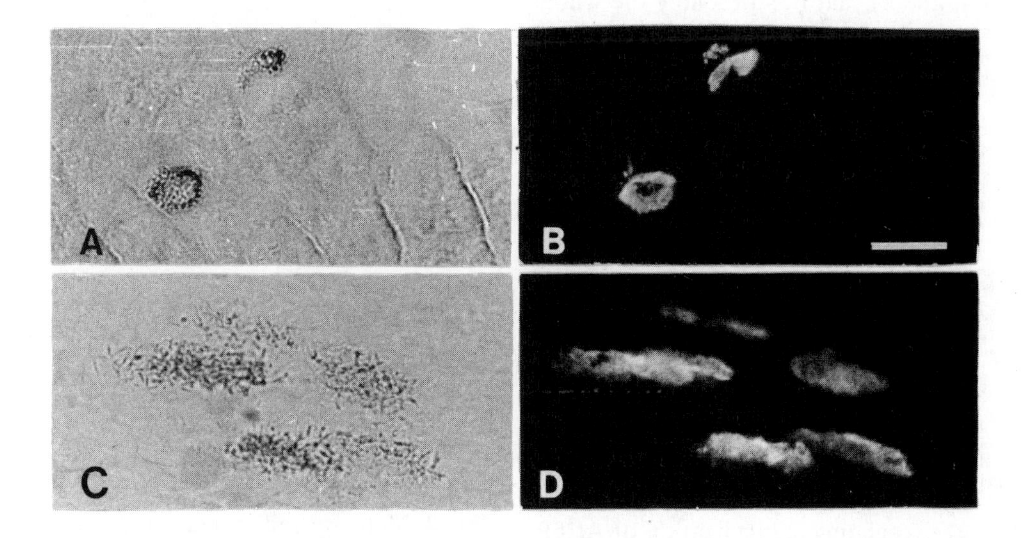

Fig. 1. Histochemical localization of AChE (left side) and AChR (right side) in normal (A, B) and two weeks denervated (C, D) motor endplates of soleus muscle in 16 day old rat. Bar: 20 um.

in contrast to mature endplates which are oval, about 50 um long and display lobular appearance (cf. Sketelj et al., 1988). In denervated muscles (Fig. 1 c, d), the postsynaptic-like area bearing both AChE and AChR expanded to a length of 30-50 um along the muscle fiber surface to become ribbon-like. Similar dispersion of AChR has been described in postnatally denervated mouse muscles and it was claimed not to be due to AChR diffusion or simply muscle fiber growth (Slater, 1982). We showed, in addition, that junctional AChE also spreads after postnatal denervation. Careful examination of about 50 denervated motor endplate regions showed, within the limits of resolution of the method, fairly good coincidence of AChE and AChR microlocalization. Further, we showed by using electron microscopy that morphological differentiation of the expanding postsynaptic-like area continues in the absence of the nerve for at least four weeks after denervation, although in somewhat abortive way (Sketelj et al., 1990). Sarcolemmal infoldings containing AChE grow into the underlying sarcoplasm mimicking the secondary synaptic clefts. This indicates that after postnatal denervation of muscles a) expansion of the postsynaptic-like area cannot be fully due to unfolding of synaptic clefts, and b) that it is probably the whole postsynaptic apparatus which is being put together on a much larger area of the sarcolemma than in innervated fibers. Namely, both AChE and synaptic clefts are strictly limited to the area exactly underneath the nerve ending in normally innervated muscle fibers (Fig. 2). Expansion of the postsynaptic apparatus in denervated postnatal muscles may be a release phenomenon. Motor nerve ending, besides probably stimulating the production of synaptic components in the underlying nuclei (Fontaine et al., 1986), also inhibits accumulation of these components in the immediate vicinity of the contact site (Moody-Corbet and Cohen, 1982; Davey and Cohen, 1986). Postnatal denervation may eliminate this inhibition and allow expansion of the postsynaptic-like area along the muscle fiber surface.

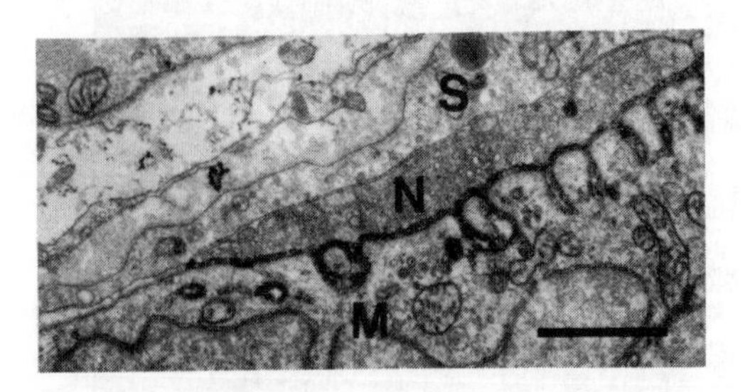

Fig. 2. Cytochemical localization of AChE (black precipitate) and distibution of synaptic clefts in a motor endplate in soleus muscle of a 17 day old rat. M - muscle, N - nerve ending, S - Schwann cell. Bar: 1 um.

B. Neural modulation vs. intrinsic muscular mechanisms in regulation of AChE molecular forms in slow and fast rat muscles

The patterns of AChE molecular forms in slow SOL and fast EDL rat muscles differ considerably (Fig. 3 a, b) as has already been

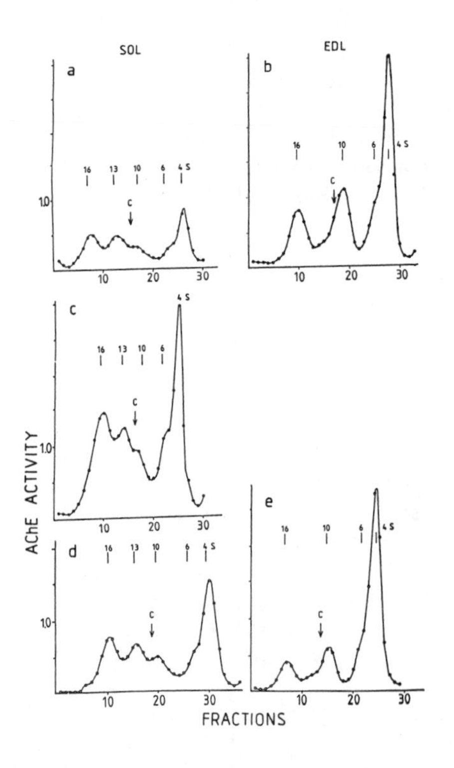

Fig. 3. Velocity sedimentation patterns of AChE molecular forms in normal (a, b), tenotomized (c) and 15 day old innervated regenerating (d, e) soleus and EDL muscles. C - catalase marker (11.3 S).

shown earlier (Groswald and Dettbarn, 1983; Lomo et al., 1985). In comparison to the EDL, the pattern of AChE molecular forms in the SOL is characterized by relatively low activities of the globular 4 S (G1) and 10 S (G4) forms and high activity of the 13 S (A8) AChE form. Both muscles, however, exibit similar patterns of AChE molecular forms at birth so that the differences arise during postnatal muscle maturation (Dettbarn et al., 1985; Sketelj et al., 1990). Neural stimulation patterns of both muscles in adult rats differ significantly (Hennig and Lomo, 1984): SOL is stimulated by tonic low frequency impulses whereas EDL stimulation is phasic and of high frequency. Experimental conditions were therefore created which are known to disturb the pattern of neural stimulation of SOL and EDL muscles in order to estimate the importance of different neural stimulation patterns for maintaining different patterns of AChE molecular forms in both muscles.

Tenotomy of the SOL muscle interrupts the stretch reflex and eliminates tonic EMG activity in SOL muscles (Vrbova, 1963). However, the typical SOL pattern of AChE molecular forms is not changed significantly 14 days after tenotomy of the muscle although specific AChE activity is increased (Fig. 3 c). Accordingly, a similar result was obtained by a special disuse model which silenced SOL motoneurons by elimination of gravity stimuli to the muscles (Dettbarn et al., 1985).

Prolonged denervation (six weeks) did not eliminate the characteristic difference regarding the distribution of the asymmetric AChE forms in SOL and EDL muscles (Fig. 4): relatively high activity of these AChE forms is preserved in the extrajunctional regions of the SOL muscle while their extrajunctional activity in the EDL is still negligeable. Accordingly, relative resistance to denervation of the asymmetric AChE forms in slow muscles of other species has been described (Bacou et al., 1982; Lai et al., 1986).

It seems therefore that permanent specific patterns of neural stimulation of SOL and EDL muscles are not necessary to maintain the differences in AChE regulation in fast and slow muscles. Specific stimulation patterns are probably not essential to institute the differences during postnatal development either since different AChE patterns in SOL and EDL appear during the first ten days after birth (Dettbarn et al., 1985; Sketelj et al., 1990) while the stimulation patterns of both muscles are still very much alike (Navarette and Vrbova, 1983). Experiments with muscle regeneration which mimics ontogenetic muscle development to a large extent (Carlson, 1978) also speak in favour of such a proposal. Fairly typical patterns of AChE molecular forms for fast and slow muscles develop in nerve intact regenerates of both muscles within ten days after the first signs of reinnervation (Fig. 3 d,e) although the SOL motoneurons are probably depressed because of degeneration of muscle spindles during muscle injury (Diwan and Rogers, 1986). Accordingly, preliminary results show that tenotomy of regenerating SOL muscle does not influence the AChE pattern.

All the results together strongly support the hypothesis that there is an essential intrinsic difference in AChE regulation between slow and fast skeletal muscle fibers, probably encoded in their myoblast lineages. Experiments to directly test this hypothesis by cross-transplantation of regenerating SOL and EDL muscles are in progress.

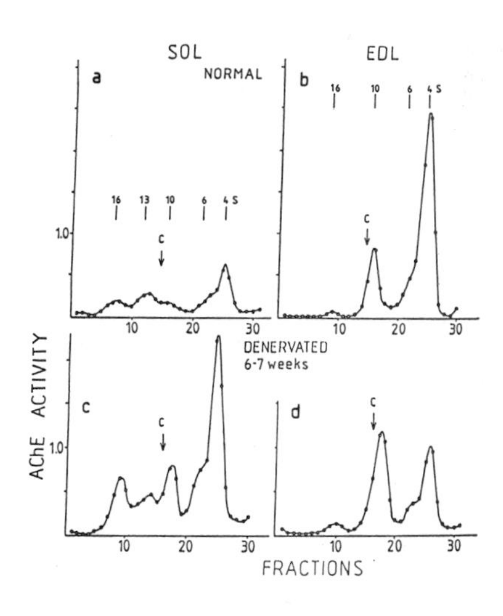

Fig. 4. Velocity sedimentation patterns of AChE molecular forms in the extrajunctional regions of normal (a, b) and 6-7 weeks denervated (c, d) soleus and EDL muscles.

Acknowledgement

The work was jointly supported by the Research Association of Slovenia and the DHHS grant JF 703.

References

Bacou, F., Vigneron, P., Massoulie, J., *Nature*, **1982**, *296*, 661- 664.
Brzin, M., Pucihar, S., *Histochem.*, **1976**, *48*, 283-292.
Carlson, B.M., *Physiol. Bohemoslov.*, **1978**, *27*, 387-400.
Carlson, B.M., Hnik, P., Tuček, S., Vejsada, R., Bader, D.M., Faulkner, J.A., *Physiol. Bohemoslov.*, **1981**, *30*, 505-513.
Davey, D.F., Cohen, M.W., *J. Neurosci.*, **1986**, *6*, 673-680.
Dettbarn, W.-D., Groswald, D., Gupta, R.C., Misulis, K.E., In *Molecular Basis of Nerve Activity*; Changeux, J.-P., Hucho, F., Maelicke, A., Neumann, E., Eds.; Walter de Gruyter: Berlin, FRG, 1985; pp 567-587.
Diwan, F.H., Milburn, A., *J. Embryol. Morph.*, **1986**, *92*, 223-254.
Do Thi, A., De La Porte, S., Koenig, *J., Biol. Cell*, **1984**, *50*, 99-102.
Fontaine, B., Klarsfeld, A., Hokfelt, T., Changeux, J.-P., *Neurosci. Lett.*, **1986**, *711*, 59-65.
Groswald, D.E., Dettbarn, W.-D., *Neurochem. Res.*, **1983**, *8*, 983- 995.
Hennig, R., Lomo, T., *Nature*, **1985**, *314*, 164-166.
Lai, J., Jedrzejczyk, J., Pizzey, J.A., Green, D., Barnard, E.., *Nature*, **1986**, *321*, 72-74.
Lomo, T., Massoulie, J., Vigny, M., *J. Neurosci.*, **1985**, *5*, 1180- 1187.
McMahan, U.J., Slater, C.R., *J. Cell Biol.*, **1984**, *98*, 1453-1473.
Moody-Corbet, F., Cohen, M.W., *J. Neurosci.*, **1982**, *2*, 633-646.
Navarrete, R., Vrbova, G., *Develop. Brain Res.*, **1983**, *8*, 11-19.
Silberstein, L., Inestrosa, N.C., Hall, Z.W., *Nature*, **1982**, *295*, 143-145.
Sketelj, J., Brzin, M., *Neurochem. Res.*, **1980**, *5*, 653-658.
Sketelj, J., Črne, N., Brzin, M., *Neurochem. Res.*, **1987**, *12*, 159- 165.
Sketelj, J., Črne, N., Brzin, M., *Neurosci Res.*, **1988**, *20*, 90- 101.
Sketelj, J., Črne, N., Brzin, M., *Cell Mol. Neurobiol.*, **1990**, (in press).
Slater, C.R., *Develop. Biol.*, **1982**, *94*, 23-30.
Sohal, G.S., *Int. J. Develop. Neurosci.*, **1988**, *6*, 553-565.
Toutant, J.-P., Massoulie, J., In *Mammalian ectoenzymes*; Kenny, P., Turner, S., Eds.; Elsevier Science Publishers B.V. (Biomedical Division): 1987; pp 289-328.
Vrbova, G., *J. Physiol.* **1963**, *166*, 241-267.
Vrbova, G., Gordon, T., Jones, R., *Nerve-Muscle Interaction*; Chapman and Hall: London, UK, 1978.

In Vivo Regulation of Acetylcholinesterase in Slow and Fast Muscle of Rat

Wolf-D. Dettbarn, Departments of Pharmacology and Neurology, Vanderbilt University, Nashville, Tennessee 37212
Douglas E. Groswald, University of Kansas - College of Health Sciences and Hospital, Kansas City, Kansas 66103
Ramesh C. Gupta, Toxicology Department, Breathitt Veterinary Center, Murray State University, Hopkinsville, Kentucky 42240
Karl E. Misulis, Departments of Neurology and Pharmacology, Vanderbilt University, Nashville, Tennessee 37212
Gary T. Patterson, State of California, Department of Food and Agriculture, Medical Toxicology Branch, Sacramento, California 95814

The mechanisms underlying regulation of acetylcholinesterase (AChE) activity in skeletal muscles have been studied for several years (Briminjoin, 1983). While the nerve is important for the maintenance and characteristics of the muscle it innervates, the relative importance of nerve invoked muscle activity, nerve derived trophic factors, and the use of the muscle such as in loadbearing is less well understood. Our interest is to assess the role of genetic factors and non genetic influences on the total phenotypic expression of AChE in skeletal muscle. In rat the fast extensor digitorum longus muscle (EDL) differs from the slow twitch soleus not only in overall AChE activity but also in contribution of individual molecular forms to total activity (Dettbarn, 1981; Groswald and Dettbarn, 1983). These differences are also expressed in the nerves innervating these muscles (Gisiger and Stephens, 1982). Therefore these two muscles offer a great opportunity to study the mechanisms that regulate AChE. In the following we will review some of our studies.

The following approaches have been chosen to study the regulation of AChE: (1) developmental differentiation of fast and slow twitch muscle, (2) the effect of denervation and reinnervation, (3) the role of muscle disuse, particularly the reduction of loadbearing, and (4) the role of the nerve as determinant in regenerating fast and slow muscle.

Developmental Differentiation of AChE Molecular Forms

In neonatal rats, all myotubes in EDL as well as soleus are undifferentiated and therefore provide the ideal beginning for studying the process that leads to the differentiation in AChE in mature muscles.

In newborn rats the enzyme activity is identical in both the EDL and the soleus muscle (Dettbarn et al., 1985). Significant differences can be observed between the first and second week. At this time the rate of enzyme increase becomes maximal but rises faster in soleus than in EDL. This rise in AChE activity is only transient in soleus.

At birth and shortly thereafter, both EDL and SOL muscles have proportionately greater quantities of the lighter AChE molecular forms. With continuing development, there is an increase in the heavier AChE molecular forms in the SOL. No such shift is seen in the EDL (Table 1).

The use of hindlimb muscles in the rat such as the soleus and EDL starts late in postnatal development. Compared to other animals, skeletal muscles in rat are not fully developed at birth. During the postnatal period, muscle fibers increase in number, hypertrophy, and their enzyme pattern changes. In addition, elimination of polyneuronal innervation is completed between days 14-16 and adult innervation patterns have been established (Redfern, 1970).

Developing muscle fibers will undergo a preprogrammed sequence of AChE synthesis unless a signal is received which will stimulate or suppress a particular protein.

The recent work by Lomo et al., 1985, demonstrates the importance of frequency pattern of nerve impulses in determining molecular form characteristics of AChE.

0–8412–2008–5/91/0071$06.00/0

Table 1. **Activities of AChE Molecular Forms During Postnatal Maturation**[1]

Days Postnatal	EDL 4S	EDL 10S	EDL 16S	SOL 4S	SOL 10S	SOL 12S	SOL 16S
1 day	41.7±9.4	33.4±6.4	24.9±2.4	32.6±7.4	30.2±1.8	19.0±4.0	19.7±2.7
1 week	55.0±0.6	23.5±0.5	21.5±0.5	31.0±5.0	23.0±3.1	15.0±1.0	25.0±2.0
2 week	49.5±4.5	28.0±1.0	22.5±5.5	29.5±1.5	14.5±0.5	13.5±0.5	42.0±1.0
3 week	51.0±1.0	22.7±1.8	26.6±1.7	22.2±3.2	13.7±0.3	24.7±1.2	39.3±2.7
4 week	50.0±0.6	27.3±3.4	22.0±2.6	18.7±0.9	22.0±2.6	32.7±0.3	26.7±3.8

Activities in AChE molecular forms from extensor digitorum longus (EDL) and soleus (SOL) muscles. AChE molecular forms were recovered from sucrose density gradients during postnatal development. Activities are expressed for each form as a percentage of the total recovered activity. Data are from three separate preparations of high speed supernatants (100,000 x g) isolated from rat EDL and SOL muscles. Values are expressed as the mean ± SEM.

Denervation and Reinnervation Induced Changes in the Regulation of AChE Molecular Forms

Three molecular forms of AChE are normally found in EDL, and the contribution to the overall AChE activity is approximately 50, 35, and 15% for the 4S, 10S, and 16S forms, respectively, and approximately 25, 20, 25, and 30% for the 4S, 10S, 12S, and 16S forms in the SOL (Groswald and Dettbarn, 1983a).

As shown in Table 2, following nerve crush the EDL exhibits decreases in all three forms. On the other hand, SOL muscle exhibited selective increases in the AChE activity associated with the 4S and 10S forms and decreases in the 12S and 16S forms. With reinnervation, only a gradual increase is evident in the EDL muscle. However, transient increases to several times the contralateral controls was apparent for all four molecular forms of AChE in the SOL muscle (Groswald and Dettbarn, 1983b).

The Role of Disuse in the Regulation of AChE

Many functional properties of skeletal muscle are known to be altered by deprivation of neural influences. Elimination of these influences by denervation results in a variety of changes in rat muscles which include

Table 2. **Activities of the Molecular Forms of Acetylcholinesterase Expressed as a Percent of the Contralateral Controls**[1]

Time after crush (weeks)	N	Extensor Digitorum Longus 4S	Extensor Digitorum Longus 10S	Extensor Digitorum Longus 16S	Soleus 4S	Soleus 10S	Soleus 12S	Soleus 16S
1	3	11.0± 2.3	39.5± 9.6	29.6±16.3	140± 30	85± 47	36± 10	28± 18
2	5	29.9±12.2	20.8± 7.9	22.0±10.6	222± 47	157± 10	55± 17	81± 32
3	3	45.6± 6.7	21.2± 2.2	30.5± 1.3	687±173	310± 10	248± 24	439± 39
4	5	83.1± 5.7	37.9± 9.9	177.8±37.3	419±101	257± 57	324±200	504±146
5	3	83.8± 5.0	49.8±16.8	94.1±27.6	155± 85	262±158	219± 94	165±101
6	5	114.2± 8.1	50.1±10.8	96.4±10.0	157± 7	168± 32	186± 49	165± 23

[1]Comparisons of the AChE activity from EDL and SOL muscles after denervation for each molecular form. Peaks from the distributions between experimental muscles were compared with the peaks from their contralateral controls, to give the percentage of contralateral controls. Between three to five determinations (N) were made for each postoperative period, and there were three muscles for each group. Values are the mean ± SE.

atrophy and decreased AChE activity. Some of these changes occur in a progressive manner rather than immediately which raises the question of the role of disuse versus trophic factors in the regulation of muscle functions.

We have used a recently developed technique, hindlimb suspension, to investigate the effects of no loadbearing on the regulation of AChE (Gupta et al., 1985). The advantages in using this hindlimb suspension technique to study skeletal muscle changes due to disuse are: (1) maintenance of intact neural influences, (2) allows full range of joint movements, (3) removes weightbearing function, and (4) only isotonic contractions occur.

No change in the total number of fibers in the EDL or SOL was found between control and suspended rats, although a generalized fiber atrophy was apparent in the SOL.

The AChE activity of the EDL from suspended rats did not significantly change compared to controls (Table 3). However, AChE activity in the SOL from suspended rats was significantly increased during the 2nd and 3rd week of suspension. When the individual molecular forms of AChE in the SOL were assayed, an increase of activity of all four major molecular forms was found; whereas, in the EDL no significant change was observed. The contribution of the individual forms to the total activity was not different from that of control muscles (Table 4).

This data provides additional evidence that regulation of AChE in use and disuse of a given muscle may be different in EDL and soleus muscles. No loadbearing in the antigravity muscle such as the soleus appears to stimulate the synthesis of AChE without modifying the contribution of the individual molecular forms.

Role of Nerve in the Regulation of AChE in Regnerating Muscle

In this study we have investigated the role of neural influences on muscle fiber types and AChE regulation during regeneration of the EDL and the soleus muscles of the rat. Innervated and denervated EDL and SOL were injected _in situ_ with 1 ml of 0.5% bupivacaine (BPVC, Marcaine) solution to produce complete muscle fiber degeneration. Histochemical ATPase activity, AChE activity and its molecular forms were determined on the regenerating muscles for up to four weeks. Complete fiber destruction was found within three days after the BPVC injection in both innervated and denervated muscles. By day 7 the onset of muscle regeneration was evident in all treated muscles. In the innervated regenerating muscles both the EDL and SOL showed initially increases in the type II (fast) fibers. With progressing regeneration the population of type I (slow) fibers increased in SOL to that found in the

Table 3. Effect of Hindlimb Suspension-Induced Muscle Disuse on Acetylcholinesterase Activity in Rat Slow Soleus and Fast Extensor Digitorum Longus Muscle[1]

Muscle	Animal	Duration of Suspension		
		1 Week	2 Weeks	3 Weeks
Soleus	Control	285.9±33.4	266.4±14.1	275.7±23.0
	Experimental	358.3±45.1 (125)	605.0±26.3[2] (227)	726.1±58.7[2] (263)
Extensor digitorum longus	Control	510.9±43.4	466.2±30.6	467.0±39.0
	Experimental	450.1±49.3 (-12)	504.5±28.3 (+8)	393.7±24.3 (-16)

SOURCE: Reprinted with permission from ref. 7. Copyright 1985.

[1]Each value is the x ± SE nmol acetylcholinesterase hydrolyzed/whole muscle protein/h obtained from 6 to 15 animals. Numbers in parentheses represent percentage change compared with nonsuspended control.

[2]A statistically significant difference ($P < 0.05$) compared with control value at each corresponding time interval.

Table 4. Activities of Acetylcholinesterase Molecular Forms After 3 Weeks of Muscle Disuse[1]

	Soleus		Extensor Digitorum Longus	
	Control	Disuse	Control	Disuse
4S	30±2	33±2	51±3	41±9
10S	19±2	20±3	27±3	36±6
12S	23±1	25±2	-	-
16S	28±3	22±3	22±1	24±3

SOURCE: Reprinted with permission from ref. 7. Copyright 1985.
[1]Data are from four separate experiments and activities are expressed for each form as percentage of total recovered activity.

normal muscle and the EDL remained populated mainly by type II fibers similar to those found in normal EDL. Acetylcholinesterase activity in both the innervated EDL and SOL initially decreased with fiber degeneration followed by an increase in soleus to above that of control with return towards normal levels. No qualitative differences in the AChE molecular forms pattern in the innervated regenerating muscles, compared to controls, were found during regeneration. In the denervated regenerating muscles, both the EDL and SOL were repopulated with type II fibers and only 1.6% of type I fibers were found in the soleus. Acetylcholinesterase activity of the denervated regenerating muscles remained below the level for controls. Qualitatively, the AChE molecular form pattern in the regenerating denervated SOL changed towards that of the EDL pattern (Table 5).

These experiments clearly indicate that both muscles start out with a uniform fiber population at day one and differentiate into regular patterns by four weeks while innervated. In the absence of innervation, however, the regenerating soleus showed a totally different pattern of fiber distribution shifting from a type I muscle to a IIB muscle after two weeks. By the end of the fourth week the fiber profile has changed to 100% IIA fibers. From this it can be concluded that while both muscles start out histochemically as undifferentiated fast muscles (high ATPase) it is the innervation that determines the ultimate characteristic fiber type distribution of the overall muscle profile. EDL seemed to be less dependent on innervation for its fiber type characteristics during regeneration than the soleus. This muscle in the absence of innervation never develops the typical fiber type distribution or AChE molecular forms profile seen in mature innervated soleus.

Denervation at the onset of regeneration does not change the

Table 5. Effect of Denervation on the Activities of AChE Molecular Forms in Regenerating Rat Soleus Muscle. Activities are expressed for each form as a percentage of the total recovered activity

	AChE Molecular Forms[1]			
	4S	10S	12S	16S
Control	21	19	37	23
2 Weeks Denervation	44	18	17	21
2 Weeks BPVC[2]	28	13	33	26
2 Weeks BPVC and Denervation	41	43	0	16

[1]See Table 2 for legend.
[2]Muscles were directly injected with 1 ml of 0.5% bupivacaine (BPVC, marcaine) solution to produce total muscle fiber degeneration. Muscles were denervated at time of injection.

development of mature pattern characteristics in the fast EDL, but totally prevents the appearance of molecular form pattern typical for mature muscle in regenerating soleus. Indeed the soleus has developed the pattern of the fast muscle with major contribution of the lighter molecular form such as 4S and 10S. Thus denervation at the onset of regeneration prevents the development of slow fiber characteristics such as type I fibers and molecular forms of AChE in the slow soleus (Sketelj et al., 1986). The initiation of slow muscle characteristics does not appear to be a simple dependence on nerve muscle contact, but involves the development of functional activity such as loadbearing, postural and antigravity movements. This is supported by the demonstration of the importance of frequency and pattern of impulses determining AChE molecular form patterns especially in slow muscle (Lomo et al., 1985). The transition from fast to slow after birth or during regeneration is a progressive change. How is this transition effected? Do new population of cells appear or do existing cells switch from fast to slow muscle characteristics? Does functional activity alter the genes being transcribed to switch from fast to slow characteristics?

Regenerating EDL muscle, whether becoming reinnervated or not, has retained some information regarding fiber type and distribution of its AChE molecular forms that is not retained in denervated soleus, since the regenerating denervated soleus in contrast to the innervated muscle does not retain its fiber type nor its characteristic molecular form pattern for AChE but indeed resembles a regenerating denervated EDL muscle with a preponderance of fast type IIA fibers.

During differentiation suppression of information must take place to promote characteristics of specialized cells. During development the interaction between nerve and muscle may have an influence on gene expression. This can be caused by direct contact, growth factor, or by the activity of the muscle. A common regulatory gene initiates the synthesis of AChE. The differentiation between EDL and soleus is then the result of interaction between the genetic progress and muscle. Trophic factors may initiate differentiation. They can be produced by the target organ or the neuron.

References

Brimijoin, S.; Progr. Neurobiol., **1983**, 21, 291-322.

Dettbarn, W-D.; Exp. Neurol., **1981**, 74, 33-50.

Dettbarn, W-D., Groswald, D.E., Gupta, R.C., Misulis, K.E. In: Molecular Basis of Nerve Activity; Eds., Changeux, Hucho, Moulicke, Neuman; W. de Gruyter & Co., Berlin-New York, 1985, 567-588.

Gisiger, V., Stephens, H., Neurosci. Lett., **1982**, 31, 301-305.

Groswald, D.E., Dettbarn, W-D., Neurochem. Res., **1983a**, 8, 983-995.

Groswald, D.E., Dettbarn, W-D., Exp. Neurol., **1983b**, 79, 519-531.

Gupta, R.C., Misulis, K.E., Dettbarn, W-D., Exp. Neurol., **1985**, 89, 622-633.

Lomo, T., Massoulie, J., Vigny, M., J. Neurosci., **1985**, 5, 1180-1187.

Redfern, P.A., J. Physiol., **1970**, 209, 701-709.

Sketelj, L., Crne, N., Brzin, M., Neurochem. Res., **1986**, 12, 159-165.

Physiological Regulation of Tetrameric (G_4) Acetylcholinesterase in Adult Mammalian Skeletal Muscle

Hugo L. Fernandez, Dept. Veterans Affairs Med. Ctr., Kansas City, MO 64128; Dept. Physiology, Univ. of Kansas, Kansas City, KS 66103, USA
Cheryl A. Hodges-Savola, Dept. Veterans Affairs Med. Ctr., Kansas City, MO 64128; Univ. of Kansas Center on Aging, Kansas City, KS 66103, USA

This work addresses the physiological regulation of acetylcholinesterase (AChE) isoenzymes in adult mammalian skeletal muscles. It focuses on defining the mechanism(s) underlying the selective increase in G_4 AChE induced by elimination or enhancement of neuromuscular activity. Our findings document the differential control exerted by motor nerves on individual AChE isoforms and, most importantly, demonstrate the direct participation of acetylcholine receptor (AChR) activation in modulating the production of G_4 AChE in fast-twitch muscles.

The importance of motoneurons in the regulation of adult skeletal muscle AChE has long been acknowledged, yet the precise operation and relative contribution of the neuromuscular factors (agents) involved have not been fully elucidated (Fernandez and Donoso, 1988). Further complicating this problem is the polymorphic nature of AChE and the selective subcellular localization of its multiple oligomers, each of which may serve a distinct physiological function in cholinergic transmission (Toutant and Massoulie, 1988). Although some evidence exists in support of the differential regulation of individual AChE isoforms, it is often assumed that the metabolism of these molecules is uniformly controlled via the operation of a single set of neuromuscular agents. This assumption is based on the qualitatively similar changes exhibited by AChE isoforms (i.e., either increases or decreases in activity depending on muscle type and animal species) in response to certain experimental perturbations (Rotundo, 1987). Recently, however, we have identified two preparations in which the activity of the globular G_4 tetramer in adult rat hindlimb fast-twitch muscles dramatically increases, whereas those of other AChE isoforms either remain unchanged (exercise-enhanced motor activity; Fernandez and Donoso, 1988) or decrease (short-term denervation; Gregory et al., 1989). As described below, we have taken advantage of these experimental paradigms to explore how muscle AChE isoenzymes are influenced by changes in motor activity and to define which synaptic transmission-related factor(s) serves as the molecular signal(s) that triggers such selective G_4 AChE responses.

Enhanced Motor Activity

Exaggerated neuromuscular use or disuse invariably results in certain functional modifications that involve the muscle's membranous elements and eventually its contractile apparatus (Eisenberg, 1985). Since these generalized modifications may be preceded by more subtle changes in synaptic transmission-related molecules, we explored whether a non-invasive method of increasing neuromuscular use (i.e., treadmill exercise) elicited any AChE isoform changes in fast- and/or slow-twitch muscles from adult rats (Fernandez and Donoso, 1988). AChE isoenzyme activities were evaluated after the animals were systematically subjected to low-intensity treadmill running (daily sessions of 1 h/day over 1-30 days at a treadmill speed of 8.5 m/min, with 1 min sprints at 15 m/min given every 10 min). This protocol was selected because it effectively increases motor activity without producing drastic changes in the structural or biochemical composition of skeletal muscle fibers. Indeed, it does not alter the activities of oxidative and glycolytic enzymes, nor the metabolism of muscle contractile proteins (Holloszy and Booth, 1976). Under these conditions, a single exercise session was sufficient to trigger a significant increase in G_4 AChE activity in fast-twitch gracilis muscles without affecting other AChE isoforms, or the muscles' wet weight and total protein content. Such an elevation in the membrane-associated tetramer became even more apparent after 3 treadmill sessions, and persisted at increased levels (≈ 185% of control) through at least 30 days. This early and selective enzymatic change was detected throughout the muscle, but was slightly more pronounced in endplate as compared to non-endplate regions. Results similar to those shown in Table 1 were also

Table 1. *Effects of exercise on gracilis muscle endplate AChE isoform activities*

	G_1+G_2	G_4	A_{12}
Control	1330± 41	777± 50	332±29
Exercised	1254±115	1367±376*	308±63
% Control	94±10	176±28	93±22

Values (pmol/min/gradient) are the means ± SD for 3 control and 3 exercised (7 days) 12 mo-old male Fischer 344 rats. * Greater than control ($p < 0.01$).

obtained for the predominantly fast-twitch anterior tibialis muscle, whereas no changes in AChE isoforms occurred in the slow-twitch soleus. These and additional findings led us to conclude that the G_4 AChE increment represented an adaptive response to enhanced neuromuscular use, rather than to generalized circulatory or hormonal effects. In this context, it must be noted that animal locomotion, as promoted by treadmill running, selectively affects the pattern of neuromuscular activity in distinct sets of muscles; thus, phasic motor units of gracilis and tibialis muscles are affected to a greater extent than are the tonic units of postural soleus muscles. Allowing for some technical differences in the treadmill regimens used (mild exercise versus endurance training), other investigators have recently confirmed most of the foregoing findings (Jasmin and Gisiger, 1990). Altogether, these experiments are consistent with the hypothesis that G_4 AChE is influenced by a combination of neuromuscular agents/factors which is distinct from that affecting other AChE isoforms. More specifically, they indicate that nerve-evoked muscle activity serves a primary role in the regulation of G_4 AChE in mature fast-twitch muscles (Fernandez and Donoso, 1988).

Short-Term Denervation

Although the effects of denervation on skeletal muscle AChE have been extensively examined, only a few reports refer to the enzyme's individual isoforms and primarily focus on the long-term response of the asymmetric A_{12} oligomer (Brimijoin 1983). Surprisingly, less attention has been given to the response of G_4 AChE even though this isoenzyme is closely related to the muscle's dynamic properties (Bacou et al., 1982; Groswald and Dettbarn, 1983) and, as noted above, is particularly sensitive to experimentally-induced alterations in motor activity. Thus, we re-examined the effects of denervation on AChE isoforms, with particular emphasis placed on identifying early changes in the G_4 oligomer (Gregory et al., 1989). Young adult rat gracilis muscles were unilaterally denervated and 24-96 h thereafter AChE isoforms were evaluated in both experimental and contralateral control preparations. Results showed that denervation induced pronounced changes in AChE iso-form activities; indeed, after 24 h significant declines were detected in G_1+G_2 (≈60% of control) and A_{12} (≈70%) AChE activities, whereas the G_4 tetramer remained unchanged. In turn, while the G_1+G_2 and A_{12} isoforms gradually continued to decay (≈25% at 96 h), G_4 AChE increased by 36 h (≈140%), reached an apparent maximum (≈180%) at 42 h, and subsequently returned to control levels (≈ 60 h) which were maintained at least through 96 h. Additional experiments indicated that the elevation in G_4

AChE was present throughout the gracilis muscle, but its magnitude was greater in the motor endplate regions. Fig. 1 illustrates an example of the changes in AChE isoforms 48 h post-denervation. The possible relationship

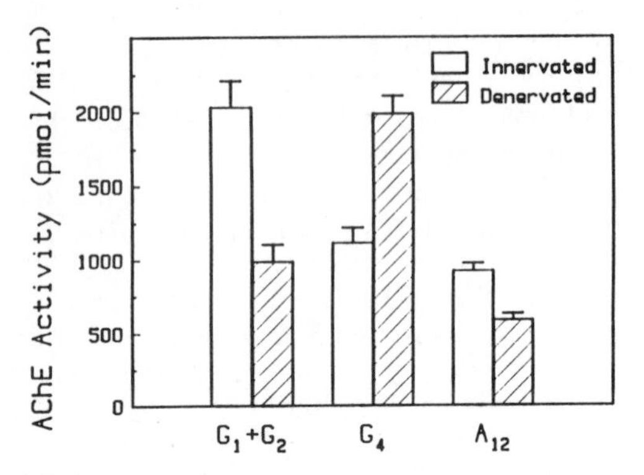

Fig. 1. Effects of denervation (48 h) on 3 mo-old Sprague-Dawley rat gracilis muscle endplate AChE isoforms (pmol/min/gradient) extracted under high salt conditions (1:35 wt/vol; 50 mM Tris-HCl, pH 6.8, 1 M NaCl, 5 mM EDTA, and 1% Triton X-100) and separated in 5-20% sucrose gradients (260,000 g_{max}, ≈18h, 4° C). AChE isoforms were quantified by adding the enzymatic activities under corresponding sedimentation profile peaks; values are the mean ± SEM of 4 experiments.

between the above changes and alterations in certain aspects of muscle metabolism was also explored following either enzymatic inactivation by diisopropylfluorophosphate (DFP) or inhibition of protein synthesis by cycloheximide (CHX). As expected, the denervation-induced increase in G_4 AChE was observed after concurrent denervation and AChE inactivation by DFP, but not following treatment with CHX. Other experiments involved treatment of mus-cles, *in situ,* with AChE inhibitors which allow differentiation between intra- and extracellular AChE isoform pools. Results showed that the G_4 increase was more prominent in the extracellular compartment of muscle endplate regions. Further analyses also indicated that the G_4 increment could not be entirely accounted for by the concomitant decay in asymmetric isoform activities, nor by a denervation-induced deficit in the rate of the tetramer's secretion. From these and other data we concluded that the G_4 AChE response to short-term denervation primarily reflected a transient increase in the assembly of these tetramers from newly synthesized (catalytically active and/or inactive) precursors (Gregory et al., 1989).

Role of ACh and AChR in the Regulation of G4 AChE

At first sight, our findings from the treadmill exercise experiments appear to be in contrast with those involving short-term muscle denervation, i.e., augmentation or elimination of nerve-evoked muscle activity elicited analogous elevations in G_4 AChE. These observations, however, can be reconciled by considering that the G_4 increase induced by exercise represents an adaptive reaction of fast-twitch muscles to the resulting enhanced frequency of interaction between nerve-released ACh and postsynaptic AChRs (Fernandez and Donoso, 1988). In turn, quantal and non-quantal spontaneous ACh release persists for several hours after nerve transection and temporarily exceeds normal levels before ultimately subsiding (Stanley and Drachman, 1986; Zemkova et al., 1987). Since this phenomenon occurs prior to the G_4 AChE response, and ACh or its analogues have been shown to promote the synthesis of total AChE in cultured muscle cells (Walker and Wilson, 1976), it is reasonable to postulate that the transient excess of spontaneously released ACh and its subsequent interactions with AChRs stimulate the enhanced production of the AChE tetramer. On the other hand, the effects of denervation could simply result from the development of spontaneous muscle contractions (fibrillation). To test these possibilities, AChE isoforms were assayed immediately after recording muscle electromechanical activity at various times following nerve transection. Fibrillation potentials were first detected 48 h post-denervation (5-15 potentials/s), their frequency increased between 48 and 72 h, reached a maximum by 72 h (150-200 potentials/s), and persisted at such elevated levels through at least 168 h. Thus, the response of G_4 AChE (as well as that of other isoen-

Table 2. *AChE isoforms in quiescent and fibrillating gracilis muscle endplate regions*

	n	G_1+G_2	G_4	A_{12}	Fibrill.
Control	10	2201±240	1187±156	972±64	-ND-
Den (36h)	7	1228±236	2463±397	760±57	-ND-
Den (96h)	3	873±163	1300± 64	263±39	1430±50

Values are the mean ± SEM of AChE isoforms (pmol/min/gradient) and fibrillation (fibrill.) frequencies (potentials/10 sec) in denervated (Den) and innervated contralateral (pooled controls) muscle samples from 3 mo-old male Sprague-Dawley rats (n = number of animals; ND = non-detectable). All Den values are different from control ($p < 0.01$) except for Den (96h) G_4 AChE.

zymes) to denervation preceded by at least 12 h the onset of fibrillation which was well-established only at a time when the G_4 increment was declining. In other words, as indicated in Table 2, dramatically elevated levels of G_4 AChE were observed in the absence of fibrillation (36 h post-denervation), whereas such enzymatic levels had subsided to control values in the presence of vigorous fibrillation (96 h). Accordingly, whether nerve-evoked (innervated muscle) or spontaneous (denervated muscle), the muscle's contractile activity alone cannot explain our results. Next, the possible involvement of ACh-AChR interactions in the denervation-induced G_4 AChE response was studied by specifically and irreversibly blocking AChRs via intramuscular injection of α-bungarotoxin (α-BTX) immediately before (≈ 30 min) obturator nerve transection. Analyses of AChE isoforms after 48 h showed (Fig. 2) that while this treatment did not affect G_4 AChE in innervated preparations, it completely blocked the denervation-induced increase in this isoform. Other experiments showed that when receptor blockade was initiated 24 h after denervation, it had little to no effect on G_4 AChE, i.e., elevated levels of the tetramer were observed at 48 h. These and additional data indicate that unobstructed ACh-AChR interactions are essential for the G_4 AChE response to take place, but once this response is triggered (within the first 24 h of denervation), it occurs regardless of the functional status of AChRs.

In summary, our findings not only support the differential regulation of AChE isoenzymes by neuromuscular transmission-related events, but demonstrate the direct role of AChR activation in modulating the production of G_4 AChE in fast-twitch muscles. While our studies do not address the precise nature of the neurogenic substance(s) involved in such receptor activation, the well-known specificity of α-BTX's action at transmitter binding sites strongly suggests that this substance is ACh. Accordingly, data from experiments involving enhancement (exercise) or suppression (denervation) of nerve-evoked muscle activity can be readily explained by

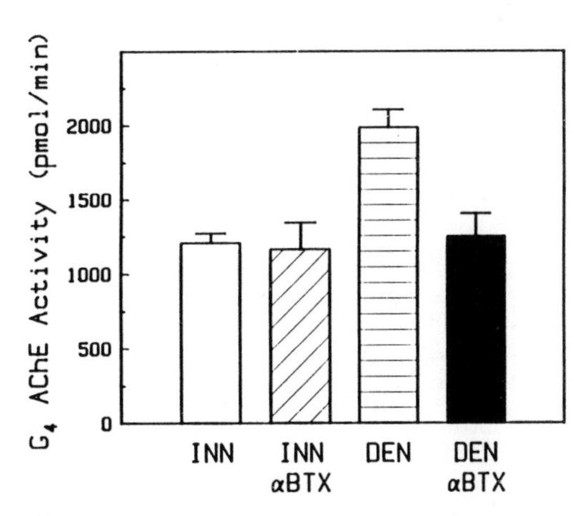

Fig. 2. Effects of denervation (DEN; 48 h) and/or α-bungarotoxin (αBTX) on 3 mo-old Sprague-Dawley rat gracilis muscle endplate G_4 AChE (pmol/min/gradient). Values represent the mean ± SEM of 6-8 experiments.

our conclusion that an increased frequency of interaction between ACh and AChRs signals the muscle to specifically produce more G_4 AChE.

Acknowledgments

We thank E. Gregory for technical assistance. This work was supported by the Medical Research Service of the US Dept. of Veterans Affairs, the National Institute on Aging, and the University of Kansas Center on Aging.

References

Bacou, F., Vigneron, P., Massoulie, J., *Nature*, **1982**, *296*, 661-664.

Brimijoin, S., *Prog. Neurobiol.*, **1983**, *21*, 291-322.

Eisenberg, B.R., *J. Exp. Biol.*, **1985**, *115*, 55-68.

Fernandez, H.L., Donoso, J.A., *J. Appl. Physiol.*, **1988**, *65*, 2245-2252.

Gregory, E.J., Hodges-Savola, C.A., Fernandez, H.L., *J. Neurochem.*, **1989**, *53*, 1411-1118.

Groswald, D.E., Dettbarn, W.-D., *Neurochem. Res.*, **1983**, *8*, 983-995.

Holloszy, J.O., Booth, F.W., *Annu. Rev. Physiol.*, **1976**, *38*, 273-291.

Jasmin, B.J., Gisiger, V., *J. Neurosci.*, **1990**, in press.

Rotundo, R.L., In *The Vertebrate Neuromuscular Junction*; Salpeter, M.M., Ed.; Alan R. Liss: New York, NY, **1987**, 791-808.

Stanley, E.F., Drachman, D.B., *Brain Res.*, **1986**, *365*, 289-292.

Toutant, J.P., Massoulie, J., *Handbook Exper. Pharmacol.*, **1988**, *86*, 225-265.

Walker, C.R., Wilson, B.W., *Neurosci.*, **1976**, *1*, 508-513.

Zemkova, H., Vyskocil, F., Edwards, C.A., *Pflugers Arch.*, **1987**, *409*, 540-546.

The Pool of G_4 Acetylcholinesterase Characterizing Rodent Fast Muscles is Differentially Regulated by the Predominant Type, Dynamic or Tonic, of Natural Activity

Victor Gisiger*, Département d'Anatomie et Centre de Recherche en Sciences Neurologiques, Université de Montréal, C.P. 6128, Succ. A, Montréal (Québec) Canada H3C 3J7.
Bernard J. Jasmin, Shauna Sherker and Phillip F. Gardiner, Sciences de l'Activité Physique, Université de Montréal, C.P. 6128, Succ. A, Montréal (Québec) Canada H3C 3J7.

Following chronic elevation of neuromuscular activity, as achieved by training programs, the G_4 content of fast muscles selectively increased or decreased according to whether the actual activity the muscles were called upon to perform was predominantly dynamic or tonic. These data strongly confirm that, in innervated mature fast muscles of rodents, their characteristic G_4 pool is subject to a specific regulation, independent from that controlling the asymmetric forms. The results strengthen the view that G_4 plays in fast muscles a specific and essential functional role, definitely distinct from that of A_{12}. The G_4 alterations induced by training are likely to modulate the ACh synaptic background concentration and therefore, endplate excitability.

Fast muscles of rodents characteristically differ from their slow-twitch counterparts by exhibiting high levels of G_4, the tetrameric form of AChE (see refs. in Jasmin, 1990). According to a recent report (Gisiger, 1988), this additional pool is concentrated at the perijunctional sarcoplasmic reticulum of the muscle fibers, where it forms a cojunctional compartment embedding the neuromuscular junctions in an AChE-rich environment. Furthermore, whereas innervation plays a fundamental role in establishing and maintaining the pools of both A_{12} and G_4 (Brimijoin, 1983), converging lines of evidence indicate that, once established, the pool of G_4 is particularly sensitive to the functional demands placed upon the muscle (see refs. in Jasmin, 1990). The ensemble of these facts has led to the proposal that the G_4 content of a muscle depends more on the type of activity it actually performs than on the influence of its motor innervation (Gisiger, 1988).

In the present study, we tested this hypothesis by examining the effect of a chronic elevation in motor activity on the content in AChE molecular forms of functionally antagonist rat muscles. Neuromuscular activity was enhanced by running and swimming training programs, i.e. 2 exercises which entail clearly distinct types of muscle activity.

Materials and Methods

Running training: Female Sprague-Dawley rats (180-200 g) were trained to run on a motor driven treadmill (Quinton Instruments, Seattle, WA) at a speed of 27m/min with a 10% grade. Running time was progressively increased by 10 min every fourth day so that by the seventh week, the animals ran 120 min per exercise session. Swimming training: The swimming regimen consisted of swimming twice a day in a tub filled with water kept at 33°C. Swimming time was gradually increased until the animals swam 90 min twice a day by the third week. Each training program lasted approximately 12 weeks. Control rats were cage-confined.

Tissue preparation and homogenization as well as analyses of AChE and citrate synthase were performed as described in detail elsewhere (Jasmin, 1990).

Results

The effect of the prolonged increase in neuromuscular activity on the AChE molecular forms has been studied in five different rat hindlimb muscles. They included four predominantly fast muscles, i.e. gastrocnemius

(GAST), plantaris (PL), tibialis anterior (TA) and extensor digitorum longus (EDL), as well as the slow-twitch soleus (SOL). On a functional basis, these 5 muscles belong to 2 antagonist groups: GAST, PL and SOL act as ankle extensors while TA and EDL operate as ankle flexors (Gruner, 1980a,b).

General effects of training.
Following the 12 week training programs, the mean body weight of the trained rats was very similar to that of controls (about 290 g). The trained muscles showed no obvious signs of hypertrophy and, accordingly, the protein concentration of the extracts obtained from control and trained muscles was generally unchanged. Only two muscles showed a slight protein increase: running trained TA (14%) and swimming trained EDL (28%) ($P < 0.001$). The muscle's adaptation to enhanced activation was evaluated by measuring the activity of the mitochondrial enzyme citrate synthase. Our running training program raised the activity of this enzyme by 35-38% ($P < 0.02$) in GAST, PL, EDL and SOL, and by 9% in TA, whereas the swimming program induced a generally smaller, although still significant, elevation of the citrate synthase activity in all 5 muscles (13-18%; $P < 0.05$). Such increases are similar to those which are usually reported for, respectively, the running (Holloszy, 1976) and swimming (Hickson, 1984) training programs used in this study.

Effect of the training programs on AChE.
The impact of training varied widely according to the type of exercise, the molecular forms as well as the properties and function of the muscles. However, the most distinctive effect of training was seen on G_4 of fast muscles, in terms not only of its extent, but also of its direction.

Running training. This program induced marked changes in the G_4 content (activity per muscle) of the fast muscles which were opposite in the ankle extensors and flexors (Fig. 1). G_4 selectively increased by 66% and 54% in GAST and PL ($P < 0.001$), respectively, with no significant change in the level of the other forms, except for a slight A_{12} increase in PL ($P < 0.05$) (Fig. 2). In contrast, the ankle flexors showed a 30% to 40% ($P < 0.05$) decrease in their G_4 content, which was accompanied in TA by a slight increase of the other forms, including A_{12} (Fig. 2). In the case

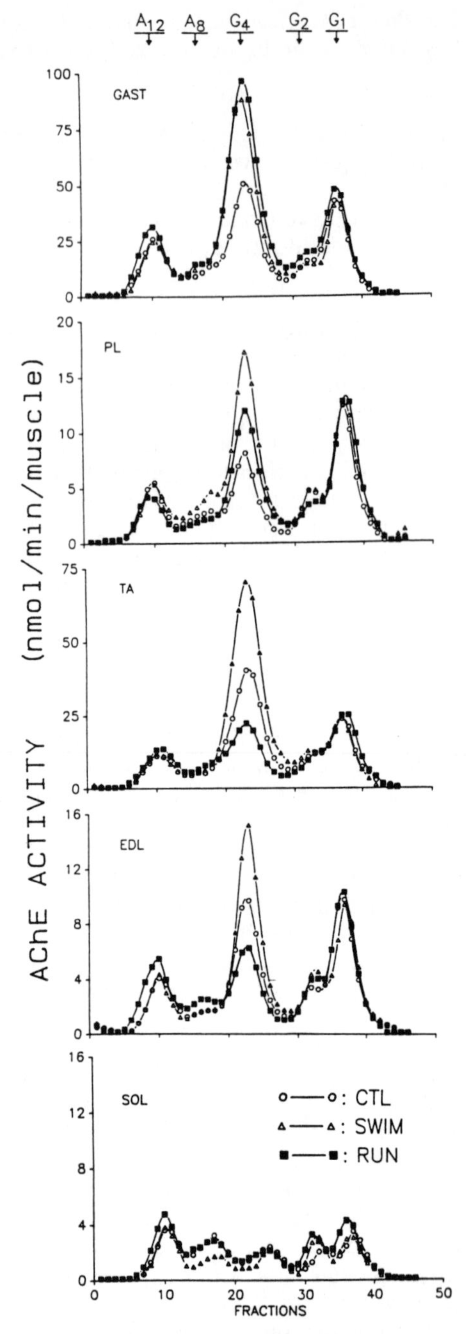

Fig. 1. Effect of running and swimming training programs on AChE molecular forms of functionally antagonist muscles of the rat hindlimb. Shown are actual distributions that are representative of the average content in AChE molecular forms observed in samples of each control (n= 15-24) and trained (n= 8-14) muscle.

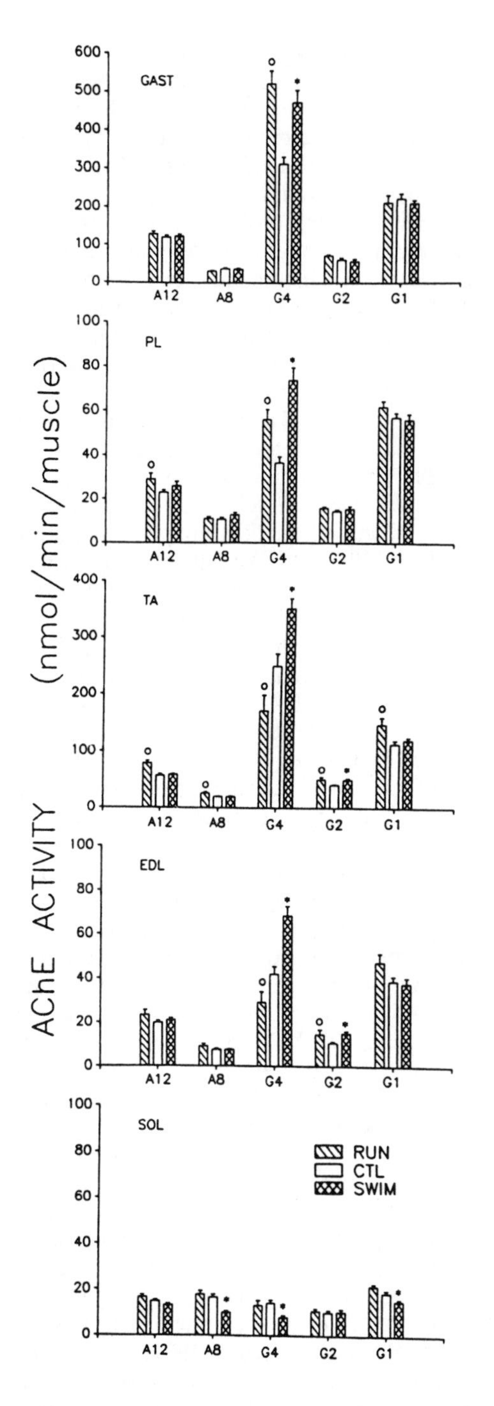

Fig. 2. Activities per muscle of the AChE molecular forms in control and trained (running and swimming) muscles. Values are $\bar{X} \pm$ S.E.M.; o and * denote statistically significant differences from control muscles ($P<0.05$). The sample size is the same as in Fig. 1.

of the slow-twitch soleus, running training resulted in small, nonspecific changes which affected most of the molecular forms (Fig. 2). Swimming training. This program induced a marked elevation of G_4 in all fast muscles including both ankle extensors and flexors (Fig. 1). G_4 was increased by 40% to 60% ($P< 0.005$) in GAST, TA and EDL, and by 100% in PL (Fig. 2). Actually, the impact of swimming was even more selective than that of running: among all the other molecular forms, only G_2 was slightly increased in TA and EDL (Fig. 2). In the case of SOL, swimming reduced A_8 and G_4 to about 60 % of control values ($P< 0.005$). It should be stressed however that, although significant, these reductions involved only low amounts of AChE activity (Fig. 2).

Discussion

Our training programs selectively affected the G_4 content of fast muscles, with almost no effect on the asymmetric forms, as did the much milder exercise regimen used by Fernandez and Donoso (1988). This highly selective G_4 adaptation to enhanced activity exhibited by intact muscles is especially striking when compared to the concomitant alterations of both asymmetric forms and G_4 displayed by denervated muscles in response to direct stimulation (Lomo, 1985). However, the most significant finding yielded by this study is the observation that running training induced a marked reduction of G_4 in the ankle flexors, opposing both the concomitant tetramer elevation in the functionally antagonist ankle extensors and the G_4 elevation displayed by these same ankle flexors following swimming training. During running, the ankle extensors generate a significant part of the propulsive force throughout the stance phase (Gruner, 1980a). Therefore, the relationship between the G_4 elevation displayed by GAST and PL and the dynamic role they play during running appears straightforward. As for the ankle flexors, the fact that they exhibited an increased level of citrate synthase activity indicates that they were significantly sollicited during running training. EMG data support this conclusion (Gruner, 1980a; Roy, 1985). Therefore, the G_4 decrease cannot be ascribed to a reduction in neuromuscular activity. Moreover, no significant difference has been found between the pattern as well as the rate of activation by motoneurones of fast ankle extensors and flexors (Zajac, 1980). The determinant difference seems to be the peculiar biomechanical action of the ankle flexors. During running, TA and EDL flex the foot and then maintain this flexion throughout the swing phase while the hindlimb is recovered and protracted forward by the action of other muscles (Gruner, 1980a). Consequently, during running, the ankle flexors contract for most of the time against the inertial torque generated at the ankle by acceleration of the protracted hindlimb, counteracting stretching forces (Grillner, 1975; Rasmussen, 1978). By contrast, during swimming, the ankle flexors contract against the water resistance, as do the extensors, and both groups of muscles exhibit a clear dynamic activity (Gruner, 1980b). This latter situation resulted in a marked G_4 increase in TA and EDL. Taking into account the biomechanical data, the results of the present study support the conclusion that the G_4 content of a fast muscle is high or low according to whether the actual activity the muscle is called upon to perform is predominantly dynamic or tonic.

Functional impact of the G_4 alterations induced by training.

Especially important is the strong confirmation brought by this report that, in innervated mature fast muscles, their characteristic G_4 pool is subject to a specific regulation, completely distinct from that controlling the asymmetric forms (Gisiger, 1983; Massoulié, 1982). The highly differentiated response exhibited by the G_4 pool to changes in muscle activity together with the independence of the G_4 and A_{12} regulations support the proposal that G_4 fulfills a specific and essential function, definitely distinct from that of A_{12} (Gisiger, 1983; Gisiger, 1988). This conclusion is strengthened by converging evidence indicating that the two AChE end-products G_4 and A_{12} are assembled along two separate biosynthetic pathways (Toutant, 1987) and occupy distinct localizations at the surface of mature innervated muscle fibers, i.e. mainly intrajunctional for A_{12} versus mainly perijunctional for G_4. Indeed, whereas no less than 70% of all the asymmetric forms present in rodent muscles are clustered within the endplates (Dreyfus, 1983; Fernandez, 1984), by contrast, the G_4 molecules do not exhibit any specific affinity for the junctional structures (Toutant, 1987) and, accordingly, they appear to distribute themselves fairly uniformly over the sarcolemmal membrane both within and around the endplates (Gisiger, 1988; Stephens and Gisiger, unpublished observations). This marked distribution divergence makes it unlikely that the G_4 pool specific of fast muscles may significantly contribute to the rapid hydrolysis of ACh achieved by the asymmetric forms within the junctions. In contrast, the cojunctional compartment of G_4, in particular the G_4 molecules externalized around the endplates, as both membrane-bound and secreted molecules, occupy a perfect strategic position in order to very effectively remove ACh molecules diffusing away from the postsynaptic membrane (Gisiger, 1982-83; Wathey, 1979). In this respect, the AChE changes induced by training appear substantial enough to significantly modify the concentration of G_4 molecules around the endplates and consequently, the hydrolysis rate of ACh evading the junction-bound AChE. Accordingly, the G_4 alterations induced by training are expected to affect the extent of ACh accumulation around the endplates during the trains of impulses, and therefore the background level of ACh within the junctions. Such changes should be of consequence since even slight increases in the synaptic background concentration of transmitter significantly depress the excitability of the motor endplate by facilitating desensitization of the nicotinic receptors (Changeux, 1981). In this view, an increase of the muscle G_4 content mediated by elevation of dynamic activity, of the magnitude reported here, is likely to enhance the ability of the endplate to effectively respond to high frequency activation. The group of Magazanik has shown that the slow time course of endplate potentials in SOL, which constitutes an important limiting factor in the frequency of synaptic transmission, is directly related to the low AChE content characterizing this slow-twitch muscle (Magazanik, 1979). Interestingly, on the basis of their electrophysiological data, Magazanik and collaborators (Giniatullin, 1986) have proposed that, in addition to the classical one, the functional role of muscle AChE also consists of eliminating the traces of preceding activity in order to avert desensitization. At the light of this convergent and complementary approach, the selective and opposite G_4 regulation exhibited by fast muscles in response to enhanced dynamic versus tonic activity, which we report here, appears as an especially appropriate and powerful adaptation.

[This work was supported by grants from NSERC and FCAR (Team grant)]

References

Brimijoin, S., *Prog. Neurobiol.,* **1983**, *21,* 291-322.

Changeux, J.P., *Harvey Lect.,* **1981**, *75,* 85-254.

Dreyfus, P.A., Rieger, F., Pinçon-Raymond, M., *Proc. Natl. Acad. Sci. USA,* **1983**, *80,* 6698-6702.

Fernandez, H.L., Inestrosa, N.C., Stiles, J.R., *Neurochem. Res.,* **1984**, *9,* 1211-1230.

Fernandez, H.L., Donoso, J.A., *J. Appl. Physiol.,* **1988**, *65,* 2245-2252.

Giniatullin, R.A., Bal'tser, S.K., Nikol'skii, E.E., Magazanik, L.G., *Neirofiziologiya,* **1986**, *18,* 645-654.

Gisiger, V., Stephens, H.R., *Gif Lectures in Neurobiology. J. Physiol. (Paris),* **1982-83**, *78,* 720-728.

Gisiger, V., Stephens, H.R., *J. Neurochem.,* **1983**, *41,* 919-929.

Gisiger, V., Stephens, H.R., *J. Neurosci. Res.,* **1988**, *19,* 62-78.

Grillner, S., *Physiol. Rev.,* **1975**, *55,* 247-304.

Gruner, J.A., Altman, J., Spivack, N., *Exp. Brain Res.,* **1980a**, *40,* 361-373.

Gruner, J.A., Altman, J., *Exp. Brain Res.,* **1980b**, *40,* 374-382.

Hickson, R.C., Overland, S.M., Dougherty, K.A., *J. Appl. Physiol.,* **1984**, *57,* 1834-1841.

Holloszy, J.O., Booth, F.W., *Annu. Rev. Physiol.,* **1976**, *38,* 273-291.

Jasmin, B.J., Gisiger, V., *J. Neurosci.,* **1990**, *10,* 1444-1454.

Lomo, T., Massoulié, J., Vigny, M., *J. Neurosci.,* **1985**, *5,* 1180-1187.

Magazanik, L.G., Fedorov, V.V., Snetkov, V.A., *Prog. Brain Res.,* **1979**, *49,* 225-240.

Massoulié, J., Bon, S., *Annu. Rev. Neurosci.,* **1982**, *5,* 57-106.

Rasmussen, S., Chan, A.K., Goslow, G.E. Jr., *J. Morph.,* **1978**, *155,* 253-269.

Roy, R.R., Hirota, W.K., Kuehl, M., Edgerton, V.R., *Brain Res.,* **1985**, *337,* 175-178.

Toutant, J.-P., Massoulié, J., In *Mammalian Ectoenzymes,* **1987**, A.J. Kenny and A.J. Turner, eds., pp. *289-328,* Elsevier, Amsterdam.

Wathey, J.C., Nass, M.M., Lester, H.A., *Biophys. J.,* **1979**, *27,* 145-164.

Zajac, F.E., Young, J.L., *J. Neurophysiol.,* **1980**, *43,* 1221-1235.

The Long-Term Effects of Acetylcholinesterase Inhibition at the Frog Neuromuscular Junction

Lev G. Magazanik, Vladimir A. Snetkov, Sechenov Institute of Evolutionary Physiology and Biochemistry, Academy of Sciences of the USSR, 44 M.Thorez pr., Leningrad, 194223, U.S.S.R.

The effects of anticholinesterase drugs on the amplitude and the time course of the miniature end-plate currents (MEPCs) were studied in the isolated frog neuromuscular preparation. It was shown that relatively small increase of amplitude and prominent initial prolongation of MEPCs were followed by slow restoration of parameters. The desensitization of cholinoreceptors as a possible mechanism of the phenomenon observed was investigated. Experimental data and results of MEPC modeling speak in favor of this hypothesis.

There are many evidences that the accumulation and long-lasting presence of ACh molecules in the synaptic cleft is the main result of AChE inhibition. It provokes in turn many simultaneous processes (sustained postsynaptic depolarization, potentiation, desensitization, presynaptic failures etc.) which lead to transmission blockade (Magazanik, 1989; Magazanik, Giniatullin, 1989). The paper presented is devoted to long-lasting changes of end-plate currents (EPCs) and their possible molecular nature.

Changes of EPC shape induced by AChE inhibition.

Although both the initial MEPC shape and their changes induced by AChE inhibition vary greatly in different neuromuscular junctions (Magazanik et al., 1984), the increase of EPCs decay time constant (τ) is much more pronounced than the amplitude enhancement. There were no evidences obtained that AChE inhibition influences open channels lifetime (Katz, Miledi, 1973). The extent of decay lengthening (τ_1/τ_o) correlates with quantity and density of synaptic AChE active sites measured in biochemical and cytochemical investigations (Barnard, Lyles, Pizzey, 1982). Evidently, the main effect of any anticholinesterase drug on neuromuscular junction is the increase of the lifetime of free ACh molecules in the synaptic cleft. It must alter the time course of ACh-activated ion channels gating, i.e MEPC shape. Katz and Miledi proposed that MEPC decay lengthening after AChE inhibition is a result of repetitive binding of ACh molecules to ACh receptors and related buffering of ACh diffusion from the synaptic space.

0–8412–2008–5/91/0086$06.00/0

MEPC lengthening induced by AChE inhibition is transient

As it was mentioned above, the 15-30 min treatment with either prostigmine (6 μM) or organophosphorus inhibitor, armine (1 μM), induced an increase of MEPC amplitude and τ. However, the maximal prolongation of MEPC decay was followed by the slow and continuous fall of τ without pronounced changes of MEPC amplitude (see Potapyeva, Snetkov, Magazanik, this volume) (Fig. 1). This fading out of anti-AChE drug effects may not be a result of the restoration of AChE activity, since repetitive treatment with drug had no influence on τ evolution. We proposed that the phenomena observed may be a result of the prolongation of the lifetime of free ACh molecules into the synaptic cleft favoring the desensitization of ACh receptors. This hypothesis was supported by the fact that the decrease of temperature to 11°C prevented MEPC shortening completely, and maximal values of τ and amplitude remained stable during several hours.

On other hand, desensitization can be promoted by the application of exogenous ACh. Actually, in the experiments on anti-AChE-treated preparations after relatively short-term (10-15 min) bath application and following washout of 2 μM ACh amplitude and τ dropped to 74 ±3% and 49±3%, respectively (n=6). During further washout in contrast to rapid recovery of amplitude, τ restored very slowly. At a time when the amplitude was recovered completely, τ reached only 61±4% (n=6) of the control level (Fig. 2.); in most experiments τ did not reach the initial value during the long washing period (1-2 hours). This than in control (τ_1/τ_0=4.4) (Fig. 3.). During permanent presence of CCh the effect of prostigmine was even less prominent ($\tau_1/\tau_0 \approx 1.4$). In any way no significant change of MEPC amplitude was observed.

The modeling of the effects of AChE inhibition

The quantitative evaluation of relative contribution of different processes in the MEPC generation

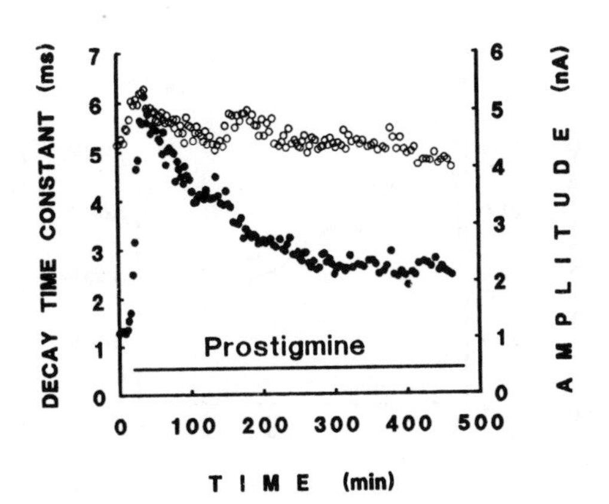

Fig. 1. Long-term effects of prostigmine (6.6 μM) on MEPC parameters. Open circles - amplitude, filled circles - decay time constant.

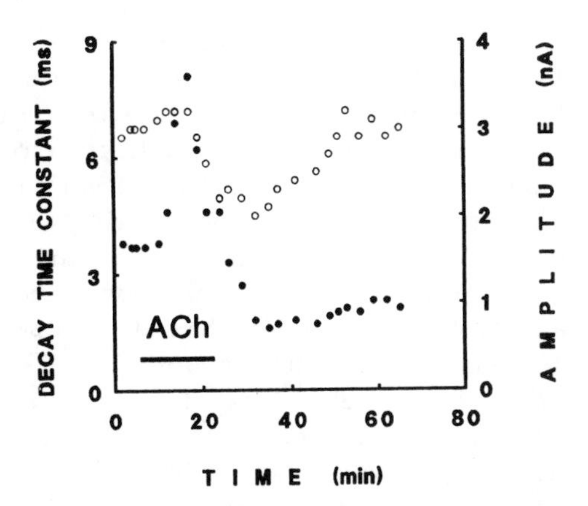

Fig. 2. The effects of application and washout of 2 μM ACh on MEPC parameters. Symbols used - as in Fig. 1.

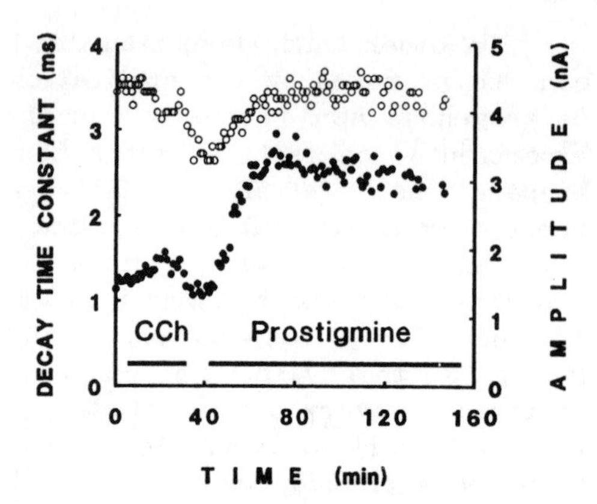

Fig. 3. The decreasing of the effects of AChE inhibition after pretreatment with 10 μM CCh. Symbols used - as in Fig. 1.

(Rosenberry, 1979; Whatey, Lester, Nass, 1979, Nigmatullin et al., 1984) is necessary for interpretation of anti-AChE drugs action. A set of differential equations describing several simultaneous processes has been solved for this purpose. As a result the functions representing the time course of concentrations of the all contributors (AChE active sites; resting, single- and double-bound states of the ACh receptor; free ACh etc.) were obtained. This approach allowed the adequate simulation of normal MEPC (AChE is active), its temperature and potential dependences. However, some difficulties appeared when the effect of AChE inhibition was simulated: in model a large increase of τ corresponded to much greater increase of MEPC amplitude than it was experimentally observed. To solve this problem and to obtain a more flexible model we use now another approach based upon description on microscopic level of stochastic behavior of all individual molecules involved. This model allows to change the kinetics schemes, geometry, initial distribution of contributors etc.

One of the most important results of modeling was a presentation of the time course of free ACh concentration (Fig. 4). It is clear that if AChE is active it relation of MEPC amplitude and τ changes could not be the result of reduction of resting ACh receptors number. Actually, the same shortening of MEPC decay (to 61%) induced by α-bungarotoxin on the anti-AChE-treated preparations was accompanied by amplitude fall to 66±5% (n=6). We supposed that the desensitized ACh receptors can serve as traps for free ACh (Magazanik et al., 1990). In comparison with resting receptors the desensitized ones do not open the channels but have about 20 times higher affinity to ACh (Cohen, Strnad, 1987). During an exposure to ACh the number of desensitized receptors increased and during the washing period these receptors lost bounded exogenous ACh (and thereby became ready to catch quantal ACh) much more rapidly than converted to the resting state. Hence, increase of the number of desensitized receptors (traps) after exposure to ACh leads to qualitatively same results as ACh hydrolysis by AChE.

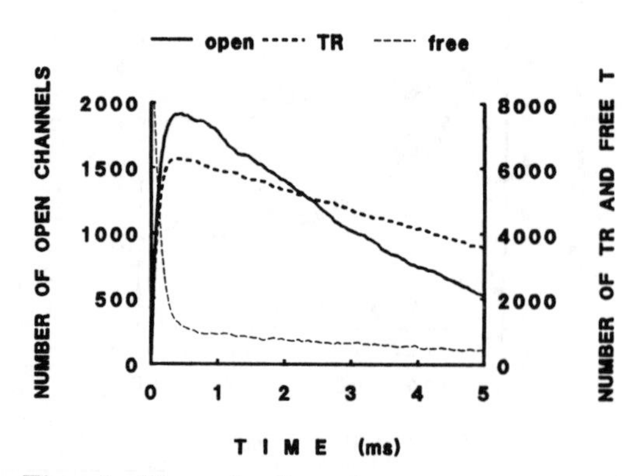

Fig. 4. The simulated time course of double-bounded (open), single-bounded (TR) receptors and free transmitter (T) after AChE inhibition.

Prominent shortening of MEPC without decrease of amplitude was observed earlier in prostigmine-treated muscle during the action of desensitization-promoting drug proadifen (Giniatullin et al., 1989).

A some number of desensitized receptors may exists on postsynaptic membrane under normal conditions (Katz, Thesleff, 1957; Cohen, Strnad, 1987). To evaluate the contribution of pre-existing desensitized receptors in effects of the AChE inhibition the action of non-hydrolysable agonist, carbacholine (CCh), was studied. After the 15 min treatment with 10 μM CCh followed by 15 min washing MEPC were changed negligibly, but AChE inhibition by prostigmine led to much smaller increase of τ ($\tau_1/\tau_0 \approx 2$) hydrolyzes the whole ACh of quanta in several tens of microseconds. After AChE inhibition free ACh concentration falls in two steps. The fast initial decrease is due to the binding of the main part of ACh to resting receptors which leads to the formation of their single- and double-bounded states, i.e. MEPC amplitude. During the following slow phase all receptor states are in equilibrium with free ACh in any point of postsynaptic area and therefore MEPC decay reflects directly the reduction of ACh molecules number. Thus any way of ACh removal in addition to ordinary diffusion should lead to the MEPC decay shortening.

An attempt to simulate the effect of free ACh trapping on MEPC shape was made in order to describe the results of experiments presented in Fig. 1 and 2. In particular, 20% of resting receptors were replaced with desensitized ones which differed by ability to bind ACh molecule for time much longer than MEPC duration. This replacement went to 2-fold shortening of MEPC decay if AChE was absent. Corresponding MEPC amplitude fall was 30% vs. less than 20% in experiments. This difference seems to be not critical and may be ascribed to some processes not included into modeling.

Conclusion

Unfortunately, there are no precise information on the changes of all synaptic processes taking place in the time scale of hours after AChE inhibition. But the main body of evidences speaks in favor of suggestion that all these changes are due to retention of ACh in bound form. There are several sites capable for trapping of ACh differed in affinity. Binding to these sites and especially to desensitized receptors can partly compensate the lack of enzymatic ACh removal and thereby attenuate slowly the changes of postsynaptic currents.

References

Barnard, E.A., Lyles, J.M., Pizzey, J.A., *J.Physiol.*, **1982**, *331*, 333-354.

Cohen, J.B., Strnad, N.P., In *Molecular Mechanisms of Desensitization to Signal Molecules*, T.M.Konjin et al., Eds., NATO ASI Series, Springer, N.Y.,1987, vol.**H6**, pp. 257-273.

Giniatullin, R.A., Khamitov, G., Khazipov,R., Magazanik, L.G., Nikolsky, E.E., Snetkov, V.A., Vyskocil,F., *J. Physiol.*, **1989**, *412*, 113-122.

Katz, B., Miledi, R., *J.Physiol.*, **1973**, *231*, 549-574.

Katz, B., Thesleff, S., *J. Physiol.*, **1957**, *138*, 63-80.

Magazanik, L.G., In *Neuromuscular junction*, L.C.Sellin, R.Libelius and S.Thesleff, Eds., Fernstrom Found. Ser., Elsevier, Amsterdam, 1989, vol. **13**, pp. 259-272.

Magazanik, L.G., Fedorov V.V., Giniatullin, R.A., Nikolsky, E.E., Snetkov, V.A., In *Cholinesterase*, M.Brzin, E.Barnard and D.Sket, Eds., Walter de Greuter, Berlin, 1984, pp. 229-242.

Magazanik, L.G., Giniatullin, R.A., *Progress in Zoology,* **1989**, *37*, 197- 208.

Magazanik, L.G., Snetkov, V.A., Giniatullin, R.A., Khazipov, R.N., *Neurosci. Lett.*, **1990** (in press).

Nigmatullin, N.R., Snetkov, V.A., Nikolsky, E.E., Magazanik, L.G., *Neurofiziologia*, **1988**, *20*, 390-398 (in Russian).

Rosenberry, T.L., *Biophys.J.*, **1979**, *26*, 263-289.

Wathey, J.C., Nass, M.M., Lester, H.A., *Biophys.J.*, **1979**, *27*, 145-164.

Acetylcholinesterase in Damaged and Regenerating Neuromuscular Junctions

Lili Anglister* and **Brigitte Haesaert**, Department of Anatomy and Embryology, Hebrew University-Hadassah Medical School, Jerusalem 91010, Israel

Acetylcholinesterase (AChE) in skeletal muscle is highly concentrated at the neuromuscular junction, where it is found in the synaptic cleft, associated with the synaptic portion of the myofiber's basal lamina. We used surgical manipulations in frog muscles to examine the individual contribution of myofibers and nerve terminals to synaptic AChE. We demonstrate that the nerve, in the absence of myofibers, produces AChE that becomes adherent to synaptic basal lamina, and carries high concentrations of the A_{12} AChE molecular form. Denervated adult myofibers do not produce synaptic AChE and A_{12} form unless they are regenerating.

In vertebrate skeletal muscles AChE is concentrated at neuromusclular junctions (Couteaux, 1955; Salpeter et al., 1972; Fig. 1a). Some of the surface enzyme at the synaptic cleft is probably associated with the myofiber's or neuronal plasma membranes (for reviews, Massoulie and Bon, 1982; Brimijoin, 1983; Silman and Futerman, 1986; Rotundo, 1987). But, a substantial fraction of the synaptic enzyme is tightly associated with the synaptic cleft basal lamina (Hall and Kelly, 1971; Betz and Sakmann, 1973; McMahan et al., 1977; Anglister and McMahan, 1984). If muscles are damaged in ways that spare the basal lamina sheaths, new myofibers develop within the sheaths and the damaged axons regrow to form new neuromuscular junctions at the original syanptic sites. AChE accumulates at the regenerating neuromuscular junctions, and as at normal ones, is found in the synaptic cleft, adherent to synaptic basal lamina. The studies we describe were aimed at learning the individual roles motor nerve, muscle and basal lamina play in producing and directing the accumulation of AChE at synaptic sites.

Contribution of Nerve to Synaptic Acetylcholinesterase

Various lines of evidence indicate that nerve may play an indirect role: By inducing activity in the myofibers it triggers them to produce synaptic AChE. Thus denervated myofibers that are stimulated electrically (Lomo and Slater, 1980; Lomo et al., 1985), are ectopically innervated (Weinberg and Hall, 1979), or are regenerating (Anglister and McMahan, 1985) reaccumulate AChE at the denervated synaptic sites. A fundamental question addressed in this study was whether nerve can directly provide AChE to the synaptic cleft (Anglister, 1987).

The preparations we studied were damaged frog muscles, in which axons were innervating empty basal lamina sheaths, devoid of myofibers. Cutaneus pectoris muscles were exposed in anasthesized frogs and were cut inbetween axonal branches, without damaging motor nerves and their terminals. This caused degeneration and removal in the animal of all myofibers from their basal lamina sheaths while leaving intact motor axons, nerve terminals and synaptic basal lamina sheaths. The operated frogs were X-irradiated to prevent regeneration of myofibers. In these preparations nerve terminals persist at almost all the synaptic sites on the sheaths and show no discernible differences in structural features from normal (Yao and McMahan, 1984). At the time of operation muscles were exposed to DFP or MSF (10mM, 1h, RT), which block irreversibly and inactivate all AChE activity detected by either histochemical or biochemical assays. Five weeks later the

0–8412–2008–5/91/0091$06.00/0

preparations were examined for the appearance of newly formed AChE. Histochemical staining revealed an arborized pattern, characteristic of neuromuscular junctions, which co-localized with the staining for nerve terminals (not shown).

Examination at the Electron microscope of a cross section through a stained arbor revealed that AChE reaction product was concentrated adjacent to the nerve terminals, obscuring synaptic basal lamina (Fig. 1b), as in the intact neuromuscular junction (Fig. 1a). Since all original AChE had been inactivated and myofibers had been removed, the newly formed AChE at the synaptic sites must have been produced by the persisting axon terminals. Further studies indicate that this AChE became associated with the extracellular matrix: 1) The innervated sheaths preparations were denervated after the enzyme had been produced and accumulated at nerve terminal sites (30 days after the original operation). Histochemical examination of AChE was performed 2 weeks later, when all the cellular components have been degraded and removed. The enzyme remained adherent to the synaptic basal laminae although nerve terminals had been removed. 2) Much of the newly formed AChE was dissociated by collagenase which is known to remove extracellular matrix components including AChE from neuromuscular junctions (e.g. Anglister and McMahan, 1985) without affecting nerve terminals.

These results demonstrate that motor nerves are capable of producing some of the synaptic AChE at the neuromuscular junctions and that the AChE becomes a component of synaptic basal lamina.

Molecular Forms of AChE Provided by Motor Axons Innervating Myofiber Basal Lamina Sheaths

Acetylcholinesterase (AChE) consists of asymmetric or globular molecular forms that can be distinguished by their sedimentation coefficients (Massoulie and Bon, 1982). In frog muscle the major asymmetric form is a 17.5S AChE, while most globular forms are 4-6S AChE (Fig. 2A, also Nicolet et al., 1986). We analyzed the innervated basal lamina sheaths preparation, described above, to determine which of the molecular forms of AChE may be contributed by the nerve terminals to the synaptic sites (Anglister and Haesaert, 1988).

The preparations were examined at 30 days after surgery and inactivation of the original AChE. At that time all synaptic sites on the vacated basal laminae were occupied by nerve terminals and showed histochemically the presence of AChE, newly formed by nerve. Biochemical analysis by sucrose gradient centrifugation (Fig. 2B) revealed that 30-50% of this AChE was a 17.5S species; an equal fraction migrated as 4-6S molecules. In innervated region of intact frog muscle only 15-20% AChE consisted of 17.5S species, while most (60-70%) were 4-6S forms (Fig. 2A).

The high 17.5S content in the innervated sheaths originated from the motor axons: In the nerve bundle to CP most of the enzyme (usually more than 50%) was 17.5S AChE, while 4-6S species were a minor component (10-20%) (Fig. 2C). Similar composition was found in the brachial nerve from which the nerve to CP branched. Ligation experiments (not presented here) showed that the 17.5S migrated by fast axonal transport in both directions, more was carried in the anterograde than in the retrograde direction. Moreover, some of the 17.5S AChE in the innervated sheaths preparation, could be inhibited by echothiophate, a cholinesterase inhibitor, under conditions in which it does not permeate cells (e.g., Younkin et al., 1982), thus indicating that this fraction of the enzyme resided extracellularly.

Altogether, the results show that the motor nerve to the cutaneus pectoris muscle, produces all forms of AChE, but carries high concentrations of 17.5S AChE. This is also reflected in the high content of this form in the innervated basal lamina preparation, in the absence of myofibers. It is most likely that some of the 17.5S AChE is destined for extracellular synaptic localization or for trophic purposes.

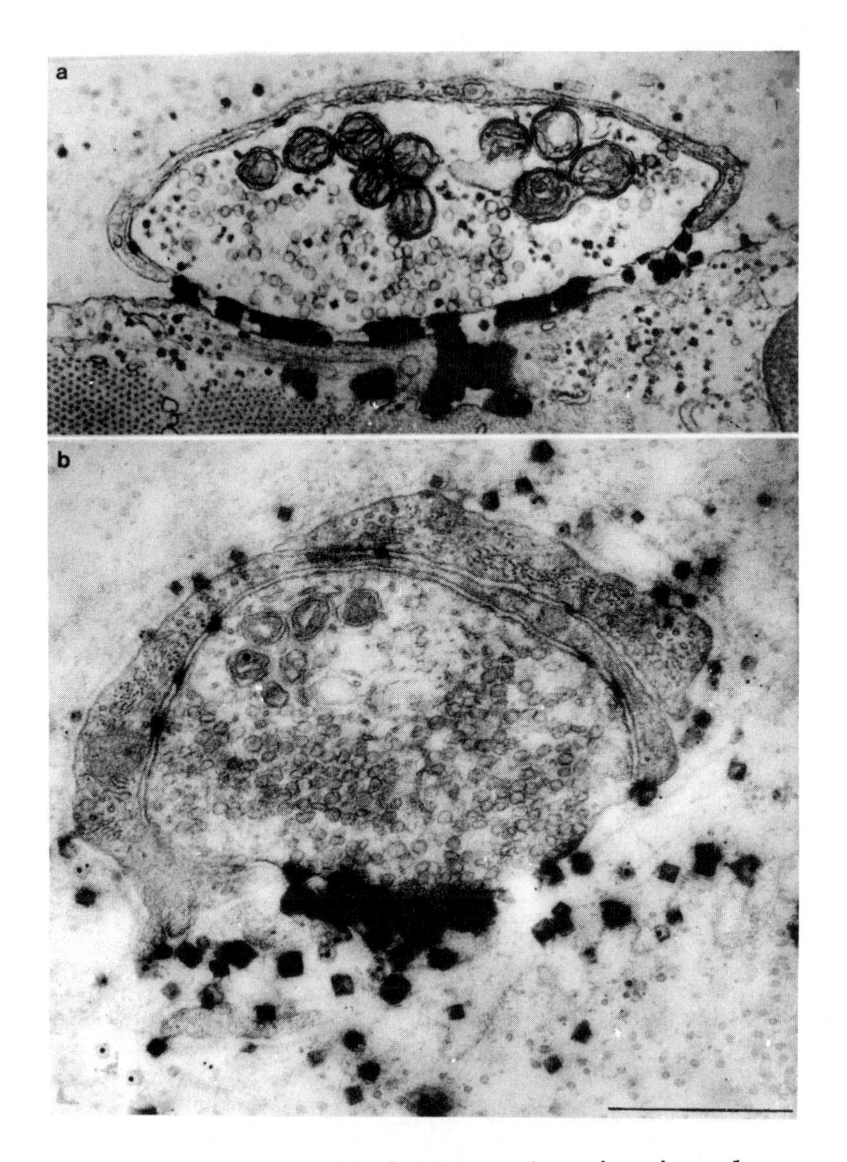

Fig. 1. Electron micrographs of cross sections through muscle preparations histochemically stained for AChE (Karnovsky, 1964). (a) Neuromuscular junction in a normal muscle. Crystals of AChE stain are concentrated in the synaptic cleft and junctional folds, obscuring myofiber basal lamina. (b) Innervated basal lamina sheaths 5 weeks after surgery and irreversible inactivation of all original AChE, and after new AChE has been produced. AChE stain is highly concentrated on the surface of the nerve terminal where it is positioned against the persisting myofiber basal lamina, obscuring that part of the sheaths. Bar, 1 um.

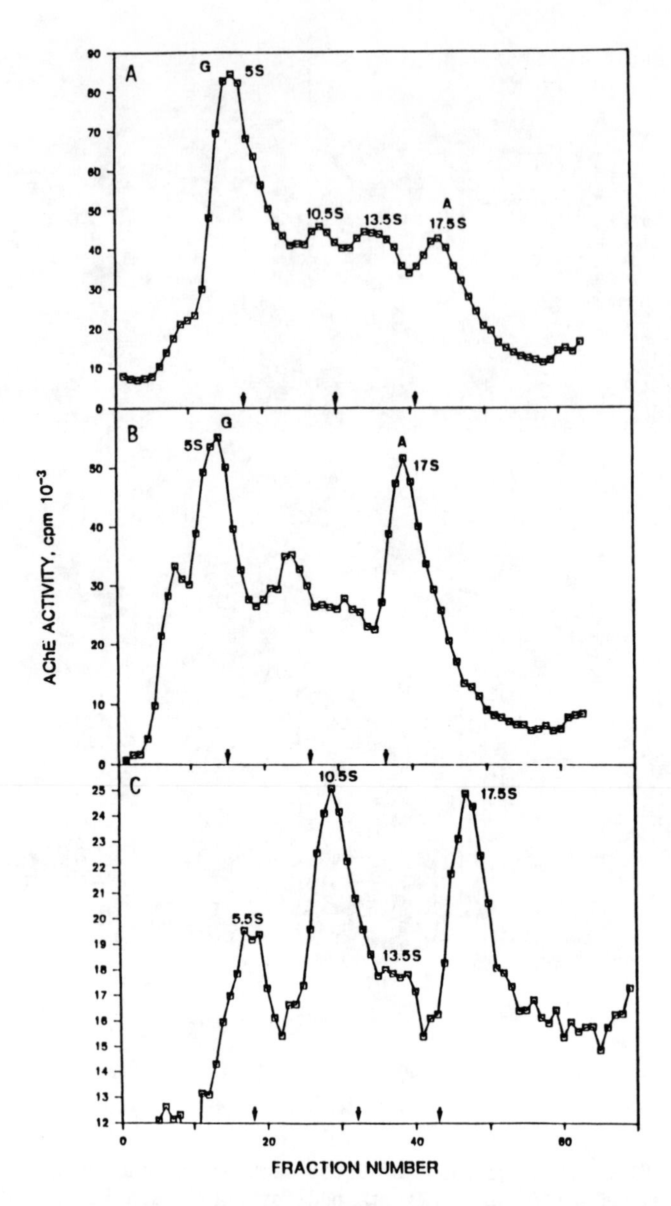

Fig. 2. Sucrose gradient analysis for AChE molecular forms extracted from frog Cutaneus pectoris neuromuscular preparations. A)normal muscle. B) innervated basal lamina sheaths. C) axons innervating that muscle. Samples, extracted in buffer containing high salt concentrations and detergent were layered on 5-20% gradient and spinned for 6.5hrs at 54,000 rpm, 4 C. Arrows indicate sedimentation of standard protein markers: alkaline phosphatase, catalase, β-galactosidase (from left to right). AChE determined by a radiometric assay (Johnson and Russel, 1975).

AChE in Denervated Frog Muscle

Denervation of frog muscle causes a reduction in the surface AChE, at the neuromuscular junctions, as indicated by Cholinesterase histochemical staining and physiological response (Birks et al., 1960). Yet, some staining appears at the denervated sites (Fig. 3a). AChE at the frog neuromuscular junction is tightly adherent to synaptic basal lamina: It persists associated with it even in the absence of both myofibers and nerve terminals (McMahan et al., 1977; Anglister and McMahan, 1983). Thus, it was possible that the AChE producing the reaction product at the denervated synaptic site, was left there, as a remenant of the intact junction. In order to learn whether the denervated muscle was producing synaptic AChE, in the absence of innervation, we irreversibly inactivated all the muscle's AChE (using MSF, conditions described above) immediately following muscle denervation. We searched the preparation for histochemical staining of AChE at the denervated synaptic sites. Almost no staining was detected (Fig. 3b). Even at 8 weeks after denervation there was still only a background level of AChE reaction product. This leads to the conclusion that the denervated muscle did not produce AChE or didnot put the AChE on the surface at the denervated synaptic site.

Different result was obtained when we examined denervated regenerating muscle. In that experiment muscles were damaged by crushing them, in a way that caused disintegration of the myofibers, sparing the basal lamina sheaths. Myofibers were allowed to regenerate but reinnervation was deliberately prevented. As before, original AChE was irreversibly inactivated during surgery. The regenerating myofibers produced new AChE, as demonstrated by histochemical staining of the muscle (Fig. 3c). The new AChE preferentially accumulated at points where the plasma membrane of the new muscle fibers was apposed to the region of the basal lamina that had occupied the synaptic sites at the original neuromuscular junctions (Anglister and McMahan, 1985). The newly formed enzyme became incorporated into the synaptic basal lamina. These results demonstrate that regenerating myofibers, unlike intact denervated myofibers, produce synaptic AChE in the absence of nerve. The accumulation of the enzyme is directed by molecules in the synaptic basal lamina that play a role in organizing postsynaptic differentiation in the regenerating muscle (Reist et al., 1987).

Biochemical analysis of denervated frog muscle for AChE molecular forms, provides results that are consistent with the histochemical observations described above. Denervated intact muscle did not produce asymmetric 17.5S AChE, but rather the globular tetramer. The regenerating muscle, however, produces significant amounts of the asymmetric enzyme, even when it is maintained chronically denervated. Thus the regenerating muscle, possibly like a developing muscle, is in a different regulatory state of its AChE biosynthesis, assembly and incorporation to the cellular structures.

Conclusion

Frog myofibers and nerve terminals can produce synaptic cleft AChE. Its appearance in the regenerating neuromuscular junctions is directed by synaptic basal lamina. Nerve - muscle - matrix interactions and additional factors play a role in regulating the production and maintenance of the synaptic enzyme, and are a subject for further research.

Acknowledgement

We are grateful to Mrs R. Cohen for expert technical assistance. This study was supported by grants from the US-Israel Binational Science Foundation, Israel Academy of Sciences, Bat Sheva de Rothscild Foundation and the Bruno Goldberg Endowment to L.A.

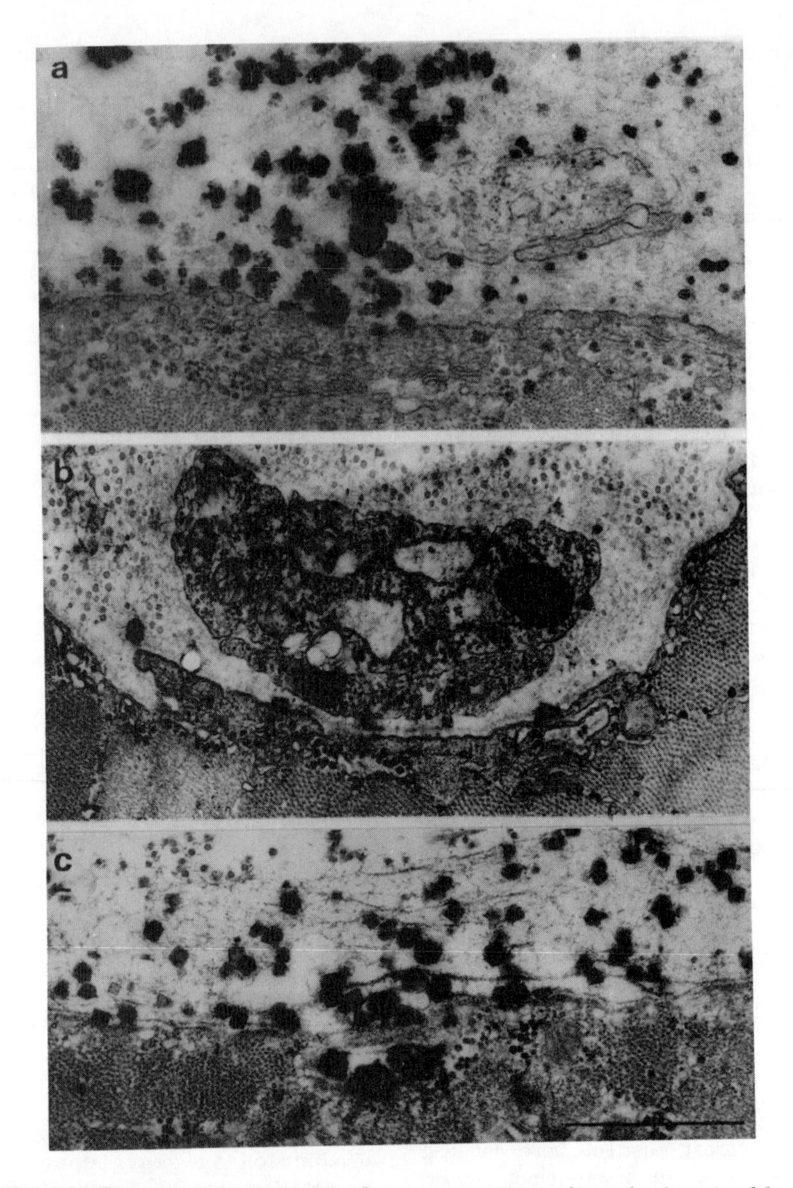

Fig. 3. Electron micrographs of cross sections through chronically denervated synaptic sites in frog muscle preparations, histochemically stained for AChE, 30 days after denervation. (a) Denervated muscle. (b) Muscle was treated with MSF to inactivate irreversibly original AChE, at the time of denervation (see text for details). (c) Regenerating muscle, that had been denervated and MSF treated, as in (b), but at the same time myofibers had been damaged to cause first their degeneration and later their regeneration from dividing satellite cells (Anglister and McMahan, 1985). Note the presence of phagocytic processes and possibly remenants of the axons at the denervated sites in (a) and (b). Bar, 1 um.

References

Anglister, L. Neurosci. Abstr., **1987**, 13, 1211.

Anglister, L. and Haesaert, B., Neurosci. Abstr., **1988**, 14, 165.

Anglister, L. and McMahan, U.J. In: Basement Membrane and Cell Movement (Ciba Foundation Symposium 108), 1984, pp. 163-182.

Anglister, L. and McMahan, U.J. J. Cell

Betz, W. and Sakmann, B. J. Physiol., **1973**, 230, 673-688.

Birks, R., Katz, B. and Miledi, R. J. Physiol. (Lond.), **1960**, 150, 145-168.

Brimijoin, S. Progress in Neurobiology, **1983**, 21, 291-322.

Couteaux, R. Int. Rev. Cytochem., **1955**, 30, 874-886.

Hall, Z.W. and Kelly, R.B. Nature New Biol., **1971**, 232, 62-64.

Johnson, C.D. and Russel, R.L. Anal. Biochem., **1975**, 64, 229-238.

Karnovsky, M.J. J. Cell Biol., **1964**, 23, 217-232.

Lomo, T. and Slater, C.R. J. Physiol., **1980**, 303, 191-202.

Lomo, T., Massoulie, J. and Vigny, M. J. Neurosci., **1985**, 5, 1180-1187.

Massoulie, J. and Bon, S. Ann. Rev. Neurosci., **1982**, 5, 57-106.

McMahan, U.J., Sanes J.R. and Marshall, L.M. Nature, **1977**, 271, 172-174.

Nicolet, M., Pincon-Raymond, M. and Rieger, F., J. Cell Biol., **1986**, 102, 762-768.

Reist, N.E., Magill, C. and McMahan, U.J. J. Cell Biol., **1987**, 105, 2457-2469.

Rotundo, R.L. In: The Vertebrate Neuromuscular Junction, Alan R. Liss, Inc., 1987, pp. 247-284.

Salpeter, M.M., Plattner, H. and Rogers, A.W. J. Histochem. Cytochem., **1972**, 20, 1059-1068.

Silman, I. and Futerman, A.H. Eur. J. Biochem., **1986**, 170, 11-22.

Weinberg, C.G. and Hall, Z.W. Dev. Biol., **1979**, 68, 631-635.

Yao, Y.-M. and McMahan, U.J. Neurosci. Abstr., **1984**, 10, 1085

Younkin, S.G., Rosenstein, C., Collins, P.L. and Rosenberry, T.L. J. Biol. Chem., **1982**, 257, 13630-13637.

REGULATION AND DISTRIBUTION OF ACETYLCHOLINESTERASE MOLECULAR FORMS *IN VIVO* AND *IN VITRO*

François-Marie Vallette *°, Sabine De la Porte ** and Jean Massoulié *

***Laboratoire de Neurobiologie, E.N.S, 46 rue d'Ulm, 75230 Paris Cedex 5, France.**
****Laboratoire de Neurobiologie Cellulaire, Université de Bordeaux I, avenue des Facultés, 33405 Talence Cedex, France.**

°Corresponding author (Fax : (33) 1 43 29 81 72)

In this paper, we discuss the biogenesis of the different molecular forms of cholinesterases, with special attention to the most complex ones: the collagen-tailed, asymmetric forms (A forms). Each tissue exhibits a specific pattern of AChE molecular forms which varies with the evolution of its interactions with other elements. During the past years we studied the regulation of AChE biosynthesis, both *in vivo* and *in vitro*, in muscles and in central and peripheral nervous tissues of birds and rodents.

Considerable attention has been focussed on the role of nerve-muscle interactions in the distribution and the regulation of the AChE molecular forms (reviewed in Massoulié and Bon, 1982; Toutant and Massoulié, 1988). It is well known that denervation causes drastic changes in AChE activity and in the pattern of the molecular forms. In the rat, the A forms disappear almost completely 2 weeks after denervation, whereas the globular forms decrease more slowly. In chicken, denervation leads to the loss of A_{12} forms but the total activity increases (increase due to the globular forms). In rabbit, both total activity and A_{12} forms increase in slow muscles. The effect of denervation thus depends on the species and on the fast or slow type of the muscle investigated (see Toutant and Massoulié, 1988).

Biosynthesis and distribution of AChE molecular forms during myogenesis *in vivo* and *in vitro*

1) Role of the nerve

AChE activity can be observed very early in vertebrate embryos. Intense patches of AChE activity can be detected histochemically at neuro-muscular contacts immediately after the first nerve-muscle interactions. The A forms appear at this stage, i.e at 14-d in the posterior leg muscles of rat embryos and 5-d *in ovo* in chicken leg muscles.

We investigated the expression of A forms of AChE in two different myogenic territories, the limb buds and the dermomyotomes of chick and quail embryos *in ovo* (Vallette *et al.*, 1987).

0–8412–2008–5/91/0098$06.00/0

The dermomyotome is the part of the somite that gives rise to dermis and muscles. In 3-d quail embryos, the A forms are present in the mononucleated, morphologically undifferentiated cells which constitute the dermomyotome at this stage. In contrast, in chick embryos, these forms are not observed in the dermomyotomes, indicating a very important difference between these two closely related species. The latter territory, however, produces A forms after differentiation into myotubes in culture, in both quail and chicken embryos (Vallette *et al.*, 1987).

To avoid any neural influence on the biosynthesis of A forms, the neural crest and/or the neural tube (i.e. the origin of motor nerves) were removed at 2-d *in ovo*. These operations did not suppress the capacity of dermomyotomal cells to express A forms, when analyzed 24 hr later. This result shows that under certain conditions, mononucleated cells can produce the complex A forms without any physical contact with nervous elements.

We obtained additional evidence that the expression of A forms by muscle cells *in vivo* does not always require innervation, by studying heterologous cocultures of chicken myotubes and rat neuronal cells or rat myotubes and chicken neuronal cells. The specific origin of the AChE could be identified with polyclonal antibodies raised against pure rat or chicken AChE. We found that the focalized enzyme, present in the numerous patches, predominantly originated from the muscle. By selective immunoprecipitation of AChE of each species, we could not detect any influence of nervous elements on the production of muscle AChE molecular forms contents or *vice-versa* (De la Porte *et al.*, 1986).

The major event observed in neuromuscular cocultures was the formation of patches of AChE, which always seemed to coincide with the neuromuscular contacts. AChE patches are, in this respect, more specifically associated with synaptic contacts than the clusters of acetylcholine receptors, which are found all along the muscle fibers.

No difference was detected in the content of AChE molecular forms before and after mild treatment of rat neuromuscular cocultures with collagenase, in spite of the complete removal of AChE patches (Vallette *et al.*, 1990). Thus, the AChE molecules in these structures represent only a small part of the total AChE activity, and present an extreme sensitivity to collagenase, probably due to their molecular insertion.

Using permeant and non permeant inhibitors of AChE, we analyzed the role of the nerve on the repartition of the enzyme between the extra- and intracellular compartments (see below). We found that the nerve has little influence on the repartition of the enzyme in the two compartments (Vallette *et al.*, 1990).

2) Role of muscular activity

In vivo, Lømo *et al.* (1985) showed that electrical stimulation of a denervated rat soleus induced a marked and rapid increase of AChE activity. The pattern of molecular forms depended strongly on the mode of stimulus. A "fast" mode, mimicking the normal activity of a fast muscle (short bursts of high frequency) resulted in a pattern of AChE forms resembling that of a rapid muscle, whereas a "slow" mode (constant low frequency) resulted in a pattern resembling that of the normal slow soleus muscle. This result illustrates the important role played by electrical stimulation in the biosynthesis of AChE.

Several experiments using murine cells *in vitro* suggest that the level of A forms is regulated by ionic fluxes, rather than by the muscular contractions (rewieved in Toutant and Massoulié, 1988). In rat primary muscular cultures, the level of A forms decreases upon treatment with TTX (a blocker of Na^+ channels), whereas the addition of veratridine (a drug which maintains these channels in the open state) results in a marked increase of AChE activity, and in particular of the A forms (De la Porte *et al.*, 1984). A direct influence of muscle contractions can be ruled out, since both drugs paralyze the cultures. Ca^{++} ions have been suggested to play a key role in the regulation of the biosynthesis of AChE (Rubin, 1985).

We examined the role of TTX and veratridine in quail and chicken myotubes *in vitro*, and found that veratridine increased the proportion of A forms; this effect was reproduced by a similar depolarisation induced by KCl. TTX treatment, on the other hand, did not affect the pattern of AChE molecular forms (F.M. Vallette and J. Massoulié, in preparation). Fernandez-Valle and Rotundo (1989) recently reported that TTX treatment led to the disappearance of A forms in quail muscles in culture. This discrepancy is probably due to a difference in the stage of maturation of the cells used. The secretion of AChE molecules was however affected in the same manner in both cases: TTX treatment increased the secretion of AChE molecules, while veratridine or depolarization induced by KCl had the opposite effect (Rotundo, this volume; F.M. Vallette and J. Massoulié, in preparation).

Secretion of AChE; distribution of molecular forms between cellular compartments

In chick muscles the rate of secretion of AChE is extremely high (Rotundo and Fambrough, 1980). The secreted AChE corresponds mostly to non amphiphilic G_4 and G_2 forms (F.M. Vallette and J. Massoulié, in preparation).

By combination of membrane permeant and non permeant inhibitors of AChE, it is possible to determine the extra- or intracellular localization of the different AChE forms. These studies were performed mostly in myogenic or neuronal cultures (Lazar and Vigny, 1980; Brockman *et al.*, 1982; Rotundo, 1984; Vallette *et al.*, 1986) and demonstrated that all forms exist in both compartments. This implies that the complex A forms are assembled intracellularly, as directly confirmed by Rotundo (1984), who showed that this process takes place in the Golgi apparatus (see Figure 1). The A forms, as well as the G_4 molecules, are however preferentially located extracellularly, e.g. in muscle cells.

In a neuroblastoma cell line (T_{28}), G_4 is mostly exposed extracellularly while G_1, the other main form in these cells, is essentially intracellular (Lazar and Vigny, 1980). In chromaffin cells, we showed that part of the non amphiphilic G_4 is contained in granules, and thus may be secreted by exocytosis, together with catecholamines (Bon *et al.*, 1990).

Active and inactive pools of AChE

Studying the metabolism of ^{35}S-methionine labelled AChE in quail muscle cultures, Rotundo (1988) found that more than 80% of the newly synthesized molecules were rapidly degraded, without acquisition of enzymatic activity.

The importance of a rapid intracellular degradation was also illustrated in a study of the metabolism of active AChE, in the murine neural cell line, T_{28}. Incorporation of aminoacids labelled with heavy isotopes made it possible to distinguish pre-existing light molecules from newly synthesized heavy molecules (Lazar *et al.*, 1984). This study showed i) that AChE was first synthesized as an inactive precursor; ii) that both G_1 and G_4 molecular forms existed under several metabolic pools, with different turn over rates; iii) and that most of the active G_1 molecules in these cells were degraded, less than 20% being processed into G_4 or secreted. The molecular nature of the different pools of each oligomeric form, as well as their metabolic relationships, are still unknown.

Comparing the intensity of immunostaining and enzymatic activity, we found that chicken tissues contain an important proportion of inactive AChE molecules (F.M. Vallette, J.M. Chatel, Y. Frobert, J. Grassi and J. Massoulié, in preparation). Such molecules are poorly retained on a N-methylacridinium Sepharose affinity column, under conditions which allow the quantitative binding of active AChE (Vallette *et al.*, 1983). Sedimentation analyses of low salt soluble (LSS) and detergent soluble (DS) fractions showed that the AChE activity corresponds mostly to G_2

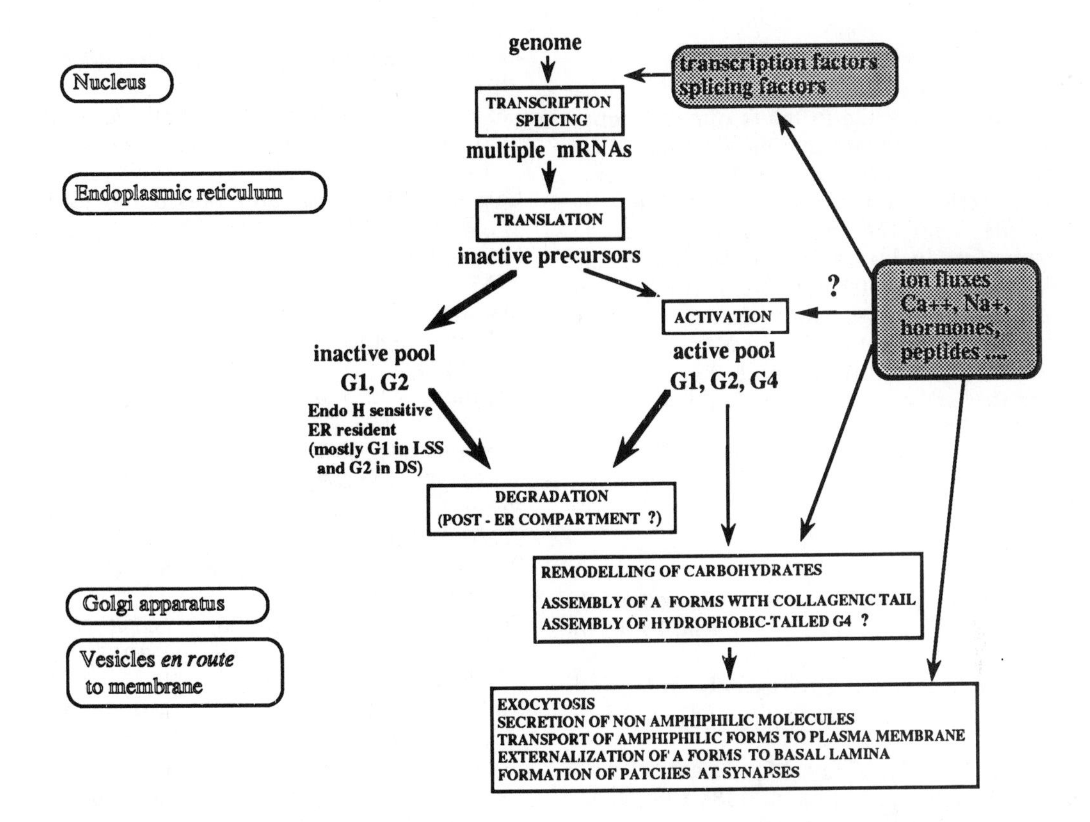

Figure 1

Biosynthesis maturation and transfert of AChE forms.

The left side of the diagram indicates the subcellular localization of the various maturation processes. The shaded boxes (right) indicate the suggested pathways by which environmental conditions modify these biosynthetic processes. Analyses of neuroblastoma cultures (Lazar *et al.*, 1984), chicken muscle cultures (Rotundo, 1988) and chicken tissues *in vivo* (Vallette *et al.*, in preparation) suggest that the inactive pool and the pathways leading to internal degradation may be quantitatively predominant.

and G_4 forms. Inactive AChE is preferentially extracted in the LSS fraction and corresponds to monomers and dimers (G_1 being predominant in LSS and G_2 in DS). Treatment with endoglycosidases H or F demonstrated that monomers and dimers carry high-mannose carbohydrate chains while tetramers contain complex carbohydrate chains, suggesting that the inactive enzyme is most likely a resident of the endoplasmic reticulum. The physiological meaning of the presence of a large fraction of inactive AChE both *in vivo* and *in vitro* is unknown.

Conclusion

In muscles, the nerve influences the total AChE activity, the distribution of molecular forms, and their localization at synapses. It is possible that the acquisition of catalytic activity and the level of the inactive pool may be controlled by muscular activity. Other unknown factors are probably involved in these regulations.

Acknowledgements

We would like to thank our colleagues who were involved in this work and Lisa Oliver for critical reading of this manuscript. This work was supported by grants from the Institut National de la Santé et de la Recherche Médicale, the Muscular Dystrophy Association of America, the Association Française contre les Myopathies, the Centre National de la Recherche Scientifique and the Direction des Recherches et Etudes Techniques.

REFERENCES

Bon, S., Bader, M.F., Aunis, D., Massoulié, J. and Henry, J.P., *Eur. J. Biochem.* **1990**, in press.

Brockman, S.K., Przybylski, R.J. and Younkin, S.G., *J. Neurosci.* **1982**, *2*, 1775-1785.

De la Porte, S., Vigny, M., Koenig, J. and Massoulié, J., *Dev. Biol.* **1984**, *106*, 1337-1342.

De la Porte, S., Vallette, F.M., Grassi, J., Vigny, M. and Koenig, J., *Dev. Biol.* **1986**, *116*, 69-77.

Fernandez-Valle, C. and Rotundo, R.L., *J. Biol. Chem.* **1989**, *264*, 14043-14049.

Lazar, M. and Vigny, M., *J. Neurochem.* **1980**, *35*, 1067-1079.

Lazar, M., Salmeron, E., Vigny, M. and Massoulié, J., *J. Biol. Chem.* **1984**, *259*, 3703-3713.

Lømo,T., Vigny, M. and Massoulié, J., *J. Neurosci.* **1985**, *5*, 1180-1187.

Massoulié, J. and Bon, S., *Annu. Rev. Neurosci.* **1982**, *5*, 57-106.

Massoulié, J. and Toutant, J.P., *Handbook of Experimental Pharmacology*, **1988**, *86*, 167-224.

Rotundo, R.L. and Fambrough, D.M., *Cell* **1980**, *22*, 595-602.

Rotundo, R.L. *Proc. Natl. Acad. Sci. (USA).* **1984**, *81*, 479-483.

Rotundo, R.L., *J. Biol. Chem.* **1988**, *263*, 19398-19406.

Rubin, L.L., *Proc. Natl. Acad. Sci. (U.S.A)* **1985**, *82*, 7121-7125.

Toutant, J.P. and Massoulié, J., *Handbook of Experimental Pharmacology* **1988**, *86*, 225-265.

Vallette, F.M., Marsh, D.J., Muller, F., Massoulié, J., Marçot, B. and Viel, C., *J. Chromatog.* **1983**, *257*, 285-296.

Vallette, F.M., Vigny, M. and Massoulié, J., *Neurochem. Int.* **1986**, *8*, 121-133.

Vallette, F.M., Fauquet, M. and Teillet, M.A., *Dev. Biol.* **1987**, *120*, 77-84.

Vallette, F.M., De la Porte, S., Koenig, J., Massoulié, J. and Vigny, M., *J. Neurochem.* **1990**, *54*, 915-923.

Regulation of AchE Expression in Cultures of Rat Sympathetic Neurons

Michel Weber*, **Brigitte Raynaud and Simone Vidal**, Laboratoire de Pharmacologie et Toxicologie Fondamentales, Toulouse, France

Although a large amount of knowledge has been gained on the expression of AchE in muscle cells, comparatively little is known on its regulation in neurons. For such a study, primary cultures of sympathetic neurons from new-born rats present distinct advantages:

- this homogeneous population of post-mitotic neurons can be cultured for many weeks in the virtual absence of non-neuronal cells;

- the expression of neurotransmitter phenotypic traits is regulated by several extra-cellular signals or by the long-term depolarization of the neurons with high K^+ medium. It is thus posssible to determine if the expression of AchE in these cultures is correlated with the expression of a particular neurotransmitter phenotype.

In such cultures, we have studied the regulation of the expression of acetylcholine and catecholamine synthesizing enzymes, as well as AchE molecular forms by Nerve Growth Factor, by high K^+ medium and by a partially purified choline acetyltransferase inducing factor from muscle conditioned medium (Weber, 1981; Weber et al, 1985); this factor is most probably identical to a cholinergic differentiation factor purified to homogeneity by Fukada (1985). The recent molecular cloning of this factor established its identity with the previously characterized Leukemia Inhibitory Factor (Yamamori et al, 1989).

Molecular forms of AchE of dual-function sympathetic neurons

Sympathetic neurons cultured in a medium non-permissive for the proliferation of non-neuronal cells and containing 5% rat serum as well as an optimal concentration of NGF acquire many characteristics of mature noradrenergic neurons. In particular, a rapid development of the three noradrenergic synthesizing enzymes (tyrosine hydroxylase (TOH), Dopa decarboxylase (DDC) and dopamine-β-hydroxylase (DBH)) is observed. In addition, these neurons express a low, but significant level of CAT activity, and are appropriatly called dual-function neurons.

Under these culture conditions, the neurons express a high level of AchE, composed of the A12 form (up to 30%), G4 form (45-60%) and G1-G2 forms (10-25%) (see Figure 2). These neurons thus contain a distinctly higher percentage of A12 than observed in the rat SCG (Gisiger et al, 1978). The possible role of the preganglionic fibers in the repression of A12 expression *in vivo* is discuted below. Sympathetic neurons in culture also release large amounts of AchE, as the activity released in 48 hrs corresponds to 130-200% of the cell associated activity. The percentage of the released molecular forms reflects that of the cell associated forms; in particular, the neurons release large amounts of 16S AchE. The identity of the released 16S form as A12 AchE was verified by its sensitivity to collagenase form III ,which characteristically shifted the sedimentation coefficient from 16 to 18S, and by its aggregation properties at low ionic strengh (Ferrand et al, 1986).

The cellular localization of AchE forms was determined by two methods (Ferrand et al, 1986):

- in the first method (Taylor et al, 1981; Inestrosa et al, 1981), neurons are treated with the impermeable and reversible inhibitor BW 284c51 and with DFP, a permeable, but irreversible inhibitor. Under these conditions, the catalytic sites exposed at the cell surface are protected by BW from DFP poisoning, and can be reactivated by washing.

- in the second method (Lazar and Vigny, 1980), external sites are selectively and irreversibly inactivated by phospholine, which, at 4°C, penetrates cells very slowly.

These two methods gave comparable results and showed that 60-70% of the total AchE activity is exposed at the neuron surface, a relatively high percentage compared to other neuronal cells (see Ferrand et al, 1986). The same experiments demonstrated that 60% of the A12 form is exposed at the cell surface, as well as 80% of the G4 form and 60% of the G1-G2 forms. These neurons thus contain a sizeable pool of internal A12 form (Figure 1).

The AchE molecular forms in sympathetic neurons also differ in their solubilization properties: at least 76% of the external A12 form can be solubilized at high ionic strength, but 44-50% of the internal A12 form is only solubilized in the presence of Triton X100. In contrast, 78% of the external G4 form is solubilized by Triton at low ionic stength, whereas about 50% of the internal G4 form can be solubilized at high ionic strength. As far as the G1-G2 forms are concerned, 50% of the neuron surface activity, and 100% of the internal activity are solubilized by high salt.

0–8412–2008–5/91/0103$06.00/0

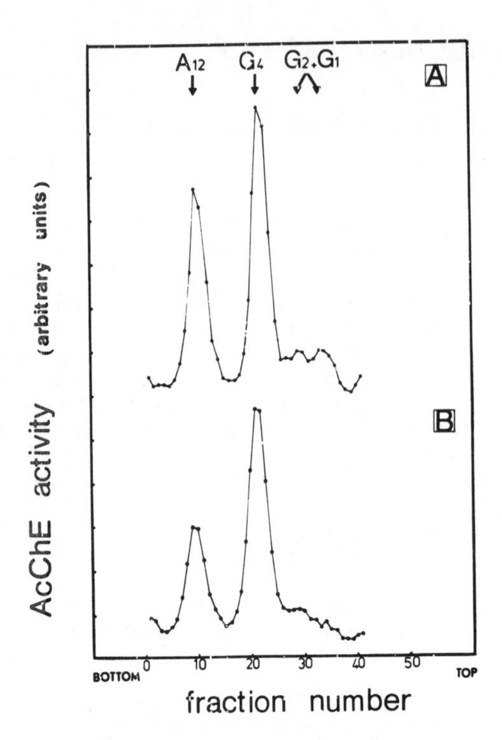

Fig. 1: cellular localization of AchE molecular forms in cultured sympathtic neurons. Neurons were grown for 20 days in the presence of 370 ng/ml NGF and 5% rat serum. Sister cultures were treated with 0.75 mM BW and 50 μM DFP (panel B) or with BW alone (panel A). After washing, the molecular forms of AchE were analyzed on sucrose gradients. Catalytic site which are not inactivated by the treatment with the two inhibitors (panel B) are exposed at the cell surface. (Reprinted from Ferrand et al., 1986)

Regulation of AchE expression by Nerve Growth Factor

NGF increases the survival of cultured sympathetic neurons in a dose-dependent manner. Moreover, this factor increases CAT and TOH activities per surviving neurons. A Northern blot analysis performed with RNA extracted from 15 day old cultures demonstrated that NGF increases TOH mRNA level (Raynaud et al., 1988). As the induction of TOH activity by NGF is insensitive to inhibitors of RNA polymerase (Rohrer et al., 1978; Hefti et al., 1982), our data suggest a post-transcriptional control of TOH mRNA level by NGF.

When primary cultures of sympathetic neurons grown in the presence of a varying concentration of 7S NGF (20-370 ng/ml) are compared, AchE activity per neuron is increased about 2-fold. This effect is distinctly smaller that the increases in CAT and TOH activities observed under the same conditions (40- and 5-fold, respectively). The increased AchE activity caused by high NGF concentrations is mostly due to a large induction of the A12 form, the abundance of which rises from an undetectable level (46 ng/ml 7S NGF) to 45% of the total activity (370 ng/ml) (Figure 2).

It has been previously reported that NGF induces AchE in PC12 cells and in cultured

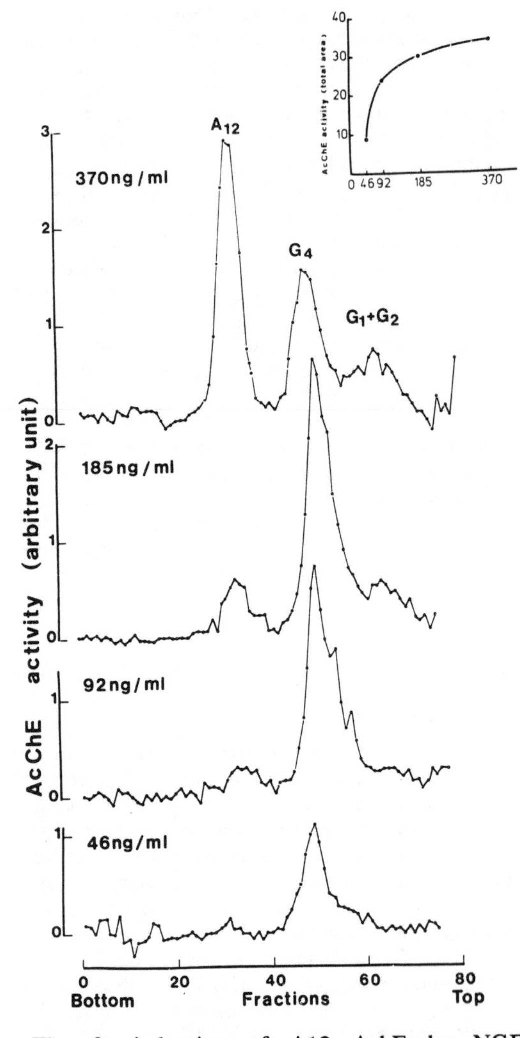

Fig. 2: induction of A12 AchE by NGF. Sympathetic neurons were cultured for 15 days in the presence of the indicated concentrations of 7S-NGF. The survival of the neurons was optimal at 92 ng/ml and higher, and was reduced 2-fold in the presence of 46 ng/ml. Inset: total AchE activity per culture. (Reprinted from Raynaud et al., 1988)

chromaffin cells. This effect is fully sensitive to RNA polymerase inhibitors, suggesting that NGF enhances AchE gene transcription (Greene and Rukenstein, 1981; Acheson et al., 1984). Moreover, NGF induces A12 AchE in PC12 cells from an undetectable level to 5-13% of the total activity (Rieger et al., 1980; Inestrosa et al., 1981, 1985).

Regulation of AchE expression by a cholinergic differentiation factor

We have partially purified a glycoprotein (app. M_r 43,000) from muscle conditioned medium, which increases up to 100-fold the development of CAT activity in cultured sympathetic neurons (Weber, 1981; Swerts et al, 1983). This effect can be at least in part attributed to an increase in CAT mRNA level, evoking the possibility that this factor induces the transcription of CAT gene (Brice et al, 1989). Moreover, this factor represses 2-4 fold the expression of the catecholamine synthesizing enzymes TOH, DDC and DBH (Swerts et al, 1983). Again, the decrease in the expression of TOH is accompanied by a similar deficit in TOH mRNA level, presumably due to transcriptional control (Raynaud et al, 1987).

As this factor triggers the cholinergic differentiation of sympathetic neurons, it was surprizing to find (Swerts et al, 1984) that it actually decreases AchE expression 2-3 fold. This decrease is the result of a similar decrease in the number of immunoprecitable enzyme molecules (Figure 2). This observation is particularly intriguing, as cholinergic sympathetic neurons in the cat are particularly rich in AchE (Sjöqvist, 1963; Lundberg et al, 1979). Such a correlation, however, does not seem to exist in the rat. Nevertheless, the relevance of the inhibition of AchE expression by a CAT inducing factor in culture to the *in vivo* situation remains unclear (see Landis and Keefe, 1983; Leblanc and Landis, 1986).

Most strikingly, the partially purified cholinergic factor totally blocks the expression of A12 AchE, whereas the development of the globular forms is affected to a lesser extent (Figure 3). This is not the result of an increased release of the A12 form, as its secretion was also undetectable in cultures maintained in the presence of the factor. These data evoke the possibility that the factor regulates the splicing of the primary transcript of the gene. The fact that the mRNAs coding for the catalytic subunits of the A12 and G4 forms might be generated by the same exon splicing pattern (see P. Taylor et al., this volume) makes this possibility unlikely. Alternatively, the factor could interfere with the

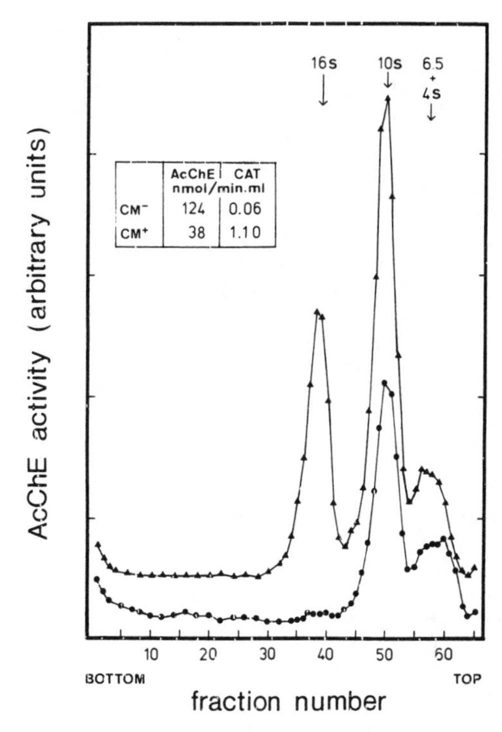

Fig. 3: effects of muscle conditioned medium on the expression of AchE molecular forms. Neurons were cultured for 12 days in the presence (dots) or absence (triangles) of muscle CM, a source of the cholinergic neuronal differentiation factor. (Reprinted from Swerts et al., 1984)

assembly of the A12 form in the Golgi, or inhibit the synthesis of A12 collagen-like tail.

The effects of the factor on the expression of A12 AchE are reversible (Raynaud et al., 1987b): when the factor is withdrawn from the medium after 5-20 days, A12 AchE develop at a rate identical to that observed in cultures which were never exposed to the factor. Such a treatment stops the further development of CAT, and increases the expression of TOH (Vidal et al., 1987). Inversely, the delayed treatement of the cultures with the factor at days 10-15, i.e. when the percentage of A12 has reached about 30%, causes a decrease of this form to the low level observed in cultures treated with the factor from the beginning. This treatment rapidly induces CAT expression, and depresses the rate of TOH development.

Regulation of AchE expression by neuronal depolarization

The long term depolarization of cultured sympathetic neurons by high K^+ does not affect their survival or their growth, as measured by their content in protein and RNA. However, high K^+ medium decreases by 80% the development of CAT activity, both in the presence or absence of the cholinergic factor. Moreover, this medium stimulates the development of TOH activity and TOH mRNA level 2-3 fold (Raynaud et al, 1987a,b). As far as CAT and TOH are concerned, the cholinergic differentiation factor and neuronal depolarization have opposite effects. However, the long-term depolarization of the neurons, as the cholinergic factor, *decreases* the expression of AchE about 5-fold. In addition, high K^+ medium decreases the expression of A12 form, although it is not totally blocked as observed with the cholinergic factor. Again, the effects of depolarization on A12 AchE are reversible in a few days (Raynaud et al., 1987b).

It is thus striking that both strongly cholinergic neurons (grown in the presence of the CAT inducing factor) and strongly noradrenergic neurons (grown in high K^+ medium) are characterized by relatively low levels of AchE, and particularly by a decreased development of A12 AchE. Is a high level of A12 AchE characteristic of dual-function neurons?

It is also intriguing to note that the depolarization of cultured quail muscle cells by veratridine leads to an actual increase of A12 form, rather than a decrease observed in our neuron cultures (Rotundo et al., this volume). The reasons for this discreapancy are not unknown; they may originate from species differences, or a differential regulation of A12 expression by internal Ca^{++} in nerve and muscle cells.

To what extent does the long term depolarization of cultured neurons reproduce the effects of neuronal activity triggered by preganglionic fibers *in vivo*? The stimulation of preganglionic fibers is known to cause a trans-synaptic induction of TOH mRNA, an effect similar to our observations in culture. On the other hand, the section of the preganglionic fibers at birth depresses the development of TOH, but does not induces CAT expression in the denervated SCG. Moreover, the preganglionic denervation causes a 2-fold decrease in AchE activity *in vivo* (Gisiger et al, 1978), whereas our culture work would suggest an opposite effect. Nevertheless, the denervation of the adult SCG increases A12 AchE 2-fold. In agreement with our culture work, this last result suggest that neuronal activity represses the expression of the asymmetric form. Taken together, these data suggest that the long-term depolarization of sympathetic neurons in culture only imperfectly reproduces the effects of neuronal activity triggered by preganglionic fibers *in vivo*.

The role of Ca++ channel in neurotransmitter plasticity

We wondered if the effects of high K^+ medium on the expression of CAT, TOH and AchE were relayed by a Ca^{++} entry, and which types of voltage-activated Ca^{++} channels were involved in these regulations (Vidal et al., 1989). Indeed, two types of V-sensitive Ca^{++} channels, named N-type and L-type, have been characterized in cultured rat sympathetic neurons (Perney et al., 1986; Plummer et al., 1989). N- and L-channels differ in their gating properties and their pharmacological sensitivities; in particular, only L-type channels are sensitive to the so-called Ca^{++} antagonists in the dihydroperidine (like nitrendipine), diphenylalkylamines (desmethoxyverapamil) or diphenylbutylpiperidine (fluspirilene) series. N-channels are under normal circonstances specifically involved in catecholamine release by sympathetic neurons; the function of L-channels remains unclear, although their slow inactivation under a prolonged depolarization might significantly increase internal Ca^{++}.

None of the inhibitors of the L-type channels tested have any effect on the expression of neurotransmitter metabolizing enzymes by neurons cultured in low (5 mM) K^+ medium. However, nitrendipine (3 μM) or fluspirilene (1 μM) totally reversed the induction of TOH and the repression of CAT caused by high (40 mM) K^+ medium. Moreover, the inhibition of AchE expression caused by high K^+ medium was also totally reversed by these two inhibitors, and could thus be attributed to a Ca^{++} entry specifically through channels of the L-type (Figure 4). On the other hand, the decrease in the expression of AchE caused by the cholinergic differentiation factor was unaffected by these drugs, suggesting that this factor does not depress AchE development by raising internal Ca^{++}. Ca^{++} antagonists like nitrendipine also antagonized the adverse effect of high K^+ medium on A12 AchE development.

These experiments thus suggest that the effects of a long term depolarization of sympathetic neurons on CAT, TOH and AchE expressions are caused by a Ca^{++} entry through L-type channels. Although others explanations remain possible, these results evoke the intriguing possibility that only Ca^{++} entrering through these channels can affect the expression of phenotypic traits, and possibly gene expression.

Conclusions

We have studied the expression of AchE by cultured sympathetic neurons in relation with

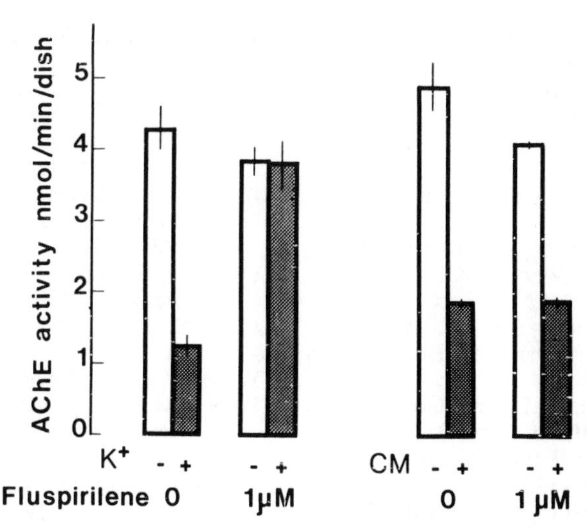

Fig. 4: Fluspirilene, a Ca^{++} antagonist, antagonizes the depression of AchE expression caused by long term K^+-depolarization, whereas the depression caused by muscle CM is unaffected. Left panel: neurons were grown for 8 days in a medium containing either 5mM K^+ (clear bars) or 35 mM K^+ (stippled bars). Half of the cultures were simultaneously treated with 1μM Fluspirilene, as indicated. Right panel: neurons were cultured in the absence (clear bars) or presence (stippled bars) of muscle conditioned medium.(Reprinted from Vidal et al., 1989).

their neurotransmitter phenotype. NGF stimulates the expression of both choline acetyltransferase and tyrosine hydroxylase, and increases AchE development about two-fold. This effect is mostly due to a induction of A12 AchE. A cholinergic factor purified from muscle conditioned medium induces choline acetyltransferase and represses tyrosine hydroxylase; this factor actually decreases AchE development. In particular, the expression of A12 AchE is blocked in a reversible manner by this factor. Raising internal Ca^{++} by K^+-mediated depolarization represses choline acetyltransferase and induces tyrosine hydroxylase; under this condition, the expression of AchE, and in particular of its A12 form, were also impaired.

There is therefore no correlation between a high rate of AchE development and the expression of noradrenergic or cholinergic phenotypic traits by cultured rat sympathetic neurons. On the other hand, there is a differential regulation of the assymetric and globular forms of AchE by NGF, by the cholinergic neuronal differentiation factor and by internal Ca^{++}. The origins for the particular mode of regulation of A12 AchE remain unclear; they might involve the differential splicing of the pre-mRNA, the regulation of the collagen-like tail, or the assembly of the assymmetric form in the Golgi apparatus.

The analysis of AchE promoter (see Taylor et al., this volume) opens new avenues in understanding AchE gene regulation. In particular, the AP1 site present in this promoter might play a key role in the regulation by internal Ca^{++} (Sheng and Greenberg, 1990).

References

Acheson, A.L., Naujoks, K. and Thoenen, H. *J. Neurosci.*, **1984**, *4*, 1771-1780

Brice, A., Berrard S., Raynaud B., Ansieau S., Coppola T., Weber M.J. and Mallet J. *J. Neurosci. Res.*, **1989**, *23*, 266-273

Ferrand C., Clarous D., Delteil C. and Weber M.J. *J. Neurochem.*, **1986**, *2*, 349-358.

Fukada, K. *Proc. Natl. Acad. Sci. U.S.A.*, **1985**, *82*, 8795-8799

Gisiger, V., Vigny, M., Gautron, J. and Rieger, F. *J. Neurochem.*, **1978**, *30*, 501-516

Greene, L.A. and Rukenstein, A. *J. Biol. Chem.*, **1981**, *256*, 6363-6367

Hefti, F., Gnahn, H., Schwab, M.E. and Thoenen, H. *J. Neurosci.*, **1982**, *2*, 1554-1566

Inestrosa, N.C., Matthew, W.D., Reiness, C.G., Hall, Z.W. and Reichardt, L.F. *J. Neurochem.*, **1985**, *45*, 86-94

Inestrosa, N.C., Reiness, C.G., Reichardt, L.F.and Hall, Z.W. *J. Neurosci.*, **1981**, *1*, 1260-1267

Landis, S. and Keefe, D. *Dev. Biol*, **.1983**, *98*, 349-372

Leblanc, G. and Landis, S. *J. Neurosci.*, **1986**, *6*, 260-265

Lundberg, JM., Hökfelt, T., Schultzberg, M., Uvnäs-Wallenstein, K., Kohler, C. and Said, S.I. *Neurosci.*, **1979**, *4*, 1539-1559

Perney, T.M., Hirning, L.D., Leeman, S.E. and Miller, R.J. *Proc. Natl. Acad. Sci. U.S.A.*, **1986**, *83*, 6656-6659

Plummer, M.R., Logothetis, D.E. and Hess, P. *Neuron*, **1989**, *2* 1453-1463

Raynaud, B., Clarous, D., Vidal, S., Ferrand, C. and Weber, MJ *Dev. Biol.*, **1987b**, *121*, 548-558

Raynaud, B., Faucon-Biguet, N., Vidal, S., Mallet, J. and Weber, MJ. *Dev. Biol.*, **1987a**, *119*, 305-312

Raynaud, B., Faucon-Biguet, N., Vidal, S., Mallet, J. and Weber, MJ. *Development*, **1988**, *102*, 361-368

Rieger, F., Shelanski, M.L. and Greene, L.A. *Dev. Biol.*, **1980**, *76*, 238-243

Rohrer, H., Otten, U. and Thoenen, H. *Brain Res.*, **1978**, *159*, 436-439

Sheng, M. and Greenberg, M.E. *Neuron*, **1990**, *4*, 477-485

Sjöqvist, F. *Acta Physiol. Scand.*, **1963**, *57*, 339-351

Swerts, JP., Le Van Thaï, A. and Weber, MJ. *Dev. Biol.*, **1984**, *103*, 230-234

Swerts, JP., Le Van Thaï, A., Vigny, A. and Weber, MJ. *Dev. Biol.*, **1983**, *100*, 1-11

Taylor, P.B., Rieger, F., Shelanski, M.L. and Greene L.A. *J. Biol. Chem*, .**1981**, *256*, 3827-3830

Vidal, S., Raynaud, B. and Weber, M.J. *Mol. Brain Res.*, **1989**, *6*, 187-196

Vidal, S., Raynaud, B., Clarous, D. and Weber M.J. *Development*, **1987**, *101*, 617-625

Weber M. *J. Biol. Chem.*, **1981**, *256*, 3447-3453.

Weber M.J., Raynaud B. and Delteil C. *J. Neurochem.*, **1985**, *45*, 1541-1547.

Yamamori, T., Fukada. K., Aebersold, R., Korsching. S., Fann, M-J. and Patterson, P.H. *Science*, **1989**, *246*, 1412-1416

Effect of β-Endorphin and β-Endorphin 1-27 on the Localisation of Acetylcholinesterase in Rat Skeletal Muscle

Sharon Hughes Lawrence W. Haynes and Margaret E. Smith*, Department of Physiolology, University of Birmingham B15 2TT, U.K. (Fax: 021-414-6924)

In previous work, immunoreactivity for β-endorphin was demonstrated in developing motoneurones (Haynes et al, 1982) and evidence was reported for a role for β-endorphin in the control of AChE molecular forms during the development of the neuromuscular system (Haynes et al, 1984). It was also shown that both β-endorphin and its C-terminally truncated derivative β-endorphin 1-27 are released by embryonic spinal cord in vitro (Haynes et al, 1984). Here we provide evidence for a role for β-endorphin, β-endorphin 1-27 and the β-endorphin C-terminal dipeptide, glycylglutamine, in the control of (AChE) at the developing neuromuscular junction.

Myotubes were cultivated from rat gastrocnemius muscle of 20-21 days gestation and the AChE molecular forms were separated by sucrose density gradient sedimentation (Haynes et al, 1984) and assayed by the method of Ellman et al. (1961). In addition, cultures were stained for AChE with the Karnovsky-Roots method at pH 7.5.

In control cultures activity of several forms of AChE, including the A12 form, could be separated and the enzyme reaction product was detectable in fine deposits on or near the sarcolemma. However, when myotubes were cultivated in the presence of β-endorphin (30nM for 7-12 days), specific activity of AChE was higher, and an increased predominance of the globular (precursor) forms of the enzyme was found. Only small amounts of the A12 AChE were present. In these cultures enzyme product was largely located internally over the nuclear envelope and Golgi complex.

When myotubes were cultured in the presence of β-endorphin 1-27 (500nM for 7 days), the relative proportions of the different forms of AChE were similar to those seen in the controls but the peptide induced a perinuclear accumulation of AChE in a few of the cells, and in these a focal localisation of the enzyme activity was seen (Haynes and Smith, 1990).

In other experiments myotubes were cultivated from rat embryos at the earlier time of 16 days gestation. The cultures were treated with glycylglutamine (600 nM) for 5 days, and this increased the activity of all forms of the enzyme, particularly the A_{12} form (Haynes and Smith, 1985b), and many myoblasts stained intensely for AChE. The dipeptide had no effect on the activity or distribution of AChE in myotubes cultured from more mature embryos.

The results indicate that β-endorphin, β-endorphin 1-27 and glycylglutamine may be independently involved in AChE regulation during the development of the muscle fibre. It is possible that the peptides exert their influence at different times during development.

Supported by the Medical Research Council and the International Spinal Research Trust.

Ellman, G.L.; Courtney, K.D.; Andres, V. Jnr; Featherstone, R.M. Biochem. Pharmacol. **1961**, 7, 88-95.

Haynes, L.W.; Smith, M.E. Biochem. Soc. Trans. **1985b**, 13, 174-175.

Haynes, L.W.; Smith, M.E. Experientia. **1990**, 46, 211-213.

Haynes, L.W.; Smith, M.E.; Smyth, D.G. J. Neurochem. **1984**, 42, 1542-1551.

Haynes, L.W.; Smyth, D.G.; Zakarian, S. Brain Res. **1982**, 232, 115-128.

0–8412–2008–5/91/0109$06.00/0

Electrical Stimulation Prevents Curare-related effects on Acetylcholinesterase Accumulation and on its Molecular-form Distribution in Slow and Fast Muscles of the Chick Embryo

A. KHASKIYE, G. SUIGNARD-KHASKIYE, D. RENAUD C.N.R.S.-U.R.A. 1340, Faculté des Sciences et des Techniques, 2, rue de la Houssinière, F-44072 Nantes Cédex 03, France

The effects of curarization and electrical stimulation upon the embryonic changes in acetylcholinesterase (AChE) accumulation, AChE-specific activity and molecular-form distribution were studied in slow-tonic anterior latissimus dorsi (ALD) and in fast-twitch posterior latissimus dorsi (PLD) muscles of the chick embryo. In control muscles, synaptic AChE accumulation first appeared at day 12 of incubation in both muscles and they were intensively stained at day 17. The AChE spots were smaller and fainter labelled in ALD muscle than in PLD muscle. From day 11 to day 17 of incubation, the AChE-specific activity decreased, while the relative proportion of asymmetric A12 and A8 forms increased.

Curare was injected over the chorioallantoic membrane from day 8 to day 17. Repetitive injections of curare resulted in a decrease in the AChE-specific activity, in the accumulation of the synaptic AChE and in the proportion of AChE asymmetric forms (Fig. 1). These effects affected more severely the slow ALD than the fast PLD muscle. These results are in agreement with the work of Gordon et al. (1974), according to which the high concentration of AChE at the neuromuscular junction of ALD muscle fails to develop and is very much reduced in PLD muscle in the presence of curare.

Direct electrical stimulation at a relatively high frequency (40 Hz) of curarized ALD and PLD muscles resulted in an increase in the AChE asymmetric form proportion, in these series, they were accumulated in an equivalent amount to the control muscles. A low frequency (5 Hz) stimulation of curarized muscles resulted in a dominance of globular forms (Fig. 1). These results are fully consistent with our previous work (Khaskiye and Renaud, 1988), in which the distribution of AChE molecular forms in PLD muscle depended on the rhythm of evoked muscle activity. This is also the case in rat soleus muscle, in which slow and fast activity patterns induce differences in expression of AChE molecular forms (Lømo et al., 1985). Both patterns of stimulation partly prevented the loss in synaptic AChE accumulations due to the muscle curarization. Moreover, in curarized-stimulated PLD muscle, AChE spots appeared to be more numerous than in curarized or normal PLD muscles and they were differently distributed.

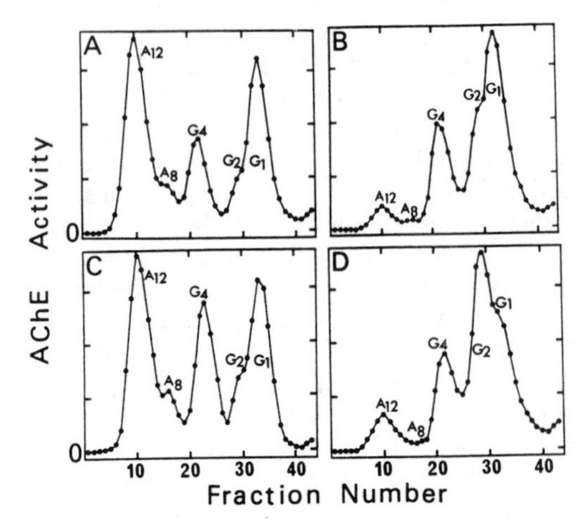

Fig. 1. Distribution of AChE molecular forms in chick embryo ALD muscle at day 17 of incubation. A, control muscle; B, curarized muscle; C, curarized muscle stimulated at a fast (40 Hz) rhythm from day 10 to day 17; D, curarized-stimulated muscle at low (5 Hz) rhythm.

These results suggest that in chick embryo ALD and PLD muscles, the muscle activity and the rhythms of this activity are strongly involved in the AChE evolution.

References

Gordon, T., Perry, R., Tuffery, A.R., Vrbova, G., *Cell Tissue Res.*, **1975**, *153*, 13-25.

Khaskiye, A., Renaud, D., *Differentiation*, **1988**, *39*, 28-33.

Lømo, T., Massoulié, J., Vigny, M., *J. Neurosci.*, **1985**, *5*, 1180-1187.

0–8412–2008–5/91/0110$06.00/0

Changes in the Neuromuscular Junction of Rat Soleus Muscle induced by Disuse Atrophy

Francis Bacou, INRA Physiologie animale, Place Viala, 34060 Montpellier Cedex 1
Maurice Falempin, Yvonne Mounier, Laboratoire de Physiologie des Structures Contractiles, Université des Sciences et Techniques de Lille, 59655 Villeneuve d'Ascq Cedex, France.

The functional properties of neuromuscular junctions (NMJ) were studied on normal Soleus young (three months) Wistar male rats and on atrophied Soleus of animals placed two or four weeks in hypokinesia-hypodynamia conditions using the hindlimb suspension (HS) model (Morey, 1979).

The amplitudes of m.e.p.p.s. vary considerably in disuse conditions and the histogram appears bimodal. After four week HS, the m.e.e.p. frequency significantly decreases; moreover, the electrophysiological characteristics of the post synaptic membrane show significant increase in input resistance (R_{in}) while resting potential (RP), specific membrane resistance (Rm), space constant (λ) and time constant (τm) are either unchanged or not significantly modified.

Cholinesterase staining of the NMJs is higher in atrophied Soleus, although a similar arrangement was observed in both atrophied and control muscles. Changes in acetylcholinesterase (AChE) activity and its molecular form pattern has been studied on normal or HS Soleus and EDL. AChE activity is higher in experimental Soleus and its polymorphism is not altered (Gupta et al., 1985). On the contrary, AChE activity decreases in experimental EDL, and its polymorphism is markedly affected: the percentage of G4 forms significantly increases and that of G1/G2 forms decreases.

These results suggest that changes in the NMJ of atrophied rat Soleus appear after suspension-induced disuse. The changes in the m.e.p.p. amplitudes and frequency after HS could be explained by the higher R_{in} and by morphological modifications of the axonal endings as already described (Baranski & Marciniak, 1979; D'Amelio et al., 1988). However, changes in the polymorphism of AChE observed only on EDL attest a different regulation in the biosynthesis of AChE molecular forms in fast-twitch and slow-twitch muscles.

References

Baranski, S., Marciniak, M., *Aviat. Space Environ. Med.*, **1979**, *50*, 14-17.
D'Amelio, F., Daunton, N.G., Fast, T. Grindeland, R., *Proceeding of Symposium on Neuroscience*, **1988**, 204.
Gupta, R.C., Misulis, K.E., Dettbarn, W.D., *Exp. Neurol.*, **1985**, *89*, 622-633.
Morey, E.R., *Bioscience*, **1979**, *29*, 168-172.

0–8412–2008–5/91/0111$06.00/0

Molecular Forms of Acetylcholinesterase in Nuclei isolated from Fast and Slow Rabbit Muscles

Francis Bacou, Pierre Vigneron, Jacques Dainat
INRA Physiologie animale, Place Viala, 34060 Montpellier Cedex 1, France.

Denervated rabbit muscles produce increased levels of Acetylcholinesterase (AChE), with highly variable patterns of molecular forms depending on the type - fast or slow- of muscles (Bacou et al., 1982). The histochemical analysis of Tennyson et al. (1977) showed that after denervation of rabbit muscles, the AChE activity increases mostly at intracellular sites and is largely associated with the sarcoplasmic reticulum. Here we report further biochemical data on nuclei isolated from control or one month denervated fast (Semimembranosus accessorius; SMa) or slow (Semimembranosus proprius; SMp) muscles of the rabbit.

Nuclei were isolated according to Dainat et al. (1984). AChE activities are higher in denervated muscles (3.32 mU. mg prot.$^{-1}$ /2.65 in control SMa, and 4.95 mU. mg prot.$^{-1}$/ 0.92 in control SMp). Similar differences are obtained when AChE activities are plotted as mU. mg DNA^{-1}. The molecular form patterns of nuclei AChE is similar to that observed in muscles. In control SMa isolated nuclei, the percentage of asymmetric (A) forms corresponds to 47% of the total AChE activity and 9% in denervated SMa nuclei. In control SMp nuclei, A forms represent 45 % of the AChE activity and 51 % in denervated SMp nuclei.

The present observations show that it is very likely that a fraction of the synthetized molecular forms, including the collagen-tailed forms, remain intracellular and associated with the nuclear part of the sarcoplasmic reticulum. One example of such intracellular localization of collagen-tailed forms has been reported in the case of motor nerves, where these molecules are transported by the fast axonal flow (Di Giamberardino and Couraud, 1978). However, the function of AChE in the sarcotubular system, and particularly the sarcoplasmic reticulum related to the nuclei, remains to be elucidated.

References

Bacou, F., Vigneron, P., Massoulié, J., *Nature*, **1982**, *296*, 661-664.

Dainat, J., Bressot, C., Bacou, F., Rebière, A., Vigneron, P., *Mol. Cell Endocr.*, **1984**, *35*, 215-220.

Di Giamberardino, L., Couraud, J.Y., *Nature*, **1978**, *271*, 170-172.

Tennysson, V.M., Kremzner, L.T., Brzin, M., *J. Neuropath. Exp. Neurol.*, **1977**, *36*, 245-275.

0–8412–2008–5/91/0112$06.00/0

Tetrameric (G_4) Acetylcholinesterase from Skeletal Muscle as Affected by Advancing Age and Changes in Motor Activity

Cheryl A. Hodges-Savola, Research Service, Dept. Veterans Affairs Med. Ctr., Kansas City, MO 64128; and Univ. of Kansas Center on Aging, Kansas City, KS 66103, USA
Hugo L. Fernandez, Research Service, Dept. Veterans Affairs Med. Ctr., Kansas City, MO 64128; and Dept. of Physiology, Univ. of Kansas, Kansas City, KS 66103, USA

Skeletal muscles exhibit remarkable plasticity in response to changes in motor activity, i.e. they undergo metabolic and regulatory modifications which particularly affect synaptic transmission-related molecules. Since alterations in these molecules may partly explain the slowing of motor reactions and reduced synaptic efficiency associated with advancing age, we studied the response of AChE to enhanced motor activity during aging.

Experiments were performed on sedentary 3-33 mo-old male Fischer 344 rats and age/wt matched animals subjected to treadmill exercise (8.5 m/min, 1 min sprints at 15 m/min every 10 min, for 1 hr/day, for 5 days). This paradigm elicits a specific increase in rat fast-twitch muscle G_4 AChE (Fernandez and Donoso, 1989).

All AChE isoforms in gracilis muscle endplate regions from sedentary rats similarly declined after 6 mo, attained minimal levels by ≈ 9-15 mo, and partly recovered after 21 mo. Non-endplate regions and tibialis muscles showed analogous changes, but the soleus and diaphragm maintained constant AChE levels through 24 mo. Denervation (48 h) did not affect gracilis G_4 AChE through 28 mo, but induced reductions in G_1+G_2 and A_{12} AChE which became less evident with age. Gracilis AChE isoform recovery from inactivation by DFP was unaltered by advancing age, whereas the exercise-induced elevation of G_4 AChE diminished from ≈200% of control (6 and 12 mo) to 170% (24 mo), and was indistinguishable from control by 30 mo (Fig 1).

These data support the importance of neuromuscular activity in regulating muscle G_4 AChE and show that aging reduces the muscle's capacity to respond to enhanced motor activity. This effect does not necessarily result from a decline in the muscle's intrinsic ability to "produce" G_4 AChE, but rather from deficits in this isoform's "regulation".

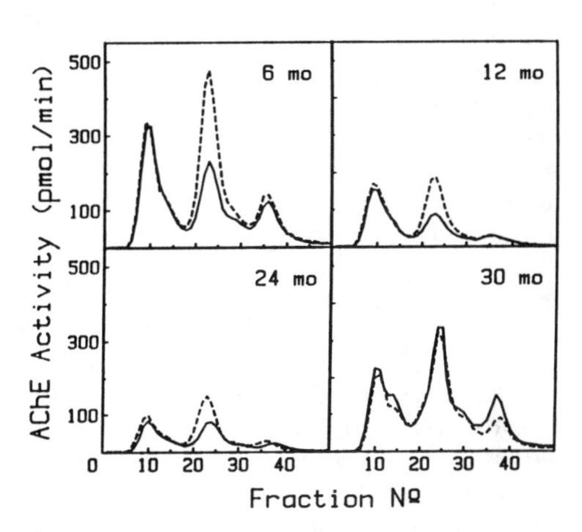

Fig. 1. Velocity sedimentation pro- files of AChE (pmol/min/gradient; Gregory et al., 1989) in gracilis muscle endplate samples from seden- tary (—) and exercised (---) rats (6, 12, 24, and 30 mo). Each profile represents the average of 3-5 exper-iments. Menisci are to the left and major activity peaks depict (left to right) G_1+G_2, G_4, and A_{12} isoforms.

Supported by the Medical Research Service US Dept. of Veterans Affairs and the University of Kansas Center on Aging.

References

Fernandez, H.L., Donoso, J.A., *J. Appl. Physiol.*, **1989**, *65*, 2245-2252.

Molecular Forms of Acetylcholinesterase in Innervated Regenerating Fast and Slow Rat Muscles

J. Sketelj*, N. Črne and M. Brzin, Institute of Pathophysiology, School of Medicine, 61105 Ljubljana, Yugoslavia

Muscle ontogenetic development and maturation is largely repeated during its regeneration after injury. Muscle regeneration therefore offers a suitable experimental model for studying the regulation of AChE in immature muscle cells under the influence of the motor nerves during early reiinervation period.

Nerve-intact muscle regenerates were prepared by ischemic-toxic injury of slow soleus (SOL) and fast extensor digitorum longus (EDL) muscles of the rat. Muscle regenerates were isolated during the period of six weeks after injury. Junctional and extrajunctional regions were sometimes separated after histochemical visualization of AChE in the motor endplates. Molecular forms of AChE were analysed by velocity sedimentation in linear sucrose gradients.

Effects of innervation on AChE regulation revealed by changes of its velocity sedimentation patterns can be classified as early or late. Already a few days after innervation, the pattern of AChE molecular forms in the innervated SOL regenerate was different from that in a denervated regenerate: the activity of the 13 S (A8) AChE form was significantly increased. The pattern became similar to that in normal SOL during the next week. No significant early changes of AChE molecular forms can be observed in innervated EDL regenerates but later the domination of the 10 S AChE which takes place during the second week after injury in noninnervated regenerates was prevented by innervation. Virtual extinction of the 16 S AChE form in the extrajunctional regions of the EDL regenerates and its decrease, together with the decrease of 4 S AChE form in the SOL, can be looked upon as late effects of innervation, needing 4-6 weeks to be completed. By then, both the patterns and the distribution of AChE molecular forms in innervated EDL and SOL regenerates became indistinguishable from those in normal mature muscles.

Changes observed in AChE regulation in innervated regenerating muscles are similar to developmental changes of AChE patterns in postnatal SOL and EDL muscles, although not identical. Very early differentiation of AChE molecular forms patterns in regenerating SOL and EDL muscles, probably preceding establishment of normal neural stimulation of both muscles, supports the hypothesis that intrinsic differences in AChE regulation mechanisms encoded in the myoblast lineages of both muscles basically determine the patterns of AChE molecular forms in slow and fast muscles.

0–8412–2008–5/91/0114$06.00/0

Acetylcholinesterase and Other Synaptic Components in the Myotendinous Junctions of Rat Muscles

A.Trampuž, N. Črne and J. Sketelj*, Institute of Pathophysiology, School of Medicine, 61105 Ljubljana, Yugoslavia

The presence of AChE in myotendinous junctions (MJT) has been described long ago but its function, if any, is still unknown. AChE localization in MTJ of slow soleus (SOL) and fast extensor digitorum longus (EDL) muscles of the rat was reexamined, its molecular forms were analysed by velocity sedimentation and its possible coincidence with two other components of the neuromuscular junction, acetylcholine receptor (AChR) and lectin Dolichos biflorus agglutinin receptor (DBAR) was investigated. Triple histochemical staining allowed visualization of all three synaptic components on the same tissue section. Thus, characteristics of local interactions of muscle cell with two other cells, i.e. neuron and fibroblast, can be studied as well as the regulation of individual synaptic components.

AChE could not be detected in the MTJ of fast EDL muscles. On the other hand, most but not all of the MTJ in the SOL muscle accumulate significant amounts of AChE. The pattern of AChE molecular forms in the MTJ region of SOL did not differ significantly from that in the extrajunctional regions in general: the asymmetric AChE forms, especially the 13 S (A 8) form, represented about 40 % of total activity. Inability of the adult EDL muscle to synthesize the asymmetric AChE forms extrajunctionally may be the reason that AChE does not accumulate in its MTJ. Accordingly, in 17-day old rats, the asymmetric AChE forms are still produced in the extrajunctional regions of the EDL muscles, and AChE is present in their MTJ. AChR and DBAR are not especially concentrated in the MTJ of SOL muscles containing AChE. Three weeks after SOL muscle denervation, AChE activity in the MTJ is decreased but still detectable whereas AChR is not yet perceptibly concentrated in the MTJ although its synthesis in extrajunctional regions of the muscle fibers is enhanced.

The results demonstrate that individual components which appear together at the neuromuscular junction can be synthesized and accumulated in an independent manner by the muscle cell at some other type of muscle cell junction.

Focal Accumulations of Acetylcholinesterase and Other Synaptic Components in Non-Innervated Regenerating Rat Muscles

N. Črne, J. Sketelj*, and M. Brzin, Institute of Pathophysiolgy, School of Medicine, 61105 Ljubljana, Yugoslavia

Two types of focal accumulations of AChE occur in non-innervated regenerating muscles: 'junctional' in places of former motor endplates, and 'extrajunctional' in other regions of the myotubes. Their coincidence with two other components of the motor endplate, acetylcholine receptor (AChR) and receptor for a lectin Dolichos biflorus agglutinin (DBAR), was investigated. Muscle regeneration was induced by ischemic-toxic injury of rat soleus (SOL) and extensor digitorum longus (EDL) muscles.

In 8-day old regenerates of both muscles, 94 % of junctional AChE accumulations coincided with identical accumulations of AChR, whereas those without AChR seemingly had no contact with regenerating myotubes. Most (90 %) of extrajunctional AChE accumulations also accumulated AChR but only 70 % of AChR patches showed AChE activity in the SOL, and 60 % in the EDL regenerates. Faint presence of DBAR was seen in junctional AChE accumulations including those without AChR, but not in extrajunctional ones.

In 17-day old muscle regenerates all examined junctional AChE accumulations contained AChR but in 10 % of them microlocalization of AChE and AChR was not completely identical. A higher percentage of extrajunctional accumulations contained only one synaptic component (AChE or AChR) than at 8 days. In some accumulations displaying both components the microlocalization of AChE and AChR was not identical. In SOL regenerates, 62 % of extrajunctional AChE accumulations coincided exactly with a patch of AChR, in 15 % of them localization of both components was not identical, and 23 % had no detectable AChR. About half (53 %) of AChR patches coincided with AChE accumulation of identical shape, in 10 % localization was not quite identical, and 37 % of them had no detectable AChE. A similar situation was found in the EDL regenerates except that the percentage of AChR patches lacking AChE activity was even higher (54 %). DBAR could not be detected in either junctional or extrajunctional AChE accumulations in regenerates of this age.

It seems that postsynaptic-like specializations arising in non-innervated regenerating slow or fast muscles may get transformed with time due to selective accumulation, transposition or elimination of their components. DBAR accumulation in postsynaptic specializations may need the presence of motor nerve endings.

0–8412–2008–5/91/0116$06.00/0

Acetylcholinesterase in Junctional and Extrajunctional Regions of Denervated Slow and Fast Muscles of the Rat

N. Črne and J. Sketelj*, Institute of Pathophysiology, School of Medicine, 61105 Ljubljana, Yugoslavia

The patterns and the distribution of AChE molecular forms in slow soleus (SOL) and fast extensor digitorum longus (EDL) muscles of the rat differ significantly. The longterm effects of denervation on AChE molecular forms in junctional and extrajunctional regions of both muscles were investigated in order to reveal the role played by either neuronal influences or intrinsic muscular mechanisms in maintaining these differences. Molecular forms of AChE were analysed by velocity sedimentation in linear sucrose gradients.

One week after denervation of the EDL muscles, specific activity of the 4 S AChE, and especially that of the 16 S AChE form, in the junctional region of the muscles decreased much more than activity of the 13 S AChE, while a significant relative increase of the 9 S AChE form (A4) occured. In the extrajunctional regions, specific activity of the 4 S AChE form dropped precipitously whereas the asymmetric AChE forms were nearly undetectable, like in the normal EDL. In the denervated SOL muscles, AChE pattern in the junctional region resembled that in the EDL, but extrajunctionally the asymmetric AChE forms were still present, like in the normal SOL, although their relative and absolute activity was decreased.

Three weeks after denervation, the asymmetric 16 S and 13 S AChE forms were quite conspicuous in the junctional regions of both muscles. They accumulated also in the extrajunctional regions of the SOL muscles but could not be detected extrajunctionally in the EDL.

Relative and absolute activities of the asymmetric AChE forms in the extrajunctional regions of SOL muscles were much higher than in the EDL also at six weeks after denervation. In the SOL, the 16 S form became predominat asymmetric AChE form instead of the 13 S form. In contrast, activities of the 4 S AChE forms, which are very different in normal muscles, became comparable.

Differences in the distribution of the asymmetric AChE forms which are characteristic for the normal EDL and SOL muscles are therefore preserved or even accentuated after prolonged denervation. It seems that intrinsic differences between the two muscles stemming either from different properties of their myoblast lineages or from irreversible maturational effects of innervation, affect AChE regulation in the two muscles in addition to immediate neuronal influences.

0–8412–2008–5/91/0117$06.00/0

Effect of Botulinum Toxin Type A on the Appearance of Organophosphate-Induced Lesions and AChE Recovery in Skeletal Muscles

Dušan Sket*, Dušanka Čuček, Miro Brzin, Institute of Pathophysiology, School of Medicine, 61105 Ljubljana, Yugoslavia
Wolf-D. Dettbarn, Department of Pharmacology, School of Medicine, Vanderbilt University, Nashville, TN 37232, USA

The induction of severe morphological lesions of muscle fibers is one among various toxicological effects of organophosphates (c.f. Laskowski and Dettbarn, 1977). The lesions can be attenuated by muscle inactivity, such as in cases of inhibition of AChR or muscle denervation.

In skeletal muscles Botulinum toxin, type A (BoTx) inhibits both, the evoked and spontaneous quantal release of ACh but nonquantal release is only partially blocked (Doležal et al., 1983), creating an experimental condition different from that existing during AChR blockade or denervation. Therefore, a possible lesion-preventing effect of BoTx was examined by unilaterally injecting BoTx (5 U) into the hind limb of the rat. Four days later, lesion-inducing dose of organophoshate (DFP s.c.) was applied and after additional 24 h, the soleus and extensor digitorum longus muscles of the BoTx injected side were isolated and endplate regions were processed for E.M. examination (c.f. Dekleva et al., 1989). As controls, rats injected with DFP only were used. The recovery rate of AChE was studied in parallel by assaying muscle AChE activity the first and the seventh day after DFP application in rats with or without BoTx treatment. AChE activity was measured in homogenates of muscles using radiometric method (Lewis and Eldefrawi, 1974).

E.M. examination of endplate muscle regions has revealed that BoTx like denervation completely blocks the appearance of lesions. This finding suggests that the spontaneous nonquantal release of ACh is obviously ineffective in producing changes in muscle fibers, leading to lesions.

The results on AChE recovery rates in muscles of DFP treated animals appear somewhat unexpected in that BoTx inhibits AChE recovery almost completly, compared to the substantial recovery rate in contralateral, BoTx noninjected leg. Since denervation does not prevent the recovery of irreversibly inhibited muscle AChE (Kiauta et al., 1977), the muscle inactivity apparently can not explain the absence of neosynthesis of AChE after BoTx in DFP-treated rats.

References

Dekleva, A., Sket, D., Sketelj, J., Brzin, M., *Acta Neuropathol.*, **1989**, *79*, 183-189.
Doležal, V., Vyskočil, F., Tuček, S., *Pfluegers Arch.*, **1983**, *397*, 319-322.
Kiauta, T., Brzin, M., Dettbarn W-D., *Exp. Neurol.*, **1977**, *56*, 281-288.
Laskowski, M.B., Dettbarn, W-D., *Ann. Rev. Pharmacol. Toxicol.*, **1977**, *17*, 387-409.
Lewis, M.K., Eldefrawi, M.E., *Anal. Biochem.*, **1974**, *57*, 588-592.

Stimulatory and Inhibitory Effects of Ethanol on Acetylcholinesterase Associated to Membranes Derived from Sarcoplasmic Reticulum

Juan Cabezas-Herrera and **Cecilio J. Vidal***, Dpto. de Bioquímica y Biología Molecular, Facultad de Biología, 30071, Espinardo, Murcia, Spain.

The effects of ethanol on membrane-bound and soluble AChE from sarcoplasmic reticulum (SR) were studied. Treatment of membranes with 2.5-12.5% v/v ethanol produced a stimulation of the AChE activity and inhibition thereafter, the enzyme remaining bound to the membranes. Addition of increasing concentrations of ethanol to Triton-solubilized AChE led to its progressive inhibition. The inhibition was reversible or irreversible depending on the solvent concentration. Kinetic studies showed that the solvent behaved as a competitive or mixed inhibitor when added to hydrophilic or amphiphilic forms, respectively.

Isolation of muscle microsomes, solubilization of AChE, enzymatic assays and sedimentation analyses were performed as described earlier (Vidal et al, 1987).

The effect of ethanol on microsomal AChE is shown in Fig.1. As the alkanol concentration rose from 5 to 15% v/v, the enzyme activity increased but higher concentrations produced an inhibition of the enzyme. The biphasic effect of ethanol on membrane-bound AChE was abolished by adding Triton X-100 to the microsomes. The soluble enzyme was progressively inhibited by ethanol (Fig. 1). The effect of the alkanol on the individual molecular forms (4.5, 10.5 and 13.5 S) showed a similar pattern of inhibition. This indicated that interactions of ethanol with the enzyme were not dependent on its state of aggregation. Reversible inhibition was shown up to 12.5% v/v, and irreversible thereafter. In the range of reversible interactions, ethanol behaved as a competitive inhibitor of the hydrophilic 10.5 and 13.5 S forms, but led to a mixed inhibition with the amphiphilic 4.5 S forms.

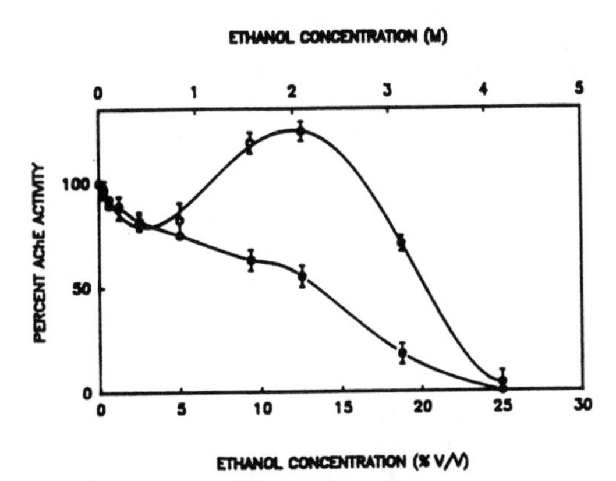

Fig. 1. Effect of increasing concentrations of ethanol on membrane-bound (o) and Triton-solubilized (•) AChE. Samples were incubated without (100%) and with the solvent for 1 hr and aliquots assayed for AChE activity.

Ethanol has been shown to stimulate the activity of the calcium pump of SR (Melgunov et al, 1988). Whether the stimulation of the membrane-bound AChE by ethanol is triggered by a direct action on the enzyme or by the overall fluidising effect of the alkanol on membranes, is under current research.

A reversible inhibitory action of ethanol on AChE indicates that the solvent is accommodated in the active or regulatory site of the enzyme.

This work has been supported by the FISss of Spain (89/514). J.Cabezas is a holder of scholarship from the Instituto de Fomento de la Region de Murcia.

References

Melgunov, V.I., Jindal, S., Melicova, M.P., *FEBS Lett*, **1988**, 227, 157-160.

Vidal, C.J., Muñoz-Delgado, E., Yague-Guirao, A., *Neurochem. Int.*, **1987**, 10, 329-338.

0–8412–2008–5/91/0119$06.00/0

Localization of Acetylcholinesterase in Membranes of Sarcoplasmic Reticulum of Skeletal Muscle

M. Dolores Cánovas-Muñoz, Encarnación Muñoz-Delgado and Cecilio Vidal*, Dpto. Bioquímica y Biología Molecular, Facultad de Biología, 30071 Murcia, Spain.

The preferential localization of AChE in SR membranes of rabbit muscle has been investigated. Integrity and orientation of the vesicles was assessed by measuring their inulin-inaccessible space and calcium-loading capacity. The orientation of AChE in the vesicles was established by means of irreversible and reversible inhibitors which are poorly and freely soluble in lipid. Further insights into the localization of the enzyme was gained by investigating the interactions of concanavalin A and the antibody AE1 with AChE in intact membranes. The data are consistent with an external orientation (cytoplasmic side) of AChE in the membrane.

Isolation of SR membranes and measurements of AChE, Mg^{2+}ATPase and Ca^{2+}ATPase activities were performed as described earlier (Vidal et al, 1987). The values of inulin-inaccessible space, 4 ± 0.5 μl/mg protein, (Hidalgo *et al*, 1977) and capacity to accumulate Ca^{2+}, 130-150 nmol Ca^{2+}/mg protein, established the intactness and correct orientation of the vesicles.

About 17% of the enzyme was active after incubation of the membranes with 25 μM echothiopate, a poorly-soluble in-lipid inhibitor (internal enzyme). Furthermore, about 78% of total enzyme was not inhibited by 10 μM DFP in samples preincubated with 10 μM BW284c51, a reversible and non-permeant inhibitor (external enzyme).

Application of Con A to the membranes, followed by solubilization and sedimentation of the complexes led to a recovery of 13% of the enzyme activity in the supernatant.

Sedimentation profiles of the molecular forms of AChE isolated from microsomes, before and after incubation of the membranes with AE1 revealed the formation of immunocomplexes whether AE1 was applied to the membranes before solubilization or added to the soluble AChE (Fig. 1).

This work has been supported by a grant (89/514) from the FISss of Spain. M.D.C. is a holder of a Scholarship from DGICYT of Spain.

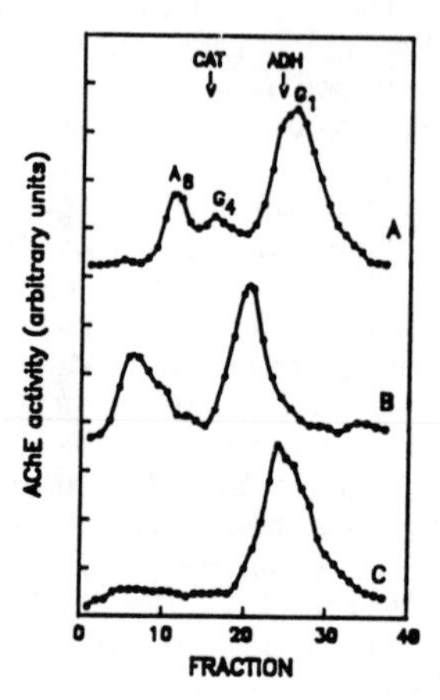

Fig. 1. Sedimentation patterns of molecular forms of AChE before (A) and after incubation of the Triton-solubilized AChE with AE1 (B). Alternatively, membranes were first treated with AE1, washed and resuspended in the extraction medium (C). (CAT, 11.4 S; ADH, 4.8 S).

References

Hidalgo, C., Ikemoto, N. *J. Biol. Chem.* **1977**,252, 8446-8454.

Vidal, C.J., Muñoz-Delgado, E., Yagüe-Guirao, A. *Neurochem. Int.* **1987**, 10, 329-338.

0–8412–2008–5/91/0120$06.00/0

Distribution of Acetylcholinesterase in Membranes Constituting the Sarcotubular System of Skeletal Muscle

Francisco J. Campoy and Cecilio J. Vidal*. Dpto. Bioquímica y Biología Molecular, Facultad de Biología, 30071, Espinardo, Murcia, Spain.

Sucrose gradient subfractionation of a crude microsomal preparation led to three distinct fractions (F_1-F_3). F_1 is rich in T-tubules, F_2 in longitudinal sarcoplasmic reticulum (SR) and F_3 in terminal cisternae of SR, as established by electron microscopy, PAGE and measurement of enzyme markers of surface membrane and SR. All membranes contained a significant amount of AChE. From each membrane, a variable extent of solubilization of AChE was achieved with Triton X-100. Main molecular forms of 4.5S (G_1), 10.5S (G_4) and 13.5S (A_8) occurred in all membranes, but in different proportions.

A crude microsomal fraction rich in SR was isolated from white rabbit muscle (Cánovas-Muñoz et al., 1990). By centrifugation on sucrose gradients (Salama et al., 1984) three membrane fractions were obtained. Electron microscopy studies, SDS-PAGE and enzyme markers showed that the minor band at 25 % w/w sucrose (F_1) was rich in T-tubules. The peak at 33 % (F_2) contained membranes of longitudinal SR. The band at 38 % (F_3) was mainly composed of terminal cisternae of SR.

F_1, F_2 and F_3 contained 2.5, 1.7 and 1.8 μmol . h^{-1} . mg^{-1} (U/mg) of AChE, respectively, increasing to 3.9, 2.4 and 2.7 U/mg with Triton X-100. Extraction of F_1, F_2 and F_3 with 0.5 mg/mg protein Triton X-100 solubilized a 20%, 55% and 40 % of total activity in a first step (S_1), and 36%, 19% and 16% in a second one (S_2). The profiles of molecular forms can be seen in Fig. 1. The extent of solubilization of AChE with Triton X-100 was different for each type of membrane and this suggests a differential way of attachment to the membranes. In all types of vesicles, the most abundant molecules are amphiphilic G_1 forms (Cánovas-Muñoz et al., 1990).

Financial support was obtained from the FISss of Spain, grant 89/514. J. Campoy is a holder of a scholarschip from DGICYT.

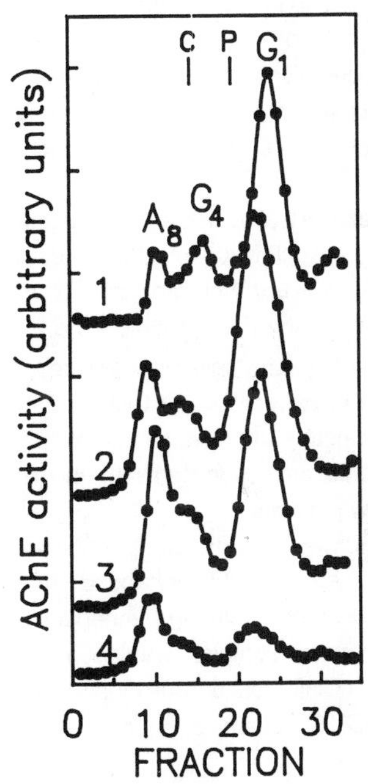

Fig. 1. AChE molecular forms solubilized with Triton X-100. Profiles 1-3 show the forms solubilized in a first step (S_1) from F_1, F_2 and F_3, respectively. The forms obtained in a further extraction (S_2) of F_3 are displayed in profile 4, those of F_1 and F_2 being similar (C, 11.4S; P, 6.1S).

References

Cánovas-Muñoz, M.D., Campoy, J., Muñoz-Delgado, E., Vidal, C.J., *Biochim. Biophys. Acta*, **1990**, (in press).

Salama, G., Abramson, J., *J. Biol. Chem.*, **1984**, *259*, 13363-13369.

Histochemical Study of Non-Specific Cholinesterase Activity in PNS as a Tool for the Recognition of Relationships Between Axon and Schwann cells

Petr Dubovy, Department of Anatomy, Medical Faculty, Masaryk University Brno, 662 43 Czechoslovakia

The present study was undertaken to determine non-specific cholinesterase (nCHE) activity in peripheral nerve structures of the rats during development, degeneration and regeneration. The nCHE activity in Schwann cells at various experiments suggested that the nCHE molecules might be involved in the relationships between axons and their Schwann cells.

The nCHE molecules are synthesized by all Schwann cells of developing peripheral nerves. The synthesis of these molecules is repressed in Schwann cells starting to produce the myelin sheath and continues in non-myelin-forming Schwann cells. The histochemical findings indicate that the nCHE molecules are secreted and become an integral component of the extracellular matrix of Schwann cells.

The recovery of nCHE activity after axotomy and subsequent application of OMPA revealed that the sythesis of nCHE in Schwann cells of adult rats is independent on the neurotrophic factors.

Cryo-injury of Schwann cells results in disappearance of extracellular nCHE activity. Thus, the nCHE molecules attached to the extracellular matrix components are associated with the presence of viable Schwann cells.

All Schwann cells of peripheral stumps of damaged nerves exhibit enhanced nCHE reactivity up to 12 days. During reinnervation the axonal growth cones contain invaginations and vesicles filled with nCHE reaction product indicating the uptake of nCHE molecules. Four weeks after ligature tied above the site of crush, the axons are loaded by accumulated nCHE end product in distal segments removed near to ligature. Reactive Schwann cells of peripheral stumps are probable source of nCHE molecules transported retrogradely in regenerating axons.

The ligatures tied to intact nerves caused also peripheral accumulation of nCHE reactivity. The axonal profiles in sensory nerve endings as well as neuromuscular junctions contained invaginations and small vesicles filled with nCHE reaction product. The extracellular matrix of Schwann cells surrounding sensory and motor terminals is a source of transported nCHE molecules in intact sciatic nerves.

Histochemical detection of nCHE activity indicating reactive Schwann cells is effective methodical approach for the study of Schwann cell migration. For example, the segments of denervated peripheral stumps damaged by cryo-treatment are settled by Schwann cells migrating from both intact ends. Migrating Schwann cells without their basal laminae exhibit nCHE reaction product on the plasma membrane.

0–8412–2008–5/91/0122$06.00/0

Characterization of Acetylcholinesterase Secretion in Mouse Neuroblastoma Cells

Biagioni S., Bevilacqua P., Carucci C., Scarsella G.* and Augusti-Tocco G., Dipartimento di Biologia Cellulare e dello Sviluppo, Universita' "La Sapienza" P.le A. Moro 5, 00185 Roma, Italy.

Acetylcholinesterse (AChE) release has been observed in both the central and peripheral nervous system *in vivo*, and in cultured neurons and muscle cells (Greenfield, 1985; Rotundo, 1987). The presence of AChE in non-cholinergic circuits and its reported peptidase activity (Chubb et al., 1980; Small, 1989) have suggested possible functions different from that exerted in cholinergic synapses. Moreover the ability of several AChE containing cells to release selectively a particular form (the G4 form) of the enzyme in the extracellular space may be relevant for its function as a neuromodulator (Greenfield, 1985; Toutant & Massoulie', 1988).

Together with several proteins of undefined function AChE is released into the culture medium by mouse neuroblastoma cells. Previous results obtained in our laboratory indicate that a neuroblastoma x glioma hybrid line, capable of establishing cholinergic contacts, release lower amounts of AChE when compared with the parental mouse neuroblastoma line, which is defective in establishing functional contacts (Melone M.A.R. et al., 1987). This observation may suggest a possible inverse correlation between release of AChE and the establishment of cholinergic contacts. On the basis of these considerations it appeared of interest to further characterize AChE release in the parental line.

Since AChE release in NGF treated PC12 cells has been related to fiber outgrowth (Lucas & Kreutzberg, 1985) we have examined the effect of drugs interfering with cytoskeletal structures on AChE release in the mouse neuroblastoma line N18TG2. Cells were grown in DMEM supplemented with 2% Ultroser (IBF), to avoid interference with serum cholinesterase. Prostaglandin (PGE_1 10 μM) were added to the medium as a culture condition promoting fiber outgrowth.

Cytochalasin B does not affect AChE release in neuroblastoma cells, while nocodazole significantly reduces the amount of released AChE of about 50% after 3 hours. No difference was observed in the rate of AChE release under conditions promoting fiber outgrowth, suggesting that nocodazole mainly alters the transport of secretory vescicles from the Golgi apparatus to the cell surface.

The effect of K^+ and Ca^{++} on the enzyme release was also studied, in order to reveal the possible existence of a regulated component. High K^+ level (100 mM, final concentration) in the culture medium induces an increase in the AChE secretion. The potassium induces a fast and temporary increase (about 50%); in fact, 20 min. after the increase of K^+ level in the media, AChE release is restored to a rate similar to that previous K^+ elevation. These effects were observed both in differentiated and undifferentiated cells and only in presence of Ca^{++}. On the other hand preliminary results obtained in the presence of Ca^{++} ionophore A23187 in the culture medium, apparently does not modify the secretion of the enzyme. These data suggest the existence of a regulated secretion which coexists with a basal level of constitutive release. At the same time they do not allow a precise evaluation of the role of Ca^{++} in the regulation of AChE secretion.

(Supported by funds from CNR and Ministero P.I.)

References

Chubb I.W., Hodgson A.J., White G.*Neuroscience*, **1980**, *5*, 2065-2072.

Greenfield S., *Neurochem. Int.*, **1985**, *7*, 887-901.

Lucas C.A., Kreutzberg G.W., *Neuroscience*, **1985**, *14*, 349-360.

Melone M.A.R., Longo A., Taddei C., Augusti-Tocco G., *Int. J. Devl. Neuroscience*, **1987**, *5*, 417-428.

Rotundo R.L., in *The Vertebrate Neuromuscular Junction;* Salpeter M.M., Ed.; Alan R. Liss: New York, U.S.A.: 1987; 247-284.

Small D.H., *Neuroscience*, **1989**, *29*, 241-249.

Toutant J.P., Massoulie' J., *Handbook of Experimental Pharmacology;* Whittaker V.P., Ed.; Springer Verlag: Berlin-New York, Germany, 1988, Vol. 86, 225-265.

AChE AND BChE ACTIVITIES DURING DIFFERENTIATION OF RETINAL PIGMENT EPITHELIAL CELLS IN CULTURE

Salceda, R., Sánchez, G., León-Cázares, J.M. Instituto de Fisiología Celular, Universidad Nacional Autónoma de México, México D.F., México

Acetylcholinesterase has been involved in cell differentiation (Drews 1975, Miki et al 1980). Vollmer and Layer (1987) found in cultures of retina and pigment epithelial cells (RPE) a correlation between BChE and AChE activities and cell proliferation and differentiation, respectively. Therefore we studied the incorporation of ^{3}H-thymidine, melanin content and AChE and BChE activities of RPE in culture in order to explore the relation between AChE and cell proliferation and differentiation.

Cells were growth in TC 199 culture medium supplemented with 10% heat-inactivated fetal calf serum. Cells were seeded at 4 x 10^4 cells in a well culture plate and incubated at 37°C. At different periods of growing ^{3}H-thymidine was added. Cells were scraped off with 10% trichloroacetic acid, transferred to scintillation vials and radioactivity measured in a liquid scintillation counter. Melanin content was measured as described by Siegrist and Eberle (1986). Cells were incubated according to Karnovsky and Roots (1964) for the histochemical visualization of AChE and BChE activities. AChE activity was measured according to Ellman (1961).

RPE cells in culture showed an AChE activity of 10 nmolas/mg protein/min. BChE activity was 2 times lower than that of AChE. Both activities were found to be present in the culture medium. AChE activity increased during the days of culture, parallel to the cell differentiation as indicated by ^{3}H-thymidine incorporation and melanin content. Histochemistry studies showed the AChE only in the differentiated cells. Analysis of the molecular forms of AChE revealed the presence of 19S, 16S, 10S, 6.5S and 4.5S in RPE cells. However only the 4.5S and 10S were found in the culture medium. These results support the hypothesis that AChE can be considered as a very early marker for cell differentiation.

References

Drews, U. Prog Histochem Cytochem. 1975, 7,1.

Ellman, G.L. Biochem Pharmacol. 1961, 7, 88.

Karnovsky, M.J., Roots, L.J. Histochem Cytochem. 1964, 12, 219.

Miki, A., Atoji, Y., Mizoguti, H. Acta Histochem Cytochem. 1981, 14, 641.

Siegrist, W., Eberle, A.N. Anal Biochem. 1986, 159, 191.

Vollmer, G., Layer, P.G. Cell Tissue Res. 1987, 250, 481.

0–8412–2008–5/91/0124$06.00/0

Contribution of an Inactive Precursor to the Renewal of Acetylcholinesterase Activity in Rat Sympathetic Neurons

Bertrand Tavitian, Raymonde Hässig and Luigi Di Giamberardino, INSERM U334, S.H.F.J., CEA, 91406 Orsay, France.
Martine Verdière-Sahuqué and Danièle Goudou, INSERM U 153, 17 rue du Fer-à-Moulin, 75014 Paris, France.

The aim of this study was to measure the relative contributions of 1) protein synthesis and 2) activation of an inactive precursor to the renewal of intracellular AChE in neurons. For this purpose, cycloheximide (CX), an inhibitor of protein synthesis, and MPT (methylphosphonothiolate), an organophosphorous irreversible inhibitor of AChE, were added in various combinations to primary cell cultures of rat superior cervical ganglion, grown in the presence of cytosine arabinoside, in order to kill non-neuronal cells. In these cultures, AChE is present in the sympathetic neurones themselves, and essentially stable over 48 hour period.

Intracellular AChE activity was inhibited up to 95% 15 minutes after addition of 7.5×10^{-7} M MPT to the cultures. Washing the cells and allowing them to recover in MPT-free medium led to a recovery of 40% of the control activity in 17 hours. Cells grown in 5×10^{-4} M CX contained only 66% of the control AChE activity after 17 hours. Since protein synthesis was below detectable levels in those conditions, this suggests that 34% of intracellular AChE was lost (degraded or secreted) during that time. Cells grown in the presence of CX after an initial exposure to MPT for 15 minutes still recovered AChE activity to 16% of control levels. However, if the cultures were exposed to MPT a second time 3 hours after the first exposure, no AChE activity recovery was observed during the same 17 hours (6% of the control versus 5% just after MPT addition). Thus, AChE that had matured from an inactive precursor after the first MPT exposure was completely inhibited by the second MPT exposure.

Taken together, these results show the existence of a pool of inactive precursor of AChE in rat sympathetic neurons. This pool can mature into active enzyme in less than 3 hours and contributes to 11% of total intracellular AChE activity.

0–8412–2008–5/91/0125$06.00/0

Impact of Pesticides on the Regulation of Gonadal Function in Fish

Probodh Ghosh[1], Samir Bhattacharya[2], Shelley Bhattacharya[1], Environmental Toxicology Laboratory[1], Endocrinology Laboratory[2], Department of Zoology, Visva-Bharati University, Santiniketan-731 235, India.

Recent evidences suggest that organophosphate and carbamate pesticides produce their effects in fish by a direct neurotoxic action (Bhattacharya and Jash, 1981, Jash and Bhattacharya, 1983a, b); by secondary action mediated via hormones (Clement, 1985) and by coordinated changes in neural and hormonal functions (Guhathakurta and Bhattacharya, 1988; Ghosh *et al.*, 1989). In the present study GnRH activity and pituitary and serum GtH concentrations in *C. punctatus* were estimated by specific bioassay (Jamaluddin *et al.*, 1989) and RIA techniques (Banerjee *et al.*, 1987). These parameters were correlated with the ACh–AChE profiles in fish under both laboratory and field conditions.

Fish were exposed to 0.106 ppb of Metacid–50 (methylparathion 50% a.i., Bayer, India) and 1.66 ppm of Carbaryl (Carbaryl 50% a.i., Paushak, India) for 30 days under standard laboratory conditions. The experimental agricultural plot had an area of 78 m^2 and the volume of water approximately 6700 litres. The farmers sprayed the paddy field in such a way that the concentrations of Metacid–50 in the water amounted to 0.239 ppb and Carbaryl to 3.73 ppm after a single application. Sampling was done at day 2 and day 7 of treatment.

Under laboratory conditions, Carbaryl treatment produced a higher time dependent decrease of serum GtH level in comparison to Metacid–50. In the field, however, Metacid–50 showed 40% decrease in the serum GtH level in comparison to a 32% decrease caused by Carbaryl. Pituitary GtH content and GnRH activity, however, were inhibited to a greater extent by Metacid–50 than by Carbaryl under both field and laboratory conditions.

Under laboratory treatment of Metacid–50, the rate of inhibition of AChE progressively increased from day 2 (11% inhibition) to day 14 (53% inhibition) followed by a decreasing rate of AChE inhibition from 45% on day 21 to 20% on day 30. In Carbaryl treatment, inhibition rate was between 15% on day 2 and 61% on day 14 leading to 36% and 25% inhibition on days 21 and 30 respectively. The concurrent ACh profile had an inverse relationship with AChE activity i.e. increased inhibition of AChE leading to greater accumulation of ACh.

In the field, Metacid–50 caused 52% (day 2) to 69% inhibition (day 7); while carbaryl caused 63% (day 2) to 47% inhibition (day 7) concurrent ACh profiles varied between 140% (day 2) to 210% (day 7) in Metacid–50 exposure while 170% (day 2) to 90% (day 7) in carbaryl treatment.

Thus the prsent investigation elucidates that ACh-AChE system is a sensitive bioindicator not only of pesticidal contamination but also reflects significantly the degree of reproductive damage mediated via the cholinergic system.

References

Banerjee, P.P., Jamaluddin, Md., Bhattacharya, S., Nath, P., Kobayashi, M., Aida, K., Hanyu, I., *Indian J. Exp. Biol.*, 1987, *25*, 220-227.

Bhattacharya, S., Jash, N.B., *Com. Physiol. Ecol.*, 1981, *6*, 330-332.

Clement, J.G., *Fundam. Appl. Toxicol.*, 1985, *5*, 61-67.

Ghosh, P., Bhattacharya, S., Bhattacharya, S., *Biomed. Environ. Sci.*, 1989, *2*, 92-97.

Guhathakurta, S., Bhattacharya, S., *Biomed. Environ. Sci.*, 1988, *1*, 59-63.

Jamaluddin, Md., Banerjee, P.P., Manna, P.R., Bhattacharya, S., *Gen. Comp. Endocrinol.*, 1989, *74*, 190-198.

Jash, N.B., Bhattacharya, S., *Water Air Soil Pollut.*, 1983a, *19*, 209-213.

Jash, N.B., Bhattacharya, S., *Toxicol. Lett.*, 1983b, *15*, 349-356.

Molecular Forms of Acetylcholinesterase in Chick Eye Tissues

Milagros Ramos, Carmen Prada, Jordi Pérez-Tur, Ana Barat, Galo Ramírez*,
Centro de Biología Molecular (CSIC-UAM), Universidad Autónoma, Cantoblanco, 28049 Madrid, Spain.

We have previously described the presence of type II (high-salt/EDTA-extractable) asymmetric acetylcholinesterase (AChE) in chick 'whole retina' preparations (Barat et al, 1980). A recent paper by Martelly and Gautron (1988) ascribed these asymmetric forms (A-forms) to the retinal pigmented epithelium. Upon careful dissection of the different retinal layers and neighboring tissues, we have found that both the neural retina and the retinal pigmented epithelium contain only globular forms of AChE (Barat et al, 1990). In the case of neural retina, most of the G_4 and some of the G_{2+1} globular forms appear to be extracellular (Ramírez et al, 1989), and could be involved in synaptic function at the inner plexiform layer (Fig.1). The possible degradation of some intrinsic A-forms in the retina during the isolation procedure, although unlikely, cannot be totally dismissed at the present time (Gómez-Barriocanal et al, 1983). The choroid, together with the ciliary muscles and iris, are however rich in asymmetric enzyme, especially type II. Choroidal A-forms may be located in intrinsic muscle fibers. Most if not all the asymmetric AChE activity detected in chick 'whole retina' preparations could then be explained in terms of contamination by non-retinal eye tissues (Barat et al, 1990).

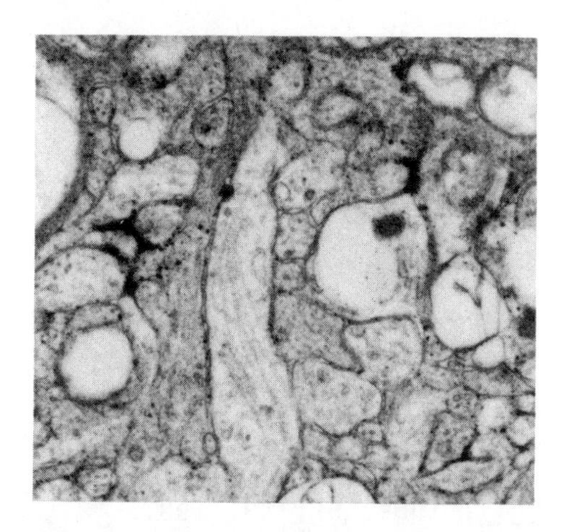

Fig.1. BW/DFP-treated chick retina showing extracellular AChE (presumably G-forms) in synaptic areas of the inner plexiform layer (x26,000). Adapted from Ramírez et al, 1989.

Supported by grants from the Dirección General de Investigación Científica y Técnica (PB87-0244), the Fundación Ramón Areces and Laboratorios Dr. Esteve, S.A.

References

Barat, A., Escudero. E., Gómez-Barriocanal, J., Ramírez G., Biochem. Biophys. Res. Commun., **1980**, 96, 1421- 1426.

Barat, A., Ramos, M., Prada, C., Ramírez, G., Neurosci. Lett., **1990**, 112, 302-306.

Gómez-Barriocanal, J., Rodríguez-Borrajo, C., Barat, A., Ramírez, G., Neurosci. Lett., **1983**, 38, 321-326.

Martelly, I., Gautron, J., Brain Research, **1988**, 460, 205-213.

Ramírez, G., Barat, A., Fernández, H., J. Neurosci. Res., **1989**, 22, 297-304.

0–8412–2008–5/91/0127$06.00/0

Surface AChE in the Avian Ciliary Ganglion

C. Gangitano, A. Del Fà, *C. Olivieri-Sangiacomo
Institute of General Biology, Catholic University, 00168 Rome

In the chick ciliary ganglion (CG), we have detected AChE activity in the cytoplasm as well as at synaptic and extrasynaptic areas of the neuronal surface. AChE reaction has been cytochemically localized according to Lewis and Shute (1969). In the present work we have focalized our attention on AChE localized at the neuronal surface, in different experimental conditions. Our data originate from investigations on preganglionic denervation, postganglionic axotomy and embryonic development of CG. During embryonic life AChE appears at the neuronal surface only after the onset of early synaptic contacts (7 d.i.), labelling the whole contour of the calyciform nerve terminals only at 15 d.i. when the calyx is morphologically mature (Olivieri-Sangiacomo et al., 1983). Besides in this period the reaction product lines increasingly larger extrasynaptic areas of the neuronal surface. As preganglionic denervation is concerned, surface AChE undergoes a progressive disappearance following the degeneration of transected nerve endings. Finally, after postganglionic axotomy the reaction markedly labels the synaptic and extrasynaptic neuronal surface and peripheral contour of nerve terminals, the cytoplasmic AChE rapidly disappearing (Gangitano et al., 1983). In conclusion in avian CG preganglionic afferences appear to exert an essential role in the maintenance of surface AChE in ganglionic neurons. Such modulating actions could be mediated by neurotrophic factors, as recently postulated in the superior cervical ganglion of the cat (Koelle et al., 1987).

References

Gangitano,C., Del Fà,A., Ardito,G., Rumi,E., Olivieri-Sangiacomo,C., Boll. Soc. It. Biol. Sper., **1983**, 7, 1005-1009.

Koelle,G.B., Sanville,U.G., Thampi, N.S., Proc. Natl. Acad. Sci. USA, **1987**, 84, 6944-6947.

Lewis,P.R., Shute,C.C.D., J. Microsc. Sci., **1969**, 89, 181-193.

Olivieri-Sangiacomo,C., Del Fà,A., Gangitano,C , Develop. Brain Res., **1983**, 7, 61-69.

Human Intestinal Mucosal Cell Cholinesterases

B. Colas, **J-P. Sine**, Laboratoire de Biochimie II, Faculté des Sciences, 2, rue de la Houssinière, 44072 Nantes Cédex 03, FRANCE
R. Ferrand, Laboratoire de Biologie du Développement, Faculté des Sciences, 2, rue de la Houssinière, 44072 Nantes Cédex 03, FRANCE

Cholinesterases are found in a large variety of non-neuronal cells. We have identified these enzymes in the intestinal mucosal cells of vertebrates belonging to different classes (Sine and Colas, 1985, 1987; Sine *et al.*, 1988, 1989) The ChE activity found in the human intestinal cells appears relatively low when compared to that of other studied mammals.

Histochemical studies show that the ChE activity is found at the apex part of the human intestinal villi whereas in other species such as rat or rabbit, it is detected all along the villi except in the crypts. Its distribution along the intestine is very similar to that found in pig with a maximum activity in the colon and a minimum one in the jejunum.

Although the AChE activity is predominant all along the intestine, the BuChE activity is also present in significant amount. The latter is relatively important in the colon. Like in the other species, only globular forms are found. Generally, the G_4 form is associated with either the G_1 form or the G_2 form (Sine *et al.*, 1988). Here, like in cat, the three forms (G_1, G_2 and G_4) are identified after sedimentation in a low-ionic strength medium without detergent.

As it is known that acetylcholine stimulates the mucus secretion, it is tempting to associate the function of ChE, present in the intestinal mucosal cells, with a regulation of this secretion. With this view, we now examine the behavior and characteristics of ChE in pathological states.

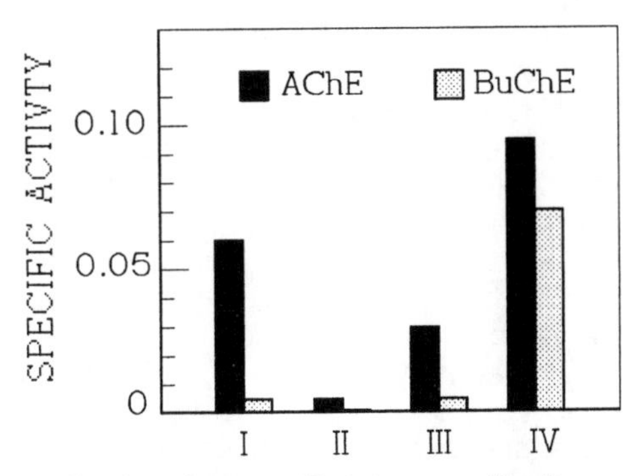

I: duodenum; II: jejunum; III: ileum; IV: colon. The AChE and BuChE specific activities are expressed as µmol of substrate hydrolyzed/h/mg of cell proteins.

References

Sine, J-P. and Colas, B., *Biochim. Biophys. Acta*, **1985**, *817*, 190-192.

Sine, J-P. and Colas, B., *Biochimie*, **1987**, *69*, 75-80, 1987.

Sine, J-P., Ferrand, R., Colas, B., *Comp. Biochem. Physiol.*, **1988**, *91C*, 597-602.

Sine, J-P., Ferrand, R., Colas, *Mol. Cell. Biochem.*, **1989**, *85*, 49-56.

Identification and Characterization of Cholinesterases in the Enterocytes

J-P. Sine, R. Ferrand, B. Colas, Faculté des Sciences, 2, rue de la Houssinière, 44072 Nantes Cédex 03, FRANCE

A cholinesterase activity was found in the enterocytes of vertebrates belonging to different classes: Mammals, Birds, Amphibia, except some Osteichthyes and all the Chondrichthyes studied (Sine *et al.*, 1988). This activity is present along the intestine villi but not in the crypt region. Its distribution, from duodenum to rectum, shows a very variable pattern according to the species.

The activity is due to either an AChE or a BuChE. Only globular forms are found. Except flounder, frog and rabbit, all the other studied species contain a G_4 form associated with either a G_1 form or a G_2 form. Both the G_1 and G_2 forms are found with the G_4 form in cat and human enterocytes. The G_1 form is always soluble. The G_2 and G_4 forms are either soluble or amphiphilic.

The presence of ChE in the enterocytes raises the question of their physiologic function. No relation could be established between, on the one hand, the presence, the localization, the activity level, the solubility and the molecular forms of the enzyme and, on the other hand, the dietary habits. However, identical molecular forms with similar properties were identified in related species (rat / mouse, chicken/quail). The study that we have performed during the chick development shows that the ChE activity differs according to the sex. The ontogeny and the regulation of the enzyme activity seem to depend on hormonal factors.

References

Sine, J-P., Ferrand, R., Colas, B., *Comp. Biochem. Physiol.*, **1988**, *91C*, 597-602.

Species	Maximum activity	AChE (μmol/h/mg	BuChE cell proteins)	Molecular forms
Dogfish		n.d.	n.d.	
Goldfish		n.d.	n.d.	
Flounder	intestine	0.30	n.d.	*G_2(100)
Eel	intestine	0.30	n.d.	G_1 (58) G_4(42)
Frog	rectum	2.18	0.60	G_2(100)
Chicken	jejunum	0.05	2.18	G_1(58) G_4 (42)
Quail	jejunum	n.d.	1.24	G_1(85) G_4 (15)
Ox	duodenum	0.05	0.71	G_1(34) G_4(66)
Pig	colon	0.06	0.39	G_1(73) G_4(27)
Sheep	rectum	0.01	0.16	*G_2(69) *G_4(31)
Rabbit	jejunum	0.82	n.d.	G_1(64) *G_2(36)
Rat	duodenum	0.04	2.65	*G_2(85) G_4(15)
Mouse	duodenum	0.05	3.10	*G_2(82) G_4(18)
Cat	jejunum	0.12	0.01	G_1(22) *G_2(21) *G_4(57)
Man	colon	0.13	0.04	G_1(35) G_2(49) G_4(16)

Figures in () give the activity percentage of each molecular form with respect to the total ChE activity. *: amphiphilic form. n.d. : not detected.

Modifications Induced by Diabetes Nellitus on Acetylcholinesterase Activity of Microvillus Plasma Membrane from Human Placenta

L. Mazzanti, R.A. Rabini, R. Staffolani, Institute of Biochemistry, University of Ancona, Ancona, Italy

C. Romanini, G. Benedetti, A.M. Cugini, Institute of Obstetrics and Gynecology, University of Ancona, Via Corridoni, Ancona, Italy

N. Cester, Department of Obstetrics and Gynecology, Ospedale Salesi, USL 12, Ancona, Italy

The human placental syncytiotrophoblast microvillus plasma membrane represents the intimate contact with the maternal blood. Acetylcholinesterase (AChE), is found in the plasma membrane of mammalian erythrocytes, but its role in microvillus plasma membrane of various cells including microvillus from human placenta is still uncertain (Ruch 1979). We have studied the AChE activity in placental microvillus membranes from 16 normal women and 16 women affected by gestational diabetes mellitus (GDM).

Materials and Methods

The study has been performed on the placentas from 16 women affected by GDM and 16 normal women at term. Microvillus plasma membranes were obtained as previously described (Withsett and Wallick 1980). Acetylcholinesterase activity was assayed in triplicate in the erythrocyte plasma membranes by Ellman, the enzymatic activity being expressed as µmol acetylthiocholine/min/ protein. Protein concentration was determined by Lowry method 1951.

Results and Discussion The GDM provokes a reduction of acetylcholinesterase activity in the microvillus plasma membrane, accompanied by a decrease of both Km and Vmax (Table l). This data is different from data obtained in erythrocyte plasma membranes during Insulin-Dependent (IDDM) and Non-Insulin-Dependent (NIDDM) diabetes mellitus (Mazzanti 1989). The different alterations might be linked to the different degree of metabolic control which was characterized by almost normoglycemic levels during GDM, unlikely IDDM and NIDDM in the reported studies (Frenkel 1980).An alternative explanation for the different membrane pattern between GDM and IDDM and NIDDM might be the peripheral action on the plasma membranes of hormones other than insulin,wich are secreted by the placenta during pregnancy (Kalkhoff 1969).The pathological implications of these abnormality deserve further studies, as the role of AChE in placenta membranes is not clear.However,this abnormal activity in GDM might reflect a general alteration in the placental membrane, with subsequent alteration of the exchange between mother and foetus. The present work was supported by a Grant of Regione Marche to Nelvio Cester

Referencee

Ellman G.L., Courtne D.K., Andres V. Jr, Featherstone R., *Biochem Pharmacol,* **1980**, 7, 88

Kalkhoff R., Richardson B.L., Beck P., *Diabetes* **1969**, 18, 153.

Lowry O.H., Rosenburg M.Y., Farr A.L., Randall R.T., *J Biol Chem* , **1951**, 193, 265

Mazzanti L., Rabini R.A., Testa I., Bertoli E., *Eur J Clin Invest,* **1989**, 19, 84

Ruch G.A., Davis R., Koelle G.B., *J Neurochem* **1979**, 26, 1189

Whitsett J.A., Wallick E.T., *Am. J. Physiol.,* **1980**, 238, E38

Table 1— Acetylcholinesterase activity in microvillus plasma membranes from human placenta in normal and GDM pregnancies. The activities are expressed as micromoles acetylcholine/min/ mg membrane protein. The kinetic parameters Km and Vmax are also presented.

	As	Km	Vmax
Controls (n= 16)	28+/- 0.4	0.131+/-0.012	9.71 +/-2.53
GDM (n= 16)	13.4+/- 0.23*	0.103+/-0.014**	6.36+/-1.52

* : p<0.01
**: p<0.05

0–8412–2008–5/91/0131$06.00/0

A Study on the Acetylcholinesterase Enzymatic Activity in Human Erythrocyte Membranes during Insulin-Dependent Diabetes Mellitus

L. Mazzanti, R.A. Rabini, G.P. Littarru, R.M. Fiorini, Institute of Biochemistry, University of Ancona, Via Ranieri, Ancona (Italy)
E. Faloia, R. De Pirro, Institute of Endrocrinology, University of Ancona, Via Ranieri, Ancona (Italy)

Physiological insulin concentrations modulate the allosteric properties and the activity of membrane-bound enzymes such as rat erythrocyte acetylcholine sterase (AchE) and Na+/K+ ATPase (Farias,1980,Unates,1979).Modifications in the AchE enzymatic activity have been observed in human erythrocyte membranes obtained from patients affected by Insulin-dependent -Diabetes Mellitus (IDDM) (Mazzanti, 1989).The process of red blood cell (RBC) aging produces a number of alterations both in the composition of the plasma membrane and in the function and activity of membrane -bound enzymes,which are partly similar to those reported in IDDM (Clark,1988). It is not clear if modifications present in erythrocyte membranes from diabetic patients are a direct consequence of alterations induced by diabetes mellitus at the plasma membrane level or if they are the consequence of an accelerated aging process induced by the disease. The aim of the present study is to investigate the influence of aging on the RBC acetylcholinesterase activity both in normal and diabetic subjects.

Patients and methods

23 insulin-dependent diabetic patients (24$\pm$43 years) and 26 healthy subjects (26$\pm$47 years) were studied. All diabetic patients showed a good metabolic control Preparation of RBC subpopulations: Erythrocytes were isolated from total blood as described (Beutler 1976) and washed three times in Hepes Buffered Stock Solution (HBSS) diluted 1:20. RBCs were separated into five subpopulations on Percoll/BSA density gradients according to the previously described method(Salvo 1982) with minor modifications.
Membrane preparation: Erythrocyte membranes from each fraction were obtained by Burton 1981 method.
AchE activity: AchE activity was measured in triplicate on erythrocyte plasma membranes according to Ellman 1961.The enzymatic activity is expressed as micromoles hydrolized of acetylthiocholine/min/protein. Protein concentration was determined by the Lowry method.

Results and Discussion

The present work demonstrated (Table 1) a marked decrease in the AchE activity during RBC aging in normal subjects confirming previous reports (Kamber 1984). On the contrary,AchE activity increased in mature RBCs from diabetic subjects. Nevertheless Diabetes Mellitus does not affect the activity of this enzyme in early mature RBCs; as a consequence,this phenomenon may not be ascribed to an effect of the disease state on plasma membrane,but it derives from an effect appearing during erythrocyte aging. On the basis of our previous work relating the AchE alterations in IDDM (Mazzanti, 1989) with membrane fluidity it might be hypothesized that the variations in the enzymatic activity appearing in diabetic patients during RBC aging may be linked to modifications in membrane fluidity developing during senescence.
The present work was supported by a Grant of Ministero Pubblica Istruzione 40%

0-8412-2008-5/91/0132$06.00/0

Table 1– Acetylcholinesterase activity of erythrocyte plasma membranes obtained from fractioned cells.The activities are expressed as micromoles of acetylcholine/min/mg membrane protein

Fractions	1	2	3	4	5
Controls (n=26)	1.54$\pm$0.19	1.09$\pm$0.12	1.04$\pm$0.12	0.73$\pm$0.15	0.52$\pm$0.16
IDD (n=23)	1.37$\pm$0.17	1.84$\pm$0.19	2.55$\pm$0.20*	3.48$\pm$0.13*	3.48$\pm$0.18*

* : $p<0.001$

References

Beutler,E.,West,C.,Blume,K.G.,J.Lab.Clin Med.,**1976**,88,328-33

Burton,G.W.;Ingold ,K.U.;Thompson,K.E . Lipids **1981**,16,946

Clark,M.R. Physiol. Rev.**1988**,68,503

Ellman,G.L.,Courtney,D.K.,Andres,V.Jr., Featherstone, R.M., Biochem. Pharmacol.,**1961**,7,88-95

Farìas,R.N.;Bloj,B.;Morero;R.D.;Sineriz F.;Trucco,R.E. Biochim. Biophys. Acta **1975**,415,231

Farìas,R.N. Adv.Lipid Res.**1980**,17,251

Kamber,E.,Poyagi,A.,Deliconstantinos,G. Comp. Biochem. Physiol.B Comp. Biochem. **1984**,77,95

Lowry,O.H.,Rosenburg,M.Y.,Farr,A.L., Randall,R.T.,J.Biol.Chem**1951**,193,265-75

Mazzanti,L.,Rabini,R.A,Testa,I,Bertoli, E.,Eur.J.Clin.Inv. **1989**,19,84-89

Salvo,G.,Caprari,P.,Samoggia,P,Mariani, G.,Salvati A.M.,Chim.Clin.Acta **1982**,122,293-297

Unates,L.E., Farias,R.N. Biochim. Biophys. Acta **1979**,568,363

GENE STRUCTURE AND EXPRESSION OF CHOLINESTERASES

Nematode Acetylcholinesterases: Diversity and Functions

Carl D. Johnson, Cambridge NeuroScience Research, Cambridge, MA 02139

Nematodes contain three, biochemically distinct classes of acetylcholinesterases (AChEs). The active subunits of the three classes appear to be the products of separate, unlinked genes. The analysis of strains carrying mutations in genes coding for the AChEs implys that the three enzymes collaborate to accomplish an essential function, presumably the hydrolysis of acetylcholine. The investigation of genetic mosaics indicates a requirement for AChE expression in muscle cells but not in neurons.

Nematodes are a highly abundant group of simple invertebrates, generally considered to have arisen early in animal evolution, prior to the split leading to the two major coelomate groups, protostomes (spiralians and arthropods) and deuterostomes (echinoderms, prechordates, vertebrates).

Over the last fifteen years, acetylcholine (ACh) functions in nematodes, including nematode AChEs have been studied in some detail, primarily by the author, R.L. Russell, A.O.W. Stretton, J.B. Rand, colleagues and collaborators.

The experiments have focussed on two species: the small, genetically tractable, soil nematode Caenorhabditis elegans and the much larger parasitic nematode Ascaris suum. C. elegans has been used for studies of the biochemistry of AChEs (Johnson and Russell, 1975; Johnson and Russell, 1983; Kolson and Russell, 1985b), of choline acetyltransferase (Rand and Johnson, 1981; Rand and Russell, 1985) and of ACh receptors (e.g. Lewis et al., 1987) and for genetic analysis of the genes which control their expression (Johnson, et al., 1981; Culotti, et al., 1981; Kolson and Russell, 1985a; Johnson et al., 1988; Rand and Russell, 1984; Rand, 1989; Lewis et al., 1987). Ascaris has been used for electrophysiological study of defined nematode neurons (e.g. Walrond, et al, 1985; review, Stretton, et al., 1985) and for single cell biochemistry experiments designed to determine which cells utilize ACh as a transmitter (Johnson and Stretton, 1985; C.D. Johnson, unpublished).

The depth and breadth of these experiments establish nematodes as a significant model for investigating both the diversity and functions of AChEs as well as the roles of ACh in the control of behavior.

The diversity of nematode AChEs: biochemical studies.

Biochemical experiments have revealed three distinguishable classes of nematode AChE activity, referred to as class A AChE, class B AChE and class C AChE (see Johnson et al., 1988). In addition, within each of the AChE classes there are multiple forms, separable on sucrose gradients and/or by ion exchange chromatography. Elongated forms of AChE containing collagen-like tails, such as have been widely observed in vertebrates, have not been found in nematodes.

All forms of AChE which are members of the same class have very similar kinetic parameters. For example, the K_M for acetylcholine is ca. 15μM for all class A forms (Johnson and Russell, 1983), ca. 80μM for class B forms (Johnson and Russell, 1983) and (remarkably) ca. 0.010μM for class C forms (Kolson and Russell, 1985b). Members of a class also have similar substrate selectivities and inhibitor sensitivities, whereas the classes differ (Johnson and Russell, 1983; Kolson and Russell, 1985b). The kinetic properties of AChE classes are conserved between C. elegans and Ascaris (Johnson and Stretton, 1980; C.D. Johnson, unpublished).

Understanding the differences in properties of the various AChE classes has facilitated the development of class-specific AChE assays capable of selectively measuring one class of enzyme in the presence of the other two classes (Fig. 1). Class-specific assays were used to perform the genetic analysis summarized below.

The diversity of nematode AChEs: genetic studies.

Each AChE class is selectively affected by mutations in one of three genetically unlinked ace (for acetylcholinesterase deficient) genes (Johnson, et al., 1981; Culotti, et al., 1981; Johnson et al. 1988). ace mutations eliminate or greatly diminish the activity of all forms of one class of AChE (Fig. 1). Mutations with selective effects on forms within a class have not be identified.

Studies of mutation frequency, of gene dosage effects and of the residual activity present in some ace mutant bearing strains suggests that ace-1 (located on the X chromosome), ace-2 (chromosome I), and ace-3 (chromosome II) are, respectively, the structural genes for AChE classes A, B, and C (Johnson, et al., 1981; Culotti, et al., 1981; Kolson and Russell, 1985a; Johnson et al., 1988).

Multiple classes of AChE, coded for by separate genes, are also observed in vertebrates, i.e. the well known distinction between vertebrate AChE and BChE. It is, therefore, important to note that none of the nematode AChE classes can be associated with a particular class of vertebrate enzyme or with the well characterized insect AChE. Rather, the nematode AChEs are most closely related one to the other, consistent with the notion that the gene duplications responsible for creating three nematode AChE genes occurred after the split of ancestral nematodes from the hypothetical common ancestor of nematodes and the more advanced coelomates.

It should also be noted that the nematode ace genes have not been cloned; whether they are homologous to vertebrate AChE and BChE and insect AChE and thereby properly a member of the ChE gene family has, therefore, not yet been determined.

The functions of nematode AChEs: analysis of mutant phenotypes.

Strains carrying any single ace mutation and ace-1;ace-3 and ace-2;ace-3 double mutant strains display no significant alteration in behavior (Johnson, et al., 1981; Culotti, et al., 1981; Johnson et al., 1988). The ace-1;ace-2 strain (which retains expression only of class C AChE) is, however, characteristically slow and uncoordinated (ACE-UNCoordinated phenotype) (Culotti, et al., 1981) and the triple mutant strain is paralyzed and dies as a larva (ACE-LEThal phenotype) (Johnson, et al., 1988). The analysis of ace mutant phenotypes thus indicates i) that no single ace gene codes for an essential function but that, nonetheless, ii) the ace genes, together perform an essential function.

Interestingly, quadruple mutant strains carrying a

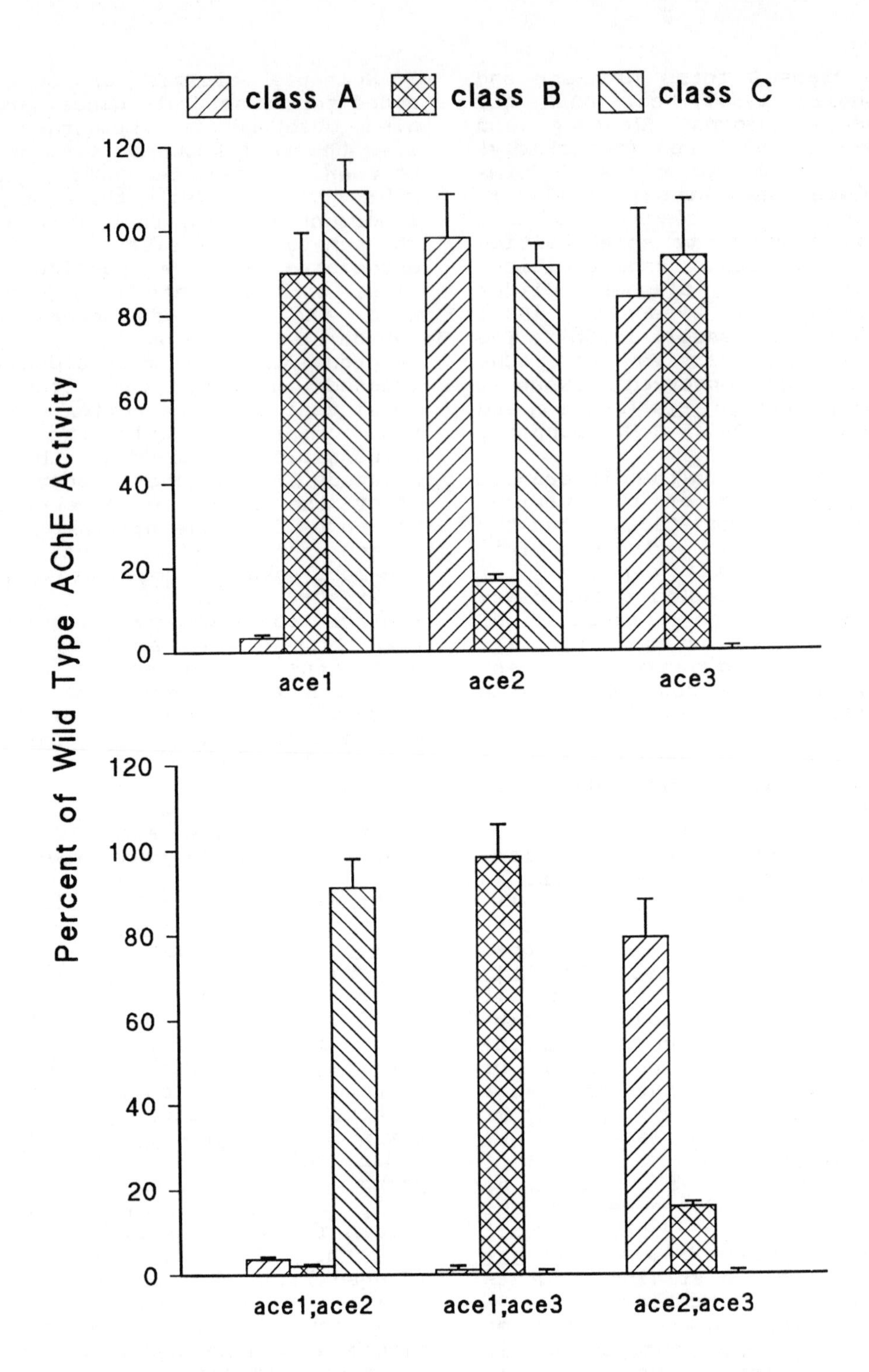

Fig. 1. Class-specific AChE activities in strains bearing ace mutations. (Adapted from data in Table 1, Johnson, et al., 1988.)

lowering-of-function mutation in the gene complex cha-1/unc-17 (the presumptive structural gene for choline acetyltransferase; Rand and Russell, 1984; Rand, 1989) in addition to mutations in all three ace genes are viable (J. Rand and C.D. Johnson, unpublished). The observation that the ACE-LET phenotype is suppressed by mutations which lower ACh synthesis, implys that the essential function performed by the ace gene products involves the hydrolysis of ACh. It further implys that the ace genes products themselves perform no other essential functions.

The functions of nematode AChEs: analysis of genetic mosaics.

The tissue and regional localization of the various AChE classes and forms have been analyzed in Ascaris (Johnson and Stretton, 1980; C.D. Johnson, unpublished). In brief, these studies show that all three classes of AChE are concentrated on muscle cells and that class C AChE is selectively located in the animal's coelomic fluid. Only small amounts of AChE are found within the nervous system.

These results are consistent with studies of genetic mosaics in C. elegans (Herman, 1984) which indicate that the loss of ace gene expression in muscle cells is responsible for both the ACE-UNC (Herman and Kari, 1985) and ACE-LET (Johnson et al., 1988) phenotypes. Somewhat surprisingly, loss of ace gene expression within all nerve cells produced no noticeable phenotype (Herman and Kari, 1985; Johnson et al., 1988). These studies thus demonstrate that the function of nerve-nerve cholinergic synapses in C. elegans does not require the expression the three ace genes in either the presynaptic or the postsynaptic cell. One explanation for this result is that ACh escapes from nematode nerve-nerve synapses exclusively by diffusion, to be hydrolyzed eventually by non-synaptic, presumable muscle-derived AChE. (There have been suggestions that ACh released at cholinergic synapses in the vertebrate brain is similiarly hydrolyzed by remote, i.e. non-synaptic, AChE.) Alternately AChE might be localized at these synapses, through the capture of enzyme synthesized and released by nearby muscle cells.

References

Culotti, J.G., von Ehrenstein, G., Culotti, M., and Russell, R.L., Genetics, **1981**, 97, 281-305.

Herman, R.K., Genetics, **1984**, 108, 165-180.

Herman, R.K. and Kari, C.K., Cell, **1985**, 40, 509-514.

Johnson, C.D. and Russell, R.L., Anal. Biochem., **1975**, 64, 229-238.

Johnson, C.D. and Russell, R.L., J. Neurochem., **1983**, 41, 30-46.

Johnson, C.D. and Stretton, A.O.W. In Nematodes as Biological Models; Zuckerman, B.M., Ed.; Academic Press: New York, NY, 1980, Vol. 1; pp. 159-195.

Johnson, C.D. and Stretton, A.O.W., J. Neurosci., **1985**, 5, 1984-1992.

Johnson, C.D., Duckett, J.G., Culotti, J.G., Herman, R.K., Meneely, P.M., and Russell, R.L., Genetics, **1981**, 97, 261-279.

Johnson, C.D., Rand, J.B., Herman, R.K., Stern, B.D., and Russell, R.L., Neuron, **1988**, 1, 165-173.

Kolson, D.L. and Russell, R.L., J. Neurogenet., **1985a**, 2, 69-91.

Kolson, D.L. and Russell, R.L., J. Neurogenet., **1985b**, 2, 93-110.

Lewis, J.A., Elmer, J.S., Skimming, S., McLafferty, S., Fleming, J., and McGee, T. J. Neurosci. **1987**, 7, 3059-3071.

Rand, J.B., Genetics, **1989**, 122, 73-80
Rand, J.B. and Johnson, C.D. Anal. Biochem., **1981**, 116, 361-371.
Rand, J.B. and Russell, R.L., Genetics, **1984**, 106, 227-248.
Rand, J.B. and Russell, R.L., J. Neurochem., **1985**, 44, 189-200.
Stretton, A.O.W, Davis, R.E., Angstadt, J.D., Donmoyer, J.E., and Johnson, C.D. Trends Neurosci., **1985**, 8, 294-300.
Walrond, J.P., Kass, I.S., Stretton, A.O.W., and Donmoyer, J.E., J. Neurosci., **1985**, 5, 1-8.

Drosophila Acetylcholinesterase Structure

Annick Mutero and **Didier Fournier**, INRA, BP 2078, 06606 Antibes, France.

Drosophila AChE has been characterized as an amphiphilic dimer linked to the membrane via a glycolipid anchor. The two active subunits are each composed of two polypeptides, 16 and 55 kDa. The 55 kDa polypeptide bears the serine of the active site, the glycolipid anchor and the cysteine responsible for the dimerization.

Acetylcholinesterase (AChE, EC 3.1.1.7) is associated with cholinergic synapses where it rapidly terminates neurotransmission by hydrolyzing acetylcholine. In contrast with vertebrates in which AChE is found at neuromuscular junctions and displays distinct molecular forms (Massoulié and Bon, 1982), most of the insects AChE is found in the central nervous system and possesses only one main form (Lenoir-Rousseaux, 1985). Although several isozymes can also be detected, they originate from the main form by protease and lipase digestion. Substrate specificity of insect AChE is intermediate between those of mammalian AChE and butyrylcholinesterase (BuChE, EC 3.1.1.8) (Gnagey *et al.*, 1987; for review of catalytic properties of insects AChE, see Toutant, 1989).

Various interests led to develop studies on insect AChE :

(1) The important quantities of enzyme which can easily be obtained from insect heads were propicious for biochemical studies.

(2) AChE is the target site for organophosphorus and carbamate classes of pesticides which react as ACh analogues and complex the enzyme. Some pest species have been found to possess altered AChE less sensitive to inhibition by these insecticides (Oppenoorth and Welling, 1976).

(3) Several mutants had been studied in *Drosophila melanogaster* which permitted to localize genetically the *Ace* locus (Hall and Kankel, 1976; Gausz *et al.*, 1986; Nagoshi and Gelbart, 1987).

Several useful tools are now available for insect AChE studies. The protein is easily purified to homogeneity by affinity chromatography (Steele and Smallman, 1976; Gnagey *et al.*, 1987), and polyclonal antibodies raised against the whole protein or against specific regions have been obtained. The gene of *Drosophila melanogaster* AChE was cloned by chromosome walking on the *Ace* locus previously localized (Bender *et al.*, 1983). The coding regions were sequenced and the protein sequence deduced (Hall and Spierer, 1986).

Functional *Drosophila* AChE has been expressed in three systems :

(1) Transient expression of *Drosophila* AChE was obtained in *Xenopus* oocytes (Fournier *et al.*, unpublished result). The protein produced in this system is active, glycosylated, but not processed for the G-PI anchor.

(2) *Drosophila Ace* cDNA was expressed in a Baculovirus-lepidoteran cells expression system (see abstract by Cerutti *et al.* in this issue). We purified and analyzed the protein produced in this system. It revealed molecular forms analogous to those found for AChE purified from *Drosophila* heads.

(3) A minigene was constructed lacking the intronic regions (Fournier *et al*, 1989) and flanked with 1.5 kb genomic sequence upstream from the start of transcription thought to contain all major promoter elements of the gene. By injection in *Drosophila* via P-mediated transformation, F. Hoffman succeeded to rescue *Ace* lethal mutants and obtained a tissue-specific expression (Hoffman *et al.*, unpublished result).

Subunit Composition

Apparent molecular weight of native AChE from insect has been estimated to 150 kDa by gel filtration, Ferguson plot or pore limit electrophoresis in the presence of anionic detergents (Steele and Smallman, 1976; Fournier *et al.*, 1987). The putative protein encoded by the *Ace* cDNA is of 70 kDa.

Analysis on SDS-PAGE of purified *Drosophila* AChE revealed two major polypeptides of 55 and 16 kDa (Fig.1). The protein is thus translated as a precursor of 70 kDa which generates by proteolytic cleavage two polypeptides. The 16 kDa polypeptide corresponds to the N-terminal end of the 70 kDa precursor when the 55 kDa polypeptide corresponds to the C-terminal end. This result was obtained by sequencing the N-terminal end of the 16 kDa polypeptide (Haas *et al.*, 1988) and by using specific polyclonal antibodies raised

0–8412–2008–5/91/0141$06.00/0

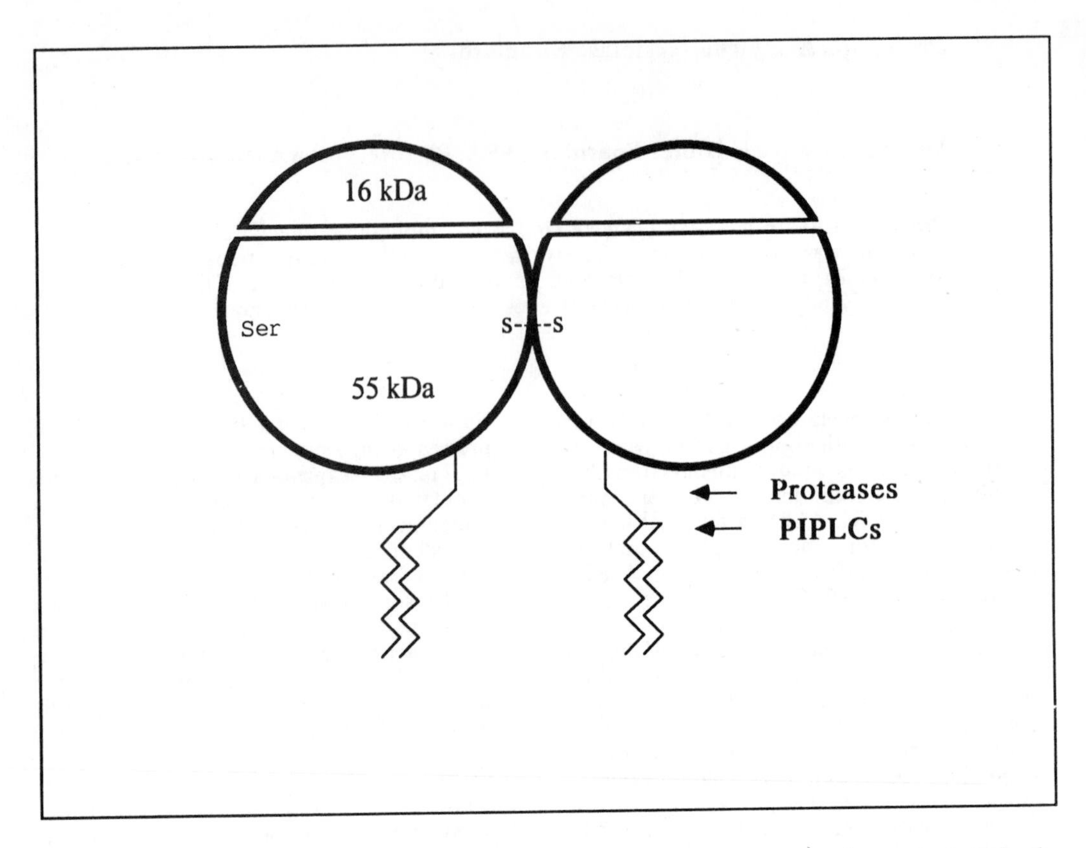

Fig. 1. A model for *Drosophila* acetylcholinesterase structure. The two 55 kDa polypeptides are linked by disulfide bond(s) while the 16 kDa polypeptides are not covalently linked.

against fusion proteins containing either N-terminus (amino acids 2-118) or C-terminus (383-534) portions of the cDNA-deduced protein sequence (Fournier *et al.*, 1988b).

In some purified extracts, a 70 kDa polypeptide is sometimes present. It was tentatively interpreted as being the precursor of the mature AChE because of its suitable molecular weight and its ability to bind [^{3}H]diisopropyl fluorophosphate (DFP) (Gagney *et al.*, 1987). But polyclonal antibodies directed against the 16 kDa or the 55 kDa subunits failed to recognize the 70 kDa polypeptide even when large amounts were present in the preparations (Fournier *et al.*, 1988b). This evidence associated with the weakness of DFP binding and the fact that this polypeptide is not covalently associated with 55 kDa and 16 kDa polypeptides supports the conclusion that this polypeptide is a purification contaminant. The precursor is absent from *Drosophila* extracts even when antiproteolytic agents are added in the extraction buffers. This suggests that the cleavage occurs *in situ* rapidly after translation.

In contrast, the 70 kDa precursor can be observed in SDS-PAGE when AChE is expressed in Baculovirus system (unpublished data). This precursor binds to [^{3}H] DFP and can be recognized by polyclonal antibodies raised against fusion proteins described above. As this form is found in extracts with weak activities, it seems that this precursor is poorly if at all active. Its presence may be explained by a saturation of enzymes involved in the proteolytic process.

The processing of a single AChE precursor into two polypeptides was not mentioned in vertebrates. When compared to vertebrates sequences, the *Drosophila* AChE presents an insertion of 33 amino acids in position 147-180 (Hall and Spierer, 1986). This additional peptide is hydrophilic and supposed to contain the cleavage site(s) of the precursor into 16 and 55 kDa subunits since its location is consistent with apparent molecular weights of the two subunits and no disulfide bond passes through this region. It looks likely that the cleavage site is not unique because two tentatives in sequencing the N-terminus of the 55kDa failed (unpublished result and Rosenberry, personnal communication). Furthermore 16 kDa and 55 kDa polypeptides sometimes appear to be composed of several bands in SDS gels.

In order to obtain some insight on the role of this hydrophilic portion of the *Drosophila* protein, a cDNA deleted for this region was constructed and used in *Xenopus* oocytes expression system. We obtained an active protein composed of one polypeptide with 65kDa of apparent molecular weigth. This result indicates that the supplementary hydrophilic polypeptide is responsible for the proteolytic cleavage found in purified extracts.

Signal Peptide

The cDNA-deduced protein sequence presents a NH_2-terminal peptide sufficiently hydrophobic to play the role of signal peptide as found in membrane associated and exported protein precursors (Hall and Spierer, 1986). It is removed from the mature protein. The N-terminal amino acid of the mature protein was determined by microsequencing of the 16kDa subunit at position Val ^{39} (Fig. 2, Haas *et al.*, 1988).

In physiological conditions, *Drosophila* AChE is anchored via a glycosyl-phosphatidylinositol (G-PI) at the outer surface of cell membranes. The expression in *Xenopus* oocytes of a cDNA lacking the C-terminal hydrophobic peptide extension that signals for the G-PI anchor attachment led to the secretion of the protein induced in the incubation medium. This secretion is due to the N-terminal signal peptide present in the construction which directs the protein into the endoplasmic reticulum.

Dimeric Protein

At all development stages, the major molecular form is an amphiphilic membrane-bound dimer (G2), but a significant proportion of amphiphilic monomer is also observed in larvae and young pupae which could represent the precursor of the G2 form (Arpagaus *et al.*, 1988). The amphiphilic dimer was shown to be converted into hydrophilic dimer and monomer on autolysis of the extract. Partial reduction of purified dimeric forms by 2-mercaptoethanol or dithiotreitol gave rise to an amphiphilic monomeric active form (Toutant *et al.*., 1988).

The 55 kDa subunit was shown to possess sulfhydryl group(s) involved in the inter-subunit linkage by two-dimensional denaturing electrophoresis, the first dimension being performed in non-reducing conditions and the second one in the presence of 2-mercaptoethanol. In these conditions, a 110 kDa polypeptide is reduced into the 55 kDa polypeptide (Fournier *et al.*, 1988b).

There are 9 cysteine residues in the *Drosophila* AChE sequence at positions 104, 131, 328, 330, 345, 480, 598 and 615 and 640. The Cys^{640} does not exist in the mature protein since it is situated in the C-terminal hydrophobic peptide replaced by the G-PI anchor.

The amino acid sequence deduced from *Drosophila* AChE gene was compared with *Torpedo* AChE (Hall and Spierer, 1986) and afterwards with human BuChE (Prody *et al.*.,1987; Lockridge *et al.*.,1987a). *Drosophila* AChE exhibits approximatively 30% overall residue identity with vertebrate ChEs. Important structural and fonctional features such as cysteine residues involved in intrachain disufide bonds or active site regions are conserved among these proteins.

The cysteine residues involved in intra-chain disulfide bonds were determined in *Torpedo californica* AChE (MacPhee-Quigley *et al.*, 1986) and the human BuChE (Lockridge *et al.*, 1987b). The two proteins exhibit three internal disulfide bonds involving cysteines in similar locations and the same number of amino acids within each disulfide loop. In peptidic sequence alignment, cysteine residues were found to be conserved in *Drosophila* AChE (Hall and Spierer, 1986). It is likely that all these proteins share an identical folding pattern. We deduced putative positions of internal disulfide bonds in *Drosophila* AChE between Cys at positions 104 and 131; 330 and 345; 480 and 598. Cys^{328} and Cys^{615} are thus available for interchain linkages (Fig.2).

Torpedo AChE and human BuChE asymmetric forms have an interchain disulfide bond near their carboxyl terminus in position 572 and 571, respectively suggesting that Cys^{615} is involved in *Drosophila*.

In order to test this hypothesis, we independently mutagenized the two free cysteines. In both cases preliminary results indicates that this two cysteines are not alone responsible for the dimerization.

G-PI Anchor

The *Drosophila melanogaster* AChE was shown to be anchored to the membrane via a G-PI anchor as previously described in human and bovine erythrocyte AChEs (Haas *et al.*, 1986; Roberts and Rosenberry, 1986; Roberts *et al.*, 1987) and *Torpedo* AChE (Futerman *et al.*, 1985). The binding of insect AChE to detergent was first shown by Arpagaus and Toutant (1985). Then, Gnagey *et al.* (1987) found ethanolamine and glucosamine in purified AChE from *Drosophila* suggesting the presence of a G-

```
MAISCRQSRV LPMSLPLPLT IPLPLVLVLS LHLSGVCGVI DRLVVQTSSG    50
PVRGRSVTVQ GREVHVYTGI PYAKPPVEDL RFRKPVPAEP WHGVLDATGL   100
SATCVQERYE YFPGFSGEEI WNPNTNVSED CLYINVWAPA KARLRHGRGA   150
NGGEHPNGKQ ADTDHLIHNG NPQNTTNGLP ILIWIYGGGF MTGSATLDIY   200
NADIMAAVGN VIVASFQYRV GAFGFLHLAP EMPSEFAEEA PGNVGLWDQA   250
LAIRWLKDNA HAFGGNPEWM TLFGESAGSS SVNAQLMSPV TRGLVKRGMM   300
QSGTMNAPWS HMTSEKAVEI GKALINDCNC NASMLKTNPA HVMSCMRSVD   350
AKTISVQQWN SYSGILSFPS APTIDGAFLP ADPMTLMKTA DLKDYDILMG   400
NVRDEGTYFL LYDLIDYFDK DDATALPRDK YLEIMNNIFG KATQAEREAI   450
IFQYTSWEGN PGYQNQQQIG RAVGDHFFTC PTNEYAQALA ERGASVHYYY   500
FTHRTSTSLW GEWMGVLHGD EIEYFFGQPL NNSLQYRPVE RELGKRMLSA   550
VIEFAKTGNP AQDGEEWPNF SKEDPVYYIF STDDKIEKLA RGPLAARCSF   600
WNDYLPKVRS WAGTCDGDSG SASISPRLQL LGIAALIYIC AALRTKRVF    649
```

Fig.2. Primary structure of *Drosophila* acetylcholinesterase. The 33 amino-acids which are absent in vertebrate cholinesterases are underlined. The arrows indicate the N-terminal end and the presumed C-terminal amino-acid of the mature protein. There are two free cysteines at positions 328 and 615. The stars mark the three amino-acids supposedly involved in the catalytic triad. The rings indicate the five potential Asn glycosylation sites.

PI anchor. This anchor was then shown to be sensitive to phosphatidylinositol-phospholipase C (PI-PLC) from *Bacillus cereus* or *Trypanosoma brucei* by Triton X-114 partitioning and electrophoresis in non denaturing gels (Fournier *et al.*, 1988a). The PI-PLC sensitivity of *Drosophila* AChE was confirmed by labeling experiment of the protein by [^{125}I] TID, a photoactivatable affinity probe specific of the lipid moiety of G-PI anchored protein which was selectively removed by PI-PLC (Haas *et al.*, 1988). In addition, the PI-PLC digestion of *Drosophila* AChE was shown to uncover a complex carbohydrate, the cross-reacting determinant antigen (CRD) using CRD-specific antibodies. The CRD antigen had originally been described on the soluble form of the variant surface glycoprotein (VSG) of *Trypanosoma brucei* (Cardoso de Almeida and Turner, 1983) and was also unmasked in other G-PI anchored proteins after PI-PLC treatment. Anti-CRD antibodies also recognized the hydrophilic dimeric form (G_2s) present in head *Drosophila* extracts suggesting the existence of an endogenous phospholipase (Fournier *et al.*, 1988a).

Protein sequence deduced from cDNA exhibits a C-terminal hydrophobic polypeptide of 30 amino acids. Such a C-terminal hydrophobic extension was shown to occur in all G-PI anchored protein precursors (see for reviews, Cross, 1987; Low, 1987; Ferguson and Williams, 1988). It is thought to be used as a temporary anchorage to the membrane before its rapid exchange to a G-PI anchor. The proof of the cleavage of the C-terminal hydrophobic peptide is given by polyclonal antibodies raised against a fusion protein containing the hydrophobic peptide sequence. These antibodies recognized the protein produced in *Xenopus* oocytes which is not processed for the G-PI anchor but failed to recognize the protein purified from *Drosophila* head. By analogy with *Torpedo* AChE, the C-terminal amino acid amide-linked to the ethanolamine is supposed to be the Cys^{615} in *Drosophila* (Haas *et al.*, 1988).

The expression of a cDNA truncated for the C-terminal hydrophobic peptide in the Baculovirus led to a water-soluble protein. This confirms the role of this sequence as signal in the anchor attachment (Cerutti *et al.*, unpublished results). Mutagenesis of the Cys^{615} residue to an Arg residue does not affect the G-PI anchorage of the protein and allows the rescue of lethal mutants.

Glycoprotein

AChE purified from *Drosophila* heads is a glycoprotein that binds to ConcavalineA-sepharose. The deduced sequence of the protein presents five potential sites of Asn-linked glycosylation at positions 126, 174, 331, 531 and 569 (Hall and Spierer, 1986). Mutagenesis of Asn^{126} or Asn^{174} to Asp residues did not drastically affect the AChE activity when expressed in *Xenopus* oocyte.

Aknowledgements - We wish to thank Drs J.P. Toutant for critical reading of the manuscript.

Arpagaus, M., Toutant, J.P., *Neurochem. Int.*, **1985**, *7*, 793-804.

Arpagaus, M., Fournier, D., Toutant, J.P., *Insect. Biochem.*, **1988**, *6*, 539-549.

Bender, W., Spierer, P., Hogness, D.S., *J. Mol. Biol.*, **1983**, *168*, 17-33.

Cardoso de Almeida, M.L., Turner, M.J., *Nature*, **1983**, *302*, 349-352.

Cross, G.A.M., *Cell*, **1987**, *48*, 179-181.

Ferguson, M.A.J., Williams, A.F., *Ann. Rev. Biochem.*, **1988**, *57*, 285-320.

Fournier, D., Cuany A., Bride J.M., Bergé J.B., *J. Neurochem.*, **1987**, *49*, 1455-1461.

Fournier, D., Bergé, J.B., Cardoso de Almeida, M.L., Bordier C., *J. Neurochem.*, **1988**a, *50*, 1158-1163.

Fournier, D., Bride, J.M., Karch, F., and Bergé, J.B., *FEBS Lett.*, **1988**b, *238*, 333-337.

Fournier, D., Karch, F., Bride, J.M., Hall, L.M.C., Bergé, J.B. and Spierer, P., *J. Mol. Biol.*, **1989**, *210*, 15-22.

Futerman, A.H., Low, M.G., Michaelson, D.M., Silman, I., *J. Neurochem.*, **1985**, *45*, 1487-1494.

Gausz, J., Hall, L.M.C., Spierer, A., Spierer, P., *Genetics*, **1986**, *112*, 65-78.

Gnagey, A.L., Forte, M., Rosenberry, T.L., *J. Biol. Chem.*, **1987**, *262*, 1140-1145.

Haas, R., Brandt, P.T., Knight, J. and Rosenberry, T.L., *Biochemistry*, **1986**, *25*, 3098-3105.

Haas, R., Marshall, T.C., Rosenberry, T.L., *Biochemistry*, **1988**, *27*, 6453-6457.

Hall, J.C., Kankel, D.R., *Genetics*, **1976**, *83*, 517-535.

Hall, L.M.C., Spierer, P., *EMBO J.*, **1986**, *5*, 2949-2954.

Lenoir-Rousseaux, J.J., *C. R. Soc. Biol.* **1985**, 179, 741-747.

Lockridge, O., Bartels, C.F., Vaughan, T.A., Wong, C.K., Norton, S.E. and Johnson, L.L., *J. Biol. Chem.*, **1987**a, *262*, 549-557.

Lockridge, O., Adkins, S., La Du, B.N., *J. Biol. Chem.*, **1987**b, *262*, 12945-12952.

Low, M.G., *Biochem. J.*, **1987**, *244*, 1-13.

MacPhee-Quigley, K., Vedvick, T.S.,Taylor, P., Taylor, S., *J. Biol. Chem.*, **1986**, *261*, 13565-13570.

Massoulié, J., Bon, S., *Ann. Rev. Neurosci.* **1982**, *5*, 57-106.

Nagoshi, R.N., Gelbart, W.M., *Genetics*, **1987**, *117*, 487-502.

Oppenoorth, F.J., Welling, W., *Insect Biochemistry and Physiology*, C.F. Wilkinson, Ed., Plenum press (N Y) 1976, 507.

Prody, C.A., Zevin-Sonkin, D., Gnatt, A., Goldberg, O., Soreq, H., *Proc. Natl. Acad. Sci. USA*, **1987**, *84*, 3555-3559.

Roberts, W.L., Rosenberry, T.L., *Biochemistry*, **1986**, *25*, 3091-3098.

Roberts, W.L., Kim, B.H., Rosenberry, T.L. *Proc. Natl. Acad. Sci. USA*, **1987**, *84*, 7817-7821.

Steele, R.W., Smallman, B.N., *Bioch. Biophys. Acta.*, **1976**, *445*, 147-157.

Toutant, J.P., Arpagaus, M., Fournier, D., *J. Neurochem.*, **1988**, *50*, 209-218.

Toutant, J.P., *Progress Neurobiol.*, **1989**, *32*, 423-446.

REGULATION OF ACETYLCHOLINESTERASE EXPRESSION IN ELECTRICALLY EXCITABLE CELLS

Richard L. Rotundo, Cristina Fernandez-Valle, Anna M. Gomez, Francesca Barton, and Philippe Leff. Department of Cell Biology and Anatomy, University of Miami School of Medicine, 1600 N.W. 10th Ave., Miami, Florida 33101

The early events in AChE biogenesis are summarized with emphasis on the assembly and fate of the AChE polypeptide chains. Recent studies on the regulation of AChE biogenesis by muscle activity suggest that post-translational mechanisms play a major role in regulating the distribution of the multiple oligomeric forms of this enzyme expressed in muscle. Furthermore, the assembly of the oligomeric AChE forms can be shown to be compartmentalized in multinucleated skeletal muscle fibers. These results are discussed in relation to the overall regulation of synaptic components at the neuromuscular junction.

The multiple forms of acetylcholinesterase expressed in vertebrates are encoded by a single gene, yet each cell type exhibits its own characteristic pattern of enzyme forms. This pattern depends not only upon the specific tissue but also on the activity state of the cells. Furthermore, different specialized plasma membrane domains of a given cell may present different versions of this enzyme as well. Research in our laboratory has focused on the mechanisms underlying the biogenesis of these multiple forms of AChE in electrically excitable cells and the molecular mechanisms underlying their regulation and localization to specific functional domains on the cell surface.

All Oligomeric Forms in Avian Nerves and Muscle are Encoded by a Single Gene:

Analysis of the two major AChE polypeptide chains expressed in avian muscle suggested the existence of allelic variants (Randall et al., 1987, Rotundo et al., 1988). Mating studies subsequently showed that the two AChE variants were co-dominant alleles, differing by approximately 10 KDa apparent MW by SDS gel electrophoresis, yet kinetically and immunologically indistinguishable. Analysis of the allelic polypeptide composition of the multiple oligomeric forms expressed in nerves and muscle cells indicated that all AChE forms were comprised of the same subunits regardless of whether they were derived from animals homozygous for either the α or ß alleles, or heterozygotes (Rotundo et al., 1988). Together with the studies from the laboratories of Drs. Taylor and Massoulié, these studies indicated that all AChE forms expressed in nerves and muscle were encoded by a single gene in these species.

Biogenesis of AChE in Muscle:

The catalytic subunit of AChE is a glycoprotein and hence translated on the rough endoplasmic reticulum. Following co-translational glycosylation, a subset of the polypeptide chains are assembled into catalytically active dimeric and tetrameric forms. These initial events in the assembly of the oligomeric forms also occurs in the rough

0–8412–2008–5/91/0146$06.00/0

endoplasmic reticulum. However, only a small fraction of the newly synthesized monomers are detected using enzymatic assays (Rotundo, 1988). The AChE subunits are transported to the early elements of the Golgi apparatus with a half-life of about 20 minutes at which time modification of the asparagine-linked oligosaccharides begins (Rotundo, 1984). Although all newly-synthesized polypeptide chains are transported to the Golgi, only about 20-30% are catalytically active. The remainder are targeted for rapid intracellular degradation by a non-lysosomal mechanism with a half-life of approximately 45 minutes (Rotundo, et al., 1988).

The catalytically active dimeric and tetrameric forms are then transported through the Golgi where they become endoglycosidase H resistant, receive terminal sugars, and are transported to the plasma membrane where they become associated with the cell surface or are secreted into the medium. In addition, a subset of catalytically active oligomers are assembled in the trans Golgi apparatus with the collagen-like tail to produce the asymmetric form (Rotundo, 1984). The entire process from synthesis to release at the plasma membrane requires about 2.5 hours, and is summarized in Figure 1.

Compartmentalization of AChE Biogenesis in Skeletal Muscle:

Recent studies from our laboratory indicate that the AChE mRNAs expressed in multinucleated muscle fibers are translated and the polypeptides chains assembled in the RER and Golgi surrounding the nucleus of transcription (Rotundo, 1989). Mosaic myotubes derived by mixing homozygous myoblasts expressing either of two allelic AChE polypeptide chains as molecular markers preferentially assembled homodimeric enzyme molecules rather than the randomly assembled heterodimeric forms expressed in heterozygous myotubes (Figure 2). These studies indicate that mRNAs

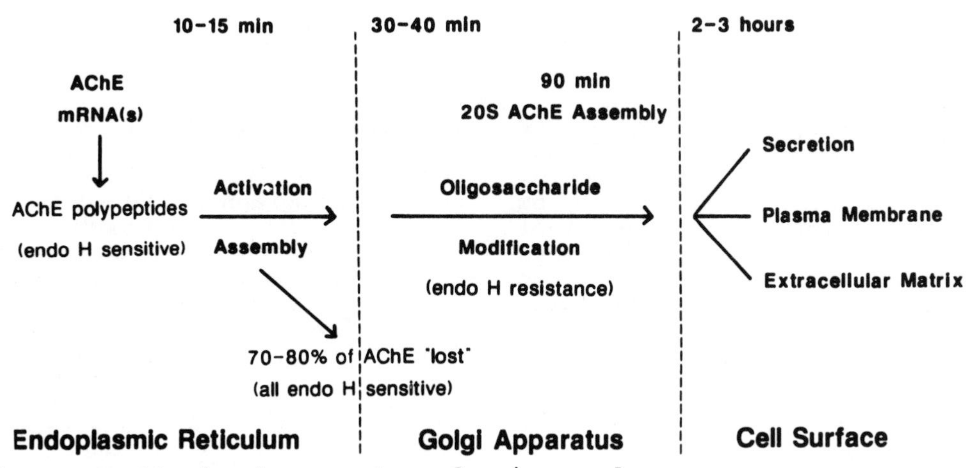

Fig. 1. Synthesis, transport, and assembly of AChE oligomeric forms in muscle.

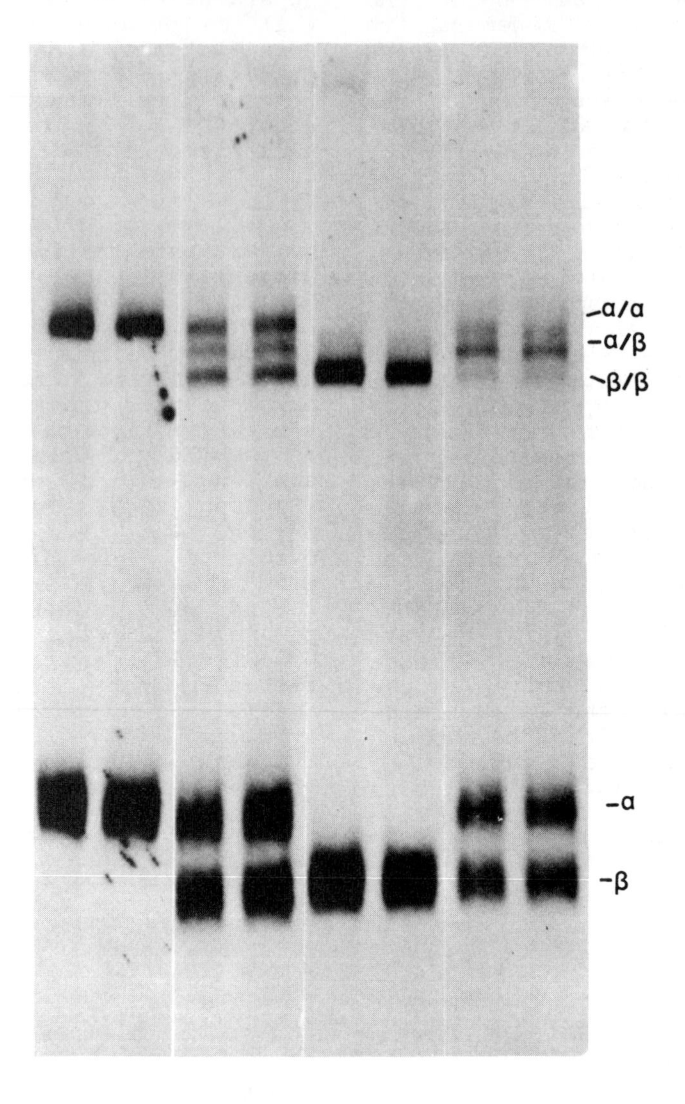

Fig. 2. Expression of AChE allelic polypeptides in tissue-cultured quail muscle cells. Muscle cultures were made from myoblasts derived from embryos of the α/α, α/β, or β/β genotypes (lanes a, b, and het.) or from mixtures of α/α + β/β myoblasts (mosaic myotubes, lane a + b). Following fusion the AChE polypeptides were labeled with ^{35}S-methionine, immunoprecipitated, and analyzed by SDS gel electrophoresis under non-reducing conditions. The gel was then fluorographed. Lower bands show monomeric AChE polypeptides; upper bands show distribution of the disulfide-bonded α-α, α-β, and β-β dimeric enzyme forms. Figure reproduced with permission from Rotundo, 1990.

encoding membrane proteins in muscle cells are preferentially translated and assembled in the vicinity of the nucleus of transcription. These observations have important implications for the local regulation of gene expression at the neuromuscular junction in that they demonstrate the very limited diffusion of locally-expressed mRNAs once transcribed.

Regulation of AChE Biogenesis by Muscle Activity:

When tissue-cultured quail skeletal muscle cells are treated with tetrodotoxin to inhibit muscle contraction, there is a 30-50% decrease in AChE activity and a complete loss of the collagen-tailed asymmetric form (Rieger et al., 1980; Brockman et al., 1984; De La Porte et al., 1984; Rubin, 1985; . This decrease is not due to changes in the rates of AChE polypeptide chain synthesis or degradation, but rather to an inhibition of asymmetric AChE assembly in the Golgi apparatus with a quantitative compensatory increase in the amount of enzyme secreted (Figure 3). In contrast, expression of AChE mRNAs increases several fold indicating that transcription and translation are oppositely regulated in this cell type

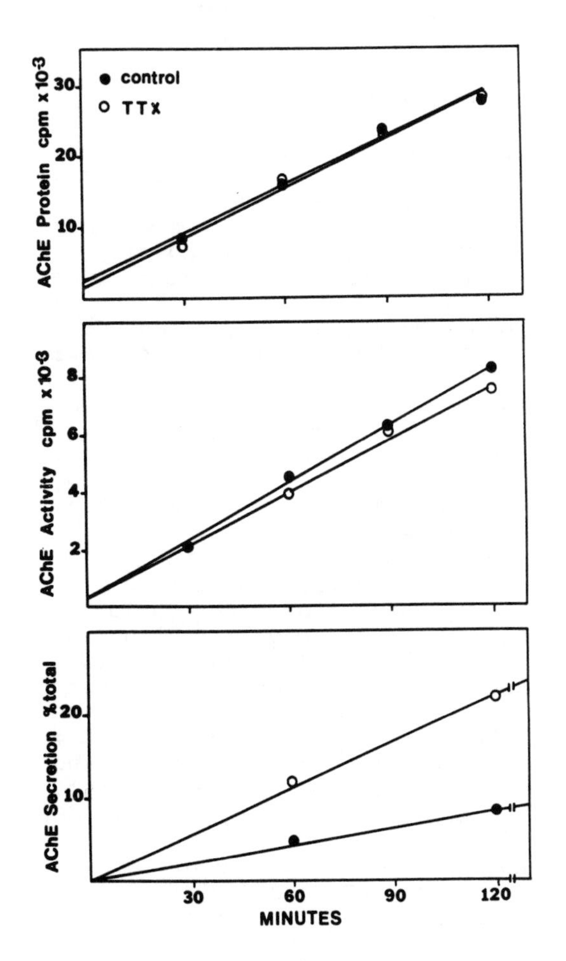

Fig. 3. Effects of tetrodotoxin (TTX) on rates of incorporation of ^{35}S-methionine into AChE protein (upper panel), appearance of catalytically active enzyme molecules (middle panel), and secretion of catalytically active AChE (lower panel) by tissue-cultured myotubes. Figure modified after Fernandez-Valle and Rotundo, 1989).

(Randall, Fernandez-Valle, and Rotundo, unpublished obs.).

Current Research and Future Directions:

Using a cDNA encoding the Torpedo AChE generously provided by Dr. P. Taylor (Schumacher et al., 1986) we have recently isolated several partial cDNAs from a quail cDNA library. A comparison of the deduced amino acid sequence with the published amino acid sequences of AChE and BuChE indicated that these cDNAs encode the catalytic subunit which associates with the collagen-like tail. Southern blot analysis of quail genomic DNA confirmed that a single gene encoding AChE exists in this species. However, Northern blot analysis indicates the existence of at least two major transcript classes in nerves and muscle of about 5.5 and 6.0 kb. At present we do not know the significance of these multiple transcripts, or whether additional ones will be found.

Preliminary studies from our laboratory show that the AChE mRNA increases dramatically during the period of myoblast fusion and skeletal muscle differentiation in tissue-cultured quail muscle cells. This period of rapid increase is then followed by one of dramatic decrease as the myotubes begin to actively contract. Incubation of the cells with tetrodotoxin, which inactivates voltage-dependent sodium channels, results in a large increase in the expression of AChE mRNA in a manner which parallels that of ACh receptor α-subunit mRNA suggesting that these two genes are coordinately regulated at the transcriptional level.

Summary:

The regulation of AChE occurs at the levels of transcription, translation, and post-translationally as well as being compartmentalized in muscle. The contributions of these multiple levels of regulation to the overall expression of AChE in skeletal muscle suggests that the AChE and acetylcholine receptors may be coordinately regulated during the development and differentiation of skeletal muscle fibers in a positive manner, and that additional levels of regulation occur subsequently resulting in divergent regulation in the differentiated state. Additional studies on the compartmentalized expression and assembly of AChE in muscle suggests that the regulation of AChE by neuronal influences and muscle activity can occur in a very limited region of the sarcoplasm, such as the neuromuscular junction

References:

Brockman, S.K.; Usiak, M.F.; and Youngkin, S.G. J. Neurosci. **1984**, 4: 131-140.

De La Porte, S.; Vigny, M.; Massoulié, J.; and Koenig, J. Dev. Biol. **1984**, 106: 450-456.

Fernandez-Valle, C.; Rotundo, R.L. J. Biol. Chem. **1989**, 264: 14043-14049.

Fernandez-Valle, C.; Rotundo, R.L. (In preparation)

Maulet, Y.; Camp, S.; Gibney, G.; Rachinsky, T.L.; Ekstrom, T.J.; Taylor, P. Neuron **1990**, 4: 289-301.

Randall, W.R.; Tsim, K.W.; Lai, J.; Barnard, E.A. Eur. J. Biochem. **1987**, 164: 95-102.

Rieger, F.; Koenig, J.; Vigny, M. Dev. Biol. **1980**,76: 358-365.

Rotundo, R.L. Proc. Natl. Acad. Sci. **1984**, 81: 479-483.

Rotundo, R.L.; Gomez, A.M.; Fernandez-Valle, C.; and Randall, W.R. Proc. Natl. Acad. Sci. **1988**, 85: 7805-7809.

Rotundo, R.L. J. Biol. Chem. **1988**, 263: 19398-19406.

Rotundo, R.L.; Thomas, K.; Porter-Jordan, K.; Benson, R.J.J.; Fernandez-Valle, C.;

and Fine, R.E., J. Biol. Chem. **1989**, 264: 3146-3152.
Rotundo, R.L. J. Cell Biol. **1990**, 110: 715-719.
Rubin, L. L. Proc. Natl. Acad. Sci. **1985**, 82: 7121-7125.
Schumacher, M.; Camp, S.; Maulet, Y.; Newton, M.; MacPhee-Quigley, K.; Taylor, S.S.; Friedman, T.; Taylor, P. Nature **1986**, 319: 407-409.
Sikorav, J.L.; Duval, N.; Anselmet, A., Bon, S., Krejci, E.; Legay, C.; Osterlund, M.; Reimund, B.; Massoulié, J. EMBO J. **1988**, 7: 2983-2993.

Cellular Expression of Murine AChE from a Cloned Brain Transcript

William R. Randall, Department of Pharmacology and Experimental Therapeutics, University of Maryland School of Medicine and Medical Biotechnology Center, Baltimore, Maryland 21201, U.S.A. FAX (301)328-3991

Catalytically active acetylcholinesterase was expressed in two cell types from a cloned mouse brain transcript encoding the asymmetric catalytic subunit. The expressed enzyme in *Xenopus* oocytes required Triton X-100 for extraction and was found only in a membrane fraction. In contrast, when the clone was expressed in 293 cells the enzyme localized in a membrane, a cytosolic and a medium fraction. The membrane-associated enzyme required detergent for extraction and interacted with it when analyzed on sucrose gradients. The cytosolic forms were extracted without detergent yet interacted with it on sucrose gradients. The secreted acetylcholinesterase forms were not detergent sensitive.

The characterization of oligomeric molecular forms of acetylcholinesterase (AChE) has been the subject of intensive research for over twenty years (Massoulié and Bon 1983, Rosenberry 1985, Rotundo 1987). The earliest studies documenting the multiple forms (Massoulié and Rieger, 1969, Wilson *et al.*, 1969) provided a foundation for later work showing precursor product relationships (Lazar *et al.*, 1984, Brockman *et al.* 1986), the localization of specific forms to different cellular compartments and studies describing their biogenesis and metabolism (Rotundo and Fambrough, 1980, Rotundo 1988). However, only recently has it been possible with the development of molecular cloning techniques to examine how the gene and its RNA transcripts may contribute to the generation of their diversity. Genomic DNA analysis and genetic lineage studies from different species suggest that all molecular forms in higher vertebrates are a product of a single gene (Sikorav *et al.*, 1987, Maulet *et al.*, 1990, Randall *et al.*, 1987, Rotundo *et al.*, 1988). Thus, the generation of diversity is a product of post-transcriptional and post-translational events. It is now well established that in the *Torpedo* electric organ, two classes of AChE, the glyco-phospholipid anchored and asymmetric catalytic subunit, are products of alternatively spliced exons encoding the carboxyl-terminal end (Gibney *et al.*, 1988, Schumacher *et al.*, 1988, Sikorav *et al.*, 1988).

In order to examine how post-translational modifications contribute to the diversity, we have chosen to express a cDNA, derived from mouse brain RNA, encoding the entire open reading frame of AChE. The principal questions are: (1) to what extent can a single transcript generate multiple molecular forms of AChE and, (2) in what cellular (or extracellular) compartments are they found? We have used two cell types as the host for the expression, the oocyte from *Xenopus laevis* and transiently transfected human 293 cells (kidney tumor derived).

cDNA Cloning of Murine Brain AChE

Four cDNA's encoding AChE were isolated from a mouse brain λgt-10 cDNA library by cross-hybridization with a cDNA probe encoding chicken AChE (homologous to bases 983-1394 in *Torpedo* λACHE-1; Schumacher *et al.*, 1986). One of these, designated MB9.1, was

0–8412–2008–5/91/0152$06.00/0

found to contain the entire open reading frame and was sequenced (Randall, W.R., *J. Biol. Chem.*, submitted, 1990). Comparison of the deduced amino acid sequence with that from *Torpedo* electric organ (Schumacher *et al.*, 1986) showed a carboxyl-terminal sequence homologous to the asymmetric type catalytic subunit (Fig. 1). The overall amino acid homology was 60%.

```
MOUSE-VRRGLRAQTCAFWNRFLPKLLS-541
TORP.-VHQRLRVQMCVFWNQFLPKLLN-533

MOUSE-ATDTLDEAERQWKAEFHRWSSY-563
TORP.-ATETIDEAERQWKTEFHRWSSY-555

MOUSE-MVHWKNQFDHYSKQERCSDL   -583
TORP.-MMHWKNQFDHYSRHESCAEL   -575
```

Fig. 1. Aligned homology of the deduced amino acid carboxyl-terminal sequence for AChE from mouse brain MB9.1 and *Torpedo californica* (TORP.) cDNA's. The sequence of the last genomic exon containing coding region is underlined.

Expression in *Xenopus* oocytes

Poly A^+ RNA, synthesized *in vitro* from MB9.1 using a poly A^+ cartridge vector, was microinjected into *Xenopus* oocytes (2ηg/oocyte) and allowed to incubate for 24 h. The oocytes produced catalytically active AChE when measured by a radiometric assay (Johnson and Russell, 1975). Suprisingly, nearly all of the activity partitioned to the pellet when oocyte membranes were prepared by four steps of repeated washing and pelleting in a "low salt" buffer (see footnote, Table 1). This membrane form was not released upon further treatment with the same buffer containing 1M NaCl indicating that it was not adhering by ionic interactions. Furthermore, a similar result was obtained using defolliculated oocytes suggesting that the enzyme was not merely trapped by an extra cell layer. We detected no enzyme activity secreted into the Barth's medium from the same oocytes. The enzyme

Table 1. Expression of a cDNA Encoding Mouse AChE in *Xenopus* Oocytes

Cell Fraction	Percent of Cellular AChE
Supernatant[§]	4
Pellet[¶]	97
Barth's Medium[@]	none detected

§ Cells were homogenized in 100μl of **"low salt buffer"**: 10mM Hepes, pH 7.5, 150mM NaCl, 5mM EDTA, 5mM EGTA, 5mM N-ethylmaleimide, 30μg/ml leupeptin, 1mg/ml bacitracin, 2mM benzamidine, 30μg/ml lima bean trypsin inhibitor, 100μg/ml aprotinin. Supernatant is from a 100,000xg centrifugation.

¶ Pellets were extracted in the "low salt buffer" containing 1M NaCl and 1% Triton X-100.

@ Barth's medium contained 8μg/ml leupeptin, 2mg/ml bacitracin, 0.04U/ml aprotinin and 2mg/ml EGTA as antiproteases.

exhibited a dose dependent inhibition with BW284c51 characteristic of mouse AChE ($IC_{50} \approx 5x10^{-8}$ M). Conversely, it was relatively insensitive to inhibition by iso-OMPA ($IC_{50} \approx 5x10^{-4}$ M). The enzyme showed a ratio of hydrolysis of acetylthiocholine to butyrylthiocholine of approximately 200 to 1 and an acetylcholine dependent inhibition of catalysis further defining its activity as AChE.

Analysis of AChE molecular forms extracted from microinjected *Xenopus* oocyte membranes using 5-20% sucrose density gradients showed a single monomeric peak at 3.5S in the presence of

Triton X-100. This form interacted with the detergent and aggregated when Triton X-100 was omitted from the gradient and removed from the sample. Thus, by these criteria, the expressed AChE in *Xenopus* oocytes was associated with a component that was inserted into the membrane.

Expression in Transfected 293 Cells

The absence of secreted enzyme from oocytes expressing mouse AChE suggested that this cell type may exhibit unusual features regarding the assembly and processing of the molecular forms. Consequently, we decided to express the same cDNA in the human cell line, 293, after transient transfection. The cDNA was inserted into an expression plasmid immediately downstream of the promoter for the human cytomegalovirus to produce a high, constituitive rate of transcription. The 293 cells within 35mm dishes were transfected using a calcium phosphate precipitation protocol (Gorman, 1985) and incubated for 48 h. prior to analysis. The cultures were partitioned into membrane, "low salt" soluble cytosolic, and medium components in the same manner as the oocytes were prepared. Unlike the oocyte, the majority of AChE activity (85%) was secreted into the medium (Table 2.). Much less activity (3%) was in the membrane fraction. However, this fraction represented a much larger proportion (22%) when only cellular AChE activity was measured (i.e. when enzyme secreted into the medium was not measured).

To further quantitate the proportion of AChE exposed to the cell surface, we used a cell surface assay originally developed by Rotundo and Fambrough (1980) for chick muscle cell cultures. Transfected cultures, cooled to 4°C to prevent secretion of AChE, were incubated in the presence of 1 ml of 0.788 mM [^{3}H]acetylcholine and aliquots (10μl) were removed to a reaction termination mixture at 5 minute intervals. To determine the rate of total cellular hydrolysis of acetylcholine, Triton X-100 was added to 1% after 15 minutes and the

Table 2. AChE Activity of Mouse cDNA Transfected 293 Cells

Cell Fraction	Percent of Total AChE
Supernatant@	12
Pellet	3
Medium	85

@ The buffers used to homogenize the cells (1 ml) and to extract the pellets (0.5 ml) were as described in Table 1.

cultures were sampled at 2 minute intervals thereafter. The percentage of cell surface activity was then calculated as the difference in slopes before and after Triton X-100 addition. The results of four experiments, each from three 35 mm dishes, gave a mean cell surface activity of 18.5%. This value compared well with the percentage of cellular AChE found in membranes using the differential extraction methods (22%) and suggested that about 20% of the enzyme within the transfected cells was localized as a membrane associated ectoenzyme.

Sensitivity to PIPLC

A glycophospholipid linked form of AChE has been previously shown to be released from bovine erythrocytes upon treatment with phosphatidylinositol specific phospholipase C (PIPLC; Futerman *et al.*, 1985, Roberts *et al.*, 1987). To determine if the expressed mouse AChE contained this modification, membranes (100μl, prepared as in Table 2 but resuspended in 40mM sodium phosphate, pH 7.0) were exposed to PIPLC at two concentrations and incubated for 30 minutes at 37°C. The suspended membranes were centrifuged (100,000xg) and the supernatant and extracted pellet were measured for AChE activity. Only minor amounts of activity were released into the supernatant (< 3%) indicating that the membrane associated form was insensitive to the treatment. It is possible that palmitoylation of the protein prevented its release from the membrane as has been shown for a glycophospholipid linked form of AChE from human erythrocytes (Roberts *et al.*, 1988). Mouse erythrocyte AChE is also insensitive to PIPLC (Futerman *et al.*, 1985). Alternatively, a PIPLC-insensitive amphipathic form, isolated from bovine brain, was shown to be anchored by a small 20 kDa polypeptide and may be analogous to the form studied here (Inestrosa *et al.*, 1987). In support, the deduced carboxyl-terminal amino acid sequence of the mouse cDNA is not homologous to the glycophospholipid form in *Torpedo* and would likely encode a form that is not attached by a phosphoinositide linkage.

Amphipathicity of the Molecular Forms

To further characterize the amphipathic nature of the expressed AChE, molecular forms contained within the membrane, cytosolic and medium fractions were analyzed by sedimentation velocity in the presence or absence of Triton X-100.

The predominant molecular forms in the extracted membranes in the presence of Triton X-100 were dimeric (6S) and tetrameric (9S) with a slight peak occurring at 13S. The dimeric and tetrameric forms shifted to higher sedimenting values and aggregated when Triton X-100 was absent from the gradients indicating that a component of these forms was lipophilic. The predominant peak in the Triton-free gradient, sedimenting at 13S, co-sedimented with a minor peak seen in the Triton-containing gradient suggesting that the minor peak was due to aggregation of the enzyme. Removing the detergent from the extracted sample by chromatography on Sephadex G-50 columns did not alter the sedimentation profiles of the Triton-free gradients. Consequently most of the samples contained 1% Triton X-100.

We detected three molecular forms in the cytosolic fraction with Triton-containing gradients. These were the monomer (3.4S), the dimer (6S) and the tetramer (9.5S). The monomeric and dimeric forms interacted with the detergent as indicated by their shift to higher sedimentation values in Triton-free gradients. However, a small peak of tetrameric form remained unchanged suggesting that it was hydrophilic in nature.

Two predominant molecular forms, the dimer (5.5S) and the tetramer (10S), were found secreted into the medium. The tetramer consistently showed a slightly higher sedimentation value than its counterpart extracted from membranes or from the cytosolic fraction - probably because it was not incorporated into a detergent micelle. Neither of these forms interacted with Triton X-100.

Summary

We have expressed catalytically active AChE from a mouse brain cDNA in two cell types and have determined its cellular location and amphipathic nature. The clone showed carboxyl-terminal amino acid homology to the asymmetric catalytic subunit from *Torpedo* tissues. AChE in microinjected oocytes was expressed as a detergent-soluble monomer and was found almost exclusively in the membrane fraction. In contrast, the AChE expressed in human 293 cells was found in three locations: associated with cell membranes, within the cell and solublized by low salt, and secreted into the medium. About 20% of the cell-associated AChE was membrane-linked as dimers and tetramers, required Triton X-100 for extraction and interacted with the detergent when analyzed by sucrose gradients. The cytosolic enzyme did not require high salt or detergent for extraction yet clearly interacted with Triton X-100 in sucrose gradients. The secreted dimeric and tetrameric forms were not detergent sensitive and quantitatively represented the predominant source of enzyme in the culture.

I would like to thank Dr. T.L. Rosenberry for his generous gift of PIPLC and Penny Bamford for her valuable technical assistance. This work was supported in part by a grant from the NIH (NS26885).

References

Brockman, S.K., Usiak, M.F., Younkin, S.G., *J. Biol. Chem.*, **1986,** *261,* 1201-1207.

Futerman, A.H., Low, M.G., Michaelson, D.M., Silman, I., *J. Neurochem.*, **1985,** *45,* 1487-1494.

Gibney, G., MacPhee-Quigley, K., Thompson, B., Vedvick, T., Low, M.G., Taylor, S.S., Taylor, P., *J. Biol. Chem.*, **1988** *263,* 1140-1145.

Gorman, C. In *DNA Cloning;* Glover, D.M., Ed.; IRL Press, Oxford, UK, 1985, Vol. II; pp 143-190.

Inestrosa, N.C., Roberts, W.L., Marshall, T.L., Rosenberry, T.L., *J. Biol. Chem.*, **1987,** *262,* 4441-4444.

Johnson, C.D., Russell, R.L., *Anal. Biochem.*, **1975,** *64,* 229-238.

Lazar, M., Salmeron, E., Vigny, M., Massoulié, J., *J. Biol. Chem.*, **1984,** *259,* 3703-3713.

Massoulié, J., Rieger, F., *Eur. J. Biochem.*, **1969,** *11,* 441-455.

Massoulié, J., Bon, S., *Annu. Rev. Neurosci.*, **1982,** *5,* 57-106.

Maulet, Y., Camp, S., Gibney, G., Rachinsky, T.L., Ekström, T.J., Taylor, P., *Neuron,* **1990,** *4,* 289-301.

Randall, W.R., Tsim, K.W.K., Lai, J., Barnard, E.A., *Eur. J. Biochem.*, **1987,** *164,* 95-102.

Roberts, W.L., Kim, B.H., Rosenberry, T.L., *Proc. Natl. Acad. Sci. U.S.A.*, **1987,** *84,* 7817-7821.

Roberts, W.L., Myher, J.J., Kuksis, A., Low, M.G., Rosenberry, T.L., *J. Biol. Chem.*, **1988,** *263,* 19398-19406.

Rosenberry, T.L., *Enzymes Biol. Memb.*, **1985,** *3,* 403-429.

Rotundo, R.L. In *The Vertebrate Neuromuscular Junction;* Salpeter, M.M., Ed.; Alan R. Liss, New York, NY, 1987; pp 247-284.

Rotundo, R.L., *J. Biol. Chem.*, **1988,** *263,* 19398-19406.

Rotundo, R.L., Fambrough, D.M., *Cell,* **1980,** *22,* 583-594.

Schumacher, M., Camp, S., Maulet, Y., Newton, M., MacPhee-Quigley, K., Taylor, S.S., Friedmann, T., Taylor, P., *Nature,* **1986,** *319,* 407-409.

Schumacher, M., Maulet, Y., Camp, S., Taylor, P., *J. Biol. Chem.*, **1988,** *35,* 19979-18987.

Sikorav, J.L., Krejei, E., Massoulié, J., *EMBO J.*, **1987,** *6,* 1865-1873.

Sikorav, J.L., Duval, N., Ansetmet, A., Bon, S., Krejei, E., Legay, C., Osterlund, M., Reimund, B., Massoulié, J., *EMBO J.*, **1988,** *7,* 2983-2993.

Wilson, B.W., Mettler, M.A., Asmundson, R.V., *J. Exp. Zool.*, **1969,** *172,* 49-57.

Rabbit Butyrylcholinesterase Gene: An Evolutionary Perspective

Arnaud Chatonnet and Omar Jbilo, Physiologie animale, INRA, place Viala, 34060 Montpellier Cedex 1, France

The study of the rabbit butyrylcholinesterase gene highlights the common features of cholinesterases. 1)The intron positions in the coding sequence are conserved in AChE and BChE genes of vertebrates. 2) Copies of incompletly spliced mRNA of ChEs are found in cDNA libraries. 3) The high A+T content of BChE gene is conserved in mammals. 4) Phylogenetic tree of the family of proteins homologous to cholinesterases can be inferred from sequence comparison and gives an idea of the gene duplications which occurred during evolution of the family.

Three genomic clones covering the entire coding sequence of the rabbit butyrylcholinesterase (BChE) gene were purified and partially sequenced (Chatonnet *et al.*, 1990). In addition a cDNA clone containing intronic sequences on both sides of exon 3 was also characterized. Southern blots of genomic rabbit DNA digested by EcoRI were hybridized with probes corresponding to the four human exons and showed only one band. These results suggest that there is a single BChE gene in rabbit, a situation similar to that found in human (Arpagaus *et al.*, 1990) and other mammalian species (Arpagaus *et al.*, this volume).

Gene Structure

As for human BChE gene, there are four exons in rabbit gene: the first exon is untranslated, the second codes for 83% of the mature protein, the third is only 167 bp long and the fourth codes for the C-terminus of the protein. The position of introns within the coding sequence is completely conserved among cholinesterases of vertebrates (Sikorav *et al.*, 1988; Arpagaus *et al.*, 1990; Maulet *et al.*, 1990). The active site serine is found in a highly conserved region. Other amino acids likely to be involved in the active site triad are conserved such as histidine 423 or 438 and aspartate 91 or 170. Aspartate 70 which is in the anionic site (McGuire *et al.*, 1989) is also conserved.

Small Interspersed Element (SINE)

A repetitive element specific of the rabbit genome was found at least twice in intronic sequences of the BChE gene. At the 3' end of one of the copies, 10 repeats of a AATAA motif was found (see Jbilo *et al.*, this volume).

cDNA Clones with Intronic Sequences

In addition to the genomic clones, one cDNA clone (BNY1) was isolated. This cDNA was unusual in that it contained intronic sequences. The insert of 1 Kb contained 167 coding bases homologous to the nucleotide sequence 1434 to 1600 of human cDNA and corresponded to exon 3 of the BChE gene. Other cDNAs of cholinesterases containing intronic sequences have been described (Fig.1). The clone Z35 of human BChE contained only the fourth exon and 100bp of the upstream intron (McTiernan *et al.*, 1987). A clone of *Torpedo marmorata* AChE lacked the last exon, the unspliced intron was processed as a normal 3' end of a mRNA (Sikorav *et al.*, 1988). This mature mRNA, if translated, would give a long sequence with a transmembrane domain that would be able to anchor the enzyme to the plasma membrane. The intronic sequence that was found in the clone of the rabbit in a position homologous to the Torpedo AChE clone could also be processed as it contains the maturation and polyadenylation signals. If translated the C-terminal end of the protein would have a hydrophobic domain of 20 amino acids that could remain in the membrane. To test the hypothesis that this region of pre mRNA might be conserved in some mature transcripts and could give rise to amphiphilic forms of BChE, a reverse transcription was performed using an oligonucleotide reverse complement of the sequence located 20 bases downstream of exon 3 at the 5' end of intron 3. The cDNA products were then amplified by PCR and are being sequenced.

Alternative Splicing of BChE Transcripts?

Thus far a single fourth exon responsible for the synthesis of the C-terminal end of the soluble form of BChE has been sequenced. Using the PCR technique with an oligonucleotide specific of

0–8412–2008–5/91/0157$06.00/0

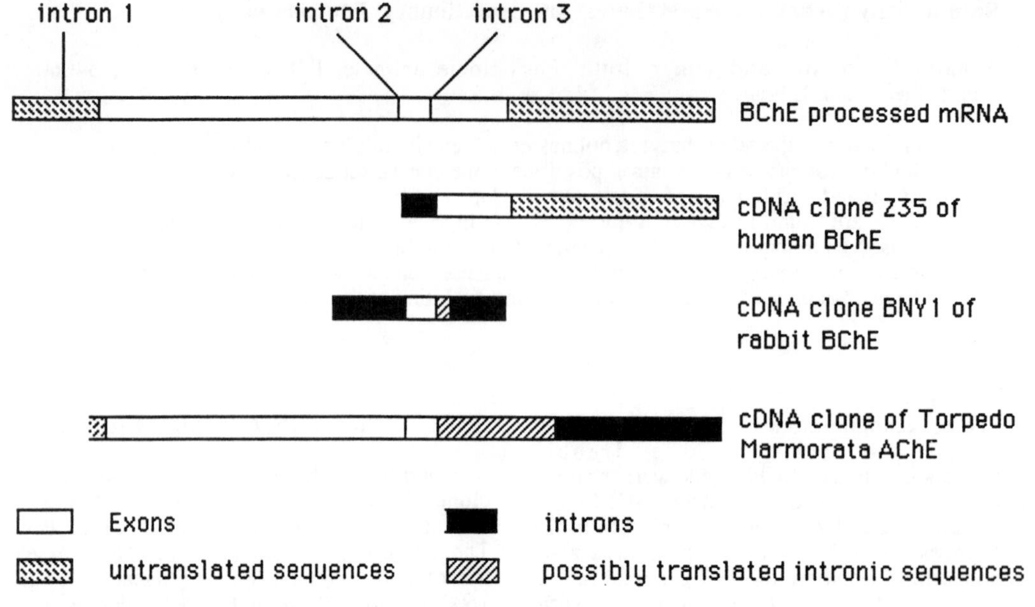

Fig. 1. Comparison of different cDNAs of Cholinesterases in which intronic sequences were found.

the end of the exon 2 and an oligo-dT (method of rapid amplification of cDNA ends, Frohman *et al.*, 1988), we tried to amplify minor transcripts which might have an alternative exon 4. We are cloning the minor bands of amplified material (see abstract by Jbilo *et al.*, this volume).

Codon Usage

The high A+T content at the third base of the codons or in untranslated sequences in the human BChE gene is also a characteristic of the rabbit gene. The percentage of G+C at this position is 30 % in rabbit and 34 % in human. This feature found in other BChE sequences of mammals (see abstract by Arpagaus *et al.*, this volume) is also found in a number of other genes. It appears that there is a nonrandom distribution of these genes as far as the content of C+G is concerned (Newgard *et al.* 1986): C+G content is low in genes highly expressed in liver and high in those expressed in skeletal muscles (Fig. 2, Table 1). Sequences of human and rabbit BChE were included as well as proteins of the cholinergic system. BChE gene (expressed in liver) are among those of lowest C+G content at the third base of the codons. All the proteins of the cholinergic system of mammals and birds so far sequenced show a high C+G content at the last base of the codons. Our hypothesis predicts that AChE of higher vertebrates should have a high C+G content at the last base of the codons and should not differ from genes highly expressed in muscle (see Camp S. *et al.* and Soreq *et al.* this volume) Although this observation may admit exceptions, we thus suggest that the C+G content could be related to the tissue- or developmental stage-specific expression of the genes. The proportion of A+T or C+G at the third base of the codons has been related to the position of the gene in larger regions of the genome with constant base composition (Bernardi *et al.*, 1985). It has been also related to the bandings of the chromosomes and to the susceptibility of the associated proteins to digestion. A correlation has also been found with the time of replication (early phase for A+T genes and late replication for C+G genes). A high proportion of housekeeping genes were found in C+G compartment (Bickmore and Sumner, 1989; Holmquist, 1989).

Phylogenetic trees of BChEs

"Maximum Parsimony" and "Neighbor Joining" methods were used to investigate evolutionary relationships between cholinesterases (see abstract by Cousin *et al.*, this volume). Published sequences of vertebrate AChEs and BChE, invertebrate AChE, non specific esterases and rat thyroglobulin were aligned and a matrix of distance deduced from the alignement. Fig. 3 shows a phylogenetic tree obtained with the Neighbor Joining method of Saitou and Nei (1987). One interesting feature of this representation is that it shows quite clearly that the duplication between ACHE and BCHE genes

Table 1. C+G content in total mRNA and at the third base of the codons for genes expressed in liver or muscle of higher vertebrates

	C+G	
Genes expressed in liver	Last base	Total
Human phosphoglycerate kinase	55.8	47.7
Human serum albumin	39	41.3
Human triose phosphate isomerase	65.7	55.4
Human haptoglobine	53.9	49.6
Human alcohol dehydrogenase	41.8	40.2
Mouse C. P-450 stéroïde 21-H.	56.2	57.6
Rat ornithine amino transferase	41	40.6
Human ornithine amino transferase	49.6	46.5
Human apolipoprotéine A-1	68	63
Rabbit liver glycogen phosphorylase	60	48.7
Human liver aldolase	58.4	50.3
Human phenylalanine hydroxylase	52.5	41.8
Rat Peroxisomal dehydrogenase	56.0	48.7
Bovine prothrombine	79.5	60.9
Human fibrinogen gamma chain	39.1	40.5
Human fibrinogen alpha chain	42.4	47.5
Human butyrylcholinesterase	34	40.2
Rabbit butyrylcholinesterase	39.4	42.3
Genes expressed in muscle		
Rat myosin heavy chain skeletal	72	54.8
Rat myosin heavy chain cardiac	85	59.4
Rat myosin light chain skeletal	61	49.4
Rabbit heavy chain cardiac	90	60.3
Rabbit muscle phosphorylase	85.8	60.3
Rabbit muscle aldolase	78.9	60.6
Rat creatine kinase	67.5	59.4
Human creatine kinase	80.1	58.5
Mouse phosphorylase kinase	71.9	50.3
Proteins of the cholinergic system		
Human nicotinic AChR γ subunit	79	61.1
Human nicotinic AChR α subunit	67	51.4
Mouse nicotinic AChR δ subunit	74.2	56.5
Bovine nicotinic AChR ε subunit	80.6	59.7
Porcine muscarinic AChR	95	66.6
Porcine Choline acetyl-transferase	70.9	56.4

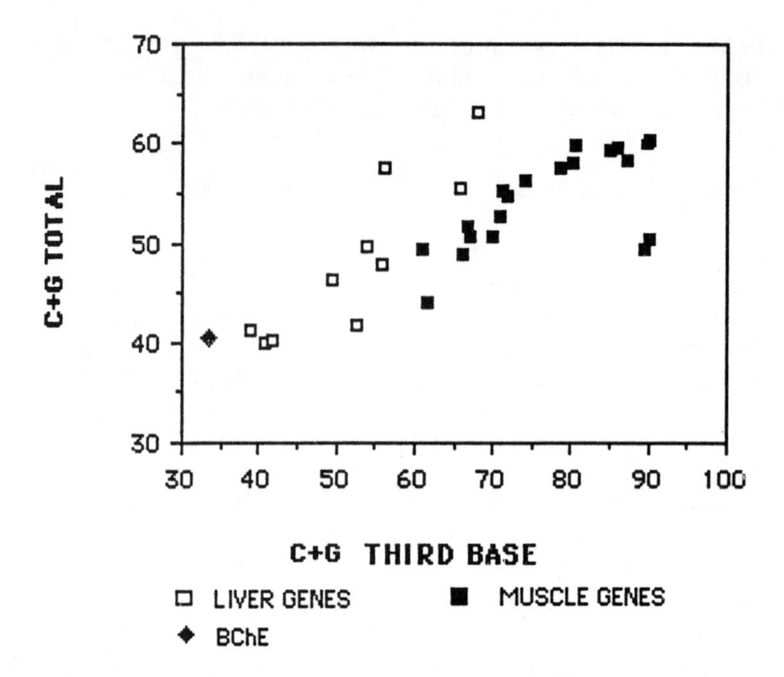

Fig.2. Percentage of C+G at the third base of codons in several mammalian genes. Open squares: genes expressed in liver (see table 1 for details); closed squares: genes preferencially expressed in muscles. Human and rabbit BChE genes are represented by filled circles.

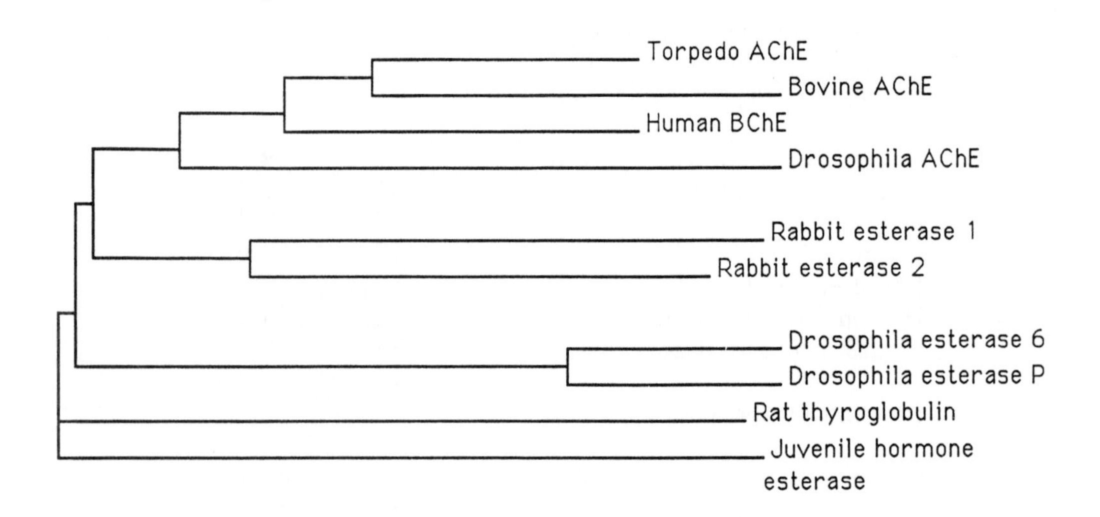

Fig.3. Phylogenetic tree of cholinesterases and related esterases constructed with the Neighbor Joining method of Saitou and Nei (1987). Peptide sequences were taken from: Doctor *et al.*, 1989; Hall and Spierer, 1986; Schumacher *et al.*, 1986; Korza and Ozols, 1988; Ozols 1989; Collet *et al.*, 1990; Hanzlik *et al.*, 1989, and Di Lauro *et al.*, 1985.

occured long after the divergence of vertebrates and insects. This observation is in full agreement with the biochemical arguments of Toutant *et al.*, 1985; Gnagey *et al.*, 1987 and Pezzementi *et al.* (this volume) who all suggested that BCHE likely appeared with the emergence of vertebrates.

Acknowledgements

We wish to thank Dr. Nei (University of Texas Houston) for kindly providing the Neighbor Joining Method program. The other programs were obtained from BISANCE at CITI 2. We thank Drs M. Arpagaus, O. Lockridge, B.N. LaDu, F. Bacou, J. Massoulié and J.-P. Toutant for support and helpful discussions. This work was supported by INRA AIP "Cholinestérases" and an "Association Française contre les Myopathies" Grant.

References

Arpagaus, M., Kott, M., Vatsis, K.P., Bartels, C.F., La Du, B.N. and Lockridge, O., *Biochemistry,* **1990,** *29,* 124-131.
Bernardi, G., Olofsson, B., Filipski, J., Zerial, M., Salinas J., Cuny G., Meunier-Rotival M. and Rodier, F., *Science,* **1985,** *228,* 953-958.
Bickmore, W.A. and Sumner, A.T., *Trends in Genetics,* **1989,** *5,* 144-148.
Chatonnet, A. and Lockridge, O., *Biochem J.,* **1989,** *260,* 625-634.
Chatonnet, A., Lorca, T., Barakat, A., Aron, E., and Jbilo, O., *Mol. Cell. Neurobiol.,* **1990,** in press.
Collet, C., Nielsen, K.M., Russel, R.J., Karl, M., Oakeshott, J.G., and Richmond,R.C.,*Mol. Biol. Evol.,* **1990,** *7,* 9-28.
Di Lauro, R., Obici, S., Condliffe, D., Ursini, V.M., Musti, A., Moscatelli, C., and Avvedimento, V.E. *Eur. J. Biochem.,* **1985,** *148,* 7-11.
Doctor, B.P., Smyth, K.K., Gentry, M.K., Ashani, Y., Christner, C.E., De La Hoz, D.M., Ogert, R.A., and Smith S.W., in: *Computer-assisted modeling of receptor-ligand interraction* Alan R. Liss, **1989,** 305-316.
Frohman, M.A., Dush, M.K., and Martin, G.R., *Proc. Natl. Acad. Sci. USA,* **1988,** *85,* 8998-9002.
Gnagey, A.L., Forte, M., Rosenberry, T.L., *J. Biol. Chem.,* **1987,** *262,* 13290-13298.
Hall, L.M.C. and Spierer, P., *EMBO J.,* **1986,** *5,* 2949-2954.
Hanzlik, T.N., Abdel-Aal, Y., Harshman, G., and Hammock, B.D., *J. Biol. Chem.,* **1989,** *264,* 12419-12425.
Holmquist, G.P. *J. Mol. Evol.,* **1989,** *28,* 469-486.
Korza, G., and Ozols, J., *J. Biol. Chem.,* **1988,** *263,* 3486-3495.
Krane, D.E. and Hardison, R.C., *Mol. Biol. Evol.,* **1990,** *7,* 1-8.
Maulet, Y., Camp, S., Gibney, G., Rachinsky, T., Ekstrom, T.J. and Taylor, P., *Neuron,* **1990,** *4,* 295-301.
McGuire, M.C., Nogueira, C.P. Bartels, C.F., Lightstone, H., Hajra, A., Van der Spek, A.F.L., Lockridge, O., and La Du, B.N.*Proc. Natl. Acad. Sci. USA,* **1989,** *86,* 953-957.
McTiernan, C., Adkins, S., Chatonnet, A., Vaughan, T.A., Bartels, C.F., Kott, M., Rosenberry, T.L., La Du, B.N., and Lockridge, O., *Proc. Natl. Acad. Sci. USA,* **1987,** *84,* 6682-6686.
Newgard, C.B., Nakano, K., Huang, P.K. and Fletterick, R.J., *Proc. Natl. Acad. Sci. USA,* **1986,** *83,* 8132-8136.
Ozols, J., *J. Biol. Chem.,* **1989,** *264,* 12533-12545.
Saitou, N., and Nei, M., *Mol. Biol. Evol.,* **1987,** *4,*406-425.
Sikorav, J.-L., Duval, N., Anselmet, A., Bon, S., Krejci, E., Legay, C., Osterlund, M., Reimund, B. and Massoulié, J., *EMBO J.,* **1988,** *7,* 2983-2993.
Schumacher M., Camp S., Maulet Y., Newton M., MacPhee-Quigley, K., Taylor S.S. Friedmann, T. and Taylor, P., *Nature* , **1986,** *319,* 407-409.
Toutant, J.-P., Massoulié, J. and Bon, S., *J. Neurochem.,* **1985,** *44,* 580-592.

Search for the molecular origins of butyrylcholinesterase polymorphism by cDNA screening, deletion mutagenesis and Xenopus oocyte co-injections

Patrick A. Dreyfus and Martine Pincon-Raymond, INSERM U153, 17 Rue du Fer-a-Moulin, 75005 Paris.

Haim Zakut, Dept. of Obstetrics and Gynecology, The Edith Wolfson Medical Center, Holon 58100, The Sackler Faculty of Medicine, Tel Aviv University

Shlomo Seidman and Hermona Soreq*, Department of Biological Chemistry, The Life Sciences Institute, The Hebrew University of Jerusalem, Israel 91904

* Corresponding author (Fax: 972-2-666804)

Screening of cDNA libraries from various fetal and adult human tissue origins resulted in the isolation of identical cDNA clones, all coding for serum butyrylcholinesterase (BuChE). These findings indicate that the human CHE gene is most probably transcribed into a single transcription product in all tissues. When injected into *Xenopus* oocytes, synthetic BuChEmRNA induced the production of an extracellular surface-associated protein displaying the characteristics of dimeric BuChE. Deletion mutagenesis experiments revealed that the entire coding sequence in BuChEmRNA is necessary for its expression in oocytes. Supplementation with total tissue mRNA's induced the appearance of molecular forms containing 4 or more catalytic subunits, and associated with the oocyte surface in a manner consistent with the prinicipal assembly patterns of cholinesterases in the native tissues.

The ubiquitious enzyme butyrylcholinesterase (EC 3.1.1.8, BuChE) is present in serum (Lockridge et al., 1987), brain (Zakut et al., 1985) and muscle (Dreyfus et al., 1983; Layer et al., 1987). BuChE displays tissue-specific polymorphism with respect to subunit assembly, hydrodynamic properties and subcellular compartmentalization.

To search for the molecular origin of this polymorphism, screening of various human cDNA libraries was carried out using both specific oligonucleotide probes (Prody et al., 1986) or a full length cDNA (Prody et al., 1987). To produce recombinant BuChE, the full length cDNA encoding human butyrylcholinesterase (Prody et al., 1987) was subcloned into the pSP64 transcription vector (Krieg and Melton, 1984) and was used as a template. Xenopus oocytes were microinjected with the transcription products using 50 nl of 0.1 mg/ml mRNA/oocyte (Soreq et al., 1984), incubated and homogenized as described (Soreq et al., 1989). Enzymatic activity was measured according to Ellman et al. (1961). Immunofluorescence detection was carried out as described (Dreyfus et al., 1989), using rabbit polyclonal antibodies prepared against Torpedo AChE, generously provided by Dr. S. Camp and P. Taylor.

Evidence for a Single Transcript

Screening of several cDNA libraries from various tissue origins (human adult liver and lymphocytes, fetal brain, muscle and liver) with the oligonucleotide and cDNA probes indicated above resulted in the isolation of several identical cDNA clones, all coding for serum BuChE, suggesting that the human CHE gene encoding this enzyme is most probably transcribed into a single transcription product in all tissues. These findings imply that alternative splicing is not responsible for the heterogeneity observed in liver, muscle and brain BuChE forms. In the absence of indications for the involvement of transcriptional control, post-transcriptional mechanisms may well be at the origin of BuChE heterogeneity. Therefore, oocyte microinjection experiments were initiated with clone-produced synthetic BuChEmRNA.

Partial Deletion Constructs

When injected into Xenopus oocytes, synthetic mRNA transcribed off a cDNA encoding human serum BuChE induced the production of a protein displaying the substrate specificity and sensitivity to selective inhibitors characteristic of native BuChE, and which clearly distinguish it from AChE (Soreq et al., 1989; Dreyfus et al., 1989). Enzymatic restriction of the complete SP64-BuChEcDNA construct was then performed to produce a series of partially deleted clones which were then similarly transcribed and expressed in oocytes (Fig. 1). Thus,

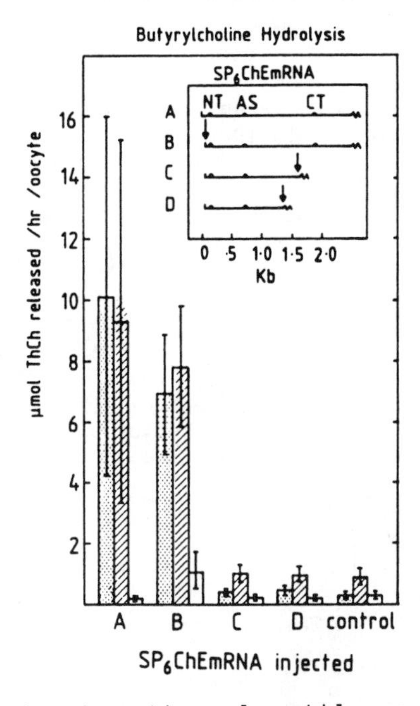

Fig. 1: The entire polypeptide sequence is essential for production of a catalytically active enzyme.

HinfI restriction of this 2450 bp long clone (Fig. 1,A) removed the 5'-end of BuChEcDNA to produce a 2380 bp long insert with only one AUG codon (B); HincII digestion produced a 1615 bp long BuChEcDNA shortened at its 3'-end which encodes a 507 amino acids long protein (C) and StuI restriction produced a yet shorter 1257 bp long insert encoding a 367 residues protein (D,see Fig. 1). The 5'-end deletion did not significantly affect BuChE synthesis in oocytes microinjected with its transcription product. In contrast, the 3'-end deletions completely abolished the production of catalytically active BuChE (Fig. 1). These experiments demonstrated that the entire sequence (574 amino acid residues) of the mature protein is essential for either production or stabilization of the catalytically active enzyme.

Molecular Form Polymorphism

Microinjected alone, synthetic BuChE-mRNA induced the formation of primarily dimeric ChE. This first level of oligomeric assembly may therefore be spontaneous, or may require a catalytic mechanism that is already available in the oocyte (Table 1).

Table 1. Molecular forms of recombinant BuChE produced in microinjected Xenopus oocytes

Co-injected mRNA	Molecular Forms
None	Dimer
Fetal Brain	Dimer + Tetramer
Fetal Muscle	Dimer + Tetramer + Tailed 16S
Fetal Liver	Monomer + Dimer

It should be noted in this respect that the detection of cholinesterase activities in Xenopus oocytes (Soreq, 1985) and high levels of BuChEmRNA in human oocytes (Soreq et al., 1987) indicate that ChE represents a natural endogenous oocyte protein. The co-injection of tissue-extracted polyA$^+$mRNAs induced higher levels of multi-subunit assembly (Table 1) - likely including the incorporation of non-catalytic subunits. It indicates that additional protein species, not available in the oocytes, are required to direct the biosynthesis of more complex molecular forms. Given the high degree of tissue-specific polymorphism of ChE molecular forms, it is not surprising to find these additional factors expressed in a tissue-specific manner.

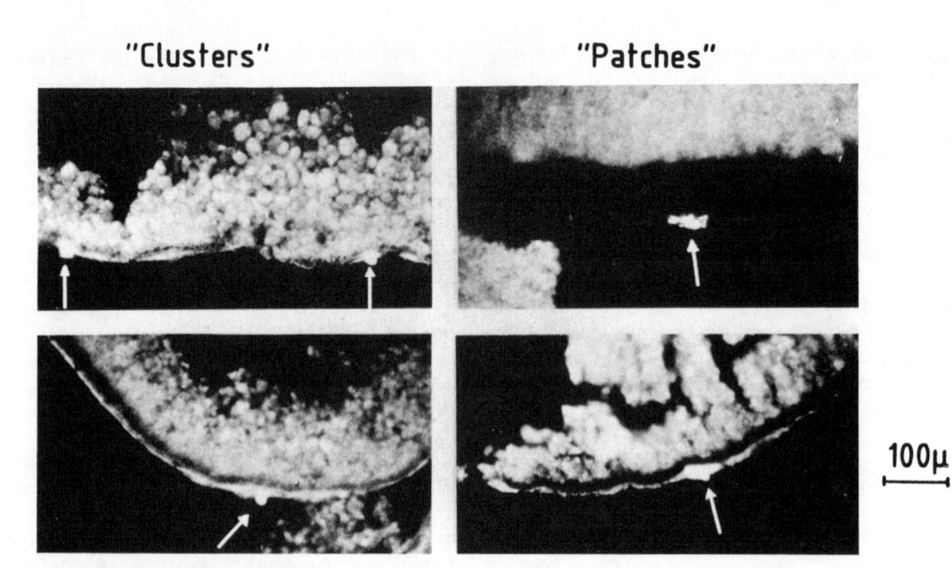

Fig. 2: Recombinant BuChE aggregates on the oocyte surface into cluster and patch structures.

Transport and Extracellular Surface Association

Immunocytochemical analysis of the oocyte-produced enzyme showed its aggregation on the oocyte surface into cluster (1 to 10 μm²) and patch (10 to 100 μm²) structures (Figure 2). About 30 minutes after injection, the immunoreactive BuChE accumulation was detectable at the external oocyte surface. Two and a half hours post-injection, fluorescent ChE clusters were bigger, more frequent and more intense than those seen at the previous injection times (Dreyfus et al., 1989). The newly expressed enzyme was detectable as intracellular immunoreactive structures (Figure 3). Tunicamycin pre-treated oocytes were capable of expressing active BuChE, but not transporting it to the cell-surface (Dreyfus et al., 1989). Supplementation with tissue-specific mRNAs induced alterations in the features of the membrane-bound aggregates (Figure 4). The relative distribution of each type of formation varied between oocytes co-

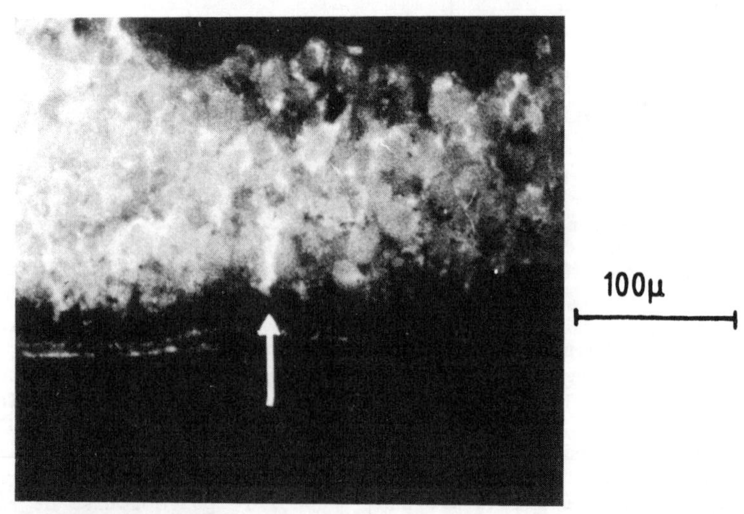

Fig. 3: Intracellular transport of recombinant newly expressed BuChE in a brain poly A⁺ mRNA co-injected oocyte; the arrow shows immunoreactive BuChE "en route" for the cell surface.

injected with mRNA from liver, brain and muscle tissues in a manner consistent with the organization of ChE in the native tissues (Figure 4) (Wallace et al., 1985; Wallace 1986). Enhanced aggregation appeared to be correlated to the appearance of molecular forms containing 4 or more catalytic subunits (Soreq et al., 1989). Both brain and muscle mRNAs increased the number of patches and clusters as compared to BuChEmRNA alone (Figure 5A). Co-injection with tissue mRNAs therefore increased both the intensity and the surface area occupied by patches and clusters indicating the induction, by tissue-specific factors of enhanced BuChE aggregation at the external surface of the oocyte (Figure`5B). The relative staining intensity of clusters and patches obtained with mucle mRNA qualitatively exceeded that obtained with brain mRNA (Figure 5B). Together, these observations imply a qualitative and/or quantitative difference between cholinesterase related mRNAs in different tissues and suggests that these mRNA pools are capable of

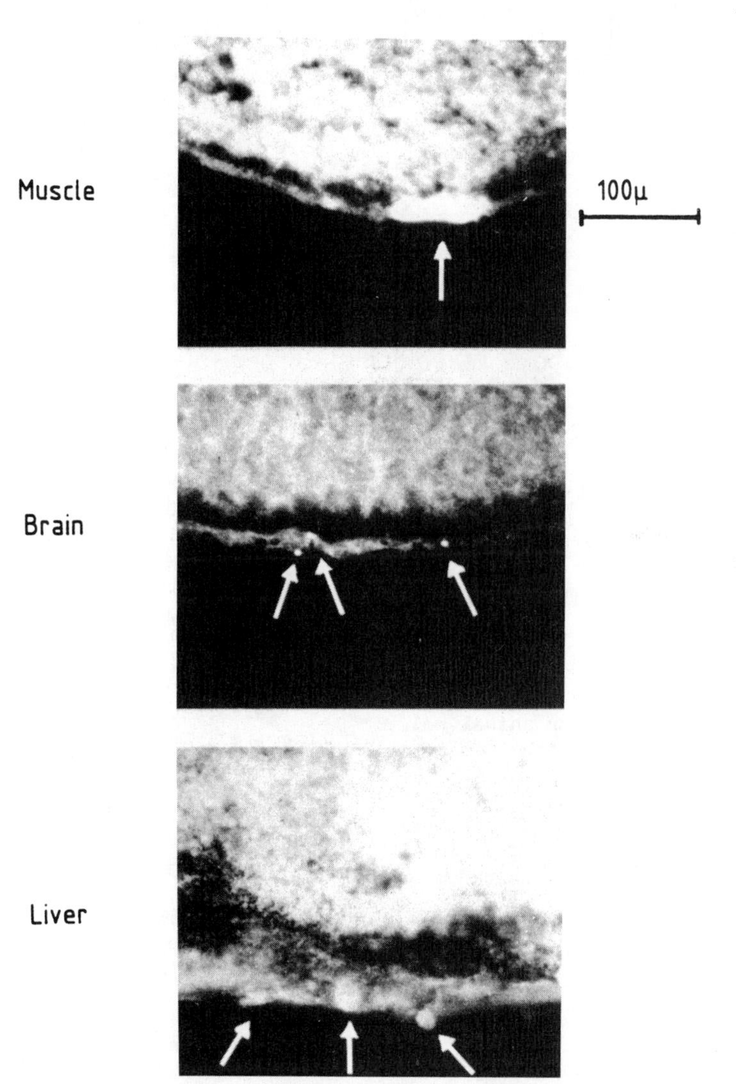

Fig. 4: Tissue-specific potentiation of BuChE assembly patterns on oocyte surface.
Muscle mRNAs induced patch formation, brain mRNAs induced numerous clusters, but fuzzy aggregates were visible with liver mRNAs.

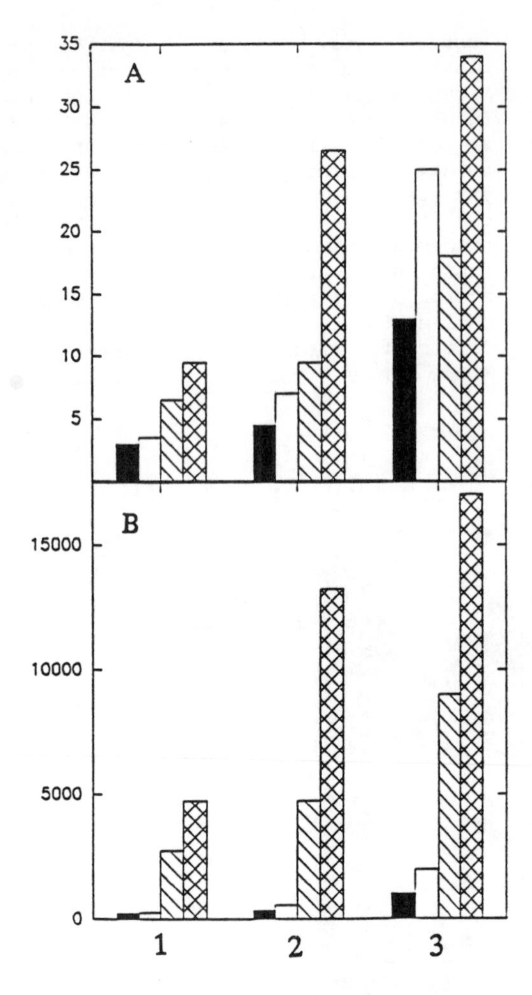

Fig.5: Immunocytochemical quantification of BuChE distribution on individual oocytes.
Black column: clusters on vegetal pole.
Open column: patches on vegetal pole.
Oblique lines: clusters on animal pole.
Crosshatched column: patches on animal pole.
1: BuChEmRNA injected alone.
2: BuChEmRNA co-injected with brain mRNAs.
3: BuChEmRNA co-injected with muscle mRNAs.
A: number of clusters and patches.
B: Total immunoreactive area in μm^2.

modulating tissue-specific usage of a single BuChEmRNA species.

These findings demonstrate that liver, muscle and brain express mRNAs encoding peptides which: i) are required for the biosynthesis of BuChE molecular forms consisting of multiple dimeric units, ii) may specify the incorporation of non-catalytic subunits, iii) direct the formation of a tissue-specific array of BuChE molecular forms, iv) direct a tissue-specific organizational pattern of BuChE at the external surface of the injected oocytes.

Acknowledgements

The contributions of Drs. Francois Rieger and Dina Zevin-Sonkin are gratefully acknowledged. Supported by INSERM (to P.A.D.), by the Association Francaise Contre Les Myopathies (to H.S. and H.Z.) and by the U.S. Army Medical Research and Development Command (DAMD 17-87-C-7169, to H.S.). We thank Drs. S. Camp and P. Taylor for antibodies.

References

Dreyfus, P.A., Rieger, F. Pincon-Raymond, M. *Proc. Natl. Acad. Sci. (USA)*, **1983**, *80*, 6698-6702.

Dreyfus, P.A., Seidman, S., Pincon-Raymond, M., Murawsky, M., Rieger, F., Schejter, E., Zakut, H., Soreq, H. *Cell. and Mol. Neurobiol.*, **1989**, *9*, 323-341.

Ellman, G.L., Courtney, D.K., Anders, V., Featherstone, R.M. *Biochem. Pharmacol.*, **1961**, *7*, 88-95.

Krieg, A., Melton, D.A. *Nucleic Acids Res.*, **1984**, *12*, 7057-7070.

Layer, P.G., Alber, R., Sporns, O. *J. Neurochem.*, **1987**, *49*, 175-182.

Lockridge, O., Bartels, C.G., Vaughan, T.A., Wong, C.K., Norton, S.E., Johnson, L.L. *J. Biol. Chem.*, **1987**, *262*, 549-557.

Prody, C., Zevin-Sonkin, D., Gnatt, A., Goldberg, O., Soreq, H., *Proc. Natl. Acad. Sci. (USA)*, **1987**, *84*, 3555-3559.

Prody, C., Zevin-Sonkin, D., Gnatt, A., Koch, R., Zisling, R., Goldberg, O., Soreq, H., *J. Neurosci. Res.*, **1986**, *16*, 25-35.

Soreq, H. *CRC Critical Reviews in Biochemistry*, **1985**, *18*, 199-238.

Soreq, H., Malinger, G., Zakut, H. *Human Reproduction*, **1987**, *2*, 689-693.
Soreq, H., Seidman, S., Dreyfus, P.A., Zevin-Sonkin, D., Zakut, H. *J. Biol. Chem.*, **1989**, *264*, 10608-10613.
Wallace, B.G. *J. Cell. Biol.*, **1986**, *102*, 783-794.
Wallace, B.G., Nitkin, R.M., Reist, N.E., Fallon, J.R., Moayeri, N.N. McMahan, U.J. *Nature*, **1985**, *315*, 574-577.
Zakut, H., Matzkel, A., Schejter, E., Avni, A., Soreq, H. *J. Neurochem.*, **1985**, *45*, 382-389.

Structure of Human Butyrylcholinesterase Gene and Expression in Mammalian Cells

Oksana Lockridge* and Bert N. La Du, Pharmacology Dept., University of Michigan Medical School, Ann Arbor, MI 48109-0626 USA
(After July 1, 1990) *Eppley Institute, University of Nebraska Medical Center, 600 S. 42nd St., Omaha, NE 68198-6805 USA

Single copy genes are very important for the Human Genome Mapping and Sequencing Project because they are used as landmarks to define position on the physical map. Our results show that butyrylcholinesterase (BCHE) is a single copy gene in the human genome as well as in other vertebrates. We have determined the structure of the human gene and have recover the gene by PCR from genomic DNA of approximately 100 individuals. PCR and DNA sequencing were used to identify nucleotide substitutions in the rare genetic variants of human BChE associated with an abnormal response to the muscle relaxant succinylcholine. When human cDNA was expressed in CHO cells >95% of activity was secreted as a tetrameric BChE having normal substrate and inhibitor specificities.

Human butyrylcholinesterase (BCHE = gene; BChE = protein) is of clinical interest because people who have genetic variants of this enzyme react abnormally to the muscle relaxant drug, succinylcholine (Kalow and Genest, 1957). For the past twenty years our laboratory's goal has been to determine the molecular basis for the genetic variants. To accomplish our goal we determined the complete amino acid sequence of the usual form of BChE (Lockridge et al., 1987), used the sequence information to find the cDNA clone (McTiernan et al., 1987), used the cDNA to find the gene (Arpagaus et al., 1990), and used information from the gene structure to determine the nucleotide substitutions in the rare genetic variants (McGuire et al., 1989; Nogueira et al., 1990). These naturally occurring genetic variants identify amino acids important for activity because most of the variants have either reduced affinity for substrate or have a reduced level of BChE enzyme in serum. Posters by Bert N. La Du, Cynthia Bartels, Christine Nogueira, Amitav Hajra, Frank Jensen, Steve Adkins and Oksana Lockridge give details of genotyping methods, genotyping results, and nomenclature. This presentation and posters by Martine Arpagaus and Steve Adkins focus on the structure of the gene and expression in cultured cells.

Four exons encode the secreted, globular G4 human BChE

Genomic clones containing exons 1 and 2, exon 3, or exon 4 were isolated by screening genomic libraries with fragments of human BCHE cDNA. Fig. 1 is a diagram of the gene showing that the human BCHE gene contains four exons and three introns. This gene structure is thought to represent the secreted, globular form of BChE because the coding sequences within exons 2, 3, and 4 have the same amino acid sequence as the BChE protein isolated from human serum.

Exon 1 is in the same reading frame as exon 2 and has an open reading frame starting at Met -69. If translation were to begin at the first methionine in exon 1 a signal peptide of 69 amino acids would be obtained. A second in frame initiator Met is located at Met -47 and would give a 47 amino acid signal peptide. A third initiator Met is in exon 2. It is possible that synthesis of BChE is initiated at several sites, and that this has a role in regulation of expression.

A transcription initiation site has not yet been determined. Therefore the exact size of exon 1 is not known, though exon 1 is greater than 120 bp. The possibility that additional

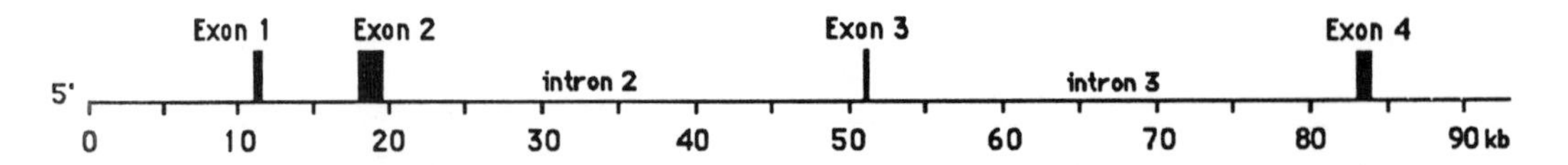

Fig. 1. Structure of the human BCHE gene. Four exons have been identified. Exons 2, 3, and 4 contain coding sequences for the globular, tetrameric butyrylcholinesterase.

exons are present on the 5' side of exon 1 has not yet been ruled out.

Maulet et al. (1990) have suggested that Torpedo ACHE may have two exons in this region.

Exon 2 contains 1525 bp. It includes a 28 amino acid signal peptide which lies within a favorable consensus sequence, AATATGC, commonly found in functional initiators. Exon 2 encodes 83% of the mature protein, from the N-terminal of the mature protein, Glu 1, to Gly 478. Exon 2 encodes the active site serine, Ser 198, the putative anionic site component, Asp 70, the potential active site histidine, His 438, 7 of the total of 9 carbohydrate chains per subunit, and two disulfide loops, Cys 65-92 and Cys 252-263.

The size and boundaries of exon 2 of human BCHE are identical to the homologous exon in Torpedo ACHE (Maulet et al., 1990). The homologous exon in Torpedo ACHE is named exon 1.

Five genetic variants of human BCHE have been mapped to exon 2. See Table 1.

Anionic site. The atypical variant has a reduced Km for all positively charged substrates and positively charged inhibitors, but a normal kcat. Atypical BChE is different from usual BChE in that it has glycine at position 70 in place of aspartic acid. This suggests that Asp 70 is a component of the anionic site (McGuire et al., 1989).

Exon 3 is a very short exon with only 167 bp. There is a disulfide bond between Cys 519 of exon 3 and Cys 400 of exon 2.

The size and boundaries of exon 3 of human BCHE are identical to the homologous exon in Torpedo ACHE.

One genetic variant, the J variant, has been mapped to exon 3. In the J variant the activity of BChE in serum is reduced by 66%. Antibody assays have shown that the amount of BChE protein is reduced, though it is not yet known whether the cause is reduced synthesis or instability of the J variant protein leading to rapid degradation. In the J variant the charged amino acid, Glu 497, is substituted by the hydrophobic residue valine.

Exon 4 contains 604 bp. It codes for the C-terminus of BChE from Gly 534 to Leu 574 and for the 3' untranslated region. Two polyadenyla- tion signals are present. Both are thought to be functional because cDNA clones were isolated terminating at each of these polyadenylation sites.

One genetic variant, the K variant, has been mapped to exon 4. The K variant has a 33% reduction in BChE activity in serum and has a reduced amount of serum protein as determined by immunological quantitation (Rubinstein et al., 1978). The substitution is from Ala 539 to Gly. The K variant is the most common genetic

Table 1. Nucleotide changes in human BCHE variants

Common name	Phenotypic description	Amino acid alteration	DNA alteration	Exon
Usual	Normal	none	none	
Atypical	Dibucaine resistant	70 Asp→Gly	nt 209 (GAT to GGT)	2
Silent-1	Silent, no activity	117 Gly→Frame shift	nt 351 (GGT to GGAG)	2
H variant	90% lower activity	142 Val→Met	nt 424 (GTG to ATG)	2
Fluoride-1	Fluoride resistant	243 Thr→Met	nt 728 (ACG to ATG)	2
Fluoride-2	Fluoride resistant	390 Gly→Val	nt 1169 (GGT to GTT)	2
J variant	66% lower activity	497 Glu→Val	nt 1490 (GAA to GTA)	3
K variant	33% lower activity	539 Ala→Thr	nt 1615 (GCA to ACA)	4

variant; it occurs in homozygous form in 1.7% of the Caucasian population and in heterozygous form in 22.6% of the population. The K variant is not associated with any abnormal drug response when it is the only variant in the human BChE gene.

One BCHE gene

Genomic blots. A set of exon probes was constructed by inserting each human exon along with flanking sequences into separate plasmids. These exon probes were hybridized to genomic blots of man, monkey, dog, rat, mouse, guinea pig, cow, sheep, pig, rabbit, and chicken. The blots showed one or at most two bands with each probe (Arpagaus et al., 1990; Masson et al., 1990; Arpagaus et al, submitted). This result was taken as strong evidence that man as well as each of these vertebrate animals contained a single gene for BCHE.

Restriction map. The restriction map of the isolated human BCHE gene was consistent with the restriction map deduced from digests of total genomic human DNA (Arpagaus et al., 1990). This supports the conclusion that there is only one human BCHE gene.

C5 (CHE2) is not on a separate BCHE locus. The C5 variant was thought to map to a second BCHE locus by Harris et al., (1963). C5 is identified by an extra activity stained band on gel electrophoresis. Eiberg et al., (1989) have very strong evidence that C5 (CHE2) maps to chromosome 2 where it is genetically linked to γ–crystallin gene cluster (CRYG). Chromosome 2 does not hybridize with BCHE cDNA (Soreq et al., 1987; Zakut et al., 1989) and therefore there is no evidence that this map point encodes a BCHE gene. Masson et al. (1990) examined the possibility that only a small percentage of the population might carry two BCHE genes by doing genomic blots of DNA from people who had the C5 band. The results of Masson et al. (1990) were consistent with the interpretation that there was only one human BCHE gene. A more likely explanation for C5 is the one offered by Scott and Powers (1974), that C5 is a combination of BChE and a second as yet unidentified protein. The BCHE gene is near transferrin on the long arm of chromosome 3 at 3q21-25 (Yang et al., 1984; Soreq et al., 1987).

PCR. To identify nucleotide alterations in the genetic variants we sequenced each exon of the BCHE gene from individuals who were known to have an abnormal BChE enzyme. Great care was taken to ensure that the DNA substitution we reported was consistently present in family members who had the genetic variant phenotype and was absent in family members who did not have the phenotype. Pedigree analysis was done to show that the trait was inherited in Mendelian fashion (see posters by C. Bartels). Genomic DNA from a single individual was amplified in 10 different PCR experiments so that Taq polymerase errors could be identified and only the inherited alterations would remain as a consistent base substitution. During this project we amplified exon 2 with many different sets of primers which hybridized inside the coding sequence. If there were more than one BCHE gene, including pseudogenes, or partial genes, we expected them to be revealed as differences in the DNA sequence. No fragment of a BCHE gene that might have been attributed to a second BCHE locus was ever retrieved by PCR. Our results were completely consistent with the existence of a single human BCHE gene.

There is agreement that the BCHE gene is located on chromosome 3. However, at this time it is not clear what other genes are hybridizing with human BCHE cDNA in chromosomal preparations where hybridization appears to occur not only on chromosome 3 but also on chromosome 16 (Soreq et al., 1987; Zakut et al .,1989). Possibly, chromosome 16 contains genes homologous to BCHE. These should not interfere with the Human Gene Mapping Project because only one BCHE gene is recovered by PCR.

Expression of human BCHE

Human BCHE cDNA was ligated into a eukaryotic expression vector containing the adenovirus major late promoter and cotransfected with the gene for dihyfrofolate reductase into Chinese Hamster Ovary cells (CHO) cells (see poster by Steve Adkins). Stable cell lines expressing BChE were created by selecting colonies resistant to methotrexate. During months of selection the BCHE gene became amplified in the DNA of the CHO cells and Southern blots revealed at least 4 copies of the human BCHE gene. Active human BChE enzyme was expressed and greater than 95% of activity was secreted into the culture medium. Secretion was consistent with the presence of a 28 amino acid signal peptide on the 5' end of the cDNA construct. The amount of BChE enzyme in the medium was sufficiently high that activity could be tested with standard spectrophotometric methods, using benzoylcholine or butyrylthiocholine as substrate. To avoid the problem of background activity from enzymes in culture medium, in particular AChE in fetal calf serum, recombinant BChE enzyme was collected into serum-free medium. The amount of recombinant BChE secreted into the medium was 125 µg per liter.

The recombinant BChE enzyme was characterized with respect to substrate and inhibitor specificity. It hydrolyzed the typical BChE substrates including benzoylcholine, proprionylthiocholine, butyrylthiocholine and alpha-naphthylacetate. It was inhibited by eserine, DFP, and paraoxon. The recombinant BChE was recognized by antibodies prepared against native human serum BChE in ELISA assays and in rocket immunoelectrophoresis. The molecular weight of the recombinant BChE was slightly lower than normal, 310,000 rather than 340,000 and this was attributed to incomplete glycosylation. The recombinant BChE had the normal tetrameric organization. This shows that the sequence in the 1825 bp cDNA fragment which included the 28 amino acid leader sequence and the 574 amino acid coding region directs secretion and assembly into a tetramer.

Supported by US Army Medical Research and Development Command Contract DAMD17-89-C-9022 (to O. L.).

Arpagaus, M., Kott, M., Vatsis,K. P., Bartels, C. F., La Du, B. N. Lockridge. O., *Biochemistry,* **1990,** 29, 124-131.

Eiberg, E., Nielsen, L. S., Klausen,J., Dahlen, M., Kristensen, M., Bisgaard,M. L. , Moller,N., Mohr , J., *Clin. Genet.* **1989**, 35, 313-321.

Harris,H., Robson, E. B. Glen-Bott , A. M. Thornton, J. A., *Nature,* **1963,** 200, 1185-1187.

Kalow, W., Genest, K. , *Can. J. Biochem. Physiol.*, **1957,** 35, 339-346.

Lockridge, O., Bartels,C. F., Vaughan,T. A., Wong,C. K., Norton , S. E., Johnson, L. L., *J. Biol. Chem.*, **1987,** 262, 549-557.

Masson, P., Chatonnet, A., Lockridge, O., *FEBS Lett.,* **1990,** 262, 115-118.

Maulet, Y., Camp,S., Gibney,G., Rachinsky,T. L., Ekstrom,T. J., Taylor, P., *Neuron* , 1990, 4, 289-301.

McGuire,M. C., Nogueira,C. P., Bartels,C. F., Lightstone, H., Hajra,A., Van der Spek, A. F. L., Lockridge, O., La Du , B. N., *Proc. Natl. Acad. Sci. USA,* **1989,** 86, 953-957.

McTiernan,C., Adkins,S., Chatonnet, A., Vaughan,T. A., Bartels,C. F., Kott,M., Rosenberry,T. L., La Du B. N., Lockridge, O., *Proc. Natl. Acad. Sci. USA* , **1987**, 84, 6682-6686.

Nogueira, C. P., McGuire, M. C., Graeser,C., Bartels,C. F., Arpagaus, M., Van der Spek, A. F. L., Lightstone,H., Lockridge,O., La Du , B. N. , *Am. J. Hum. Genet.*, **1990,** 46, 000-000.

Rubinstein, H. M., Dietz, A. A., Lubrano, T., *J. Med. Genet.*, **1978,** 15, 27-29.

Scott, E. M. , Powers, R. F., *Am. J. Hum. Genet.,* **1974,** 26, 189-194.

Soreq,H., Zamir,R., Zevin-Sonkin, D., Zakut ,H., *Hum. Genet.*, **1987,** 77, 325-328.

Sparkes, R. S., Field, L. L., Sparkes,M. C., Crist, M., Spence, M. A., James, K. Garry, P. J., *Hum. Heredity* , **1984,** 34, 96-100.

Yang, F., Lum, J. B., McGill, J. R., Moore, C. M., Naylor, S. L., Van Bragt, P. H., Baldwin, W. D., Bowman , B. H., *Proc. Natl. Acad. Sci. USA*, **1984,** 81, 2752-2756.

Zakut, H., Zamir, R., Sindel , L., Soreq, H., *Human Reprod.*, **1989,** 4, 941-946.

Differential Codon Usage and Distinct Surface Probabilities in Human Acetylcholinesterase and Butyrylcholinesterase

Revital Ben Aziz, Averell Gnatt, Catherine Prody, Efrat Lev-Lehman, Lewis Neville, Shlomo Seidman, Dalia Ginzberg and Hermona Soreq*, Department of Biological Chemistry, The Life Sciences Institute, The Hebrew University of Jerusalem, Israel 91904

Yaron Lapidot-Lifson and Haim Zakut, Department of Obstetrics and Gynecology, The Edith Wolfson Medical Center, Holon 58100, The Sackler Faculty of Medicine, Tel Aviv University

Acetylcholinesterase (AChE) and Butyrylcholinesterase (BuChE) display similar, yet distinct biochemical and immunological properties. To reveal the molecular origin for these phenomena, the coding nucleotide sequences and primary amino acid structures were compared for AChE and BuChE from man. The AChE coding sequence was found to be particularly rich in G,C residues (64%), in contrast with A,T-rich composition (63%) of the BuChE coding sequence. However, the AChE and BuChE polypeptides display a considerable level (52%) of identically aligned amino acid residues, homology which is accounted for by differential codon usage.

Molecular cloning of the human DNA sequences encoding butyrylcholinesterase (Acylcholine acyl hydrolase, EC 3.1.1.8, BuChE; Prody et al., 1986, 1987, McTiernan et al., 1987) and acetylcholinesterase (Acetylcholine acetyl hydrolase, EC 3.1.1.7, AChE, Soreq and Prody, 1989, Soreq et al., submitted) has provided the information necessary for comparative analysis of these nucleotide sequences and their translation products. Such analysis was performed with the aim of explaining the differences between the patterns of expression of these two genes and the biochemical properties of their protein products. At the level of gene expression, we wished to find out why expression of the CHE gene, encoding BuChE, is often correlated with cell division (Layer and Sporns, 1987) and precedes that of the ACHE gene encoding AChE during embryonic development (Layer et al., 1988). At the protein level, we sought to resolve the molecular determinants responsible for the antigenic differences and similarities between AChE and BuChE (see Fambrough et al., 1982, Dreyfus et al., 1988 for examples) as well as for the differences in their substrate specificities and sensitivities to selective inhibitors (Quinn, 1987, Toutant and Massoulie, 1988, Soreq and Zakut, 1990a). Finally, similarities between these two nucleotide sequences were searched for which could possibly explain the surprising phenomenon of cholinesterase gene amplification (Prody et al., 1989, Lapidot-Lifson et al., 1989, Zakut et al., 1990, Soreq and Zakut, 1990b).

1. Codon Usage Analysis

The nucleotide compositions of the coding regions of the human ACHE and CHE genes are listed in Table 1A. The sequences analyzed begin in each case with the presumptive initiator AUG codon and end with the first "nonsense" codon in the cDNAs encoding BuChE (Prody et al., 1987) or AChE (Soreq et al., submitted). Both sequences encode an "asymmetric" cholinesterase.

The AChE and BuChE coding sequences displayed almost mirror-image usage of nucleotides, with AChEcDNA being enriched in G,C-residues, and BuChEcDNA in A,T residues (Table 1A). A striking difference in the frequency of CG dinucleotides (5.9% in AChEcDNA vs. 0.8%

0–8412–2008–5/91/0172$06.00/0

TABLE 1: Nucleotide Composition and Frequencies in Human Cholinesterase genes

A. Human AChE and BuChE:sequences encoding polypeptides

AChE (2097 nucleotides)			BuChE (2325 nucleotides)		
Nucleotide	No.	%	Nucleotide	No.	%
A	370	(17.6)	A	772	(33.2)
C	682	(32.5)	C	402	(17.3)
G	656	(31.3)	G	457	(19.7)
T	389	(18.6)	T	694	(29.8)
A + T	759	(36.2)	A + T	1466	(63.1)
C + G	1338	(63.8)	C + G	859	(36.9)

B. Dinucleotide frequency :

AChE

AA	80 (3.8)	CA	127 (6.1)	GA	122 (5.8)	TA	40 (1.9)
AC	116 (5.5)	CC	266 (12.7)	GC	189 (9.0)	TC	111 (5.3)
AG	119 (5.7)	CG	124 (5.9)	GG	239 (11.4)	TG	174 (8.3)
AT	54 (2.6)	CT	165 (7.9)	GT	106 (5.1)	TT	64 (3.1)

BuChE

AA	310 (13.3)	CA	162 (7.0)	GA	157 (6.8)	TA	142 (6.1)
AC	112 (4.8)	CC	87 (3.7)	GC	77 (3.3)	TC	126 (5.4)
AG	148 (6.4)	CG	19 (0.8)	GG	114 (4.9)	TG	176 (7.6)
AT	201 (8.6)	CT	134 (5.8)	GT	109 (4.7)	TT	250 (10.8)

C. Distribution of codons in the open reading frame:

AChE

TTT	Phe	9(1.5)	TCT	Ser	3(0.5)	TAT	Tyr	3(0.5)	TGT	Cys	2(0.3)
TTC	Phe	20(3.3)	TCC	Ser	11(1.8)	TAC	Tyr	18(2.9)	TGC	Cys	6 (1.0)
TTA	Leu	0(0.0)	TCA	Ser	3(0.5)	TAA	End	0(0.0)	TGA	End	0 (0.0)
TTG	Leu	3(0.5)	TCG	Ser	2(0.3)	TAG	End	0(0.0)	TGG	Trp	17 (2.8)
CTT	Leu	3(0.5)	CCT	Pro	8(1.3)	CAT	His	2(0.3)	CGT	Arg	5 (0.8)
CTC	Leu	16(2.6)	CCC	Pro	24(3.9)	CAC	His	13(2.1)	CGC	Arg	11 (1.8)
CTA	Leu	1(0.2)	CCA	Pro	10(1.6)	CAA	Gln	3(0.5)	CGA	Arg	7 (1.1)
CTG	Leu	46(7.5)	CCG	Pro	9(1.5)	CAG	Gln	21(3.4)	CGG	Arg	14 (2.3)
ATT	Ile	1(0.2)	ACT	Thr	2(0.3)	AAT	Asn	5(0.8)	AGT	Ser	5 (0.8)
ATC	Ile	8(1.3)	ACC	Thr	7(1.1)	AAC	Asn	12(2.0)	AGC	Ser	12 (2.0)
ATA	Ile	0(0.0)	ACA	Thr	8(1.3)	AAA	Lys	3(0.5)	AGA	Arg	0 (0.0)
ATG	Met	9(1.5)	ACG	Thr	9(1.5)	AAG	Lys	7(1.1)	AGG	Arg	6 (1.0)
GTT	Val	3(0.5)	GCT	Ala	9(1.5)	GAT	Asp	8(1.3)	GGT	Gly	9 (1.5)
GTC	Val	11(1.8)	GCC	Ala	35(5.7)	GAC	Asp	21(3.4)	GGC	Gly	21 (3.4)
GTA	Val	5(0.8)	GCA	Ala	5(0.8)	GAA	Glu	3(0.5)	GGA	Gly	9 (1.5)
GTG	Val	35(5.7)	GCG	Ala	6(1.0)	GAG	Glu	31(5.0)	GGG	Gly	19 (3.1)

BuChE

TTT	Phe	30(5.0)	TCT	Ser	7(1.2)	TAT	Tyr	14(2.3)	TGT	Cys	5(0.8)
TTC	Phe	12(2.0)	TCC	Ser	5(0.8)	TAC	Tyr	6(1.0)	TGC	Cys	5(0.8)
TTA	Leu	9(1.5)	TCA	Ser	10(1.7)	TAA	End	0(0.0)	TGA	End	0(0.0)
TTG	Leu	16(2.7)	TCG	Ser	0(0.0)	TAG	End	0(0.0)	TGG	Trp	19(3.2)
CTT	Leu	14(2.3)	CCT	Pro	12(2.0)	CAT	His	9(1.5)	CGT	Arg	2(0.3)
CTC	Leu	6(1.0)	CCC	Pro	3(0.5)	CAC	His	1(0.2)	CGC	Arg	1(0.2)
CTA	Leu	4(0.7)	CCA	Pro	13(2.2)	CAA	Gln	11(1.8)	CGA	Arg	3(0.5)
CTG	Leu	6(1.0)	CCG	Pro	2(0.3)	CAG	Gln	9(1.5)	CGG	Arg	2(0.3)

Continued on next page

Table 1: Continued

ATT	Ile	13(2.2)	ACT	Thr	13(2.2)	AAT	Asn	26(4.3)	AGT	Ser	9(1.5)
ATC	Ile	6(1.0)	ACC	Thr	6(1.0)	AAC	Asn	14(2.3)	AGC	Ser	8(1.3)
ATA	Ile	12(2.0)	ACA	Thr	15(2.5)	AAA	Lys	27(4.5)	AGA	Arg	14(2.3)
ATG	Met	13(2.2)	ACG	Thr	3(0.5)	AAG	Lys	10(1.7)	AGG	Arg	2(0.3)
GTT	Val	10(1.7)	GCT	Ala	11(1.8)	GAT	Asp	17(2.8)	GGT	Gly	17(2.8)
GTC	Val	7(1.2)	GCC	Ala	10(1.7)	GAC	Asp	7(1.2)	GGC	Gly	5(0.8)
GTA	Val	8(1.3)	GCA	Ala	12(2.0)	GAA	Glu	26(4.3)	GGA	Gly	18(3.0)
GTG	Val	8(1.3)	GCG	Ala	1(0.2)	GAG	Glu	11(1.8)	GGG	Gly	7(1.2)

in BuChEcDNA) implies a particularly high potential for DNA methylation (Razin and Riggs, 1980; Weintraub et al., 1981) within the AChE coding sequence, in comparison to BuChE (Table 1B). Therefore, the conspicuously high CG dinucleotide frequency in the human AChE coding sequence may reflect a previously unforseen mode of regulation, retarding the expression of this gene where the production of BuChE would be less restricted.

The distribution of codons in the open reading frames is clearly different for the two genes. This is most apparent for the amino acids with multiple codon choices. For example, 45 of the 68 leucine residues in AChE are encoded by CTG, whereas in BuChE, only 6 of the 55 leucines translate from this codon (Table 1C). Moreover, certain codons are completely absent in the AChE gene but not in the BuChE one, like ATT and ATA for isoleucine or AGA for arginine. Thus, both the overall nucleotide composition and the third nucleotide choices for the BuChEcDNA sequence are typical of tissue specific genes (Holmquist, 1988). In contrast, the G,C-rich AChE gene resembles the α-skeletal muscle actin and the c-mos oncogene both in its exceptional nucleotide composition and in its tissue specific expression.

It is interesting to note that the proteins encoded by these two sequences have almost identical amino acid compositions (Table 2). In particular, the content of acidic and hydrophobic residues was precisely retained to be almost equal in the two proteins. The somewhat lower content of basic amino acid residues in AChE (51 as compared with 61 in BuChE) deviates from this rule, and results in a more acidic isoelectric point predicted for AChE (5.42 vs. 7.39 for BuChE).

2. Implications on Gene Amplification

Both the human AChE and BuChE coding sequences were recently found to amplify *in vivo* (Soreq and Zakut, 1990a). However, because of the very different sequence composition in these two genes, it is likely that distinct cis- and trans- elements control the initiation of RNA transcription as well as DNA replication in them. There is, consequently, no reason to attribute the observed amplification of both these discrete sequences in humans to common sequence domains or to interactions with similar DNA-binding protein(s). This leaves the acetylcholine hydrolyzing capability of cholinesterases as the sole common property which might, perhaps, explain the frequent amplification of their corresponding genes in tumor and germ cells.

Gene amplification is found under conditions where it provides a selection advantage to the cells in which it occurs (see Schimke, 1984, Stark, 1986 for reviews). The inevitable assumption would hence be that cholinesterases may provide certain cells with such an advantage, for a yet unknown reason. It was recently shown that cholinergic signalling in mammalian cells is related with cell division processes (Ashkenazi et al., 1989), supporting the above postulate. Further experiments should be performed to reveal the molecular mechanisms controlling such complex responses and explain the perpetuation of cholinesterase gene amplification.

3. Surface Probability Measurements

The polypeptide chains encoded by human AChE and BuChEcDNAs were both subjected to computerized plot structure analysis according to the Chou and

TABLE 2: Amino acid composition in complete polypeptides

AChE (613 a.a)

Ala	55	(9.0)	Leu	69	(11.2)
Arg	43	(7.0)	Lys	10	(1.6)
Asn	17	(2.8)	Met	9	(1.5)
Asp	29	(4.7)	Phe	29	(4.7)
Cys	8	(1.3)	Pro	51	(8.3)
Gln	24	(3.9)	Ser	36	(5.9)
Glu	34	(5.5)	Thr	26	(4.2)
Gly	58	(9.4)	Trp	17	(2.8)
His	15	(2.4)	Tyr	21	(3.4)
Ile	9	(1.5)	Val	54	(8.8)

Acidic	(Asp + Glu)	63	(10.3)
Basic	(Arg + Lys)	53	(8.6)
Aromatic	(Phe + Trp + Tyr)	67	(10.9)
Hydrophobic (Aromatic + Ile + Leu + Met + Val)		208	(33.9)

Molecular Weight of naked polypeptide = 67803.

BuChE (602 a.a.)

Ala	34	(5.6)	Leu	55	(9.1)
Arg	24	(4.0)	Lys	37	(6.1)
Asn	40	(6.6)	Met	13	(2.2)
Asp	24	(4.0)	Phe	42	(7.0)
Cys	10	(1.7)	Pro	30	(5.0)
Gln	20	(3.3)	Ser	39	(6.5)
Glu	37	(6.1)	Thr	37	(6.1)
Gly	47	(7.8)	Trp	19	(3.2)
His	10	(1.7)	Tyr	20	(3.3)
Ile	31	(5.1)	Val	33	(5.5)

Acidic	(Asp + Glu)	61	(10.1)
Basic	(Arg + Lys)	61	(10.1)
Aromatic	(Phe + Trp + Tyr)	81	(13.5)
Hydrophobic (Aromatic + Ile + Leu + Met + Val)		213	(35.4)

Molecular Weight of naked polypeptide = 68425.

Fassman prediction (1984). This analysis provides best-guess predictions regarding folding patterns within the analyzed proteins as well as evaluations on the probability of particular peptide domains to be located at the surface of the fully folded enzymes. Fig. 1 presents these predictions for human AChE and BuChE. As expected, the active site domain in each protein is embedded within a part of the molecule with a particularly low surface probability. In contrast, regions with higher likelihood of being at the surface appear in both enzymes at the two ends of the active site. However, in spite of the conserved positions of the 7 cysteine residues in these two proteins and the considerable amino acid sequence similarities, clear differences could be observed in the folding pattern predicted for their polypeptide chains. Thus, the low-surface probability domain in the AChE protein extends over a region longer than 300 amino acids, whereas in BuChE it is limited to about 100 residues (Fig. 1). This indicates a deeper groove for the AChE active site, and may explain its high substrate specificity as compared with that of BuChE (Quinn, 1987).

Potential N-linked glycosylation sites in both proteins appear to be close to the surface. However, the potential for glycosylation is considerably lower in AChE than in BuChE (3 vs. 7 sites; Fig. 1). This, in turn, predicts large differences between the immunological properties of AChE and BuChE, in complete agreement with previous reports (reviewed by Toutant and Massoulie, 1988). Nonetheless, the considerable similarity in the primary amino acid sequences of the two molecules (Fig. 2) explains the cross-reactivity with AChE of antibodies elicited against the recombinant BuChE fusion protein produced in E. coli (Dreyfus et al., 1988; reviewed by Soreq and Zakut, 1990b).

4. Expression of Recombinant Human Acetylcholinesterase

Microinjected oocytes of the frog *Xenopus laevis* constitute the first heterologous expression system in which recombinant cholinesterases could be produced. Studies of total mRNA microinjections (Soreq et al., 1982, 1984, 1986, Soreq, 1985, Gnatt et al., 1990), were followed by use of the oocytes for production of recombinant human BuChE (Soreq et al., 1989, Dreyfus et al., 1989). When synthesized in oocytes, cloned human cholinesterases display correct biochemical properties regarding substrate specificity and interaction with selective inhibitors. In contrast, both the assembly pattern of molecular forms and their mode of association with subcellular fractions are different from the *in vivo* patterns, and may be altered by co-injection of tissue mRNAs. This implies

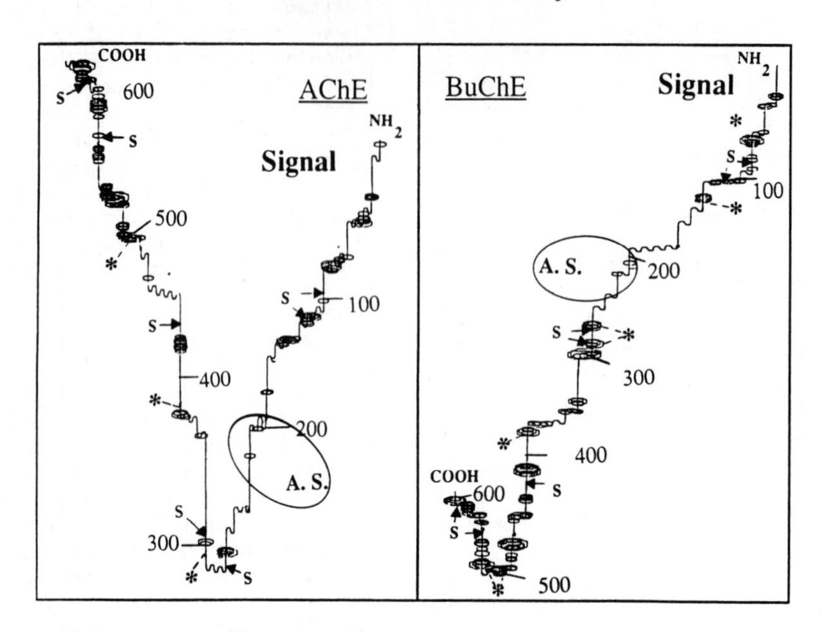

Fig. 1: The polypeptide sequences encoded by AChEcDNA and BuChEcDNA were subjected to Plot structure analysis by the Chou and Fasman prediction. Cysteine residues are noted by S- and potential glcosylation sites by stars. Peptide domains with high surface probabilities are encircled and active site regions are noted by oval signs.

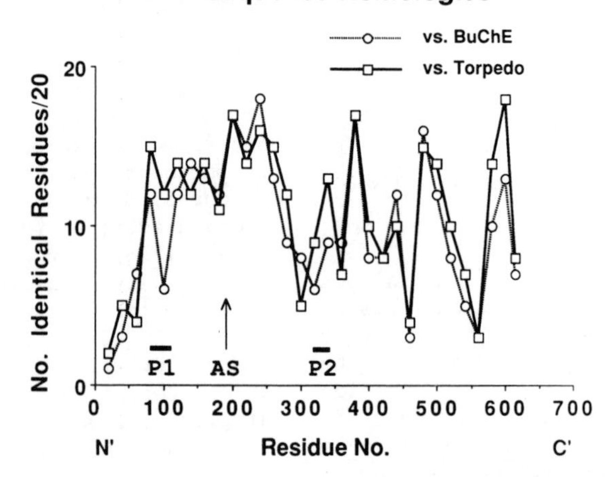

Fig. 2: AChE-specific peptides.
Amino acid similarities are plotted for human AChE as compared with human BuChE and with Torpedo AChE. Peptide domains with considerably higher homologies between AChEs are marked (P1 and P2). AS: Active site.

that faithful production of cholinesterases in large quantities will have to be performed in carefully selected cell types (Velan et a., submitted). Last, but not least, the combination of multiple scientific approaches and model systems will hopefully reveal the biological function(s) of cholinesterases in specific cell types and developmental stages.

Acknowledgements

Supported by the U.S. Army Medical Research and Development Command (contract DAMD 17-87-C-7169, to H.S.), by the Research Fund at the Edith Wolfson Medical Center (to H.Z.) and by the Association Francaise Contre les Myopathies (AFM), France (to H.S. and H.Z.). Y.L.L. and L.N. are the recipients of post-doctoral fellowships from the Levi Eshkol and the Golda Meir Funds, respectively.

References

Ashkenazi, A., Ramachandran, J., Capon, D.J., *Nature*, **1989**, *340*, 146-150.
Chou, P.Y., Fassman, G.D., *Adv. in Enzymol.*, **1978**, *47*, 45-148.
Dreyfus, P., Zevin-Sonkin, D., Seidman, S., Prody, C., Zisling, R., Zakut, H., Soreq, H., *J. Neurochem.*, **1988**, *51*, 1858-1867.
Dreyfus, P.A., Seidman, S., Pincon-Raymond, M., Murawsky, M., Rieger, F., Schejter, E., Zakut, H., Soreq, H., *Cell. Mol. Neurobiol.*, **1989**, *9*, 323-341.
Fambrough, D.M., Engel, A.G., Rosenberry, T.L., *Proc. Natl. Acad. Sci. (USA)*, **1982**, *79*, 1078-1082.
Gnatt, A., Prody, C.A., Zamir, R., Lieman-Hurwitz, J., Zakut, H., Soreq, H. *Cancer Res.*, **1990**, 50, 1983-1987.
Holmquist, G., *In chromosomes and chromatin*, Vol. II, (K.W. Adolph, ed.), CRC Press, **1988**, pp. 75-121.
Lapidot-Lifson, Y., Prody, C.A., Ginzberg, D., Meytes, D., Zakut, H., Soreq, H., *Proc. Natl. Acad. Sci. USA*, **1989**, *86*, 4715-4717.
Layer, P.G., Alber, R., Sporns, O., *J. Neurochem.*, **1987**, *49*, 175-182.
Layer, P.G., Rommel, S., Bulthoff, H., Hengstenberg, R., *Cell Tissue Res.*, **1988**, *251*, 587-595.
Massoulie, J., Toutant, J.P., *Hand. Exp. Pharm.*, **1988**, *86*, 167-193.
McTiernan, C., Adkins, S., Chantonnet, A., Vaughan, T.A., Bartels, C.F., Kott, M., Rosenberry, T.L., LaDu, B.N., *Proc. Natl. Acad. Sci. USA*, **1987**, *84*, 6682-6686.
Prody, C., Zevin-Sonkin, D., Gnatt, A., Koch, R., Zisling, R., Goldberg, O., Soreq, H., *J. Neurosci.* **1986**, *16*, 25-35.
Prody, C., Gnatt, A., Zevin-Sonkin, D., Goldberg, O., Soreq, H., *Proc. Natl. Acad. Sci. USA*, **1987**, *84*, 3555-3559.
Prody, C.A., Dreyfus, P., Zamir, R., Zakut, H., Soreq, H., *Proc. Natl. Acad. Sci. USA*, **1989**, *86*, 690-694.
Quinn, D., *Chem. Rev.*, **1987**, *87*, 955-979.
Razin, A., Riggs, A.D. *Science*, **1980**, *210*, 604-606.
Schimke, R.T., *Cell*, **1984**, *37*, 705-712.
Schumacher, M., Camp, S., Maulet, Y., Newton, M., MacPhee-Quigley, K., Taylor, S.S., Friedman, T., Taylor, P., *Nature*, **1986**, *319*, 407-409.
Soreq, H., Parvari, R., Silman, I., *Proc. Natl. Acad. Sci. (USA)*, **1982**, *79*, 830-834.
Soreq, H., Zevin-Sonkin, D., Razon, N., *EMBO J.*, **1984**, *3*, 1371-1375.
Soreq, H., *CRC Critical Reviews in Biochemistry*, **1985**, *18*, 199-238.
Soreq, H., Dziegielewska, K.M., Zevin-Sonkin, D., Zakut, H., *Cell. Mol. Neurobiol.*, **1986**, *6*, 227-237.
Soreq, H., Seidman, S., Dreyfus, P.A., Zevin-Sonkin, D., Zakut, H., *J. Biol. Chem.*, **1989**, *264*, 10608-10611.
Soreq, H., Prody, C.A., *In Computer-Assisted Modelling of Receptor-Ligand Interactions, Theoretical Aspects and Applications to Drug Design.* Eds: A. Golombeck and R. Rein, **1989**, pp. 347-359 Alan R. Liss, Inc.
Soreq, H., Zakut, H., *Pharm. Res.*, **1990a**, *7*, 1-7.
Soreq, H., Zakut, H., *Cholinesterase genes: Multileveled Regulation. Monographs in Human Genetics*, **1990b**, Vol. 13, (R.S. Sparkes,

ed.), Karger, Basel, in press.
Stark, G.R., *Cancer Surveys*, **1986**, *5*, 1-23.
Weintraub, H., Larsen, A., Groudine, M., *Cell*, **1981**, *24*, 333-341.
Zakut, H., Ehrlich, G., Ayalon, A., Prody, C.A., Malinger, G., Seidman, S., Kehlenbach, R., Ginsberg, D., Soreq, H., *J. Clin. Invest.*, **1990**, in press.

Features of Acetylcholinesterase Structure and Gene Expression Emerging from Recent Recombinant DNA Studies

Palmer Taylor, Gretchen Gibney, Shelley Camp, Yves Maulet, Tomas J. Ekström, Tara L. Rachinsky and Ying Li, Department of Pharmacology, University of California, San Diego, La Jolla, CA 92093 USA (Telefax # 619-534-6833)

Since the first primary structure of a cholinesterase was deduced, it has become clear that the enzyme defined a unique family of serine hydrolases (Schumacher et al., 1986). The sequence similarity of the cholinesterases and thyroglobulin also revealed that this family contains proteins of similar structure but without hydrolase activity (fig. 1). Molecular cloning of cholinesterases and related esterases from mammalian, lower vertebrate and invertebrate species has established several characteristics of this protein family: 1) Sequence similarity between the cholinesterases and the serine hydrolases which largely serve as proteases (i.e. trypsin and subtilisin) is limited to a small number of residues immediately around the active site serine, Ser 200, rather than global similarity within the molecule. Residue identity around the presumed catalytic histidine (His 440) is even more limited and may only be characterized by a glycine and perhaps an acidic amino acid on its C-terminal side. 2) The rank ordering of the serine and histidine within the linear sequence indicates that the cholinesterase family emerged as a consequence of convergent evolution rather than diverging from one of the other known hydrolase families. 3) Conservation of the three disulfide loops, A,B, and C, in many members of

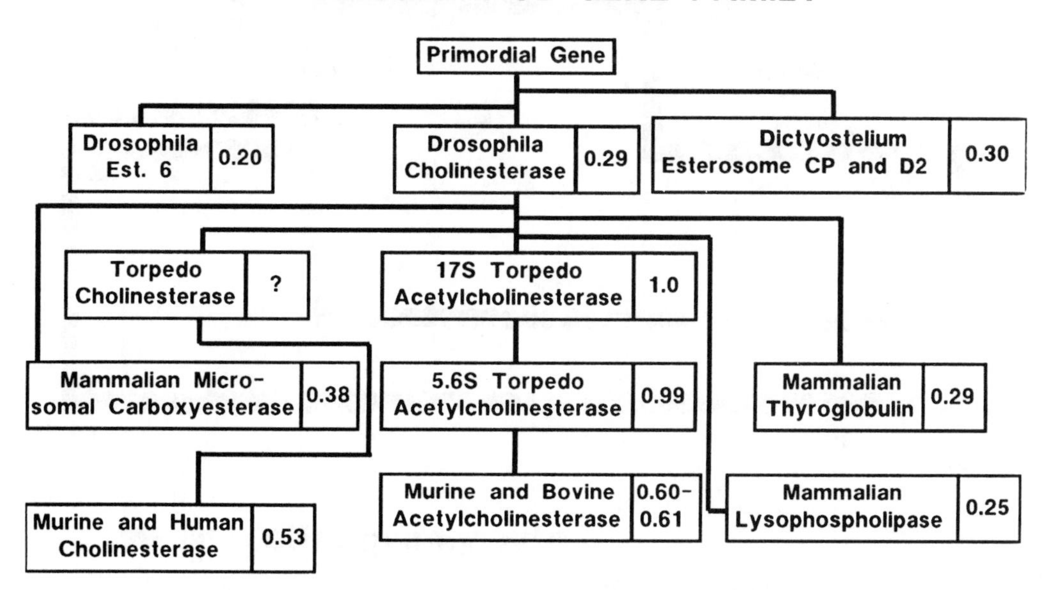

Fig. 1. The Cholinesterase Family of Proteins. The figure depicts the relationship of cholinesterases and related proteins based on overall sequence homology and evolutionary development. The numbers in the boxes represent sequence identity to Torpedo acetylcholinesterase. The selection of Torpedo is based largely on historical precident. No attempt has been made to analyze the phylogenetic relationships in terms of conservation and abundance of amino acids.

the family (MacPhee-Quigley et al., 1986) (i.e. all of the cholinesterases, thyroglobulin and certain other esterases) and loops A and B in the other esterases of the family suggests a common tertiary structure for this family of proteins (fig. 2).

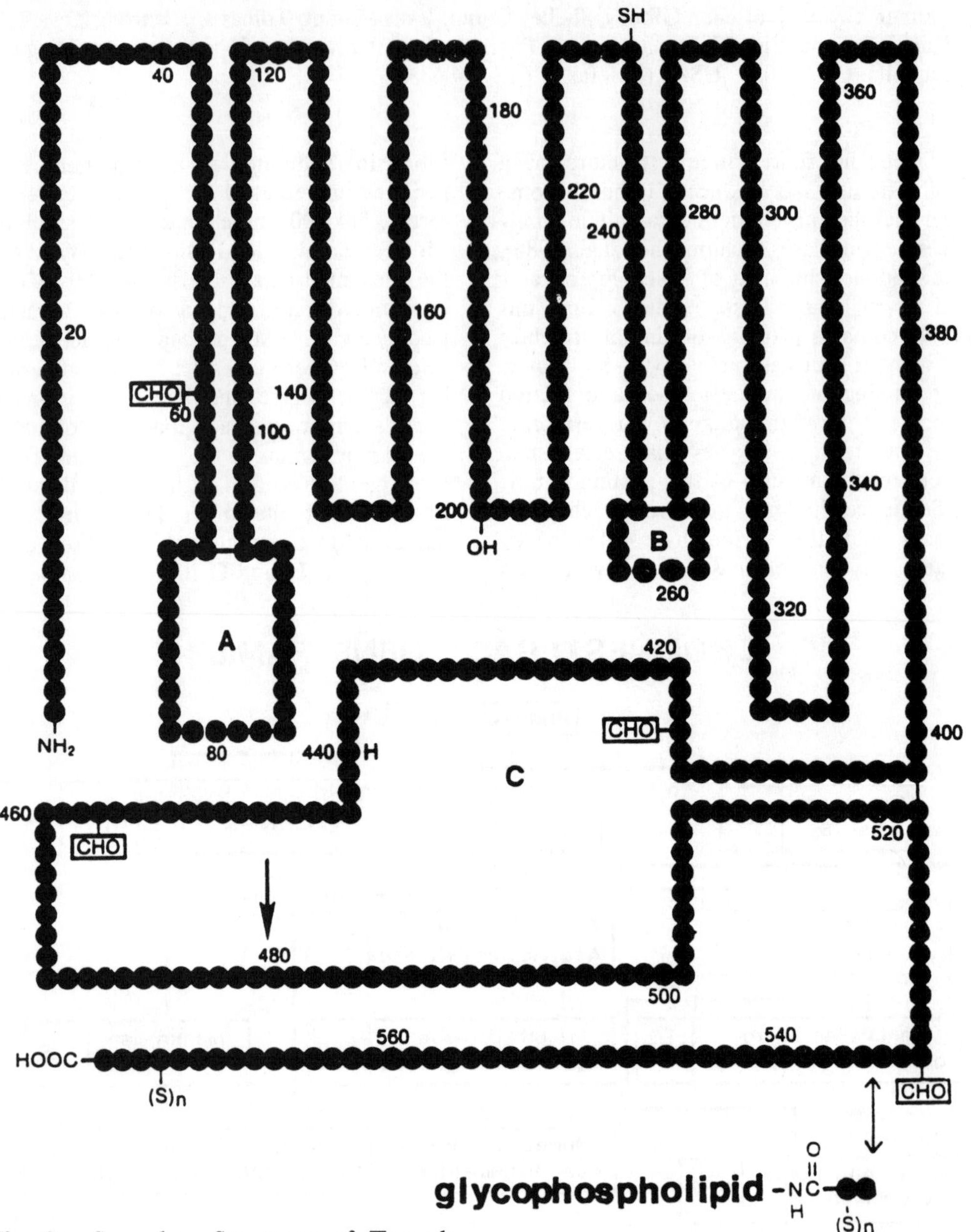

Fig. 2. Secondary Structures of Torpedo Acetylcholinesterases Based on Disulfide Bond Assignments (MacPhee-Quigley et al., 1986). The active-site serine is shown at position 200, the free cysteine at 231 and the catalytic histidine at 440. The three disulfide loops Cys 67 to Cys 94, Cys 254 to Cys 265, Cys 402 to Cys 511, are designated as A,B,C. Cys 572 bonds in intersubunit linkages. The single arrows after 479 denotes the exon 1-2 junction while the double arrow after 535 denotes the exon-intron junction of alternative mRNA processing.

Molecular Basis of Structural Diversity of the Cholinesterases

The cholinesterases can be subdivided into those containing associations of homologous subunits (homomeric) and heterologous subunits (heteromeric). The heteromeric forms result from a disulfide association of the hydrophilic form of the catalytic subunit with either a collagen-containing subunit (cf: Massoulie and Toutant, 1988) or a lipid-linked subunit (Inestrosa et al., 1987). This presumably provides a means of tethering the catalytic subunits either to the basal lamina or to the outside surface of the plasma membrane. This event occurs post-translationally at a later Golgi stage (Rotundo, 1984).

The homomeric associations give rise to dimers and tetramers which may differ substantially in their amphipathic character. This difference results from the presence of a glycophospholipid at the carboxy-terminus giving rise to an amphiphilic enzyme. Glycophospholipid attachment is a consequence of common processing events seen in such anchored proteins. It involves co-translational removal of a carboxy-terminal hydrophobic peptide and addition of a glycophospholipid at the cleavage point (Ferguson and Williams, 1989).

Amino acid sequencing of the catalytic subunits of the Torpedo enzyme species (Gibney et al, 1988) revealed the divergence point between the hydrophilic and amphiphilic species. These findings were correlated with distinct cDNA clones (Sikorav et al, 1988), S-1 nuclease digestion (Sikorav et al, 1988) and RNase digestion (Schumacher et al, 1988), which all revealed the divergence point in the encoding mRNA. Finally, the positions of the alternative exons in the DNA were established from isolation and sequencing of the genomic clones (Maulet et al, 1990) (fig. 3).

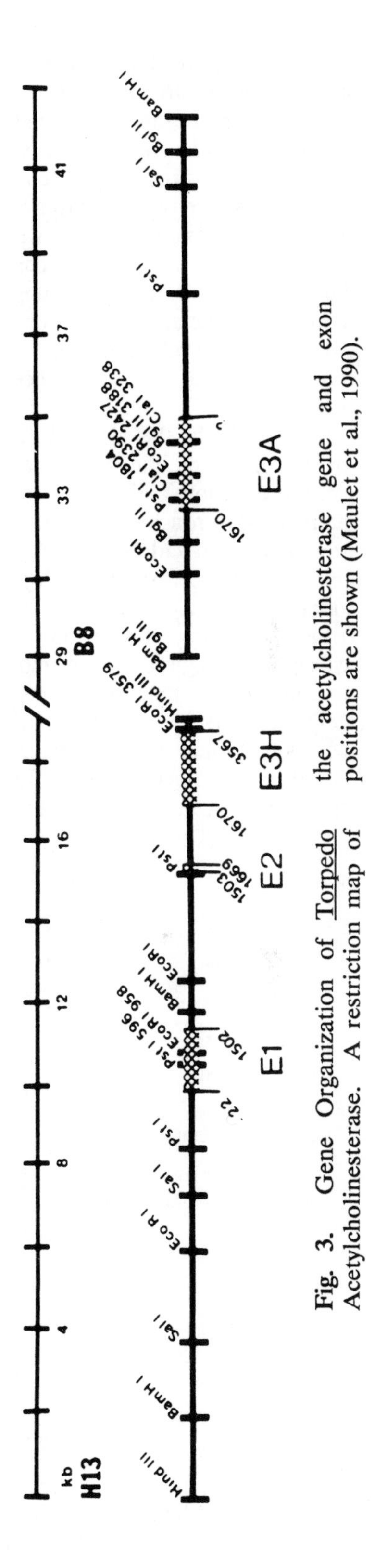

Fig. 3. Gene Organization of Torpedo Acetylcholinesterase. A restriction map of the acetylcholinesterase gene and exon positions are shown (Maulet et al., 1990).

Expression of the hydrophilic catalytic subunit leads to formation of soluble tetramer, membrane-associated tetramers linked either to a lipid-linked subunit or to each of the three triple-helical collagen containing subunits. The latter constitute the asymmetric forms associated with the basal lamina (A forms). The initial commitment to these assembled forms containing hydrophilic subunit or to the amphiphilic forms consisting of glycophospholipid-linked subunits is controlled by a mRNA processing event (fig. 4).

To determine the structural requirements for glycophospholipid attachment, we have employed genomic DNA to splice out the intervening intron between the last common exon (exon 2) and the exon encoding the unique sequence of the glycophospholipid-containing species (Gibney et al, 1990) (fig. 3). Since this exon encodes the hydrophobic region of the molecule, it is termed 3H. The mutagenesis construct yields a cDNA species containing the three exons in the open reading frame (exons 1, 2 and 3H). S1 nuclease protection establishes that single stranded DNA derived from this construct is fully protected by an existing RNA species within the cell. Transfection of the cDNA into COS-1 cells reveals expression of the sequence encoding glycophospholipid-linked form. This can be established through characterization of the gene product. The predominance of enzyme activity resides on the cell surface, as ascertained by antibody staining (fig. 5) and by extraction of the enzyme from the cell surface. At least 80% of the enzyme activity can be dissociated by treatment of the intact cell with phosphatidylinositol-specific phospholipase C. The expressed enzyme contains incorporated ethanolamine (Gibney and Taylor, 1990).

To then examine what components of the native peptide chain are necessary for processing, we deleted exon 3H from the construct. This yielded an enzyme which terminated at Thr 535 instead of Leu 575 in the hydrophilic form or Cys 537 in the processed glycophospholipid-linked form. While not catalytically active, the enzyme is synthesized in amounts equivalent to the native enzyme and secreted into the medium (Gibney et al, 1990). Hence, the sequence

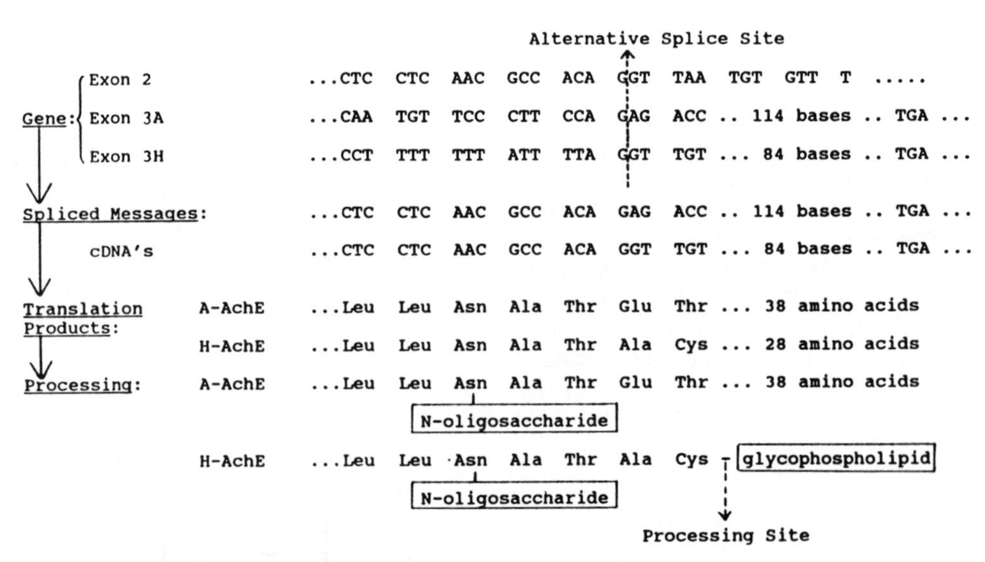

Fig. 4. Sequence of alternative mRNA processing, translation and post-translational events in the formation of the distinct carboxy termini of the catalytic subunits of Torpedo acetyl-cholinesterase (cf: Maulet et al., 1990).

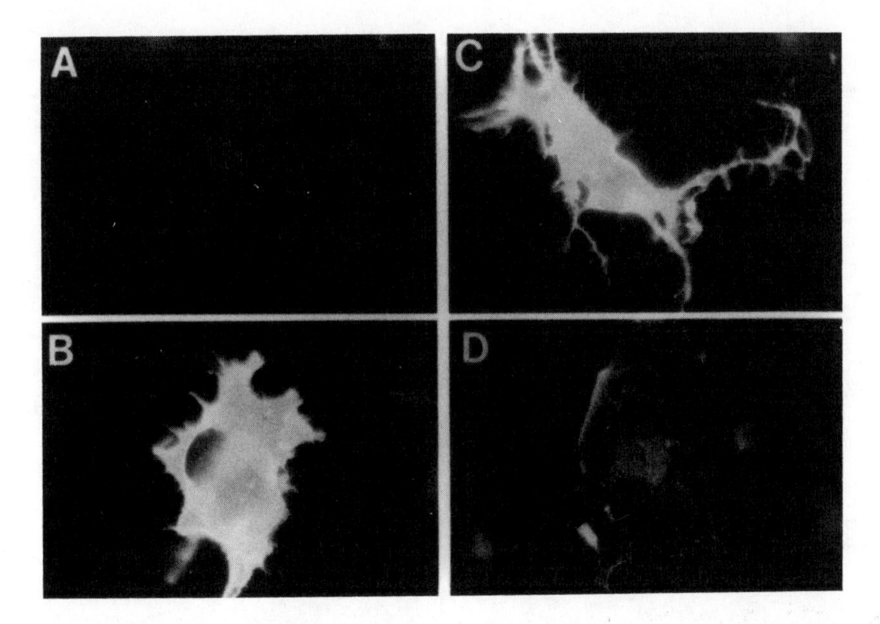

Fig. 5. Expression of Acetylcholinesterase in COS cells. A cDNA encoding the hydrophobic form of acetylcholinesterase was transfected into COS 1 cells. Expression of the enzyme on the cell surface was ascertained by reaction with a fluorescent antibody 36 hours after transfection. A) non transfected cell; B and C) cells transfected with the plasmid encoding acetylcholinesterase; D) cells transfected with the same plasmid and subsequent treatment with phosphatidyl- inositol-specific phospholipase C. (From Gibney and Taylor, 1990).

encoded by 3H is necessary for the attachment of the glycophospholipid. To establish whether exon 3H sequence is sufficient for the processing and glycophospholipid attachment, we directly ligated exon 3H to exon 1. This should yield an expressed protein which has amino acids from 480 to 535 deleted from the native sequence. Since splicing occurs within the codon triplet encoding residue 535, only residues 535 to 537 (Thr-Ala-Cys) would be common to the carboxy terminus of the deletion mutant and wild-type coded protein after processing of the 28 amino acid peptide and addition of the glycophospholipid. These residues start at position 479 in the mutant protein. This protein is also not active, but on the basis of antibody reactivity, it is expressed, incorporates ethanolamine, and is retained on the cell surface. The results of these experiments are summarized in fig. 6.

Site Directed Mutagenesis

The glycophospholipid-linked form of acetylcholinesterase is particularly well suited for mutagenesis studies because its retention on the cell surface allows for an accounting of total enzyme produced. Transient transfections do not yield sufficient enzyme for purification so we have used the ratio of catalytic activity to [^{35}S]-protein precipitated by an antibody to acetylcholinesterase to estimate specific activity. Several mutations have been created (see Table 1) and the kinetic properties of these mutants have been analyzed. The data to date have revealed essential roles for Ser200 and His440 in catalysis (Gibney et al., 1990). Mutations of Glu199 to Asp and to Gln yield interesting differences in kinetic properties of the enzyme while mutation of the Glu to His yields an enzyme with no detectable activity. Complete

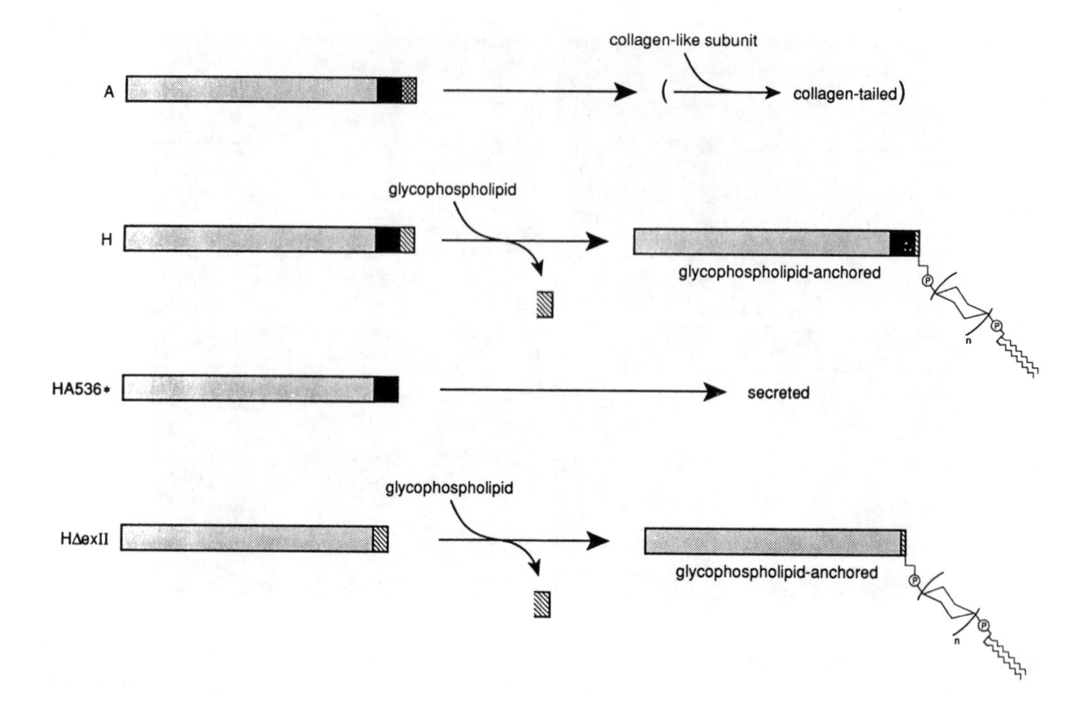

Fig. 6. Open Reading Frame Exons Used in Expression of the Asymmetric (A) and Glycophospholipid Forms (H) of Acetylcholinesterase and two mutated forms (HA536* and HΔexII). The shading denotes the individual exons from the genomic sequence (1, light; 2, dark; 3A cross hatched; 3H, diagonal). HA536 shares the same 535 amino acid sequence but lacks either carboxy terminal domain. HΔexII has the coding sequence of exons 1 (1-479) and 3H (536-565) in tandem as a result of an in frame deletion of exon 2. Residues 538 through 565 are processed upon addition of the glycophospholipid at residue 537. The results of the transfection experiments are shown above. In COS cells the A form remains intracellular, presumably because of the lack of tail unit. The H enzyme with the exon 3H are processed with addition of a glycophospholipid denoted by ethanolamine, a polymer of 6-membered rings (monosaccharides and inositol) and the diacylglycerol. Truncated HA536* is secreted into the medium. (From Gibney and Taylor, 1990).

Table 1. Mutations of Torpedo Acetylcholinesterase Which Have Been Constructed, Sequenced and Expressed

Ser_{200}	→	Val_{200}	Glu_{199}	→	His_{199}
Ser_{200}	→	Cys_{200}	His_{425}	→	Gln_{425}
Glu_{199}	→	Gln_{199}	His_{440}	→	Gln_{440}
Glu_{199}	→	Asp_{199}	His_{425},His_{440}	→	Gln_{425},Gln_{440}

characterization of these and other mutants will require enhanced production and purification of the expressed enzymes.

Alternative mRNA Processing in Mammalian Systems

Glycophospholipid-linked, asymmetric (attached to collagen-containing subunits), hydrophobic (attached to a lipid-linked subunit) and soluble forms of acetylcholinesterase have been identified in mammalian tissues, but the multiplicity of gene products makes it difficult to characterize fully all of the molecular forms of the enzyme. Accordingly, analysis of mRNA species may provide an indication of abundance of the various molecular forms of the protein. To accomplish this, we have cloned the cDNA's encoding the hydrophilic forms of murine acetylcholinesterase and butyrylcholinesterase (Rachinsky et al, 1990; see also Randall, 1990). RNase protection using antisense mRNA from the cDNA and hybridi-zation with total mRNA reveals that the cloned cDNA reflects the species in major abundance in mouse brain and muscle. In fact, no evidence for splicing at the exon 2-3 junction can be deduced in all of the murine tissues thus far examined except for cells of erythroid origin. This raises the interesting possibility that the glycophospholipid-containing species, although prevalent in fish muscle and brain (Stieger et al, 1989; Bon and Massoulie, 1989) is not a predominant species in mammalian tissues other than certain cells of hemopoietic origin. Hence, information on the genomic structure of mammalian cholinesterases and isolation of the cDNA species encoding the glycophospholipid-containing form of acetylcholinesterase will be critical to the analysis of tissue abundance and tissue specific regulation of AChE expression.

Supported by USPHS GM18360 and a contract from the DAMDC.

References

Bon, S., Toutant, J.-P., Méflah, K., Massoulie, J. *J. Neurochem.* **1988**, *51*, 776-785

Ferguson, M.A.J., Williams, A.F. *Ann. Rev. Biochem.* **1988**, *57*, 285-320.

Gibney, G., MacPhee-Quigley, K., Thompson, B., Vedvick, T., Low, M.G., Taylor, S.S., Taylor, P., *J. Biol. Chem.*, **1988**, *263*, 1140-1145.

Gibney, G., Taylor, P., *J. Biol. Chem.*, **1990**, *255*, 12576-12583.

Gibney, G., Camp, S., Dionne, M.S., MacPhee-Quigley, K., Taylor, P., **1990**, *Proc. Natl. Acad. Sci. (USA)*, in press.

Inestrosa, N.C., Roberts, W.L., Marshall, T.L., Rosenberry, T.L., *J. Biol. Chem.*, **1987**, *262*, 4441-4444.

MacPhee-Quigley, K., Vedvick, T., Taylor, P., Taylor, S.S., *J. Biol. Chem.*, **1986**, *261*, 13565-13570.

Massoulié, J., Toutant, J.P., *Handbook of Experimental Pharmacology*, V.P. Whittaker, ed., Vol. 86, Springer Verlag, Berlin, pp. 167-224.

Maulet, Y., Camp, S., Gibney, G., Rachinsky, T., Ekström, T.J., Taylor, P., *Neuron*, **1990**, *4*, 289-301.

Rachinsky, T., Li, Y., Camp, S., Newton, M., Ekström, T., Taylor, P., submitted for publication.

Randall, W., **1990** this volume.

Rotundo, R.L., *Proc. Natl. Acad. Sci. USA*, **1984**, *289*, 479-483.

Schumacher, M., Camp, S., Maulet, Y., Newton, M., MacPhee-Quigley, K., Taylor, S.S., Friedmann, T., Taylor, P., *Nature*, **1986**, *319*, 407-409.

Schumacher, M., Maulet, Y., Camp, S., Taylor, P., *J. Biol. Chem.*, **1988**, *263*, 18979-18987.

Sikorav, J.L., Duval, N., Anselmet, A., Bon, S., Krejei, E., Legay, C., Osterhind, M., Reimund, B., Massoulie, J., *EMBO J.*, **1988**, *7*, 2983-2993.

Silman, I., Futerman, A.H., *Eur. J. Biochem.*, **1987**, *170*, 11-22.

Stieger, S., Gentinetta, R. and Brodbeck, U. *Eur. J. Biochem.*, **1990**, *181*, 633-642.

Two Different Genes Encoding Cholinesterases in Chicken

Yves Maulet*, Zentrum für Molekulare Biologie, Im Neuenheimer Feld 282, D-6900 Heidelberg
Marc Ballivet, Département de Biochimie, Université de Genève, 30 quai E. Ansermet, CH-1211 Genève 4

Vertebrates express two catalytic subunits of cholinesterases encoded by different genes : acetylcholinesterase (AChE) and butyrylcholinesterase (BuChE). Their first coding exon (cE1) encodes about 85% of the polypeptide chain (Maulet et *al.*,1990).

We screened a genomic library from chicken in lambda L47 with a cDNA probe from *Torpedo californica* AChE at low stringency and obtained clones from two genes named ChEI and ChEII.

ChE I

Overlapping clones cover an entire cE1 with conserved splice sites. The gene encodes a polypeptide chain that exhibits more homology with human BuChE (67%) than with Torpedo AChE (51%). The active site serine and the histidine involved in charge relay (Gibney et *al.*,1990), the cysteins involved in disulfide bridges (Macphee-Quigley et *al.*, 1986) and part of the potential glycosylation sites are conserved.

ChE II

Two independent clones cover cE1 down to 30 nucleotides 5′to the expected donor splice site. The acceptor splice site is conserved. The exon contains an inserted sequence of 699 nucleotides which is not found in other available sequences from cholinesterases. In the conserved regions, the encoded polypeptide exhibits a higher homology with Torpedo AChE (56%) than with human BuChE (50%). The active site serine and histidine, the cysteins participating to disulfide bridges and part of the potential glycosylation sites are conserved.

The insert consists in two short motives (25 and 45 nucleotides) repeated several times. that make it a potential hot spots for mutations. No splice site is present at the boundaries of the repeats and the reading frame of the conserved sequence is not interrupted nor shifted. This region should encode a polypeptide strech of 233 amino acids. It matches the observed size difference between AChE from chicken (about 100 kDa) and those from other vertebrates (60-70 kDa). As no disulfide bridge links the conserved domains and as catalytic residues are found on each side of the insert, this finding suggests that the catalytic site is constituted by juxtaposition of two domains far apart along the polypeptide chain. The formation of a functional catalytic site would thus depend on an intrinsic affinity between the two domains. It is also possible that, in oligomeric AChE, two polypeptide chains could contribute to single catalytic subunits.

References

Maulet,Y.,Camp, S., Gibney, G., Rachinsky, T.L., Eckström, T., and Taylor, P., *Neuron,* **1990**,*4*, 289-301.

Gibney, G., Camp, S., Dionne, M.S., MacPhee-Quigley, K., Taylor,P., **1990** submitted

MacPhee-Quigley, K., Vedvick, T.S., Taylor, P., Taylor, S., *J. Biol. Chem.*, **1986**, *26*, 13565-13570.

0–8412–2008–5/91/0186$06.00/0

Search for Alternative Splicing of Butyrylcholinesterase Transcripts

Omar Jbilo and **Arnaud Chatonnet,** Département de Physiologie animale, INRA, place Viala, 34060 Montpellier Cedex 1, France
Martine Arpagaus and **Oksana Lockridge,** Department of Pharmacology, University of Michigan, Ann Arbor 48109-0626 Michigan, USA

Alternative splicing of pre-messenger RNA was shown to be responsible for the genesis of hydrophilic and amphiphilic forms of acetylcholinesterase (AChE) in Torpedo (Sikorav *et al.*, 1988; Schumacher *et al.*, 1988). The only sequences obtained so far on butyrylcholinesterase protein or cDNA and gene correspond to a hydrophilic form identical to the plasma BChE, with a C terminus translated from an exon homologous to the last exon of hydrophilic form of AChE (Lockridge et al., 1987; McTiernan *et al.*, 1987; Prody *et al.*, 1987; Jbilo *et al.*, this volume).

There is some evidence that part of BChE activity is associated with membranes. The detergent requirement for solubilisation of BChE molecules could be due to the amphiphilic nature of some BChE forms. Also a direct interaction of BChE with non denaturing detergent has been recently demonstrated (Bon *et al.*, 1988).

In order to investigate the possibility of an alternative splicing at the 3' end of BChE gene that could lead to amphiphilic forms of BChE we used PCR amplification of cDNA 3' ends. Reverse transcription was performed using an oligo dT with restriction sites at the 3' end. Amplification was performed with a primer corresponding to the restriction sites and a primer corresponding to the 3' end of exon 2 of BChE. Only cDNAs of BChE should be amplified and any type of 3' end should be represented. In rabbit and human heart liver and brain, a major band of 0.8 Kb was amplified which corresponded to the previously cloned terminal exon. Minor bands were also found which hybridized with exon 3 probes. It is not yet clear wether these bands correspond to alternative splicing or to the use of different poly-A signals.

Supported by INRA AIP "Cholinesterase" and an "Association Française contre les Myopathies" Grant.

Lockridge O., Bartels C.F., Vaughan T.A., Wong C.K., Norton S.E., and Johnson L.L. *J. Biol. Chem.,* **1987,** *262*, 549-557.

McTiernan C., Adkins S., Chatonnet A., Vaughan T.A., Bartels C.F., Kott M., Rosenberry T.L., La Du B.N. and Lockridge O. *Proc. Natl. Acad. Sci. USA* , **1987,** *84,* 6682-6686.

Prody, C.A. Zevin-Sonkin D., Gnatt A., Goldberg O. and Soreq H. *Proc. Natl. Acad. Sci. USA* , **1987,** *84*, 3555-3559.

Arpagaus M., Kott M., Vatsis K.P., Bartels C.F., La Du B.N. and Lockridge O. *Biochemistry* , **1990,** *29*, 124-131.

Schumacher M., Maulet Y., Camp S., and Taylor P. *J. Biol. Chem.* **1988,** *35*, 18979-18987.

Bon S., Toutant J.P., Méflah K., and Massoulié J. *J. Neurochem.* **1988,** *51*, 786-794.

Sikorav J.L., Duval N., Anselmet A., Bon S., Krejci E., Legay C., Osterlund M., Reimund B. and Massoulié J. *EMBO J.,* **1987,** *6*, 1865-1873.

Structure of Rabbit Butyrylcholinesterase Gene: Description of a Repetitive Short Interspersed Element (SINE) Specific of Rabbit Genome

Omar Jbilo, Thierry Lorca, Abdelhamid Barakat, and Arnaud Chatonnet,
Physiologie Animale, I N R A, place Viala, 34060 Montpellier Cedex 1, France.

The structure of the rabbit butyrylcholinesterase (BChE) gene is similar to the human BChE gene (Chatonnet *et al.* 1990, Arpagaus *et al.*, 1990) and to the *Torpedo* acetylcholinesterase gene as well (Maulet *et al.*, 1990). The identity over the coding sequence between human and rabbit BChE is 90.3% in nucleotides and 91.6% in amino acids.

Southern blots of genomic rabbit DNA digested by EcoRI were hybridized with probes corresponding to the four human exons and showed only one band. These results suggest that there is a single BChE gene in rabbit, a situation similar to that found in human (Arpagaus *et al.*, 1990).

In addition to the genomic clones, one cDNA clone (BNY1) was isolated. This cDNA was unusual in that it contained intronic sequences. A 300 nucleotide long region at the 5' end of this clone presented homologies with non coding regions of other rabbit genes. This sequence contained a short interspersed element (SINE) highly repeated in rabbit genome as detected by high hybridization of genomic southern blots. An homologous sequence was found in intron 3 upstream of exon 4, so this element was present at least twice in the BChE gene of rabbit. The same SINE has been shown recently to be a source of polyadenylation signals for some rabbit genes when this sequence is found downstream of the last exon (Krane and Hardison, 1990). In our case the SINE sequence found between exons 3 and 4 contained an uninterrupted repetition of 10 AATAA motifs.

ATCCTCCTCGATGGCCCAAGTTCTTGAGA
CCCTGCACCTGCATGGGAGACCAGGGAG
AAGCACCCAGCTCCTGGCTTCTGATTGGC
GCACGGACCGGCTGTAGTGGCCATTTGG
GGGATGAACCAAAGGAAGCAAGACCTTC
CTGTCTGTCTCTCCCTCTCACTAATTCTGC
CTGTCAAAAATAAAATAAAATAAAATAAAA
TAAAATAAAATAAAATAAAATAAAATAAAA

Fig 1. Partial sequence of a SINE element found upstream of exon 4, in rabbit BChE gene.

Supported by AIP "Cholinesterases" INRA Grant and an "Association Française contre les Myopathies" Contract.

Chatonnet, A., Lorca T., Barakat A., Aron E., and Jbilo O., *Mol. Cell. Neurobiol.* **1990,** in press.

Arpagaus M., Kott, M., Vatsis, K.P., Bartels, C.F., La Du, B.N. and Lockridge, O., *Biochemistry,* **1990,** *29*, 124-131.

Krane D.E. and Hardison R.C., *Mol. Biol. Evol.,* **1990,** *7*, 1-8.

Maulet Y., Camp S., Gibney G., Rachinsky T., Ekstrom T.J. and Taylor P., *Neuron,* **1990,** *4*, 295-301.

0–8412–2008–5/91/0188$06.00/0

A DNA Point Mutation Associated With the H-variant of Human Butyrylcholinesterase

Frank S. Jensen, Department of Anesthesia, University Hospital, Copenhagen, Denmark
Cynthia F. Bartels and Bert N. La Du*, Department of Pharmacology, University of Michigan Medical School, Ann Arbor, Michigan 48109-0626 USA

The H-variant is a quantitative variant that reduces enzymatic activity by approximately 90% (Whittaker, M. and Britten, J.J., *Hum. Hered.* **1987**, 37, 54-58). Two Danish families with this phenotype were found by the Danish Cholinesterase Research Unit and have been investigated by our laboratory. DNA from one individual heterozygous for both the H-variant and the atypical (dibucaine-resistant) variant (AH) was sequenced for the entire coding region to detect any structural abnormalities. All three exons were amplified by the polymerase chain reaction, and the two smaller exons were sequenced directly with end-labeled primers. The large exon was cloned into M13 phage and sequenced with a series of internal BCHE primers.

Family pedigree analysis indicated that the single strand carrying the H mutation did not contain the atypical mutation. Both DNA strands were completely sequenced, and a point mutation GTG->ATG (Val 142 -> Met) was found in the non-atypical strand. Subsequent sequencing of this region in an AH sibling of the above AH individual, and in two AH siblings from the other, presumably unrelated, family showed exactly the same point mutation. As yet, there is no easy way to diagnose the H-variant other than by direct sequencing of the DNA or by probability, based upon pedigree analysis. The point mutation neither removes nor creates a new RFLP (restriction fragment length polymorphism) site. Studies are in progress to relate this structural mutation to the severe quantitative reduction in BChE activity, characteristic of this phenotype.

Identification of a Frameshift Mutation Responsible for the Silent Phenotype of Human Butyrylcholinesterase in Serum (Variant Ann Arbor)

Christine P. Nogueira, Mary C. McGuire, Cynthia Bartels, Martine Arpagaus, Oksana Lockridge, and Bert N. La Du*, Department of Pharmacology, University of Michigan Medical School, Ann Arbor, Michigan 48109-0626
Abraham F.L. van der Spek, Department of Anesthesiology, University of Michigan Medical School
Harold Lightstone, Department of Anesthesiology, Metropolitan Hospital, Philadelphia, Pennsylvania

The silent phenotype of human butyrylcholinesterase is characterized by the lack of activity of the enzyme in serum. A frameshift mutation that causes this phenotype was identified in 7 individuals of two unrelated families, using the polymerase chain reaction and sequencing of DNA (white blood cells). The mutation, at position 117, changes GGT (Gly) -> GGAG (Gly+1 base). The extra base changes the reading frame from Gly 117 to a new stop codon created at position 129, upstream from the active site (Ser 198). If any protein were made, it would represent only 22% of the mature enzyme found in normal serum. Rocket immunoelectrophoresis, with alpha-naphthyl acetate to detect enzymatic activity, showed an absence of cross-reactive material as expected. Analysis of the enzymatic activities in serum agreed with the genotypes inferred from the nucleotide sequence. One additional individual with a silent phenotype did not show the same frameshift mutation. This was not unexpected since there must be considerable molecular heterogeneity involved in causes for the silent phenotype.

The mutation occurs in a tandemly repeated sequence GGT-GGT-GGT (Gly 115-117). This kind of sequence is consistently associated with a high frequency frameshift mutation in other organisms (hotspots), because of the mispairing of strands in this region (slippage mechanism). According to this model it is suggested that in the mutation Gly 117, GGT -> GGAG, G is the extra base and A is the point mutation.

Supported by NIH grant GM-27028

0–8412–2008–5/91/0189$06.00/0

Identification of Two Different Mutations Associated With Human Butyrylcholinesterase Fluoride Resistance in Serum

C.F. Bartels, C.P. Nogueira, M.C. McGuire, S. Adkins, O. Lockridge and B.N. La Du*, Department of Pharmacology, University of Michigan Medical School, Ann Arbor, MI 48109-0626 USA
H.M. Rubinstein and T. Lubrano, Medical and Research Services, Veterans Administration Hospital, Hines, IL, and Department of Medicine, Loyola University Stritch School of Medicine, Maywood, IL
A.F.L. van der Spek, Department of Anesthesiology, University of Michigan
H. Lightstone, Department of Anesthesiology, Metropolitan Hospital, Philadelphia, PA

White blood cell DNA from individuals exhibiting the fluoride-resistant phenotype of serum butyrylcholinesterase was analysed using the polymerase chain reaction followed by cloning or direct sequencing. Two different point mutations that cause this phenotype were identified. The first mutation, a transition from Thr 243 A<u>C</u>G->A<u>T</u>G(Met), was identified in an AF (heterozygous atypical/heterozygous fluoride-resistant) individual sensitive to succinylcholine and his AF mother. The mutation was called the fluoride-1 (formal name BCHE*243M). The change from the threonine residue to another amino acid in this position would result in an enzyme subunit having 8 carbohydrate chains instead of the usual 9, since Asn-X-Thr or Ser is the sequence necessary for glycosylation at the Asn site. Asn 241 would not become glycosylated.

A second mutation was found at Gly 390, where G<u>G</u>T->G<u>T</u>T(Val). This mutation was called the fluoride-2 (formal name BCHE*390V). The entire coding region of a JF (heterozygous J-variant/heterozygous fluoride-resistant) individual was sequenced and no other mutations were found, except for another point mutation that segregated with the J-phenotype. The mutation was also seen in the individual's UF and AF relatives and was subsequently found in two other pedigrees and a single individual, all exhibiting the fluoride phenotype in serum analysis. In all cases, the DNA genotypes were consistent with the phenotypes determined by serum BChE activity, dibucaine inhibition and fluoride inhibition.

Both mutations result in the loss of a restriction enzyme site (Mae II for the fluoride-1, and Hph I for the fluoride-2), so it may be possible to diagnose DNA using RFLP methods.

We gratefully acknowledge Karen James for her help in obtaining blood samples.

Supported by NIH grant GM-27028

0–8412–2008–5/91/0190$06.00/0

DNA Coding for the K Polymorphism in Linkage Disequilibrium with Atypical Human Butyrylcholinesterase Complicates Phenotyping

Cynthia Bartels*, Oksana Lockridge and Bert N. La Du, Department of Pharmacology, University of Michigan Medical School, Ann Arbor, MI 48109-0626 USA
Abraham F. L. van der Spek, Department of Anesthesiology, University of Michigan
Herbert M. Rubinstein and Tina Lubrano, Medical and Research Services, Veterans Administration Hospital, Hines, Illinois, and Department of Medicine, Loyola University Stritch School of Medicine, Maywood, Illinois

The K-variant is a quantitative variant that reduces enzyme activity by about one-third (Rubinstein HM, Dietz AA and Lubrano T, *J Med Genet*, **1978**, 15, 27-29). White blood cell genomic DNA, amplified by the polymerase chain reaction, was sequenced with 5'-labeled primers. The K-phenotype was associated with an exon 4 point mutation at nt 1615 (GCA to ACA) that changed Ala 539 to Thr. From WBC DNA of 46 phenotypically usual, unrelated people, the allelic frequencies were found to be 0.87 for GCA and 0.13 for ACA; out of the 46 individuals, one was homozygous Thr/Thr^{539} and six were heterozygous. This fits the Hardy-Weinberg equation and the 1.7% Thr/Thr^{539} value approaches the 1.3% predicted for the K-variant by Evans and Wardell from phenotyping studies (*J Med Genet*, **1984**, 21, 99-102).

DNA was obtained from members of the original two pedigrees describing the K-variant (Rubinstein). The three UK members were heterozygous Ala/Thr^{539} while the three heterozygous atypical, phenotypically AK members were homozygous Thr/Thr^{539}. It is of interest that the K-variant mutation is in linkage disequilibrium with the exon 2 mutation responsible for the atypical phenotype. When DNA from 3 homozygous atypical and 10 heterozygous atypical blood samples was sequenced, 12/13 atypical alleles (88%) were linked with Thr^{539}. This may explain the previous observation that the average atypical level of activity is reduced one-third at V_{max} compared with the average usual activity.

It is in the heterozygous atypical (UA) condition when the quantitative influence of the K-variant becomes important. Here, dibucaine and Ro2 inhibition are both determined by the relative contribution of each gene to the total amount of enzyme produced. Since most atypical alleles are linked to the K-variant, what is generally known as UA is actually U^{Ala}/A^{Thr539}, and phenotyping standards are based on this combination. Whenever this condition is varied, U^{Thr}/A^{Thr}, U^{Thr}/A^{Ala}, or U^{Ala}/A^{Ala}, the serum produced usually exhibits the AK phenotype because the dibucaine and Roche inhibition numbers are reduced . The most common of these AK combinations is U^{Thr}/A^{Thr}. The U^{Ala}/A^{Ala} combination gives borderline values for the UA versus AK distinction.

This DNA point mutation does not create or remove a restriction enzyme cutting site but can be identified by direct DNA sequencing or by the use of allele specific probes.

Supported by NIH grant GM-27028 (to B.N.L.)

0–8412–2008–5/91/0191$06.00/0

Practical Consequences of Having More Than One Mutation Within the Same Butyrylcholinesterase Gene

Bert N. La Du*, Cynthia Bartels, and Oksana Lockridge, Department of Pharmacology, University of Michigan Medical School, Ann Arbor, Michigan 48109-0626 USA

Our laboratory has identified at least 7 specific sites within the BCHE gene which determine either polymorphisms, or rate BChE variants. These intra-genetic loci are very close to each other, and one BCHE gene may carry more than one BCHE mutation. Linkage disequilibrium has been noted (see Bartels et al., poster) between atypical (dibucaine-resistant) variant and the quantitative K-variant. In another family carrying the J-variant, the affected allelic DNA strand also carried K-variant mutation.

The 7 polymorphic and variant DNA sites within the BCHE gene are being used in the search for additional structural BChE variants. Family studies and linkage analyses help to identify which allelic DNA strand must carry the unknown mutation. This information reduces the amount of DNA sequencing required, particularly when the new mutation is present only in single dose (heterozygosity) in the key pedigree members.

Application of this approach is illustrated by our current studies on the J-variant, associated with two-thirds reduction in BChE activity (Garry, P.J. et al. *J. Med. Genet.* 1976, 13, 38-42). This large pedigree also has the atypical, fluoride-2 and K-variants. These additional markers have been used to determine which allelic DNA strand must carry the J-mutation and should be sequenced.

Another consequence of having more than one variant DNA mutation on one DNA strand is that a person's BCHE genotype is no longer adequately defined by finding any two phenotype characteristics (A/K, A/F), if both are carried on one DNA strand (AK/U), the other BCHE DNA strand would be "wild type" at all of these mutation sites. Each mutation site must be examined, and its linkage to other BCHE markers suspected.

Supported by NIH grant GM-27028

Nomenclature for Human Butyrylcholinesterase Genetic Variants Identified by DNA sequencing

Oksana Lockridge*, Cynthia F. Bartels, Christine P. Nogueira, Martine Arpagaus, Steve Adkins, and **Bert N. La Du,** Pharmacology Dept., University of Michigan Medical School, Ann Arbor, MI 48109-0626 USA
*(After July 1, 1990) Eppley Institute, Universityof Nebraska Medical Center, 600 S. 42nd St., Omaha, NE 68198-6805 USA

The Human Gene Mapping Nomenclature Committee has reserved the symbol BCHE for the butyrylcholinesterase gene and the symbol ACHE for the acetylcholinesterase gene. BCHE replaces the earlier CHE1 symbol. E1 and E2 are outmoded and not in use. CHE2 will remain in use temporarily to designate the C_5 phenotype, though it is clear from the work of Arpagaus et al *(Biochemistry 29: 124-131, 1990)* and Eiberg et al *(Clin. Gen. 35:313-321, 1989)* that the CHE2 gene locus does not include a coding region for BChE, but codes for a protein that modifies butyrylcholinesterase. The BCHE gene is on chromosome 3 on the long arm q21-26 *(Soreq et al Hum. Genet 77:325-328, 1987).* Formal names for genetic variants identified by DNA sequencing include the amino acid number (amino acid #1 is the N-terminal of the mature protein) and the single letter amino acid symbol for the mutation. For example, atypical butyrylcholinesterase, which has a defective anionic site, has glycine in place of aspartic acid 70. The formal name for the atypical BCHE gene is *BCHE*70G*. Other examples are in the table below.

The fact that the BCHE gene is present in a single copy in the human genome *(Arpagaus et al 1990)* allows BCHE to be used as a landmark that defines position on the physical map of chromosome 3 for the Human Genome Mapping and Sequencing Project. We recommend use of the abbreviation BChE, with a small "h", to designate the butyrylcholinesterase protein.

Supported by US Army Medical Research and Development Command DAMD17-89-C-9022 (to O. L.) and NIH grant GM27028 (to B.N.L.).

Common name	Phenotypic description	Amino acid alteration	DNA alteration	Formal name for genotype
Usual	Normal	none	none	*BCHE*
Atypical	Dibucaine resistant	70 Asp→Gly	nt 209 (GAT to GGT)	*BCHE*70G*
Silent	Silent, no activity	117 Gly→Frame shift	nt 351 (GGT to GGAG)	*BCHE*FS117*
Fluoride	Fluoride resistant	243 Thr→Met	nt 728 (ACG to ATG)	*BCHE*243M*
Fluoride	Fluoride resistant	390 Gly→Val	nt 1169 (GGT to GTT)	*BCHE*390V*
K-variant	K-polymorphism	539 Ala→Thr	nt 1615 (GCA to ACA)	*BCHE*539T*

0–8412–2008–5/91/0193$06.00/0

Use of the Polymerase Chain Reaction for Homology Probing of Butyrylcholinesterase (BCHE) in Several Animal Species

Martine Arpagaus, Theresa A. Vaughan, Bert N. La Du, Oksana Lockridge,
Department of Pharmacology, University of Michigan Medical School, Ann Arbor, Michigan.
Patrick Masson, C.R.S.S.A., 38702 La Tronche Cedex, France.
Arnaud Chatonnet, INRA, 34060 Montpellier Cedex, France.
Michael Newton, Palmer Taylor, Department of Pharmacology, University of California, La Jolla, California, USA.

We have recently studied the structure of the human BCHE gene, and shown that the coding sequence for the protein present in serum is contained in four exons (Arpagaus *et al.*, *Biochemistry*, **1990**, *29*, 124-131). Our data unambiguously show the existence of a single BCHE locus in the human genome (see also Masson *et al.*, *FEBS Lett.*, **1990**, *262*, 115-118). We now present additional data on the BCHE gene in a series of animals (monkey, cow, sheep, dog, pig, rat, rabbit, mouse, guinea pig, chicken).

Genomic blots were hybridized with probes corresponding to each human exon. The hybridization pattern showed the existence of a single BCHE gene in every species examined. We then used the polymerase chain reaction method (PCR) on genomic DNA from these animal species. Two oligonucleotide primers were designed from the human sequence. They are located in exon 2 which contains 83% of the coding sequence for the mature protein. The amplified piece was 507 bp long; it covered the region from Asn 68 to Leu 208, thus including the active site Serine (198) and the Aspartate 70, an essential component of the anionic site (McGuire *et al.*, *Proc. Natl. Acad. Sci. USA*, **1989**, *86*, 953-957). The amplification was carried out for 30 cycles consisting of denaturation at 94°C for 1.5 min, annealing at 50 or 45°C (depending on the species) for 1.5 min, and extension at 72°C for 3 min. The amplified segments were cloned in M13mp18 and M13mp19 to be sequenced. The amplification with the human primers was successful except for rat, mouse, guinea pig and chicken DNAs. The mouse sequence was found by sequencing a genomic clone. Table 1 summarizes the number of nucleotide variations (out of 423) and amino acid changes (out of 141) in the different species as compared to the human sequence.

The active site Serine 198 of the enzyme, the presumptive anionic site Aspartate 70, the glycosylation site on Asparagine 106, as well as Cysteine 92 involved in an intrachain disulfide bridge were conserved in all sequences. Computer programs were used for the construction of phylogenic trees of mammalian BChEs (see abstract Cousin *et al.*, this volume).

Supported by US Army Medical Research and Development Command DAMD17-89-C-9022 (to O.L.) and NIH grant GM27028 (to B.N.L.).

Table 1. Number of nucleotide variations (out of 423) and amino acid changes (out of 141) in different species as compared to the human sequence

	Monkey	Sheep	Cow	Rabbit	Pig	Dog	Mouse
Nt variations	5	25	27	32	35	37	58
AA changes	0	6	6	4	6	9	12

0–8412–2008–5/91/0194$06.00/0

Phylogeny of Cholinesterases Inferred by Maximum Parsimony Method or Distance Matrix Methods (Fitch-Margoliash and Neighbor Joining Methods)

Xavier Cousin, Jean-Pierre Toutant, Omar Jbilo and Arnaud Chatonnet,
Département de Physiologie animale, INRA, place Viala, 34060 Montpellier Cedex 1, France
Martine Arpagaus, Department of Pharmacology, University of Michigan, Ann Arbor, 48109-0626 Michigan USA.

The most widely used methods for inferring phylogenies from sequence data are the maximum parsimony method and the distance matrix methods.

We first used these methods on the **nucleotide** sequences of mammalian butyrylcholinesterase (BChE) gene as determined by Arpagaus et al. (See this volume). There are 8 sequences representing five orders of mammals. We used the Phylip 2.03 package of Felsenstein (1988) for the parsimony method and the matrix method of Fitch and Margoliash (1967). We also used the more recent Neighbor Joining method of Saitou and Nei (1987) which was proved to be more powerful to give the real phylogenetic tree (Sourdis and Nei, 1988). For this method we used the program NJTREE and NJDRAW of Dr. Ferguson and Dr. Lin (U. of Texas, Houston). All methods gave similar trees (Fig.1), and the phylogeny deduced is close to phylogenies deduced from other sequence data or fossils records.

We also used the computer methods for a comparison of **amino acid** sequences of mammalian BChE, the *Torpedo* and *Drosophila* AChE, as well as sequences of non specific esterases (Est 6 and Est P of *Drosophila*, Juvenile hormone esterase of Lepidoptera, and two liver esterases of rabbit). The protein sequences were aligned and distances calculated with the Dayoff matrix. Hypotheses of gene duplications during evolution of cholinesterase family can be inferred from the trees. Specially, it appears that the divergence of AChE and BChE genes is a relatively recent event and concerns only vertebrates (see the phylogenetic tree in Chatonnet and Jbilo, this volume).

This work was supported by AIP "Cholinesterase" INRA Grant and an "Association Française contre les Myopathies" Contract. We wish to thank Pr. Nei (U. of Texas, Houston) for the Neighbor Joining Method program.

Sourdis J. and Nei M. *Mol. Biol. Evol.* **1989,** *5*, 298-311
Felsenstein J. *Ann. Rev. Genet.* **1988,** *22*, 521-565
Saitou N. and Nei M., *Mol. Biol. Evol.* **1987,** *4*, 406-425
Fitch W.M. and Margoliash E., Science **1967,** *155*, 279-284

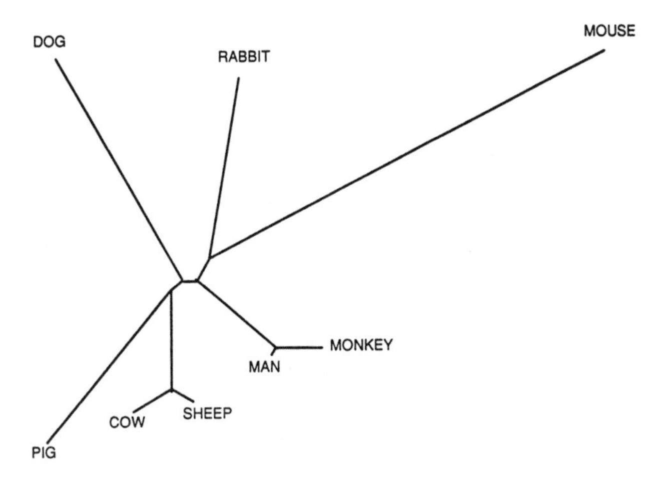

Fig. 1. Phylogeny of mammals inferred by the "Neighbor Joining Method" applied on BChE sequences

Expression of the Fluoride Variant of Human Butyrylcholinesterase in Chinese Hamster Ovary Cells

Steve Adkins, T. A. Vaughan, C. F. Bartels, B. N. La Du, O. Lockridge*
Department of Pharmacology, University of Michigan Medical School, Ann Arbor, MI 48109-0626, USA.
*(After July 1, 1990) Eppley Institute, University of Nebraska Medical Center, 600 S. 42nd St., Omaha, NE 68198-6805 USA

A plasmid vector was constructed for the expression of the fluoride variant of human butyrylcholinesterase. A cDNA containing the fluoride mutation (glycine 390 to valine) was inserted into an expression vector containing an adenovirus promoter and SV40 poly A signal. Chinese Hamster Ovary cells were cotransfected with the variant BCHE construct and a similar construct containing a mutant mouse dihydrofolate reductase cDNA. Transfected colonies were selected by nucleoside starvation and the BCHE gene was amplified by exposing the cells to increasing concentrations of methotrexate. BChE production was monitored by measuring the benzoylcholine hydrolysis activity of the culture medium. The maximum production of BChE was 50 µg/L/day secreted into serum free medium. The secreted enzyme was a tetramer.

The recombinant BChE hydrolyzed benzoylcholine, a-naphthyl acetate, butyrylthiocholine and propionylthiocholine and was inhibited by DFP (1 µM), eserine (10 µM) and paraoxon. The recombinant BChE was recognized by anti-human BChE rabbit serum. Southern blotting showed 4 copies of the human BCHE gene in one cell line and 2 copies in another cell line. The Km for benzoylcholine at 25°C was 5 µM and the Km for propionyl thiocholine at 37°C was 0.1 mM. Bartels et al (see poster) identified the Gly 390 to Val mutation as a fluoride variant. The recombinant BChE is especially valuable because it is homozygous with respect to the fluoride variant, a genotype which has been extremely difficult to detect in the population before the use of PCR and DNA sequencing.

Supported by US Army Medical Research and Development Command DAMD17-89-C-9022 (to O.L.) and NIH grant GM27028 (to B.N. L.).

Modified ligand-binding properties of butyrylcholinesterase-muteins produced in microinjected Xenopus oocytes

Lewis F. Neville, Averell Gnatt, Shlomo Seidman, Yael Loewenstein, Ruth Padan and Hermona Soreq, Department of Biological Chemistry, The Life Sciences Institute, The Hebrew University of Jerusalem, Jerusalem 91904, Israel.

Butyrylcholinesterase (BuChE, EC 3.1.1.8) (Whittaker, 1986; Soreq and Gnatt, 1987; Soreq and Zakut, 1990) has a likely biotransforming role especially in the degradation of a number of ester-containing drugs, containing the neuromuscular relaxant succinylcholine (SucCh). However, in some individuals, the presence of an abnormal or "atypical" serum BuChE whose capacity to degrade SucCh is markedly reduced, permits a greatly exaggerated neuromuscular block following SucCh administration with a resultant prolonged period of apnea of often hours duration (Whittaker, 1986). Recently with the advent of polymerase chain reaction (PCR), a single amino acid mutation of gly70 for asp70 (McGuire et al, 1989) was postulated as being the cause of "atypical" BuChE by a mechanism likely to involve deletion of the anionic site. However, expression studies pertaining to the biochemical alterations involved in transforming a normal BuChE into one presenting "atypical" characteristics are not yet available.

In order to characterize structure-function relationships within atypical BuChE, a double-mutated BuChEcDNA sequence expressed in brain neuroblastomas and glioblastomas was employed (Gnatt et al, 1990). This "unusual" BuChEcDNA encodes the amino acid substitutions asp70 to gly70 and ser425 to pro425 and was inserted into the pSP64 transcription vector. Synthetic RNA was subsequently made, injected into Xenopus oocytes and the resultant BuChE mutant (GP) characterized for its ligand-binding properties according to previously published procedures (Soreq et al, 1989). In order to delineate the precise roles of both amino acid substitutions, 2 further recombinant vectors were constructed containing the BuChEcDNA sequence encoding either the gly70 mutation (GS), or the pro425 mutation (DP). Both normal serum BuChE and the usual recombinant DS BuChE, transcribed from "normal" BuChEcDNA, served as controls. Serum and all recombinant BuChE proteins were assessed spectrophotometrically for BuChE activity, using a multiwell automated Ellman assay based upon the hydrolysis of butyrylthiocholine (BuTCh) as substrate (Dreyfus et al, 1988).

Dibucaine interacted with all muteins in a normal manner, except GP, which retained 80% activity even at 1mM dibucaine. This apparently indicated that both mutations are required to render BuChE "atypical" in reference with its dibucaine sensitivity. Inhibition curves with SucCh though, produced a different profile of the "atypical" properties of the recombinant muteins. As expected from its marked resistance to dibucaine, GP failed to bind SucCh. However, GS CHE which behaved in a normal manner with dibucaine, displayed 25 fold higher resistance to SucCh inhibition than DS.

Acknowledgements

Supported by the U.S.Army Medical Research and Development Command (Contract No. DAMD 17-87-C-7169, to H.S.) by the Association Francaise contre les Mypathies (AFM) France and by a Golda-Meir post-doctoral fellowship to L.N.

References

Dreyfus, P.A., Zevin-Sonkin, D., Seidman, S., Prody, C.A., Zisling, R., Zakut, H., Soreq, H., *J. Neurochem.*, **1988**, *51*, 1858-1867.

Gnatt, A., Prody, C.A., Zamir, R., Lieman-Hurwitz, J., Zakut, H., Soreq, H., *Cancer Res.*, **1990**, *50*, 1983-1990.

Harris, H., Whittaker, M., *Nature*, **1959**, *183*, 1808-1810.

McGuire, M.C., Nogueira, C.P., Bartels, C.F., Lightstone, H., Hajra, A., Van der Spek, A.F.L., Lockridge, O., LaDu,

0–8412–2008–5/91/0197$06.00/0

B.N. *Proc. Natl. Acad. Sci. (USA)*, **1989**, *86*, 953-957.

Quinn, D., *Chemical Rev.* **1987**, *87*, 955-979.

Soreq, H., *CRC Crit.Rev.Biochem.* **1985**, *18*, 199-238.

Soreq, H., Gnatt, A., *Mol. Neurobiol.* **1987**, *1*, 47-80.

Soreq, H., Seidman, S., Dreyfus, P.A., Zevin-Sonkin, D., Zakut, H., *J. Biol. Chem.* **1989**, *264*, 10608-10613.

Soreq, H., Zakut, H., *Cholinesterase genes: Multilevelled Regulation. Monographs in Human Genetics*, **1990**, Vol. 13, (R.S. Sparkes, ed.), Karger, Basel, in press.

Immunological Studies of the Plasma Cholinesterase Variants

V. Mortensen, A.G. Rasmussen, B. Norgaard-Pedersen, State Serum Institute, Copenhagen, Denmark
J.W. Jones, M. Whittaker, Department of Environmental Sciences, Polytechnic South West, Plymouth, England

Two monoclonal anti-cholinesterase antibodies (MAb2-1 and 2-4) have been prepared by the hybridoma technique and used as primary coating antibody on the microwell surfaces of an ELISA plate to screen the most common variants of human plasma cholinesterase (ChE). Aliquots of plasma and standard were added to each well and HRP conjugated sheep anti-ChE polyclonal antibody was used as the secondary detector (Whittaker et al, 1990). The results obtained are shown in table 1. The decreased level of cholinesterase found for $E_1^aE_1^a$ phenotypes compared with $E_1^uE_1^u$ phenotypes (37%) is greater than the 27% reported by Eckerson et al (1983). The enzymic activity for this phenotype is 43% of the activity for $E_1^uE_1^u$ individuals. The theoretical contribution of the E_1^u (50%) and E_1^a (33%) genes to the immunological level of ChE as deduced from the corresponding homozygotes are in excellent agreement with our experimental mean for the $E_1^uE_1^a$ heterozygote (84% of $E_1^uE_1^u$).

References

Whittaker, M., Jones J., Braven J., *Hum. Hered.* (in the press).

Eckersen, H.W., Oseroff A., Lockridge, O., La Du, B., *Biochem. Genet.*, **1983**, 21, 93-107.

Table 1. Comparison of enzymic activity using benzoylcholine(BzCh) and butyrylthiocholine(BTI) as substrate with the binding ratio of 2 monoclonal antibodies used in ELISA. Values are rationalized to the mean for $E_1^uE_1^u$

		Activity		MAb	
Genotype	Number	BzCh	BTI	2-1	2-4
$E_1^uE_1^u$	170	100	100	100	100
$E_1^uE_1^a$	60	75	75	85	83
$E_1^aE_1^a$	56	45	41	65	61

0–8412–2008–5/91/0199$06.00/0

Immunological Studies of the Apparent Silent Homozygotes for Plasma Cholinesterase using an ELISA Technique

M. Whittaker, J.W. Jones, Department of Environmental Sciences, Polytechnic South West, Plymouth, UK.

The antigenic activity of 37 unrelated apparently silent gene homozygotes have been examined using an ELISA technique (Whittaker et al.[a] 1990). A monoclonal anti-cholinesterase (ChE) antibody (MAb2-1 or 2-4) was used as primary coating antibody on the microwell surfaces of an ELISA plate. Aliquots of samples or standards were added to each well and HRP conjugated sheep polyclonal anti-ChE antibody was used as the secondary detector. The plasma samples were also examined by rocket immunoelectrophoresis (R.I.). A good correlation was observed in the quantitation by ELISA and R.I. Cumulative data is given in Table 1.

These groups may represent the genotypes $E_1^sE_1^s$, $E_1^sE_1^t$, $E_1^tE_1^t$ and a new genotype $E_1^xE_1^x$. Two individuals are probably segregating E_1^r - one as genotype $E_1^rE_1^r$ and the other $E_1^rE_1^x$.

Seven families segregating apparently silent gene homozygotes have been investigated by ELISA (Whittaker et al.[b] 1990). This data confirms not only the heterogeneity of the silent gene in the form of E_1^s and E_1^t, but in 3 families there is support for the segregation of the new gene E_1^x. One of the propositi in these families was the truely silent homozygote $E_1^sE_1^s$, whilst 3 propositi appear to be segregating both E_1^s and E_1^t. The remaining 3 propositi are segregating the E_1^x gene, but in all cases, as the heterozygote $E_1^sE_1^x$.

References

Whittaker, M., Jones, J.W., Braven, J., *Hum. Hered.* a and b (in the press)

Table 1 Cumulative data of enzymic activity using benzoylcholine (BzCh) and butyrylthiocholine (BTI) and the binding ratios of 2 monoclonal antibodies in ELISA. Units for BTI, R.I., MAb2-1 and 2-4 are relative to a pool of 5 $E_1^uE_1^u$ donors

Genotype	Number	Activity		R.I.	MAb	
		BzCh	BTI		2-1	2-4
$E_1^sE_1^s$	9	0.01±.01	0	0	0	0
$E_1^tE_1^t$	3	0.02±.02	8±5	20±0	15±4	21±3
$E_1^sE_1^t$	14	0.03±.03	2±2	6±6	4±4	4±4
$E_1^sE_1^x$	9	0.01±.01	0	118±29	153±34	165±38
$E_1^rE_1^r$	1	0.13	12	0	0	0
$E_1^rE_1^x$	1	0.08	7	35	120	66

Diagnosis of Human Butyrylcholinesterase Variants Using Biotinylated Oligonucleotide Probes

Amitav Hajra and **Bert N. La Du***, Department of Pharmacology , University of Michigan Medical School, Ann Arbor, MI 48109-0626

Short DNA strands hybridize very specifically to complementary DNA sequences; under proper conditions, even a single mismatched base will prevent hybridization. This high probe specificity makes it possible to detect individuals affected with genetic alterations by using short probes specific for either the normal sequence or the variant, mutated sequence. These allele-specific oligonucleotide (ASO) probes are a powerful tool to detect small DNA variations or polymorphisms and diagnose associated genetic disorders.

To date, we have used biotin-labelled ASO probes to detect five variants of human butyrylcholinesterase (BChE), all of which cause individuals to exhibit a heightened sensitivity to the drug succinylcholine. Four of these variants (atypical, fluoride-1, fluoride-2, and K variants) are the result of point mutations in the BCHE gene, while the fifth variant (silent variant) is the result of a frame shift mutation. The BCHE gene exhibits a great deal of allelic heterogeneity, and it provides an excellent demonstration of the clinical applications of nonradioactive ASO probes to the detection of genetic mutations and variants.

The procedure involves amplifying a 200 b.p. piece of the BCHE gene, containing the site of the mutation to be detected, by the polymerase chain reaction. Duplicate aliquots of the amplified DNA are dot blotted onto small nitrocellulose squares, and the dot blots are then hybridized with 19 base oligonucleotide probes biotinylated at their 5' prime end, and specific for either the normal DNA sequence or the variant, mutated, sequence. The same hybridization and wash conditions are sufficient to detect all five BChE variants. Successful probe-DNA hybridizations are detected by adding an avidin-alkaline phosphatase conjugate and performing a simple colorimetric reaction which yields an intense blue precipitate. Unsuccessful hybridizations remain colorless. Heterozygous DNA samples hybridize successfully with both probes, while homozygous samples hybridize with only the appropriate probe. The results are extremely clear and unambiguous, and can be obtained within hours of beginning the procedure.

In conclusion, short, biotinylated ASO probes can be used in a rapid and simple assay to clearly differentiate between DNA sequences that differ by a single base. Multiple variants can be detected in a single assay by using more than one set of probes. This method can detect clinically important BChE variants, even those not detectable by standard phenotyping methods (such as the K variant). This method is a means of rapidly and accurately characterizing individuals with variant forms of BChE without using radioactive materials.

0–8412–2008–5/91/0201$06.00/0

Serum Cholinesterase Polymorphism in France. An Epidemiological Survey of the Deficient Alleles

Arnaud J.*, Brun H., Llobera R., Constans J.
CRPG/CNRS: Centre de Recherche sur le Polymorphisme Génétique des Populations Humaines. CHU Purpan, 31300 Toulouse. France.

The CHE1 polymorphism is usually detected by enzymatic procedures. Four alleles codominantly transmitted have been described: CHE1*U, CHE1*A, CHE1*S and CHE1*F. Five genotypes ($E1^{aa}$, $E1^{as}$, $E1^{af}$, $E1^{ss}$, $E1^{sf}$) are associated with low cholinesterase activity and hypersensitivity to muscle relaxants.

The aim of this study is to report the CHE1 polymorphism in a large sample of the French population. Probabilities of anaesthetic risks were estimated.

A total of 2421 samples being collected from healthy unrelated adults in sixteen traditional regions of France ("Provinces Françaises").The populations concerned were living in rural areas only, their families residing in the same area for at least four generations.

We adopted an automated procedure, which is cheap and reliable.

Results and Discussion

Automated micro method: Comparison of the two methods (manual and automatic) gave identical results. Reproductibility and sensitivity were tested.The internal coefficient of variation was less than 6%.

Gene frequencies: In the central part of France, Cevennes, Limousin, Dauphiné and Auvergne show a high frequencies of the CHE1*A allele. A fifth region identified with an elevated frequency of the CHE1*A was Corse.

Higher frequencies for CHE1*A were already published in Southern part of Europe. The central focus observed in France may be explained by some gene admixture from Neolithic Southern invaders groups or immigration. Regarding the other and rare alleles, the frequencies observed are not very different from previous data.

The drug anaesthetic risk: A good correlation was described between the cholinesterase activity and apnea after the use of succinylcholine as drug for anaesthesia. In Cevennes, Limousin, Dauphine, Auvergne and Corse, the risk is very higher: 1/390 to 1/780. In others, the mean values is of the ordre 1/1870 which is not very different from published data.

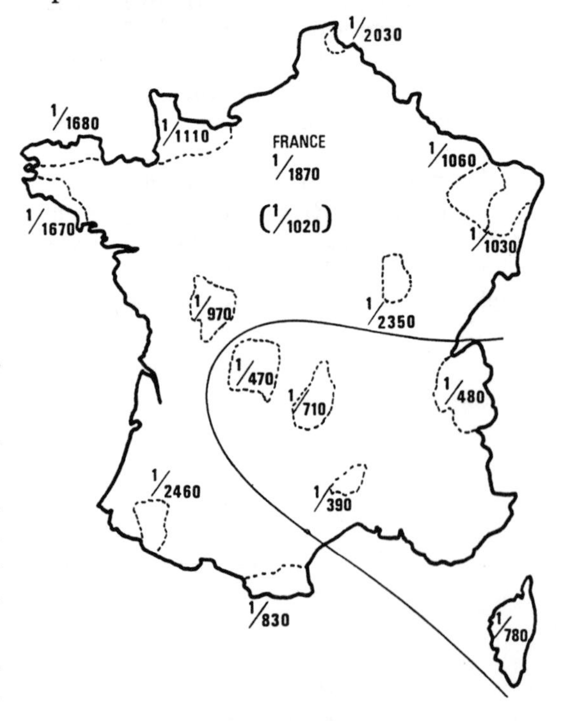

0–8412–2008–5/91/0202$06.00/0

Immunoreactive Plasma Cholinesterase Substance Concentration versus Cholinesterase Activity

Axel Brock, University of Aarhus, Department of Clinical Chemistry, Randers Centralsygehus, DK 8900 Randers, Denmark.

Substance concentrations of plasma ChE (EC 3.1.1.8) were measured in an enzyme immunoassay (Brock et al., 1990), and compared with ChE activity in 94 healthy individuals without occupational exposure to known inhibitors (six samples from each individual within a period of eight months).

ChE substance concentration

ChE substance concentration (mg immunoreactive ChE/l) was highly correlated ($r = 0.973$) to ChE activity, with a total variation corresponding to CV = 22% (mean: 5.01 mg/l, SD: 1.11 mg/l). Like ChE activity (Brock and Brock, 1990), interindividual variation of ChE substance concentration is influenced by body weight, height, sex, and ChE-1 phenotype. Estimated from a repeated measures analysis of variance, intra-individual variation of ChE substance concentration corresponds to an observed CV_{intra} = 8-8% (SD: 0.45 mg/l). Related to the actual experimental imprecision (CV = 6%), this variation implies an intraindividual variation of ChE substance concentration corresponding to a biological CV_{intra} = 6.4%

Specific catalytic ChE activity

Specific catalytic activity (kU/mg immunoreactive ChE) is influenced by the ChE-1 phenotype (U: 1.58 kU/mg, UA: 1.22 kU/mg), but neither by age, body weight, and height, nor by sex. Intraindividual variation of specific catalytic activity corresponds to an observed CV_{intra} = 6.4 % (S D.10 kU/mg). Related to the experi-mental error (CV = 6.2%), this implies an estimated biological CV_{intra} = 1.4%, which is not significantly different from zero ($F = 1.07$, $p > 0.05$).

References

Brock, A., Mortensen, V., Loft, A.G.R., Norgaard-Pedersen, B. *J. Clin. Chem. Clin. Biochem.* **1990**, 28, (in press).
Brock, A., Brock, V. *Scand. J. Clin. Lab. Invest.* **1990**, 50, (in press).

0–8412–2008–5/91/0203$06.00/0

Normal ChE-Variation: Effects by Intra- and Inter-individual Variance Components

V. Brock, Department of Ecotoxicology, Biology Institute, University of Odense, Campusvej 55, DK-5230 Odense M, Denmark

A. Brock, Department of Clinical Biochemistry, University of Aarhus, Randers Centralsygehus, DK-8900 Randers, Denmark

Assumption: Total ChE-variance for a population is the sum of inter-individual variance, intra-individual variance, and variance due to error.

Material was obtained during a period of eight months by drawing blood samples from 193 volunteers at intervals of 4-8 weeks, to obtain six samples from each individual. 131 individuals completed the full programme. None of the 193 volunteers were subject to occupational exposure to known cholinesterase inhibitors (Brock and Brock, 1990). Determination of plasma ChE activity was performed at 37°C by a continuos assay using butyrylthiocholine iodide 6 mmol/l as substrate (Brock, 1989).

Major inter-individual variance components were identified from a stepwise multiple regression analysis. 41% of total variance was explained from effects by body weight, height, sex, and ChE-1 phenotype (U or UA). Age did not influence inter-individual variance.

Intra-individual variance (which was found to be 5% of total variance) was assessed from a repeated-measures analysis of variance. Intra-individual variations (which were independent of body weight, height, sex, ChE-phenotype, and age) varied substantially from one person to another. Less than 1% of total variance was explained by experimental error.

Based on the multiple regression model a 'standardized' ChE activity may be calculated. When comparing unmatched population groups, comparisons of 'standardized' ChE activity rather than the measured ChE activity should be used to disclose a treatment effect (Table 1).

References

Brock, A., J. Clin. Chem. Clin. Biochem. **1989**, 27, 429-431.

Brock, A., Brock, V., Scand. J. Clin. Lab. Invest. **1990**, 50, 000-000.

Table 1.
Plasma ChE activity (left) and 'standardized' ChE activity (right). Model: 'Standardized' ChE = measured ChE - (weight - 75) X 0.082 + (height - 180) X 0.088 + (sex - 1) X 1.16 + (ChE phenotype - 1) X 2.78; male = 1, female = 2; ChE phenotype U = 1, UA = 2

	Mean	SD	Mean	SD	n
Whole population	8.45	1.94	8.63	1.49	192
Males	9.00	1.95	8.62	1.61	121
ChE phenotype 'U'	9.22	1.83	8.63	1.64	112
ChE phenotype 'UA'	6.36	1.50	8.58	1.36	9
Females	7.51	1.53	8.63	1.26	71
ChE phenotype 'U'	7.59	1.47	8.62	1.25	69
ChE phenotype 'UA'	4.80	1.41	8.82	1.88	2

CATALYTIC MECHANISM OF CHOLINESTERASES: STRUCTURE–FUNCTION RELATIONSHIPS OF ANTICHOLINESTERASE AGENTS, NERVE AGENTS, AND PESTICIDES

Cholinesterase and Carboxylesterase as Scavengers for Organophosphorus Agents

D.M. Maxwell*, US Army Medical Research Institute of Chemical Defense, Aberdeen Proving Ground, MD 21010-5425
A.D. Wolfe, Y. Ashani, B.P. Doctor, Division of Biochemistry, Walter Reed Army Institute of Research, Washington, DC 20307-5100

The acute toxicity of organophosphorus (OP) agents in mammals is usually attributed to their irreversible inhibition of acetylcholinesterase (AChE), an enzyme that terminates the action of acetylcholine in the nervous system (Taylor, 1985). Conventional medical treatment for OP intoxication consists of therapeutic administration of anticholinergic drugs to counteract the accumulation of acetylcholine and oximes to reactivate OP-inhibited AChE. Pretreatment with carbamates has also been used to protect AChE from OP inhibition. The inability of these pharmacological approaches to provide complete protection against OP's has led to the development of non-pharmacological approaches to protection, such as enzyme or antibody scavengers (Doctor, 1989; Dunn and Sidell, 1989).

Scavenger protection against OP's has been investigated using three enzymes- acetylcholinesterase (AChE), butyrylcholinesterase (BuChE), and carboxylesterase (CaE). CaE is an endogenous scavenger that occurs in large amounts relative to AChE and BuChE in mammals (Maxwell et al., 1988) and has been the most extensively studied OP scavenger (Boskovic, 1979; Clement, 1984; Fonnum et al., 1985; Maxwell et al., 1987). AChE is the best characterized of these esterases because of its importance in cholinergic neurotransmission (Rosenberry, 1975; Massoulie and Bon, 1982); and as an OP scavenger, AChE has the advantage that it has same specificity as the toxic site for OP's. Serum BuChE is a clinical index of OP intoxication, and purified human BuChE has been used as a clinical treatment for succinylcholine intoxication (Brown et al., 1981).

In this paper we present the results of our investigations of the ability of endogenous and exogenous scavengers to protect against OP intoxication in several species.

Materials and Methods

Animals: Male ICR mice (25-30 g), Sprague-Dawley rats (230-255), Hartley guinea pigs (330- used for *in vivo* scavenger stoichiometry measurements and LD_{50} determinations. LD_{50} values were calculated by probit analysis (Finney, 1971) of deaths occurring within 24 hr following agent administration.

Materials: Human serum BuChE was purchased from Behring Institute, FRG. AChE was purified from fetal bovine serum by the method of De La Hoz et al. (1986). Soman and VX were obtained from Chemical Research, Development and Engineering Center, Aberdeen Proving Ground, MD. 2-(O-cresyl)-4H-1:3:2-benzodioxaphosphorin oxide (CBDP) was synthesized by Starks Associates, Buffalo, NY. 7-(Methylethoxyphosphinyloxy)-1-methyl-quinolinium iodide (MEPQ) was prepared as previously described (Levy and Ashani, 1986).

Enzyme analysis: AChE and BuChE activities were determined by the method of Ellman et al. (1961) using acetylthiocholine and butyryl thiocholine as substrates, respectively. CaE activity was assayed by the method of Ecobichon (1970) using 1-naphthyl acetate as substrate.

Kinetic constants: Bimolecular rate constants were determined at pH 7.4 and 37°C by the method of Reiner and Aldridge (1967). The bimolecular rate constants were determined with racemic mixtures of chiral OP's and represent the rate constant of the fastest reacting stereoisomer.

Results and Discussion

The ability of scavengers to protect against OP's *in vivo* can be demonstrated by (1) increases in the LD_{50} of OP's and (2) measurement of OP binding to scavengers in plasma after OP challenge. Table 1 illustrates the increases in LD_{50} of VX, MEPQ and soman after administration of AChE purified from fetal bovine serum (FBS) in mice. FBS-AChE provided protection against 3.6 - 5.0 LD_{50} of these OP's. Mice contain large amounts of endogenous CaE scavenger in addition to the administered exogenous FBS-AChE. This does not influence the ability to discern the protective effect of FBS-AChE on VX and MEPQ because VX and MEPQ react poorly with CaE relative to their reaction with AChE (Raveh et al., 1989). However, soman is only 7-fold less reactive with CaE than AChE (Maxwell et al., 1988) and the endogenous CaE makes it difficult to measure the protective effect of administered FBS-AChE scavenger against soman. To overcome this problem, CBDP, a specific inhibitor of CaE, was administered to mice to eliminate the scavenging effect of endogenous CaE in mice. The dose of CBDP (2 mg/kg) that was administered to these mice has been previously demonstrated to inhibit CaE without inhibiting AChE (Maxwell et al., 1987). The scavenger effect of FBS-AChE for soman in CaE-inhibited mice was then similar to the scavenger effect of similar levels of FBS-AChE for VX and MEPQ.

TABLE 1

In Vivo Protection by FBS-AChE in Mice

Agent	FBS-AChE (nmol/kg)	CBDP (mg/kg)	LD_{50} (nmol/kg)[a] Control	FBS-AChE	Protective Ratio
VX[b]	390	---	56	201[e]	3.6
MEPQ[c]	217	---	76	310[e]	4.1
SOMAN[d]	333	2	30	151[e]	5.0

a. LD_{50} values represent only toxic P(-) stereoisomers.
b. Data taken from Wolfe et al. (1987).
c. Data taken from Ashani et al. (1989).
d. LD_{50} value for soman measured 1 hr after sc injection with CBDP.
e. Significant difference from control ($P<0.05$).

The in vivo binding of OP to scavenger can also be demonstrated by following the inhibition of plasma scavenger levels after administration of an OP. This procedure is particularly important for research when LD_{50} measurements are not possible, such as OP scavenger experiments in highly trained animals. In Fig. 1 the cumulative inhibition of 9 nmol BuChE administered to mice is demonstrated following sequential injections of small amounts (1.5 nmol/injection) of soman. All the animals in this study survived a cumulative dose of 4 LD_{50} and were behaviorally sign-free. It has been previously demonstrated that the cumulative LD_{50} resulting from several injections of small amounts of soman equals the LD_{50} of a single bolus dose of soman as long as the dosing interval does not exceed 1 hr (Sterri et al., 1980; Sterri et al., 1981). Therefore, the protection against soman produced by BuChE was a scavenger effect and not a tolerance response to cumulative doses of soman.

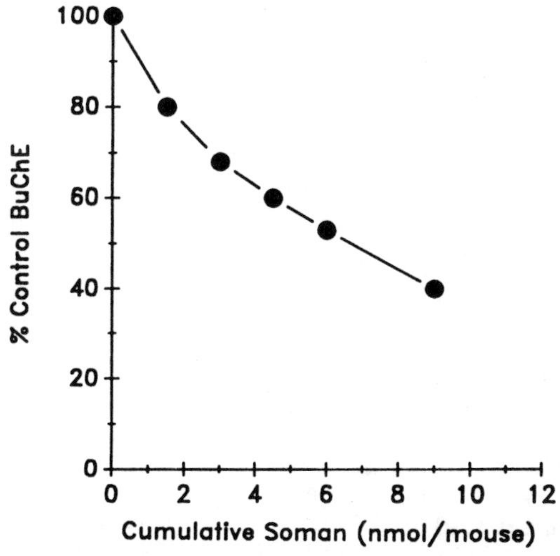

Fig. 1 In vivo stoichiometry for inhibition of plasma BuChE in mice by sequential injections with soman.

Table 2 is a comparison of the ability of AChE, BuChE, and CaE to protect against soman in mice. All three scavengers provided protection against 4.1 to 4.5 nmol/mouse of soman; however, there was a difference in the amount of scavenger necessary to provide this protection. The ratio of soman challenge to scavenger used was 0.92 for AChE, 0.75 for BuChE and 0.62 for CaE. This scavenger efficiency (soman/scavenger) decreased in the same order as the bimolecular rate constants for the reactivity of soman for each of these scavengers (AChE, 1×10^8 $M^{-1}min^{-1}$; BuChE, 7×10^7 $M^{-1}min^{-1}$; CaE, 5.2×10^6 $M^{-1}min$). AChE was the best scavenger based on scavenger efficiency (0.92) which was nearly equal to the theoretical limit of 1.0 soman molecule bound to each enzyme active site.

The importance of the plasma concentration of endogenous scavenger on individual variation in OP toxicity is shown in Fig. 2. The soman LD_{50} in aging rats correlated with the plasma level of CaE scavenger. The soman LD_{50} values were directly proportional to plasma CaE levels which varied as rats aged from 30 days to 120 days. When a similar relationship between plasma CaE and soman LD_{50} values in different species was evaluated, a direct proportionality was not observed (Fig. 3). Plasma CaE scavengers were not as effective in small animals (e.g., mice) as in larger animals (e.g., rabbits). Since the reaction between plasma scavenger and soman is dependent on both plasma CaE level and the time for CaE and soman to react, the role of circulation time on scavenger efficiency was evaluated. This evaluation demonstrated that the soman LD_{50} was

TABLE 2

In Vivo Scavenger Efficency of Enzymes for Soman in Mice

Enzyme Scavenger	Soman Dose (nmol/mouse)	Enzyme (nmol/mouse)[a] Initial	Final	Efficiency[b] (Soman/Enzyme)
AChE	4.2	8.1	3.3	0.92
BuChE	4.5	10.0	4.0	0.75
CaE	4.1	9.9	3.3	0.62

a. Calculated from enzyme level before (Initial) and after (Final) soman administration.
b. Soman dose/(Initial-Final enzyme).

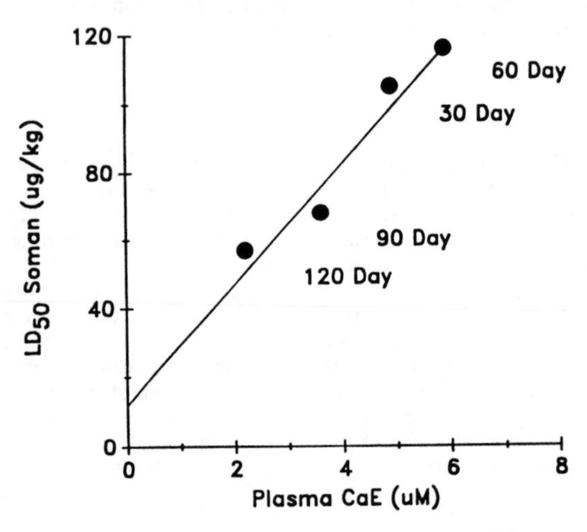

Fig. 2 Effect of plasma CaE on soman LD_{50} sc in aging rats.

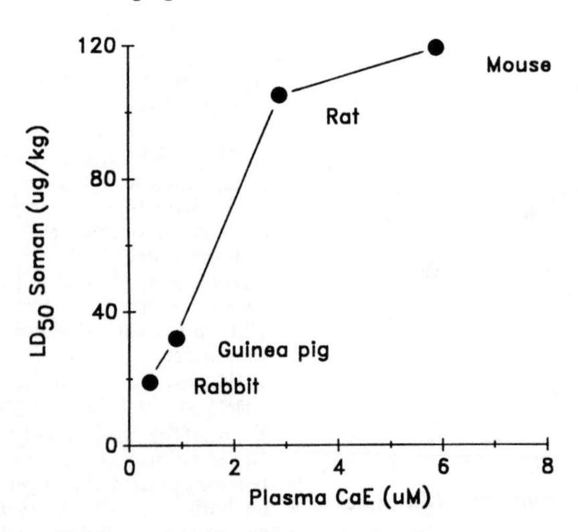

Fig. 3 Effect of plasma CaE on soman LD_{50} sc in different species.

directly proportional to the product of plasma CaE and circulation time (Fig. 4). These observations suggest that the scavenger protection observed in small animals could be equalled or exceeded in humans since the circulation time in humans (1 min) is greater than in small animals.

All of the major observations of our research are consistent with a protective action described by an irreversible bimolecular reaction between scavenger and OP. Therefore, with a combination of in vitro reactivity measurements, in vivo protection and stoichiometry experiments, and knowledge of circulation times, it should be possible to extrapolate scavenger protection against OP agents from animal models to humans.

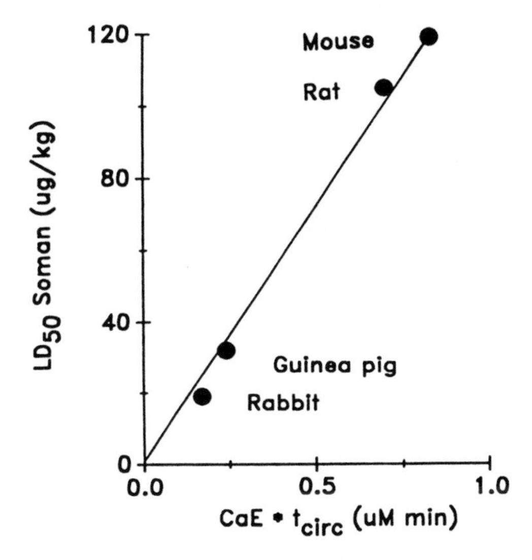

Fig. 4 Effect of the product of plasma CaE and circulation time (CaE•t_{circ}) on species variation in soman LD_{50} sc. Circulation times in each species (mouse, 0.14 min; rat, 0.24 min; guinea pig, 0.27 min; and rabbit, 0.43 min) were taken from Dedrick et al. (1970) and White et al. (1968).

References

Ashani, Y., Shapira, S., Doctor, B.P., Levy, D., Raveh, L., *Third International Symposium on Protection Against Chemical Warfare Agents*; Swedish Defence Research Establishment: Umea, Sweden, **1989**; pp 227 - 234.

Boskovic, B., *Arch. Toxicol.*, **1979**, 42, 207 - 216.

Brown, S.S., Kalow, W., Pilz, W., Whittaker, M., Worouick, C.L., *Advances in Clinical Chemistry*, **1981**, 22, 1-123.

Clement, J.C., *Biochem. Pharmacol.*, **1984**, 33, 3807-3811.

Dedrick, R.L., Bischoff, K.B., Zaharko, D.S., *Cancer Chemother. Rep.*, **1970**, 54, 95-101.

De La Hoz, D., Doctor, B.P., Ralston, J.S., Rush, R.S., Wolfe, A.D., *Life Sci.*, **1986**, 39, 195-199.

Doctor, B.P., *Third International Symposium on Protection Against Chemical Warfare Agents;* Swedish Defence Research Establishment: Umea, Sweden, **1989**; pp 225-226.

Dunn, M.A., Sidell, F.R., *JAMA*, **1989**, 262, 649-652.

Ecobichon, D.J., *Can. J. Biochem.*, **1970**, 48, 1359-1367.

Ellman, G.L., Courtney, D., Andres, V., Featherstone, R.M., *Biochem. Pharmacol.*, **1961**, 7, 88-95.

Finney, D.J., *Probit Analysis*, 3rd ed.; Cambridge University Press: Cambridge, **1971**; pp 50-124.

Fonnum, F., Sterri, S.H., Aas, P., Johnson, H., *Fund. Appl. Toxicol.*, **1985**, 5, S29-S38.

Levy, D., Ashani, Y., *Biochem. Pharmacol.*, **1986**, 35, 1079-1085.

Massoulie, J., Bon, S., *Annu. Rev. Neurosci.* **1982**, 5, 57-106.

Maxwell, D.M., Brecht, K.M., O'Neill, B.L., *Toxicol. Letters*, **1987**, 39, 35-42.

Maxwell, D.M., Vlahacos, C.P., Lenz, D.E., *Toxicol. Letters*, **1988**, 43, 175-188.

Raveh, L., Ashani, Y., Levy, D., De La Hoz, D., Wolfe, A.D., Doctor, B.P., *Biochem. Pharmacol.* **1989**, 38, 529-534.

Reiner, E., Aldridge, W.N., *Biochem. J.* **1967**, 105, 171-179.

Rosenberry, T.L., *Adv. Enzymology* **1975**, 43, 103-218.

Sterri, S.H. Lyngaas, S., Fonnum, F., *Acta Pharmacol. et Toxicol.*, **1980**, 46, 1-7.

Sterri, S.H., Lyngaas, S., Fonnum, F., *Acta Pharmacol. et Toxicol.*, **1981**, 49, 8-13.

Taylor, P. *The Pharmacological Basis of Therapeutics;* Gilman, A.G., Goodman, L.S., Rall, T.W., Murad, F., Eds.; MacMillan: New York, NY, **1985**; pp 110-129.

White, L., Haines, H., Adams, T., *Comp. Biochem. Physiol.*, **1968**, 27, 559-565.

Wolfe, A.D., Rush, R.S., Doctor, B.P., Koplovitz, I., Jones, D., *Fund. Appl. Toxicol.*, **1987**, 9, 266-270.

Significance of Plasma Cholinesterase for the Pharmacokinetics and Pharmacodynamics of Bambuterol

Leif-Å. Svensson, Research & Development Dept., AB Draco, Box 34, S-221 00 Lund, Sweden.

Bambuterol, which is an inactive carbamate prodrug of the bronchodilator terbutaline, was designed with a cholinesterase (CHE) inhibiting function in the prodrug molecule, in order to achieve the desired lung-specificity and prolonged action. Bambuterol is a good substrate for plasma CHE and a very potent and selective inhibitor of BCHE, IC_{50}= 17 nM, while the corresponding IC_{50}= 41 μM for ACHE. This paper reviews the interaction of bambuterol with cholinesterases, its behaviour in individuals with genetically impaired CHE, and its interaction with succinylcholine.

In our search for an orally active β_2-agonist bronchodilator drug with long duration of action, and with reduced systemic side effects, we observed that some lipophilic ester prodrugs of terbutaline were efficiently taken up by the lung where they bioconverted to terbutaline (Ryrfeldt, 1978). However, these prodrugs were rapidly hydrolyzed to terbutaline already during the first-pass through the gut wall (Kristoffersson, 1974). Hence, their particular lung distribution properties could not be taken advantage of.

In order to obtain sufficient first-pass- and systemic metabolic stability of the prodrug, we built into the prodrug molecule a hydrolysis brake in form of an esterase inhibitor function. The selected prodrug compound with this property is bambuterol; the **bis**-dimethylcarbamate of terbutaline (Olsson, 1984), Fig. 1. Bambuterol effectively slows down its own hydrolysis by selective inhibition of plasma cholinesterase, the characteristics of which will be further discussed below. In this way the desired first-pass hydrolytic stability was obtained, and about 65% of the absorbed bambuterol reaches the circulation intact (Nyberg, 1989). Moreover, bambuterol is at least a **prodrug-prodrug**, since its metabolism already in the gut and the liver generates several new lipophilic terbutaline carbamate prodrugs with properties similar to those of bambuterol (Svensson,1988). In fact, some of these newly formed prodrugs, the N-demethylated derivatives, may spontaneously decompose to terbutaline (Lindberg, 1989). Furthermore, very little terbutaline is formed by first-pass metabolism of bambuterol, and the terbutaline part of the prodrug molecule seems not to undergo metabolism.

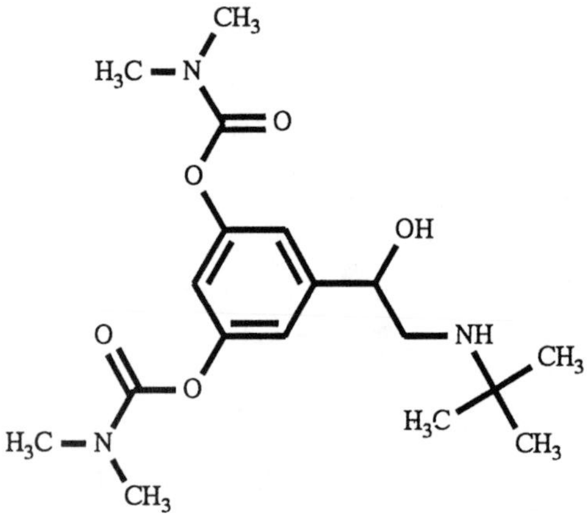

Fig. 1. Bambuterol.

Lung-uptake of bambuterol and local generation of terbutaline was demonstrated in guinea-pig isolated, perfused lungs (Ryrfeldt, 1988). A distribution study in mice revealed high lung/plasma concentration ratios for bambuterol and several of its lipophilic metabolites, even 12h after drug administration (Levin, 1989).

In asthmatic patients, an apparent enhanced distribution to the lung, results in a favorable desired effect-side effect ratio of the oral drug bambuterol. Moreover, the sustained metabolism of bambuterol allows once daily dosing.

Hydrolysis of bambuterol and selectivity of cholinesterase inhibition.

The hydrolysis of bambuterol in plasma **in vitro** was initially (0 to 5 min) a very rapid

0–8412–2008–5/91/0210$06.00/0 © 1991 American Chemical Society

process with a capacity of up to 60 pmol bambuterol/mL of plasma (Tunek, 1988). The capacity of the initial burst reaction is in good agreement with reported concentrations of BCHE in human plasma. After this, a second, slower phase occurred, with a maximal rate of around 25 pmol/h/mL of plasma. The hydrolysis of the monocarbamate however almost completely lacked the initial burst reaction. This implies that as long as bambuterol is present in appreciable amounts, the first step of hydrolysis will dominate, and little terbutaline will be formed. The effect of preincubation of human blood with 10^{-6}M of the potent cholinesterase inhibitor physostigmine resulted in the abolishment of the initial burst of the bambuterol hydrolysis, while the second slower phase was only mildly affected.

Bambuterol was found to be a selective and potent cholinesterase inhibitor with an IC_{50}= 17 nM for cholinesterase activity, while the corresponding value for acetylcholinesterase was 2400-fold higher, that is, 41 μM. The first product of hydrolysis of bambuterol, the mono-N,N-dimethylcarbamate of terbutaline also was found to be a selective, but 10-fold less potent, inhibitor of cholinesterase, probably reflecting a lower affinity for the enzyme. Due to its selective and efficient cholinesterase-inhibiting properties, bambuterol will cause a dose-dependent inhibition of these enzymes **in vivo** (Svensson, 1988). However, the chronic toxicological studies in rats and dogs have shown neither clinical nor toxicological effects that could be related to inhibition of cholinesterase activity, even though the cholinesterase activity in the high-dose group of the dog study was almost fully abolished for an entire year. Moreover, in these studies, the selectivity of the cholinesterase inhibition also was evident since there were no signs of effects resulting from inhibition of acetyl cholinesterase at doses up to 600 times a therapeutic dose of bambuterol. Also **in vivo** in man, erythrocyte acetyl cholinesterase was found to be unaffected after a high dose of bambuterol, although plasma cholinesterase was inhibited by more than 70% (Sharma, 1990).

It is believed that inhibition is caused by rapid carbamylation of a serine residue at the esteratic active site to generate an inactive carbamylated esterase intermediate which is only slowly hydrolyzed back to active esterase, i.e. k_3 in Equation 1 is rate determining. Thus, the "inhibition" is a result of a slow turnover of the catalytic cycle.

Eq. 1.

$$E + CM \underset{k_{-1}}{\overset{k_1}{\rightleftharpoons}} E\text{-}CM \xrightarrow[M]{k_2} EC \xrightarrow{k_3} E + C$$

where E= esterase, CM= carbamate substrate, M= phenolic product of hydrolysis, EC= carbamylated esterase and C= carbamic acid derivative.

The cholinesterase catalyzed hydrolysis/inhibition of bambuterol did not display any stereoselectivity since both (+)- and (-)-bambuterol gave almost identical IC_{50}-values (Svensson, 1990). Even if it is (-)-bambuterol that generates active (-)-terbutaline, bambuterol is given as the racemate in order to obtain a sufficient degree of esterase inhibition, which is the prerequisite for the desired presystemic- and systemic hydrolytic stability.

Cholinesterases inhibited by bambuterol recover their activity once exposure to the inhibitor is ceased or reduced (Tunek, 1988). Thus, inhibited ACHE and BCHE both recovered almost completely within 2.5 hr. ACHE reactivated faster than BCHE and had regained about 80% of the original activity after 60 min. The time required for regeneration of 50% of BCHE activity was 75-80 min.

The classical cholinesterase inhibitor physostigmine is a very potent, but a completely unselective inhibitor; IC_{50}= 1.3×10^{-8}M for BCHE and 1.5×10^{-8}M for ACHE. On the other hand, bambuterol is an extremely selective and potent inhibitor. This implies that the environment around the active sites in ACHE and BCHE is slightly different, which also is evident from the substrate specificity of cholinesterases in human blood presented in Table 1 (Tunek, 1988).

Table 1. Rate of hydrolysis of acetyl thiocholine, propyl thiocholine and butyryl thiocholine in human blood, **in vitro**

Substrate	Rate of hydrolysis	
	ACHE	BCHE
	$10\times^{-9}$mol/min/5 μl of blood	
AcThC	27.0	8.3
PrThC	16.2	13.8
BuThCh	1.0	14.3

We have used a number of bambuterol derivatives to further examine the substrate specificity of ACHE and BCHE (Svensson, 1990). In this study, derivatives where only the substituent at the nitrogen atom in the terbutaline part was varied by size, were used as substrates to obtain inhibition curves for ACHE and BCHE. Since k_3 is the same for all these compounds, the IC_{50}-values obtained from the inhibition curves should reflect the binding- and carbamylation steps. The results from this study clearly demonstrate that the selectivity for the two cholinesterases depend on the size (volume) of the substituent at nitrogen, see Fig. 2. Substrates with small substituents at nitrogen

R= $-CH_3$, $-CH_2CH_3$, $-CH(CH_3)_2$, $-C(CH_3)_3$, $-C(CH_3)_2CH_2CH_3$

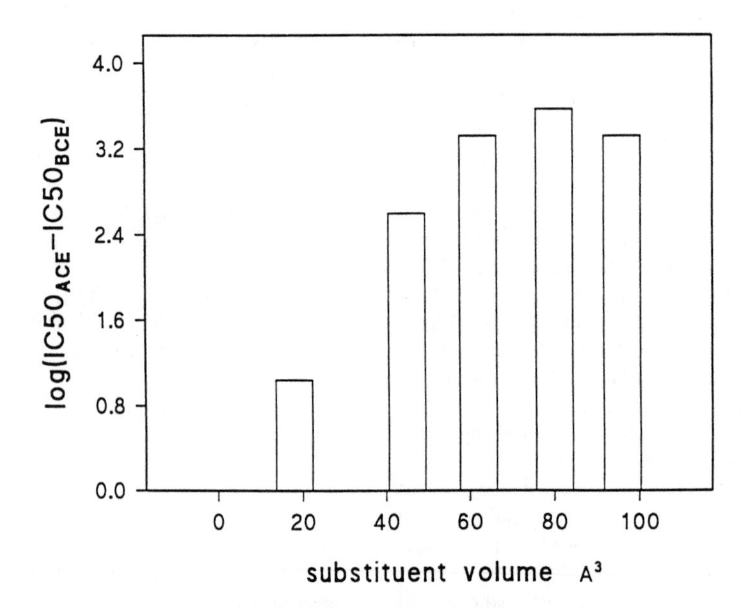

Fig. 2. Selectivity for ACHE and BCHE of a series of bambuterol derivatives as a function of the size of the substituent at nitrogen.

exhibit low selectivity, measured as $\log(IC_{50ACHE}\text{-}IC_{50BCHE})$ plotted against the volume of the nitrogen substituent, whereas the individual IC_{50}-values vary between 3.1 to 150 μM for ACHE, and for BCHE between 2.1 nM to 200 μM. But, it is evident that also other factors determine the selectivity since the racemic carbinol derivative AE 98, Fig. 3, exhibits relatively good selectivity (by a factor of about 130), although being a rather poor substrate for BCHE.

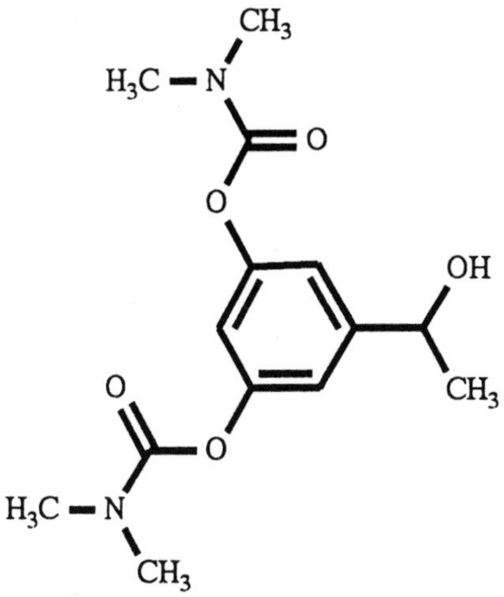

Fig. 3. The carbinol derivative AE 98.

Bambuterol and genetic variants of cholinesterase.

The behaviour of bambuterol in blood from individuals of the common, and the atypical (homozygous) phenotypes has been studied **in vitro** (Tunek, 1990). While the IC_{50}= 3 nM for the E_aE_a-type in 600-fold diluted plasma, the corresponding IC_{50}-value for the E_aE_a-type was 100-200 nM.

When 4 homozygously atypical individuals were given bambuterol 20 mg once daily for one week, the steady-state terbutaline plasma concentration curve over 24 hours showed that they were able to produce the same amount terbutaline as normal individuals. They only differ in that they have a somewhat higher peak/trough ratio in their plasma concentrations of terbutaline (Bang, 1990).

Interaction with succinylcholine.

Since plasma cholinesterase is involved in the hydrolytic inactivation of succinylcholine, bambuterol may influence the duration of succinylcholine-induced neuromuscular blockade. This was studied with clinically relevant doses, 10 and 20 mg, and a high dose, 30 mg, of bambuterol in patients, 11-14 patients/group, given succinylcholine (Fisher, 1988, Lennmarken, 1989). Bambuterol was given in the evening, **e.g.** about 12 hours before the start of surgery when succinylcholine, 1 mg/kg b.w., was given at induction of anaesthesia. Neuromuscular function was measured using nerve stimulation and time to 90% revovery of the muscle twich response was used as a measure of sufficient muscle recovery for spontaneous breathing. After 30 mg of bambuterol, the time for 90% recovery was, on average, 24.9 min, as compared with 11.2 min in the placebo group. In the other study, the time to 90% recovery was, on average, 13.5 min after bambuterol 10 mg, 15.1 min after bambuterol 20 mg, and 10.3 min in the placebo group.

Thus, a dose dependent, slight to moderate, prolongation of succinylcholine-induced neuromuscular blockade was demonstrated in patients treated with bambuterol. This is probably unimportant in most clinical situations. However, in short anaesthetic procedures, the prolonged duration of action of succinylcholine may have clinical significance.

In conclusion, although bambuterol is a potent inhibitor of plasma CHE it is a safe drug since it is also an extremely selective inhibitor of CHE with no effect on ACHE, even at high clinical doses. Individuals with genetically impaired CHE-activity also seem to hydrolyze bambuterol properly, and the interaction of bambuterol with the termination of succinylcholine activity seems to be of clinical significance only in extreme cases.

References

Ryrfeldt, Å., Nilsson, E., *Biochem. Pharmacol.*, **1978,** *27,* 301-305

Kristoffersson, J., Svensson, L.-Å., Tegnér, K., *Acta Pharm. Suec.*, **1974,** *11,* 427-438

Olsson, T., Svensson, L.-Å., *Pharm. Res.*, **1984,** *1,* 19-23

Nyberg, L., AB DRACO, *personal communication*

Svensson, L.-Å., Tunek, A., *Drug Metab. Rev.*, **1988,** *19,* 165-194
Lindberg, C., Roos, C., Tunek, A., Svensson, L.-Å., *Drug Metab. Disp.,* **1989,** *17,* 311-322
Ryrfeldt, Å., Nilsson, E., Tunek, A., Svensson, L.-Å., *Pharm. Res.,* **1988,** *5,* 151-155
Levin, E., Tegnér, K., AB DRACO, *personal communication*
Tunek, A., Levin, E., Svensson, L.-Å., *Biochem. Pharmacol.,* **1988,** *37,* 3867-3876
Sharma, M., Svensson, L.-Å., *this book*
Svensson, L.-Å., Tunek, A., Eichmuller, B., *to be published*
Tunek, A., Svensson, L.-Å., *Drug Metab. Disp.,* **1988,** *16,* 759-764
Tunek, A., Levin, E., Viby Mogensen, J., *to be published*
Fisher, D.M., Caldwell, J.E., Sharma, M., Wirén, J.E., *Anesthesiology,* **1988,** *69,* 757-759
Lennmarken, C., Staun, P., Eriksson, L.I., *Acta Anest. Scand.,* **1989,** *33,* Suppl. 91:142

"In Vivo" Determination of Striatal Acetylcholinesterase Activity by Microspectrophotometry. Physiological Modulations of the Enzyme by Various Effectors

Guy Testylier, Patrick Gourmelon, Eric Multon, Didier Clarençon and Jacques Viret.
Centre de Recherches du Service de Santé des Armées. Grenoble - BP 87
38 702 La Tronche-cedex (France)

Our knowledge about brain acetylcholinesterase (AChE) comes mainly from post-mortem biochemical assays due to the lack of well adapted technology for "in vivo" direct assay of the enzyme. However, some elements of the "in vivo" physiology of the enzyme were approached using the push-pull cannula technique (Greenfield, 1982, Taylor, 1989). Using this method, the dendritic release of AChE was studied in non-cholinergic brain structures demonstrating that the functions of this enzyme widely overpass the acetylcholine hydrolysis (Greenfield, 1984).

We have recently developed a new microspectrophotometric method which allows a colorimetric reaction inside the brain tissue of a live animal to be followed locally. We applied this technique to the measurement of AChE activity in the rat striatum using the Ellman reagent.

We present the application of this technique to the study of the "in vivo" modulations of AChE activity in various physiological situations.

Principle of the "In Vivo" Spectrophotometric Method.

"In vivo" microspectrophotometry allows a colorimetric reaction in a live animal to be monitored *in real time* and *in situ*. The physical principle of the photometric system is to illuminate by means of an optical fiber a specific site in a brain structure, to inject the colorimetric reagent at the same site and to measure, with a second optical fiber the variations of the retrodiffused light intensity.

As shown in Fig. 1 the main part of the apparatus is the miniaturized optical probe (about 150 microns in diameter) which consists of a six channel multibarrel micropipette for injections of reagents or enzyme effectors , with - at the centre - two optical fibers (30 microns) for the absorbance measurements. The incident light emitted by a mercury lamp and selected at 435 nm through a monochromator is transmitted to one of the optical fibers. The second optical fiber carries the retrodiffused light out from the analyzed site to a photomultiplier which is coupled to a microcomputer for absorbance processing.

The optical properties of the system have been investigated (Testylier, 1987), the linearity of the absorbance measurements was determined using calibrated dye solutions and the optical stability of the measuring site was checked in anaesthetized rats during a period of 12 hours .

AChE Activity Determination.

AChE activity was assayed by injection of concentrated Ellman reagent (30 or 75 mM DTNB and Acetylthiocholine). After reagent injection, we observed a rapid and linear increase of the absorbance which reached a maximum and then slowly decreased (Fig. 1). The first linear phase of the curve reflects the enzyme steady-state reaction and the initial slope is the Vmax of the enzyme. The decreasing phase of the kinetic curve results from the removal of the dye by passive diffusion and cerebral blood flow elimination. The absorbance reaches the starting base line level in from 5 to 20 minutes according to the brain structure. The maximal sample rate of the AChE assays is dependent upon this recovery time.

The specificity of the "in vivo" AChE determination was controlled by local injection of the highly specific organophosphorus inhibitor (methylphosphonothioate - MPT) through an adjacent channel of the probe (Testylier, 1987).

Ectocellular Localization of the "In Vivo" Assayed AChE.

A major question to be addressed was to know which enzyme compartment was being

0–8412–2008–5/91/0215$06.00/0

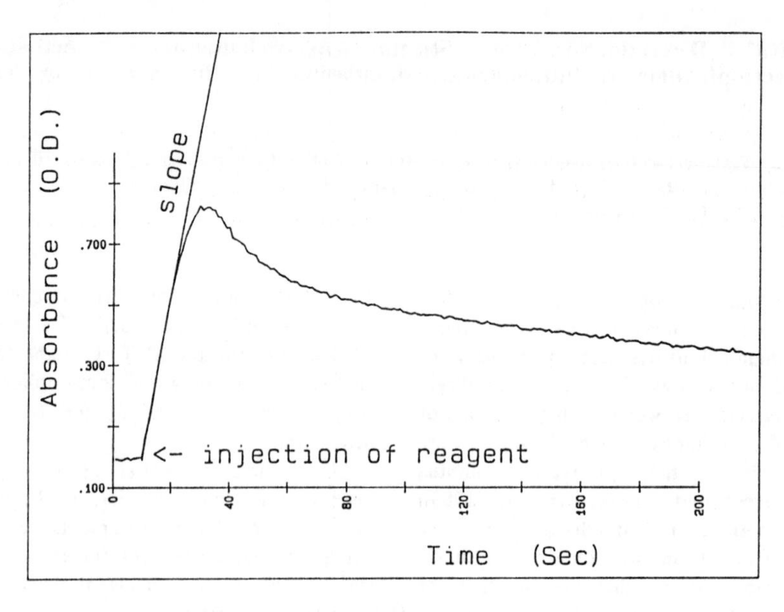

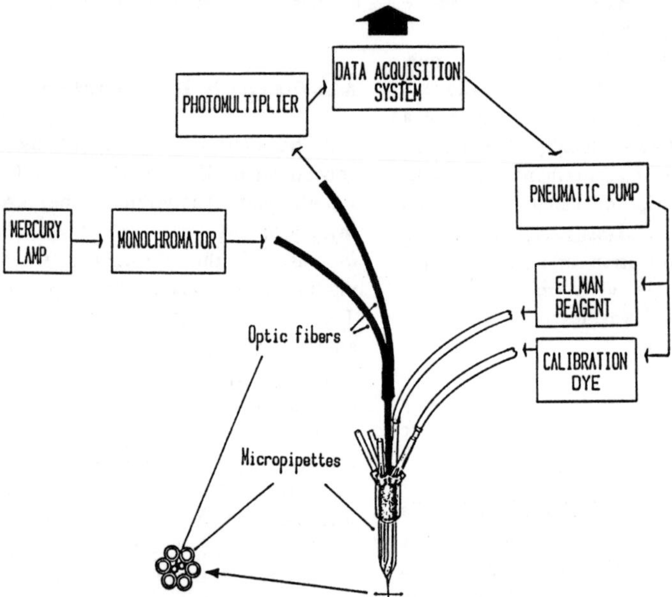

Fig. 1. Schematic diagram of the system and typical recorded curve showing the absorbance kinetic.

analyzed by the optical probe. Some authors have shown that plasmic membrane was not permeable to the Ellman reagent as observed in neuroblastoma cells (Lazar, 1980).

In an "in vitro" experiment, we verified the absence of diffusion of the Ellman reagent in the intracellular compartment in primary neuron cultures of striatum. We observed an increase of about 50% in AChE activity after the partial lysis of the striatal neurons by Triton X-100. We also checked with "in vivo" microspectrophotometric experiments that the access of the Ellman reagent was limited to the ectocellular compartment by local injections of

Triton X-100 at the measuring site. We observed an increase of the AChE activity of 20 % 10 seconds after the Triton X-100 injection.

The microspectrophotometric method thus enabled us to assay the ectocellular enzyme, the external membrane-bound AChE and the excreted soluble fraction.

Experimental Models.

We have developed two experimental animal models. The first one is an acute model where the rat is maintained under anaesthesia for several hours and the second one is a chronic model where the AChE activity is monitored over several days in an unrestrained and vigilant animal in which the optical probe is sealed on the skull.

AChE Activity in the Acute Model.

The anaesthetized model (ketamine) is characterized by a remarkable stability of the AChE activity during the 8 hours following the formation of the measuring site (Fig. 2). This

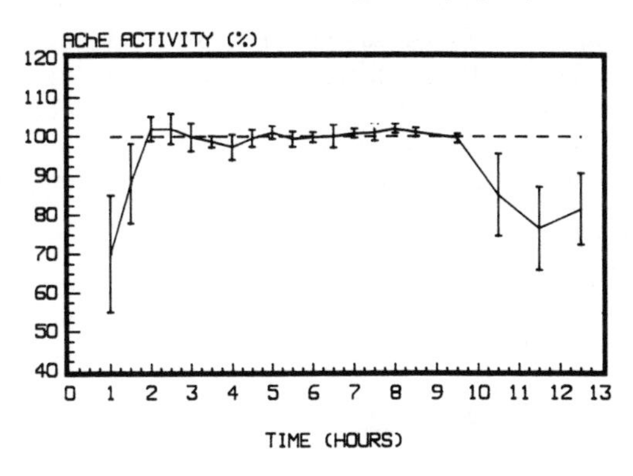

Fig. 2. Evolution of striatal AChE activity in acute model. Mean and S.D. n = 5 animals.

demonstrates the precision of the method (coefficient of variation less than 5%). The fall in the last part of the curve, due to the cumulative impregnation by ketamine, shows the operating limitations of the acute experimental model.

Histological controls of the effects of repetitive injections of Ellman reagent at the striatum level exhibited only mechanical lesions induced by the probe implantation.

Evolution of the AChE Activity in the Chronic Model.

In chronically implanted animals (Fig. 3), AChE was monitored every hour during a period of between 7 and 10 days. Brain electrical activity was continuously recorded.

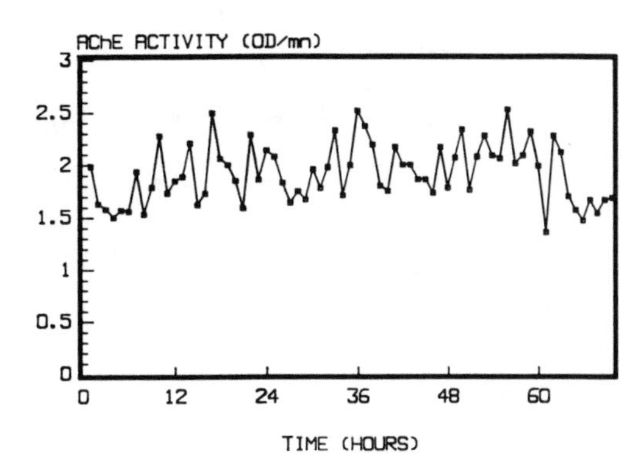

Fig. 3. Spontaneous AChE activity in a chronically implanted animal.

We always observed large and spontaneous variations of the AChE activity in this model (up to 25 %). No biological rhythm could be identified even using signal processing techniques like the autocorrelation function. Furthermore, these fluctuations were not correlated with the duration of the sleep stages (wakefulness, slow wave sleep and REM sleep).

EEG recordings (Fig. 4) did not exhibit cortical abnormalities (epileptic discharges or slowing) after the reagent injections. Moreover, when the assay was performed during slow wave sleep or sleep with spindles, no arousal reaction was observed.

Effect of Anaesthesia on the Spontaneous Fluctuations of the AChE Activity.

In order to confirm the physiological origin of the fluctuations of AChE activity which were never observed in the acute experimental model, we anaesthetized chronically implanted animals (Fig. 5). We observed a flattening of the AChE recording, occuring one hour after the induction of the anaesthesia, which remained throughout the anaesthesia and continued about 19 hours after the awakening.

We hypothesize that these spontaneous physiological fluctuations of the enzyme activity

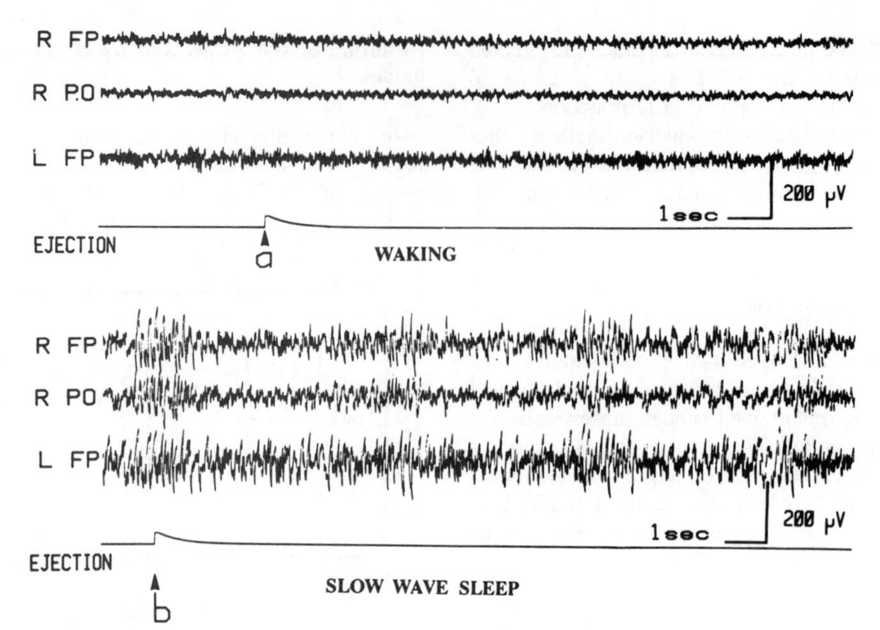

Fig. 4. Electroencephalographic recordings performed during the AChE assays in waking stage (a) or slow wave sleep stage (b).

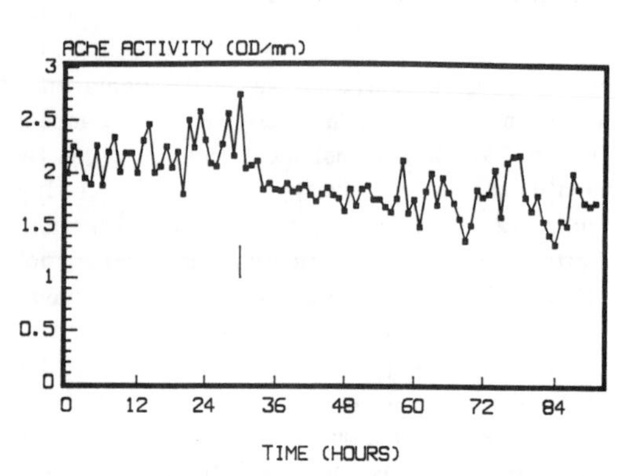

Fig. 5. Effect of anaesthesia on the spontaneous fluctuations of the AChE activity. (Ketamine : 100 mg/kg IM).

might be related to the local variations of the release of the AChE.

Physiological Responses of the AChE Activity to Local Effectors.

Effects of Electrical Stimulations.

Electrical stimulations in the measuring site were performed using two platinum wires (25 microns) inserted in two adjacent channels of the optical probe. The Fig. 6 shows the marked increasing in the AChE activity (up to 50 %) after the electrical stimulation.

It is tempting to speculate that this result could be explained by the evoked release of AChE under electrical stimulation, as observed in the perfusate of rat caudate nucleus slices (De Sarno, 1987) but the high level of the evoked response is very surprising.

Effect of Local Injection of Glutamate.

The local injection of glutamate (Fig. 7) which is known to be a major excitatory neurotransmitter of the cortico-striatal

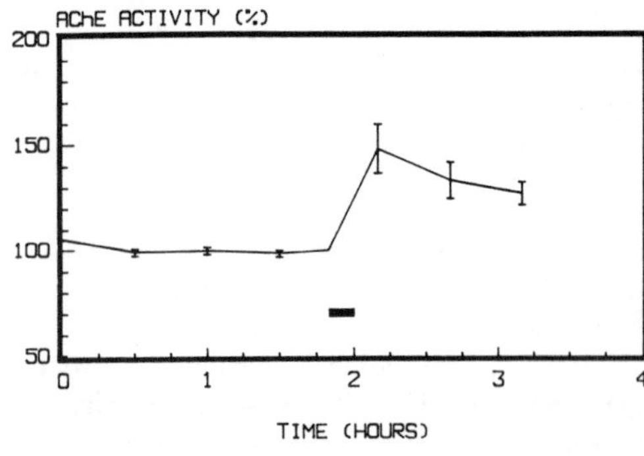

Fig. 6. Effect of local electrical stimulation on the AChE activity. Stimulation : square pulse of 4 V (3.3 microA) lasting 10 mn. Mean and S.D. n = 5 animals.

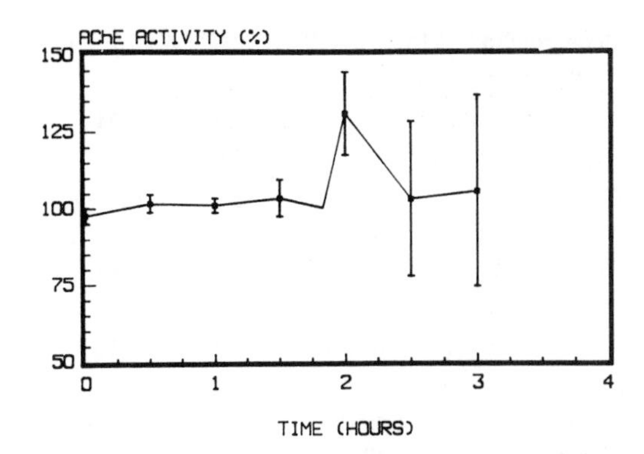

Fig. 7. Effect of local glutamate injection (60 nl, 10^{-6} M) on the AChE activity. Mean and S.D. n = 5 animals.

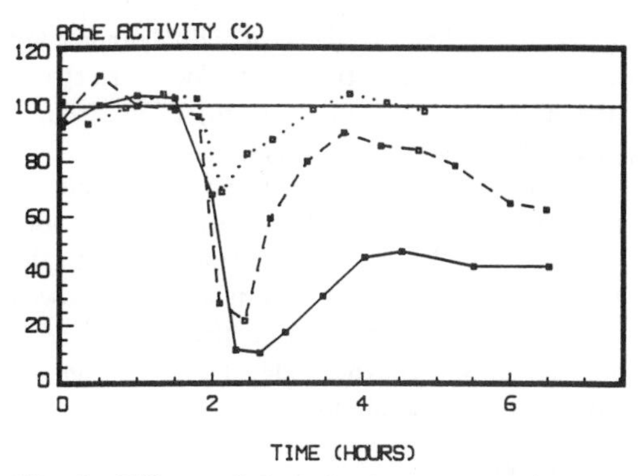

Fig. 8. Effect of intoxication by cyanide (4.5 mg/kg IV) on the AChE activity. Each curve represents the response of a single animal.

pathways, induced an increase of about 30 % in the AChE activity. It is interesting to note that an excitatory neuromediator and electrical stimulation induce a similar response. It suggests that an important fraction of the enzyme assessed by our method results from the released AChE and that its level depends on the neuronal activity in the measuring site.

Effect of Global Effectors Promoting Various Physiopathological Responses.

The "in vivo" spectrophotometric method was applied to physiopathological or neuropharmacological studies. We thus investigated the "in vivo" recovery of AChE activity after MPT intoxication by the peripheral route (see Poster), the modulation of AChE activity by a major tricyclic antidepressant (Chappey, 1990), the AChE response after a whole-body gamma irradiation (unpublished data) and the drastic decrease of the enzyme activity following death (see Poster).

We will present here an amazing result concerning the effect of a potent metabolic poison on AChE activity.

Effect of Cyanide Intoxication.

We explored the effects of cyanide, known for its cytochrome oxydase blockade properties, on the AChE activity. After intravenous injection of cyanide, at the lethal dose of 45%, we observed a rapid fall of the AChE activity, depending on the animal, which was followed by a partial or complete reactivation of the enzyme activity. Fig. 8 illustrates three typical AChE temporal profiles. The remarkable feature is the correlation, for each animal, between the amplitude of the fall and the duration of the electrical silence, which occurs in the electroencephalographic recording after the intoxication (Table 1).

Table 1

Curve	Flattening EEG Duration	Return to Normal EEG
·······	11 mn 20 s	33 mn 40 s
– – – –	12 mn 20 s	55 mn 40 s
———	22 mn 40 s	Coma

Thus, AChE activity could be a functional marker of brain activity in physiopathological situations.

Conclusion.

The "in vivo" spectrophotometric method enabled us to access to the ectocellular AChE compartment. This constitutes a great experimental advantage. Indeed, this compartment, where the enzyme is membrane-bound or soluble, is the compartment physiologically involved in the neuronal communications.

The exploration field of this technique is vast as illustrated by the above examples. Using this new methodology we showed clearly the large and spontaneous fluctuations of the enzyme activity in vigilant animals. Moreover, the level of the AChE activity seems to be dependent on the neuronal activity since local electrical stimulation or local glutamate injection strongly modulate the enzyme activity.

Finally, this method is a powerful tool to study over lengthy periods the effects of various neuropharmacological or toxic compounds on the AChE activity in the same animal. This allowed us to find out more about the physiological response of the enzyme independently from the interindividual variations.

References.

Chappey, O., Testylier, G., Gourmelon, P., Galonnier, M., Bourre, J.M., Fatome, M., Scherrmann, J.M., Viret, J., *J. Neurochem.*, **1990**, *54* (1), 333-338.

De Sarno, P., Giacobini, E., Downen, M., *J. Neurosci. Res.*, **1987**, *18*, 578-590.

Greenfield, S. A., Shaw, S. G., *Neuroscience*, **1982**, *7*, 2883-2893.

Greenfield, S.A., *Trends Neurosci*, **1984**, *7*, 364-368.

Lazar, M., Vigny, M., *J. Neurochem.*, **1980**, *35* (5), 1067-1079.

Taylor, S.J., Haggblad, J., Greenfield, S.A., *Neurochem. Int.*, **1989**, *15* (2), 199-205.

Testylier, G., Gourmelon, P., *Proc. Natl. Acad. Sci. USA*, **1987**, *84*, 8145-8149.

The Use of Acetylcholinesterase as a Universal Marker in Enzyme-Immunoassays

J.Grassi* **and P. Pradelles**, Section de Pharmacologie et d'Immunologie, Département de Biologie, CEN Saclay, 91191 Gif/Yvette Cedex, France.

AChE exerts very efficient control of cholinergic transmission because it quickly hydrolyses the neurotransmitter acetylcholine. This efficiency is largely due to its very high turnover number. As a consequence, AChE activity can be measured very sensitively under standard conditions, using routine methods. For many years, numerous studies undertaken in our laboratory have shown that AChE from electric organs of the electric eel *Electrophorus electricus* has properties ideally suited to use as a marker in enzyme-immunoassay[1]. We have demonstrated that AChE can be covalently coupled to biological molecules (including haptens, antigens and antibodies) without significant loss of enzyme activity, and that the corresponding conjugates can be used as tracers in different types of immunoassays (competitive as well as immunometric immunoassays).

These immunoassays are systematically performed using microtitre plates as support and immobilised antibodies as separation method. For competitive immunoassays, the plates are coated with a second antibody (anti-immunoglobulin antibody) so that separation occurs during specific immunoreaction. For immunometric assays, the plates are coated with a first antibody directed against a first epitope of the antigen. Solid-phase bound antigen is quantified by means of a second AChE-labelled antibody recognising a second epitope.

Purification and Measurement of AChE.

Labelling of biological molecules is performed with *Electrophorus* enzyme because it is particularly stable and has a significantly higher turnover number than other cholinesterases. In addition, milligram amounts of AChE can easily be purified from electric organs extracts (about 20 mg/kg of electric organ) by one-step affinity chromatography (Massoulié and Bon, 1976). Specific activity measured for purified enzyme (60.000 Ell.U/UA) is very close to the theoretical value calculated for pure enzyme (68.000 Ell.U/UA, Grassi, 1990). For the labelling of biological molecules, we essentially use the globular tetrameric form of the enzyme (G_4 form), which is quantitatively obtained from crude asymmetric forms by tripsin treatment (Massoulié and Rieger, 1969).

AChE activity is measured using the colorimetric method of Ellman (Ellman et al., 1961). This method allows very sensitive detection of the enzyme activity since it is possible to measure a few attomoles of AChE (10^{-18} moles) under the conditions currently used in enzyme immunoassay (see Table 1). This detection threshold is about tenfold lower than that calculated for ^{125} Iodine (2). It is also significantly lower than the detection limit calculated for other enzymes currently used in enzyme immunoassay (see Table 1). This means that conjugates prepared with AChE can be measured more sensitively than radioiodinated molecules or other enzyme conjugates, thus leading to a potential increase in precision and sensitivity for the corresponding immunoassays.

Labelling of Biological molecules.

The performance of an enzyme immunoassay is directly linked to the characteristics of the enzymatic tracer. As a consequence, the preparation of the enzyme conjugate is certainly one of the most critical steps in the development of the assay. Here, the goal is to obtain labelled molecules having both maximum immunoreactivity and specific activity and minimum non-specific binding. Generally speaking, this means preparing a conjugate for which one molecule of AChE is coupled to one molecule of immunoreactant. This implies selection of coupling reactions allowing strict control of conjugate stoichiometry. One major difficulty is to avoid reactions leading to the polymerisation of proteins (AChE, antigens or antibodies). For these reasons, as far as possible, we systematically used heterobifunctional cross-linking reagents and two-step coupling procedures.

Labelling of Antigens and Antibodies.

Antigens and antibodies are covalently coupled to AChE by the intermediary of the heterobifunctional reagent N-succinimydyl-4-(maleimido-methyl)-cyclohexane-1-carboxylate (SMCC) using procedures derived from those previously described by Ishikawa et al. (1983) for the labelling of antibodies. This method

[1] Patent number : CEA-IRF N° 83-13389 (1983)

0–8412–2008–5/91/0221$06.00/0

Table 1. Comparison of colorimetric assays for the most commonly used enzymes in EIA

Enzyme	Molecular weight (Number of catalytic subunits)	Turnover number (h^{-1} mol^{-1})	Substrates	Chromophore	Detection limit (mol)[a]
Acetylcholinesterase (*Electrophorus electricus*) EC 3.1.1.7	330000 (4)	1.8×10^8	Acetylthiocholine + 5,5'-dithiobis (2-nitrobenzoic acid)	NO_2–C₆H₃(COO⁻)–S⁻; $\varepsilon_M = 1.36 \times 10^4$ (412 nm)	1.6×10^{-18}
Alkaline phosphatase (calf intestine) EC 3.1.3.1	100000 (2)	1.45×10^7	*p*-Nitrophenyl phosphate	NO_2–C₆H₄–OH; $\varepsilon_M = 1.85 \times 10^4$ (405 nm)	1.5×10^{-17} (2.6×10^{-16})
β-Galactosidase (*E. coli*) EC 3.2.1.23	540000 (4)	2.5×10^7	*O*-Nitrophenyl-β-D-galactopyranoside	C₆H₄(OH)(NO_2); $\varepsilon_M = 4.7 \times 10^3$ (410 nm)	3.4×10^{-17} (2.6×10^{-17})
Peroxidase (horseradish) EC 1.11.1.7	40000 (1)		*O*-Phenylene diamine + H_2O_2	diaminophenazine (NH_2, NH_2); $\varepsilon_M =$? (492 nm)	3.6×10^{-18}[b] (5.9×10^{-18})

[a] Detection limit for each enzyme was calculated using the corresponding values for turnover number and ε_M. It is defined as the amount of enzyme producing an absorbance increase of 0.01 in 1 h, 0.2 ml volume, 0.5 cm pathlength. These conditions correspond to those used when EIA is performed in microtitration plates (see PRADELLES et al. 1985). Values in parentheses are those given by ISHIKAWA et al. (1983). Original values from these authors have been corrected for time, volume, and pathlength conditions in order to be consistent with our own calculations.

[b] Determined experimentally for pure crystalline enzyme (RZ = 3, Sigma, St. Louis, United States) using *O*-phenylene diamine (9.3×10^{-3} *M*) and H_2O_2 (10^{-2} *M*) as substrates.

involves the reaction of thiol groups of proteins (naturally present or previously introduced into antigens or antibodies) with maleimido groups incorporated into AChE after reaction with SMCC. Antibody labelling is generally performed using Fab' fragment in order to obtain conjugates presenting lower non-specific binding (Ishikawa et al., 1983). Antigens are thiolated before coupling by reaction of their primary amino groups with N-succinimydyl-S-acetyl-thioacetate (SATA) in neutral or alkaline media. In recent years, we have labelled more than 50 different antibodies (either polyclonal or monoclonal) and about 10 different antigens. Most of them are listed in Tables 2 and 3, respectively, together with the corresponding references. The details of the coupling procedures are given in these publications.

Labelling of Haptens.

For the labelling of haptens, a greater variety of coupling reactions is used. The different hapten conjugates prepared in our laboratory are presented in Table 4. For those haptens possessing a carboxyl group, coupling with AChE is generally achieved by reaction of the primary amino groups of the enzyme with an N-hydroxysuccinimide ester previously prepared from the hapten. For haptens possessing a primary amino group, we

Table 2 : List of the main antibody labelling performed with AChE

Antibody	Méthods	Sensitivity	References
anti-human IgE (1 monoclonal)	Fab'	0,1 mUI/ml	unpublished results
anti-human TSH (3 monoclonals)	Fab'	0,02 μUI/ml	unpublished results
anti-human Renin (1 monoclonal)	Fab'	5 pg/ml	unpublished results
anti-substance P (1 monoclonal)	whole antibody	-	Couraud et al. (1989)
anti-rabbit IgG (1 monoclonal)	Fab'	-	unpublished results
anti-rabbit IgG (polyclonal)	whole antibody		Couraud et al., (1989)
anti-mouse IgG (polyclonal)	whole antibody	-	unpublished results
anti-human $IL_{1\alpha}$ and human $IL_{1\beta}$ (20 monoclonals)	Fab'	< 5 pg/ml	Grassi et al., (1989a)
anti-human IL_2 (10 monoclonals)	Fab'	< 5 pg/ml	Grassi et al., (1989b)

Table 3 : List of the main antigen labelling performed with AChE

Antigen	Method	Sensitivity	References
aFGF (acidic Fibroblast Growth Factor)	SATA/SMCC	0,5 ng/ml	Caruelle et al., (1988)
Photosystem I components	avidin/biotin	-	Grassi et al., (1988)
human $IL_{1\alpha}$ and human $IL1\beta$	SATA/SMCC	3 ng/ml	Grassi et al., (1989a)
rat Prolactin	SATA/SMCC	0,5 ng/ml	Duhau et al., (1990)

essentially use SMCC as coupling reagent. The details of the different labelling procedures are given in the publications listed in Table 4.

Characteristics of the Enzyme Conjugates and Comparison with the Corresponding Radioimmunoassays.

Whatever the nature of the molecule to be coupled, it is always possible to achieve efficient coupling with AChE without inducing a significant loss of enzyme activity. Taking into account the great variety of enzyme-immunoassays developed during recent years, this demonstrates that AChE-labelling is of a very wide applicability. In addition, the great majority of AChE-conjugates have proved to be very stable for years when kept under suitable conditions (frozen at -20°C or lyophilised at +4°C).

In a few cases, the performances of the enzyme immunoassays were directly compared with those of the corresponding radioimmunoassays using ^{125}I-iodinated molecules as tracers. In every case, AChE-EIA appeared to be either equivalent (Renzi et al., 1987, Metreau et al., 1987, Pradelles et al., 1989, Baehr et al., 1987) or superior (Caruelle et al., 1988, Mc laughlin et al., 1987, Pradelles et al., 1985, 1990 a, b and c, Orsini et al., 1988, Porcheron et al., 1989) in terms of precision and sensitivity. From this point of view, the use of AChE conjugates constitutes a genuine advance, since it equals or improves upon the analytical performance of the immunoassay while avoiding the drawbacks associated with the use of radioactive tracers, i.e. i) no handling of radioactivity, ii) elimination of problems linked to disposal of radioactive wastes, iii) use of long-lived reagents. In addition, the combined use of enzymatic conjugates, colorimetric detection, microtitre plates and solid-phase separation allows for considerable automation of the differents steps of the assay using standard apparatus.

When compared with the other enzymes currently used in EIA, AChE presents the following advantages :

- greater detection sensitivity,
- possibility of increasing both precision and sensitivity by allowing the enzymatic reaction to proceed for very long periods (up to 40 hours),
- use of a non end-point detection method which permits continuous monitoring of the assay since it not necessary to stop the enzymatic reaction. This allows the working range of the assay (in the case of immunometric assays only) to be adapted to the level of antigen in the sample (high or low level requiring a short or long reaction time, respectively).

For all these reasons, we believe that in the years to come, AChE will prove to be a valuable and versatile tool for the sensitive and reliable determination of biological substances in basic and in clinical research.

Table 4 : List of the main hapten labelling performed with AChE

Hapten	Coupling method	Sensitivity	References
Peptides :			
Substance P	SATA/SMCC	10 pg/ml	Renzi et al., (1987)
Rat atriopeptin	SATA/SMCC	60 pg/ml	Mc Laughlin et al., (1987)
Thymulin	diazotation of a derived peptide	5 pg/ml	Metreau et al., (1987)
Thyrolibérin (TRH)	NHS ester	130 pg/ml	Grouselle et al., (1990)
Dermorphin	SMCC	30 pg/ml	Mor et al., (1989b)
Dermenkephalin	SMCC	30 pg/ml	Mor et al., (1989a)
Prostaglandins :			
Thromboxane B_2	NHS ester	10 pg/ml	Pradelles et al., (1985)
PGD_2 (méthoxamine)	"	8 pg/ml	Pradelles et al., (1985)
PGE_2	"	2 pg/ml	Orsini et al., (1988)
$PGF_{2\alpha}$	"	12 pg/ml	Pradelles et al., (1990a)
6-kéto-$PGF_{1\alpha}$	"	14 pg/ml	Pradelles et al., (1985)
Dinor TxB_2	"	8 pg/ml	Pradelles et al., (1990a)
11-déhydro-TxB_2	"	6 pg/ml	Pradelles et al., (1990a)
bicyclo PGE_2	"	20 pg/ml	Pradelles et al., (1990a)
Leukotriens :			
LTC_4	DFDNB and reverse SMCC	40 pg/ml	Pradelles et al., (1990b)
LTB_4	reverse SMCC	10 pg/ml	Pradelles et al., (1990c)
LTE_4	SMCC and reverse SMCC	20 pg/ml	Pradelles et al., (1990b)
Miscellaneous :			
cyclic AMP	DFDNB	2 pg/ml	Pradelles et al., (1989)
cyclic GMP	NHS ester	2 pg/ml	Pradelles et al, (1989)
Ecdysone	NHS ester	60 pg/ml	Porcheron et al., (1989)
Insect juvenil hormon	NHS ester	200 pg/ml	Baehr et al., (1987)
Benzyl-penicilloyl	direct reaction with AChE	300 pg/ml	Wall et al., (1990)

References

Baehr, J.C., Casas, J., Messeguer, A., Pradelles, P., Grassi, J., *Insect. Biochem.*, **(1987)**, 17(7), 929-932

Caruelle, D., Grassi,J., Courty, J., Groux-Muscatelli, B., Pradelles, P., Barritault, D., Caruelle, J.P., *Anal. Biochem.*, **(1988)**, 173, 328-339

Couraud, J.Y., Maillet, S., Grassi, J., Frobert, Y., Pradelles, P., *Methods in enzymology*, **(1989)**, 178, 275-230

Duhau, L., Grassi, J., Grouselle, D., Enjalbert, A., Grognet, J.M., submited for publication to *J. Immunoassay*, **(1990)**

Ellman, G.L., Courtney, K., Andres, V., Featherstone, R., *Biochem. Pharmacol.* , **(1961)**, 7, 88-95

Grassi, J., Frobert, Y., Lamourette, P., Lagoutte, B., *Anal. Biochem*, **(1988)**, 168, 436-450

Grassi, J, Frobert, Y., Pradelles, P., Chercuite, F., Gruaz, D., Dayer, J.M., Poubelle, P., *J. Immunol. Meth.*, **(1989a)**, 123, 193-210

Grassi, J., Frobert, Y., Pradelles, P., Dayer, J.M., Poubelle, P., *Compte rendu du VII° coloque sur les actualités en immunoanalyse, Le touquet 4-5 Octobre* **1989b**, to be published

Grassi, J., *Thése de doctorat de l'Université Paris VI*, **27 avril 1990**, Etude immunologique des cholinestérases. Utilisation de l'AChE de gymnote en tant que marqueur dans des dosages immunologiques.

Grouselle, D., Destombes, J., Barret,A., Pradelles, P., Loudes, C., Tixier-Vidal, A., Faivre-Bauman, A., *Endocrinology*, **(1990)**, 126, in press

Ishikawa, E., Imagawa, M., Hashida, S., Yoshitake, S., Hamaguchi, Y., Ueno, T., *J. Immunoassay*, **(1973)**, 4, 209-327

Massoulié, J., Rieger, F., *Eur. J. Biochem.*, **(1969)**, 11, 441-445

Massoulié, J., Bon, S., *Eur. J. Biochem.*, **(1976)**, 68, 531-539

Mc Laughlin, L., Wei, Y., Stockmann, P.T., Leahy, K.M, Needleman, P., Grassi, J., Pradelles, P., *Biochem. Biophys. Res. Com.*, **(1987)**, 144(1), 469-476

Metreau, E., Pleau, J.M., Dardenne, M., Bach, J.F., Pradelles, P., *J. Immunol. Meth.*, **(1987)**, 102, 233-242

Mor, A., Delfour, A., Sagan, S., Amiche, M., Pradelles, P., Rossier, J., Nicolas, P., *FEBS Letters*, **(1989a)**, 255, 269-274

Mor, A., Delfour, A., Amiche, M., Sagan, S., Nicolas, P., Grassi, J., Pradelles, P., *Neuropeptides*, **(1989b)**, 13, 51-57

Orsini, B., Calabro, A., Renzi, D., Pradelles, P., Fedi, P., Surrenti, C., *Clin. Chim. Acta.*, **(1988)**, 178, 305-312

Porcheron, P., Morinière, M., Grassi, J., Pradelles, P., *Insect. Biochem.*, **(1989)**, 19, 117-122

Pradelles, P., Grassi, J., Maclouf, J., *Anal. Chem.*, **(1985)**, 57, 1170-1173

Pradelles, P., Grassi, J., Chabardes, D., Guiso, N., *Anal. Chem.*, **(1989)**, 61, 447-453

Pradelles, P., Grassi, J., Maclouf, J., *Methods in enzymology*, Langone, J.J., Van vunakis, H., Eds, **(1990a)**, Academic Press, London and New York, 187, 24-34

Pradelles, P., Antoine, C., Lelouche, J.P., Maclouf, J., *Methods in enzymology*, Langone, J.J., Van vunakis, H., Eds, **(1990b)**, Academic Press, London and New York, 187, 82-89

Pradelles, P., Antoine, C., Maclouf, J., *7th Int. Conf. on Prostaglandins and related compounds.*, Florence May 28- June 1, **(1990c)**, to be published

Renzi, D., Couraud,J.Y., Frobert, Y., Nevers, M.C., Geppetti, P., Pradelles,P., Grassi, J., in *Trends in Cluster Headache*, Sicuteri, F., Vecchiet, L., Fanciullaci, M., Eds, **(1987)**, 125-134, Elsevier, Amsterdam

Wall, J.M., Yon, M., Pradelles, P., Grassi, J., *J. Immunoassay*, **(1990)**, submited for publication

Mechanism of Substrate Inhibition of Acetylcholinesterase

Elsa Reiner*, Institute for Medical Research and Occupational Health, University of Zagreb, P.O. Box 291, 41001 Zagreb, Croatia, Yugoslavia
Norman Aldridge, The Robens Institute, University of Surrey, Guildford, Surrey GU2 5XH, UK
Vera Simeon, Zoran Radić, Institute for Medical Research and Occupational Health, University of Zagreb, P.O. Box 291, 41001 Zagreb, Croatia, Yugoslavia
Palmer Taylor, Department of Pharmacology, School of Medicine, University of California San Diego, La Jolla, California 92093, USA

There are at present two hypotheses concerning the mechanism of substrate inhibition. According to one, substrate inhibition occurs when the substrate binds to the anionic site in the acetylated enzyme (Krupka and Laidler, 1961). According to the other, substrate inhibition occurs when the substrate binds to an allosteric site on the enzyme (Aldridge and Reiner, 1969 and 1972). Acetylcholinesterase (AChE) has been shown to have a peripheral (allosteric) site which can bind propidium and other reversible ligands (Taylor and Lappi, 1975; Berman et al. 1980 and 1981).

Evidence will be presented here concerning the allosteric site mechanism. The evidence is based upon kinetic studies of the effect of substrates upon reversible and progressive inhibition of AChE, and upon competition studies between substrates and inhibitors for binding to AChE using fluorescence titration with propidium.

Haloxon and coroxon (two organophosphorus compounds whose leaving group is a coumarin derivative) show a reversible component in the progressive inhibition of bovine and human erythrocyte AChE. Acetylcholine (ACh) and acetylthiocholine (ATCh) compete with the inhibitors for progressive inhibition at concentrations corresponding approximately to their respective K_m constants. However, competition for the reversible inhibition by haloxon and coroxon requires substrate concentrations close to the respective substrate-inhibition constants K_{ss}; the same is true for reversible inhibition by coumarin itself. The kinetics of reversible inhibition agrees with a theoretical equation derived upon the assumption that substrate inhibition occurs at an allosteric site of the enzyme and that haloxon, coroxon and coumarin compete with the substrates for binding to that site (Aldridge and Reiner, 1969 and 1972; Reiner and Simeon, 1975; Simeon et al., 1977 ; Reiner, 1984; Radić et al., 1984). In contrast to the above inhibitors, edrophonium and tetramethylammonium (TMA) compete with ACh and ATCh for reversible inhibition of AChE at substrate concentrations corresponding approximately to their K_m constants, indicating that these inhibitors bind to the catalytic site (active centre) of the enzyme.

ACh, ATCh, haloxon and coumarin were shown to dissociate propidium from the peripheral site of *Torpedo californica* AChE (Radić, 1988). Measurements were made by back-titration of propidium after the catalytic site had been completely phosphorylated by DFP. The concentrations of ACh and ATCh required for propidium displacement corresponded to their K_{ss} constants. This result confirmed that substrate inhibition occurred at the peripheral site of AChE. Displacement of propidium by haloxon and coumarin confirmed that these inhibitors also bound to the peripheral site of the enzyme. The K_{ss} constants derived

0–8412–2008–5/91/0227$06.00/0

from back-titration of propidium and from enzyme activity measurements were: 24 mM and 25 mM for ACh, 15 mM and 23 mM for ATCh. The enzyme/inhibitor dissociation constants derived from back-titration of propidium and from enzyme inhibition measurements were: 22 μM and 16 μM for haloxon, 26 μM and 116 μM for coumarin. The propidium dissociation constants for the DFP-modified enzyme and for the native enzyme (derived from inhibition kinetics) were 0.72 μM and 0.13 μM respectively.

Propidium was also displaced from the DFP-modified AChE by edrophonium and TMA, but the concentrations required for the displacement were significantly higher than the concentrations of edrophonium and TMA required for enzyme inhibition. The enzyme/inhibitor dissociation constants derived from back-titration of propidium and from enzyme inhibition measuements were: 590 μM and 0.30 μM for edrophonium, 113 mM and 4.0 mM for TMA. These results confirmed that AChE inhibition by edrophonium and TMA was primarily due to interaction at the active centre.

References

Aldridge, W.N., Reiner, E., *Biochem. J.*, **1969**, *115*, 147-162.

Aldridge, W.N., Reiner, E., *Enzyme Inhibitors as Substrates. Interaction of Esterases with Esters of Organophosphorus and Carbamic Acids*, Frontiers of Biology, Volume 26; North Holland, Amsterdam, 1972;pp 1--328.

Berman, H.A., Becktel, W., Taylor, P., *Biochemistry*, **1981**, *20*, 4803-4810.

Berman, H.A., Yguerabide, J., Taylor, P., *Biochemistry*, **1980**, *19*, 2226-2235.

Krupka, R.M., Laidler, K.J., *J. Am. Chem. Soc.*, **1961**, *83*, 1445--1447.

Radić, Z., *Ph. D. Thesis*, University of Zagreb, Yugoslavia, **1988**

Radić, Z., Reiner, E., Simeon, V., *Biochem. Pharmacol.*, **1984**, *33*, 671-677.

Reiner, E., In *Cholinesterases. Fundamental and Applied Aspects;* Brzin, M., Barnard, E.A., Sket, D., Eds.; Walter de Gruyter: Berlin, **1984**; pp 37-44.

Reiner, E., Simeon, V., *Croat. Chem. Acta*, **1975**, *47*, 321-331.

Simeon, V., Kobrehel, Dj., Reiner, E., *Croat. Chem. Acta*, **1977**, *50*, 331-334.

Taylor, P., Lappi, S., *Biochemistry*, **1975**, *14*, 1989-1997.

Function of the Peripheral Anionic Site of Acetylcholinesterase

Harvey Alan Berman, Kathryn Leonard, and **Mark W. Nowak**, Department of Biochemical Pharmacology, State University of New York at Buffalo, Buffalo, New York 14260

This chapter concerns electrostatic interactions at play on the surface of acetylcholinesterase (AchE) and the consequences of those interactions on enzyme catalysis. Our studies focus on the kinetics of covalent reactivity of AchE with respect to uncharged and cationic substrates and phosphonate inhibitors, and spectroscopic methods for assessing the magnitude and symmetry of electrostatic charge distribution on the enzyme surface.

Topography of the Acetylcholinesterase Subunit

The subunit of AchE contains an active center in which a nucleophilic residue, serine-200 (Schumacher, *et al.*, 1986), is separated by at least 4.7A from an anionic region. The nucleophilic serine covalently reacts with esters of acetic-, carbamic-, phosphonic- and phosphoric acids, while the anionic site serves to attract cationic portions of these molecules. Recently, our lab examined the topography surrounding the active center by employing a series of enantiomeric methylphosphonothioates containing an asymmetric phosphorus atom (Berman and Leonard, 1989; Berman and Decker, 1989). These studies indicate the presence of a second binding region that is situated within 4.7 A of the reactive serine and confers affinity for hydrophobic alkyl groups.

Kinetic (Changeux, 1966) and equilibrium studies (Taylor and Lappi, 1975) have indicated the presence of an additional site, a peripheral anionic site, topographically remote (>20 A) from the active center (Berman, *et al.*, 1980). A variety of structurally diverse cations, including *d*-tubocurarine, gallamine, and propidium, associate at this site and cause, in Changeux's words, a "bewildering array" of patterns of reversible inhibition at the active center. As suggested by Taylor and Lappi (1975), the peripheral site, rather than representing a structurally-defined, singular entity, may be more appropriately viewed as a matrix of overlapping anionic sites, thereby accounting for the relaxed requirements in accommodating structurally diverse ligands.

Function of the Peripheral Anionic Site as an Electrostatic Sensor

We propose that the peripheral anionic site functions as an electrostatic sensor of the immediate ionic milieu, providing an intramolecular mechanism for conserving enzyme covalent reactivity within a narrow range. In support of this idea are observations that the kinetics of acetylthiocholine turnover, as exemplified by k_{cat}, K_M, and k_{cat}/K_M, are reasonably constant, falling within a narrow 2-3-fold range over a wide range of ionic strength (Table 1). Similar behaviour is observed for covalent inhibition of AchE by methylphosphonyl thiocholines with respect to analogous kinetic parameters k_p, K_D, and k_i ($=k_p/K_D$).

Table 1. Influence of Ionic Strength on Covalent Reactivity of AchE

A. Hydrolysis of acetylthiocholine

[NaCl] (M)	Ks (μM)	$k_{cat} \times 10^{-5}$ (min^{-1})	$k_{cat}/Ks \times 10^{-9}$ (M^{-1}-min^{-1})
0.50	120	3.5	2.9
0.20	56	1.9	3.5
0.0	23	1.2	5.9

B. Irreversible inhibition by *S*P-CHMP-thiocholine

[NaCl] (M)	K_D (μM)	k_p (min^{-1})	$k_i \times 10^{-8}$ (M^{-1}-min^{-1})
0.50	1.1	150	1.4
0.20	0.88	170	2.1
0.0	0.35	140	4.0

As a noteworthy contrast, ligand association at the peripheral site and the active center, and the resulting modes of inhibition, are highly dependent on the ionic strength of the bulk aqueous medium (Taylor and Lappi, 1975; Berman and Decker, 1986a). Moreover, while the peripheral anionic site exerts long-range effects on active center reactivity, ligand occupation of the peripheral anionic site fails to preclude subsequent occupation

0–8412–2008–5/91/0229$06.00/0

of the active center by cationic ligands such as N-methylacridinium or edrophonium.

Covalent Reactivity of AchE with Respect to Uncharged and Cationic Methylphosphonothioates

R:
1. $-CH_3$
2. $-n C_5H_{11}$
3. $-CH_2CH_2N^+Me_3$

Covalent inhibition of AchE from *Torpedo californica* by structurally related methylphosphothioates (**1-3**) was assayed by measuring residual activity by the method of Parvari, *et al.* (1983). In all cases inhibition of AchE by the uncharged cycloheptyl methylphosphonothioates followed exponential behaviour. As summarized in Table 2, inhibition by **1** was slowed in the presence of edrophonium but not in the presence of decamethonium or propidium. Inhibition by the uncharged agent **2** was slowed by both edrophonium and decamethonium, but not by propidium. Inhibition of the cationic methylphosphonyl thiocholine **3** was slowed in the presence of edrophonium, decamethonium, and propidium; the reaction rates were at least 25-fold slower than those in the absence of ligand. Cation antagonism of **3** far exceeded that seen for the uncharged agents **1** and **2**. Hence, the extent of antagonism varied with both the individual cationic ligand and the nature of the methylphosphonothioate. While these studies characterized reactivity of the *S*P-enantiomer, it is noteworthy that identical results were obtained with the *R*P-enantiomer.

Reversible Noncovalent Inhibition of AchE Hydrolysis of Acetylthiocholine and 7-Acetoxy-4-Methylcoumarin

Non-covalent inhibition was examined by measuring the capacity of cationic ligands to inhibit AchE hydrolysis of acetylthiocholine ($AcSch^+$) and 7-acetoxy-4-methylcoumarin (7AMC). A typical reciprocal plot for inhibition of AchE by edrophonium is shown in Fig. 1A. As seen from the slope-*versus*-[I] replot, edrophonium displayed *linear mixed* inhibition. Linear mixed inhibition was observed also for propidium and decamethonium. Analysis of the y-intercept replots afforded estimates of αK_I, where α measures the difference in the inhibitor affinity for free enzyme and enzyme-substrate complex (Table 3). The higher value of α seen for edrophonium indicated that this ligand displays a higher affinity for free enzyme than for the Michaelis complex; the lower values of α seen for decamethonium and propidium indicated that these ligands show comparable affinity for free enzyme and the Michaelis complex.

Table 2. Influence of Reversible Cationic Ligands on Reaction of *S*P-Cycloheptyl Methylphosphonothioates with Acetylcholinesterase

Ligand	*S*P-CHMP-SMe	*S*P-CHMP-S*n*Pe	*S*P-CHMP-thiocholine
NONE			
k_{obs} (min^{-1})	2.6	7.4	1.9
k_i (M^{-1}-min^{-1})	$1.3x10^5$	$2.1x10^5$	$2.2x10^7$
DECAMETHONIUM			
C/L	1.1	9.4	25
EDROPHONIUM			
C/L	19	14	39
PROPIDIUM			
C/L	1.6	2.1	34

Kinetics were determined in a 0.01 M sodium phosphate buffer, pH 7.0, containing 0.2 M NaCl. The inhibition rate constant, k_{obs} (min^{-1}) was obtained with the methylphosphonothioate in at least 40-fold excess over enzyme, and at concentrations 0.25 times the dissociation constant: *S*P-CHMP-SMe ($2.0x10^{-5}$ M), *S*P-CHMP-S*n*Pe ($3.6x10^{-5}$ M), *S*P-CHMP-thiocholine ($8.6x10^{-8}$ M). The apparent bimolecular inhibition constant, k_i, was derived as k_{obs}/[methylphosphonothioate]. C/L denotes the ratio of k_{obs} in the absence (C, control) and presence (L, ligand) of non-covalent cationic ligands.

Inhibition of uncharged substrates was determined by examining AchE hydrolysis of 7AMC. Hydrolysis of this synthetic substrate produces 7HMC which is readily detectable by measurement of the linear evolution of fluorescence at 450 nm (λ_{ex}=360 nm). Edrophonium was a *linear competitive* inhibitor of 7AMC hydrolysis. Inhibition by propidium was also competitive, in strong contrast to the nearly noncompetitive behaviour against $AcSch^+$, but the slope-*versus*-[I] replot was distinctly nonlinear (Fig. 1B). The value for K_I from plots of 1/Δslope-*versus*-1/[I], as shown in Fig. 1C (cf. Segel, 1975) was calculated to be 1.2 x 10^{-6} M (α =6.0).

Spectroscopic Examination of NBD-Aminoethyl Methylphosphono-AchE

The influence of ionic strength on AchE conformation was assessed by examining the dependence of fluorescence of an environmentally-sensitive probe covalently linked at the AchE active center. AchE was labelled with NBD-AE-MPF to form the covalent conjugate NBD-AE-MP-AchE (Berman, *et al.*, 1985). With increasing concentrations of NaCl the intensity at 530 nm (λ_{ex}=470 nm), increased without any accompanying change in spectral position (Fig. 2). The dependence of NBD-aminoethyl fluorescence on Na^+

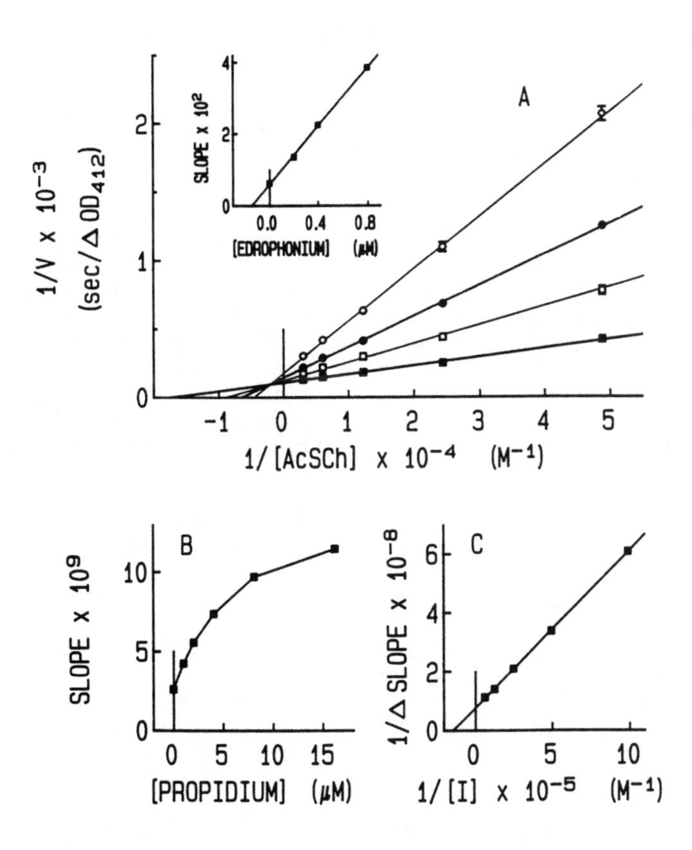

Fig. 1. Reversible Inhibition of AchE Hydrolysis of $AcSch^+$ and 7AMC. *Panel A:* Double reciprocal plots of reaction velocity *vs* acetylthiocholine concentration in the presence and absence of edrophonium. (■), No edrophonium; (□), 0.20; (●), 0.40; (O), 0.80 μM. The value for Ks was calculated to be 5.6x10^{-5} M. The *inset* presents a replot of the reciprocal plot slopes *vs* inhibitor concentration, and is indicative of linear mixed inhibition. The inhibition constant for edrophonium is calculated to be 0.15 μM. *Panel B*: Replot of the reciprocal plot slopes *vs* propidium concentration determined for inhibition of 7AMC hydrolysis. *Panel C*: Transformation of the data in *Panel B* in the form of 1/Δslope-*vs*-1/[I], affording an inhibition constant for propidium of 1.2 μM.

Table 3. Reversible Inhibition by Cationic Ligands of Acetylcholinesterase Hydrolysis of $AcSch^+$ and 7AMC

$AcSch^+$		
LIGAND	Pattern (α)	K_I (μM)
Edrophonium	Linear mixed (7.1)	0.15
Decamethonium	Linear mixed (3.2)	0.56
Propidium	Linear mixed (1.6)	1.1
7AMC		
LIGAND	Pattern (α)	K_I (μM)
Edrophonium	Linear competitive	0.14
Decamethonium	*	0.87
Propidium	Nonlinear competitive (6.0)	1.2

Inhibition was determined in a 0.01 M sodium phosphate buffer, pH 7.0, containing 0.2 M NaCl.

* The intersection point occurred in the upper right quadrant. The slopes replot was linear and the inhibition constant reported represents the [I]-intercept of this plot.

concentration was saturable, with a maximal increase of 100-150 percent occurring at concentrations greater than approximately 30 mM (Fig. 2, *inset*). Similar behaviour was observed for KCl. In the presence of $MgCl_2$ or $CaCl_2$ the fluorescence of NBD-AE-MP-AchE was *reduced*.

Discussion

Site-selective cationic ligands do not cause comparable degrees of antagonism of the methylphosphonate reaction. Similarly, a single ligand does not cause equal antagonism of all methylphosphonates. While all ligands effectively antagonize reaction of the methylphosphonyl thiocholine (**3**), the distinctions among the different ligands are most evident with respect to the kinetic behaviour of the uncharged methylphosphonothioates (**1**, **2**).

Decamethonium antagonism of **2** but not **1** indicates a clear steric effect related to the dimensions of the thiol leaving group (-SR). These findings are not compatible with competitive antagonism at a single class of sites, and require that *aromatic* ligands associate at a locus separate from that for *n*-alkyl mono- and bisquaternary ligands. These findings will be discussed elsewhere. For the present discussion it is of interest that these observations provide a kinetic index establishing the presence in the active center of a heterogenous population of cation binding sites, and substantiate earlier conclusions derived from equilibrium measurements (Berman and Decker, 1986b).

For the peripheral site ligand propidium, antagonism of covalent reaction at the active center shows an absolute dependence on the electrostatic nature of the covalent reactant. Propidium shows little capacity to antagonize inhibition by the uncharged methylphosphonates (**1**, **2**), but inhibits reaction by the methylphosphonyl thiocholine (**3**) with the same efficacy as the aromatic active-center selective cation edrophonium.

Ligand Exclusion Kinetics

Linear inhibition of $AcSch^+$ by peripheral site ligands signifies that the reaction velocity can be driven to zero by sufficiently high inhibitor concentrations. This finding, typical of the inhibition seen for acetylcholine (Mooser and Sigman, 1974), indicates that the ternary complex formed with a cationic substrate is inactive. *Nonlinear* inhibition of 7AMC by propidium signifies that high concentrations of inhibitor do not drive the reaction velocity to zero. Such nonlinear inhibition arises because ES and ESI remain productive and continue to undergo covalent reaction. For propidium, the nonlinear competitive inhibition of 7AMC hydrolysis reflects an increase in Ks ($\alpha > 1$) without effect

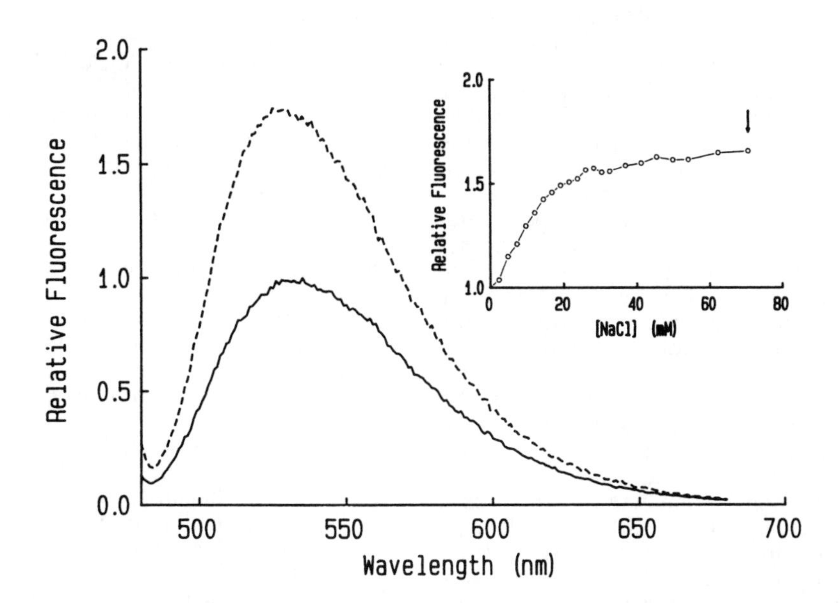

Fig. 2. Fluorescence Emission Spectra of NBD-Aminoethyl Methylphosphono-acetylcholinesterase in the Absence and Presence of NaCl. NBD-aminoethyl methylphosphonofluoridate was allowed to react with AchE; the covalent conjugate was isolated by gel filtration and dialyzed against a 0.01 N Tris-Cl buffer, pH 8.0, in the absence of NaCl. Emission spectra were obtained upon excitation at 470 nm in the absence (—) and presence (----) of NaCl. *Inset*: Fluorescence intensity at 530 nm (λ_{ex}=470 nm) *vs* NaCl concentration. Arrow indicates point at which emission spectrum was obtained.

on VMAX, since product arises from both ES and ESI with equal facility. These findings demonstrate a distinguishing feature of peripheral site occupation on AchE reactivity in that, as exemplified by the actions of propidium, peripheral site occupation blocks covalent reactivity of *cationic* but not *uncharged* acetyl ester substrates.

Electrostatic Regulation of Covalent Reactivity

The behaviour of peripheral site ligands stands out with respect to the antagonism of cationic and uncharged acetyl esters and methylphosphonates. Linear inhibition of $AcSch^+$ hydrolysis parallels the linear inhibition of acetylcholine (Mooser and Sigman, 1974) and the substantial antagonism of methylphosphonyl thiocholines. Nonlinear inhibition of 7AMC hydrolysis parallels the negligible antagonism of uncharged methylphosphonates. Since ligand association at the peripheral anionic site more effectively precludes AchE reactivity with cationic rather than uncharged substrates and methylphosphonates, peripheral site occupation appears to alter reactivity through an electrostatic interaction with the net negative active center.

In view of the sensitivity of non-covalent ligand association to changes in ionic strength and the near-constant covalent reactivity under conditions of varying ionic strength, the selective actions of peripheral site ligands on covalent reactivity of cationic agents indicates a potential role for the peripheral anionic site in conserving enzyme reactivity in an inconstant ionic environment. Such a function is anticipated to be of considerable importance in neuromuscular transmission since the synaptic cleft contains rapidly varying concentrations of mono- and divalent cations, acetylcholine, and its charged hydrolysis products choline and acetate (Kuffler, *et al.*, 1984). An acute dependence of AchE catalysis on ionic composition of the synapse, leading to large excursions in catalytic efficiency, would be incompatible with the requirement for rapid removal of acetylcholine from the synaptic cleft (Hartzell, *et al.*, 1975). The proposal that the peripheral anionic site serves as an electrostatic sensor of the ionic environment provides an intramolecular mechanism for accommodating transient changes in ionic composition of the medium without altering the kinetics of substrate turnover.

Through what mechanism can ligand occupation of the peripheral site alter catalysis at the active center? As shown above, fluoresence of the covalent NBD-aminoethyl methylphosphonyl conjugate at the active center is readily altered with subtle changes in concentration and composition (Na^+ *versus* Mg^{++}) of the ionic medium. Since mono- and divalent cations exert *opposite* effects with respect to altering fluorescence of a covalently linked environmentally-sensitive probe, these actions can be due only in part to ionic strength of the medium and must also reflect the influence of direct ion binding at the enzyme surface. These spectroscopic studies clearly demonstrate that conformation of AchE is highly responsive to subtle changes in ionic composition of the medium. The electrostatic charge carried by the substrate, net negative charge on the enzyme, and the ionic composition of the bulk aqueous medium therefore combine to play a role in determining covalent reactivity of AchE.

References

Berman, H.A.; Decker, M.M. *J. Biol. Chem.* **1986a**, *261*, 10646-10652.

Berman, H.A.; Decker, M.M. *Biochim. Biophys. Acta* **1986b**, *872*, 125-133.

Berman, H.A.; Decker, M.M. *J. Biol. Chem.* **1989**, *264*, 3951-3956.

Berman, H.A.; Leonard, K. *J. Biol. Chem.* **1989**, *264*, 3942-3950.

Berman, H.A.; Yguerabide, J.; Taylor, P. *Biochemistry* **1980**, *19*, 2226-2235.

Berman, H.A.; Olshefski, D.F.; Gilbert, M.; Decker, M.M. *J. Biol. Chem.* **1985**, *260*, 3462-3468

Changeux, J.-P. *Mol. Pharmacol.* **1966**, *2*, 369-392.

Hartzell, H.C.; Kuffler, S.W.; Yoshikami, D. *J. Physiol. (London)* **1975**, *251*, 427-463.

Kuffler, S.W.; Nicholls, J.G.; Martin, A.R. *From Neuron to Brain*; Sinauer Associates, Inc.: Sunderland, MA, 1984; pp 207-261.

Mooser, G.M.; Sigman, D.S. *Biochemistry* **1974**, *13*, 2299-2307.

Nolte, H.-J.; Rosenberry, T.L.; Neumann, E. *Biochemistry* **1980**, *19*, 3705-3711.

Parvari, R.; Pecht, I.; Soreq, H. *Anal. Biochem.* **1983**, *133*, 450-456.

Quinn, D.M. *Chem. Revs.* **1987**, *87*, 955-979.

Schumacher, M.; Camp, S.; Maulet, Y.; Newton, M.; MacPhee-Quigley, K.; Taylor, S.S.; Friedmann, T.; Taylor, P. *Nature* **1986**, *319*, 407-409.

Segel, I.H. *Enzyme Kinetics*; J. Wiley and Sons: New York, NY, 1975; pp 161-226.

Taylor, P; Lappi, S. *Biochemistry* **1975**, *14*, 1989-1997.

Changes in the Catalytic Activity of Acetylcholinesterase upon Complexation with Monoclonal Antibodies

Y. Ashani, A. Bromberg and D. Levy*, Israel Institute for Biological Research, Ness–Ziona, Israel.
M. K. Gentry, D. R. Brady and B. P. Doctor, Division of Biochemistry, Walter Reed Army Institute of Research, Washington, DC 20307–5100, USA

Two inhibitory monoclonal antibodies, 25B1 and 6H9, were raised against organophosphoryl (OP) conjugates of acetylcholinesterase (AChE) purified from fetal bovine serum (FBS) and were characterized with respect to their anti–AChE properties. Although residual enzyme activity of AChE of both enzyme:antibody conjugates was less than 3%, 6H9:AChE demonstrated considerably different kinetic behavior than 25B1:AChE in terms of rate of reaction with organophosphorus inhibitors of AChE. Results show that these inhibitory mAbs may be suitable for probing amino acid sequences associated with the catalytic activity of FBS–AChE.

The isolation of inhibitory monoclonal antibodies (mAb) to acetylcholinesterase (EC 3.1.1.7) has provided useful molecular probes to study polymorphism and tissue distribution of AChE (Fambrough et al., 1982; Mintz and Brimijoin, 1985), as well as the relationship between AChE structure and function (Brimijoin et al., 1985; Sorensen et al., 1987). In a recent report (Ashani et al., 1990), a mAb (25B1) raised against diisopropylfluorophosphate (DFP)–inhibited AChE has been extensively characterized with respect to its anti–AChE properties. More recently, a second inhibitory mAb (6H9) has been isolated and partially characterized. This mAb was raised against $[CH_3P(O)(OC_2H_5)]$–AChE, which was obtained by using 7–[(methylethoxyphosphinyloxy]–1–methylquinolinium iodide (MEPQ). Despite the similarity in anti–AChE properties of the two mAbs, the inactivated conjugates 25B1:AChE and 6H9:AChE demonstrated considerably different kinetic behavior toward phosphorylation by either DFP or MEPQ.

In this report we describe results which demonstrate differences between the residual nucleophilicity of AChE from conjugates 25B1:AChE and 6H9:AChE. It appears that both mAbs may be used to probe AChE regions which are associated with enzyme activity.

Materials and Methods

MEPQ was prepared as described previously (Levy and Ashani, 1986). 1–Methyl–2–hydroxyiminoethyl) pyridinium methanesulphonate (P_2S) was obtained from Sigma. 1,1′–Trimethylenebis–[4–(hydroxyiminomethyl)pyridinium] dibromide (TMB_4) was prepared according to Poziomek et al. (1958). AChE was purified from fetal bovine serum according to the procedure of De La Hoz et al. (1986). Enzyme activity was 400 units/nmol of active site. Monoclonal antibodies 25B1 and 6H9, both IgG, were elicited in mice with DFP–inhibited and MEPQ–inhibited FBS–AChE, respectively. Monoclonal antibodies were amplified in mouse ascitic fluids and purified by Protein A affinity chromatography. Measurements of enzyme activity were carried out according to the Ellman procedure using acetylthiocholine as substrate (Ellman et al., 1961). $[^3H]$DFP radiolabeling of free and mAb–bound AChE and displacement of radioactivity by TMB_4 were determined by gel filtration as described previously (Ashani et al., 1990). The release of the fluorescent chromophore 7–hydroxyquinolinium methiodide (7–HQ) from MEPQ was monitored (ex: 400 nm; em: 509 nm) in 5 mM phosphate, pH 8 (Levy and

0–8412–2008–5/91/0235$06.00/0

Ashani, 1986). Regeneration of enzyme activity from mAb:AChE or from mAb-bound phosphorylated AChE following reactivation was carried out by diluting complexed enzyme into 1M guanidine (Gdn)-HCl (Ashani et al., 1990) containing 2 mM procainamide (pH 7.6, 25°C). Solid-phase binding (ELISA) of mAbs to either AChE or OP-AChE conjugates was detected with a horseradish peroxidase-labeled goat antibody to mouse IgG+IGM (H+L). Competition between mAbs 6H9 and 25B1 for binding to FBS-AChE was determined using horseradish peroxidase-labeled 25B1.

Results

More than 97% of the enzymic activity of FBS-AChE was inhibited by mAb 6H9. A similar degree of inhibition was observed with mAb 25B1. Pseudo first-order plots (not shown) of enzyme activity did not produce straight lines despite a large excess of mAb. The rate of inhibition of FBS-AChE activity by mAb 6H9 was also determined in the presence of edrophonium, propidium (10 μM each), P_2S (1 mM), or TMB_4 (0.2 mM). Only propidium, which was previously demonstrated to bind at the putative peripheral anionic site (Taylor, 1975), interfered, i.e., retarded the rate of inhibition of AChE by 6H9 (Fig. 1). Titration of FBS-AChE with mAbs showed that one molecule of either mAb inactivated approximately one to two active sites of FBS-AChE. The stoichiometry observed for the inhibition of AChE by either mAb did not change in the presence of the cationic ligands mentioned above. Solid-phase immunoabsorbance assays showed that heat denaturation of FBS-AChE caused considerable loss of binding of both 25B1 and 6H9 to the perturbed antigen. Competition experiments using ELISA revealed that 25B1 and 6H9 share, in part, similar antigenic determinant(s).

To further characterize mAb-bound AChE, the extent of radiolabeling of both 25B1:AChE and 6H9:AChE by [^{3}H]DFP was determined following 4 h of incubation with tritiated DFP at 25°C (not shown). 25B1:AChE failed to show significant binding of radioactivity when more than 97% of enzyme activity was preinhibited by mAb. In contrast, the extent of binding of [^{3}H]DFP to 6H9:AChE, in which AChE activity was also preinhibited

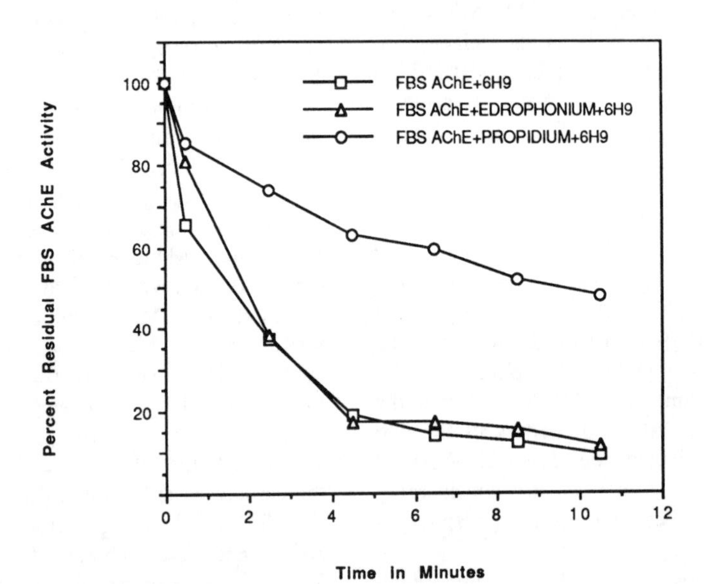

Fig. 1. Effect of edrophonium and propidium on the time course for inhibition of FBS-AChE by mAb 6H9. AChE and 6H9 were 1 nM and 2nM, respectively (5 mM phosphate, pH 8.0, 25°C). Edrophonium and propidium were both present at 10 μM. These concentrations approximated 90- and 30-fold the dissociation constants of edrophonium and propidium, respectively.

(>97%), amounted to 70–80% of that obtained for free enzyme. Increasing the time of incubation with [^{3}H]DFP to 24 h reduced the amount of radiolabeling of 6H9:AChE to 25–30% of that of free enzyme, while no corresponding changes were observed for similarly treated 25B1:AChE (not shown). This loss of radioactivity may be attributed to dealkylation of the radiolabeled isopropyl residue attached to the phosphoryl moiety of the inhibited enzyme (aging reaction). To further demonstrate differences between the nucleophilicity of the two mAb–bound AChE conjugates, the release of the fluorescent–leaving group 7–HQ from MEPQ was monitored in the presence or absence of AChE and mAb:AChE. 7–HQ was quantitatively released in less than 10 seconds from MEPQ by either free AChE or the same amount of AChE which had been preincubated for 20 h with sufficient mAb 6H9 to inactivate 97% of enzyme activity, prior to reacting mAb–bound enzyme with MEPQ (Fig. 2). In contrast, although the same amount of 25B1:AChE released the expected amount of 7–HQ, the rate of reaction between 25B1–bound enzyme and MEPQ was decreased by more than 1000–fold. The bimolecular second–order rate constant obtained by computer–fitted analysis of the experimental points shown in Fig. 2, was $2.1x10^5$ M^{-1} min^{-1}, whereas the bimolecular second–order rate constant for inhibition of FBS–AChE by MEPQ was $3.4x10^8$ M^{-1} min^{-1}. Neither mAb changed the rate of the spontaneous hydrolysis of MEPQ.

In view of the marked differences between the nucleophilicity of 25B1– and 6H9–bound FBS–AChE for OPs, the ability of oxime reactivators to displace the OP moiety of mAb–bound OP–AChE was measured. In one set of experiments the rate of displacement of enzyme–bound radioactivity from [^{3}H]DFP–AChE, complexed to either 25B1 or 6H9, was monitored in the presence of 1 mM TMB_4. Irrespective of the mAb used, the rate of displacement of radioactivity in the presence of 1 mM TMB_4 (pH 8.0) was significantly lower for mAb–bound DFP–AChE than in the absence of mAb (not shown). In the presence of 1 mM TMB_4, 6H9–bound DFP–AChE was more rapidly converted to a form which was resistant to further release of radioactivity than 25B1–bound DFP–AChE (Fig. 3). This observation may be associated with mAb–enhanced aging of DFP–AChE, which appeared to be more pronounced for the 6H9:AChE conjugate. This explanation is consistent with results obtained from radiolabeling experiments of 6H9:AChE (see above). After 48 h of incubation with

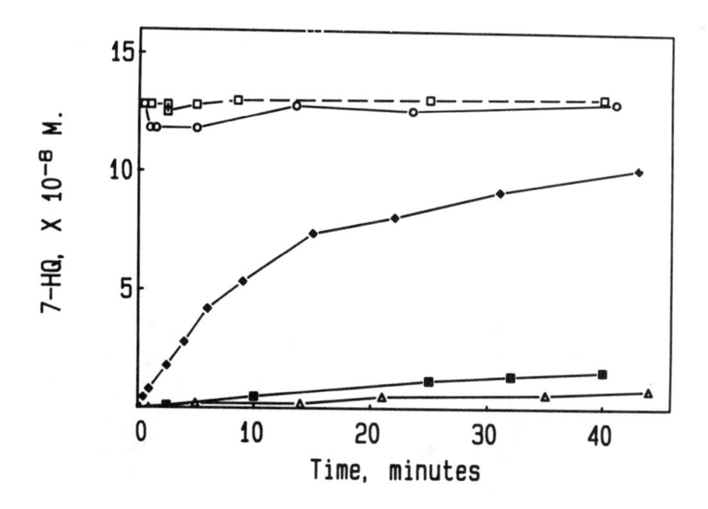

Fig. 2. Release of 7–HQ from MEPQ with and without the presence of mAb:AChE conjugates (Δ , spontaneous hydrolysis of MEPQ, subtracted). Antibody:AChE conjugates were prepared by incubation of a 3– to 4–fold stoichiometric excess of mAb with FBS–AChE for 20 h. Final concentrations of free AChE (□), 6H9–AChE (○), 25B1:AChE (◊) and diethylphosphoryl–AChE (■) were 0.16 μM each. Final MEPQ concentration was 0.25 μM.

TMB_4, all protein–bound radioactivity was released from the corresponding conjugate with 25B1, whereas the radioactivity of 6H9–bound [^{3}H]DFP–AChE decreased to 25–30% after 24 h and remained unchanged after an additional 24 h of incubation (Fig 3).

To correlate displacement of radioactivity from [^{3}H]DFP–inhibited AChE with possible concomitant regeneration of enzyme activity from mAb–bound OP–AChE, it was necessary to dissociate dephosphorylated enzyme from its complex with either 25B1 or 6H9, since the complexes were enzymatically inactive. At various time intervals, the reactivation mixture of mAb:[paraoxon–inhibited AChE] and either P_2S or TMB_4 was diluted 100–fold into 1.0 M Gdn–HCl containing 2 mM procainamide–HCl. Control experiments showed that dissociation of the complex mAb:AChE occurred, as evidenced by progressive restoration of enzyme activity until a steady state of AChE activity was obtained. The rate of approach and the level of the observed steady state depended on the type of mAb used and the final concentration of both mAb and AChE in diluted Gdn solution. Using the present protocol we were able to estimate the extent to which mAbs 6H9 and 25B1 retarded the reactivation of paraoxon–inhibited AChE. The half–life times for reactivation of mAb–bound, paraoxon–inhibited AChE in the presence of 1 mM P_2S were 8, 7 and 100 min for non–bound, 6H9–bound and 25B1–bound enzyme, respectively. Retardation of the reactivation of paraoxon–inhibited AChE by 6H9 and 25B1 was significantly higher for TMB_4 than for P_2S.

Discussion

The similarity observed in the anti–AChE properties of monoclonal antibodies 25B1 (Ashani et al., 1990) and 6H9 suggests that both mAbs may be directed, in part, against the same region of FBS–AChE. It appears that mAbs 6H9 and 25B1 are directed against a conformational epitope which is comprised of antigenic determinants located in close proximity to the active site of FBS–AChE. Both epitopes may encompass part of the putative peripheral anionic site. Thus, the observation that the two mAb:AChE complexes displayed significantly different properties in terms of their reaction with phosphorylating agents and different

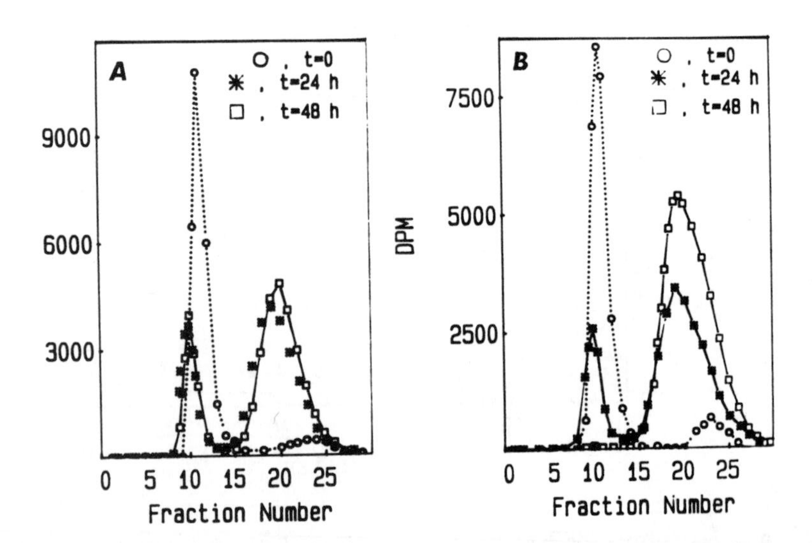

Fig. 3. Displacement of radioactivity from 6H9–bound (Panel A) and 25B1–bound (Panel B) [^{3}H]DFP–AChE. [^{3}H]DFP–treated FBS–AChE was incubated 20 h with either mAb and then subjected to 1 mM TMB_4 (pH 8.0). Protein–bound and free radioactivity were determined by gel filtration (Bio–Rad P–6).

stabilities when the corresponding mAb:OP–AChE conjugates were exposed to oxime reactivators is of considerable importance. Of particular interest is the finding that loss of activity of AChE upon binding to 6H9 was not accompanied by a significant decrease in the nucleophilicity of the complexed enzyme toward phosphorylating agents.

We hope that binding of these monoclonal antibodies to synthetic octapeptides spanning the entire sequence of FBS–AChE will allow us to probe specific amino acids which participate in the catalytic activity of AChE.

References

Ashani, Y., Gentry, M.K., Doctor, B.P., *Biochem.,* **1990,** *29,* 2456–2463.

Brimijoin, S., Mintz, K.P., Pedergast, F.G., *Mol. Pharmacol.,* **1985,** *28,* 539–545.

De La Hoz, D., Doctor, B.P., Ralston, J.S., Rush, R.S., Wolf, A.D., *Life Sci.,* **1986,** *39,* 195–199.

Ellman, G.L., Courtney, D., Andres, V., Featherstone, R.M., *Biochem. Pharmacol.,* **1961,** *7,* 88–95.

Fambrough, D.M., Engel, A.G., Rosenberry, T.L., *Proc. Natl. Acad. Sci., USA,* **1982,** *79,* 1078–1082.

Levy, D., Ashani, Y., *Biochem. Pharmacol.,* **1986,** *35,* 1079–1085.

Mintz, K.P., Brimijoin, S., *J. Neurochem.,* **1985,** *45,* 289–292.

Poziomek, E.J., Hackley, B.E., Steinberg, M., *J. Org. Chem.,* **1958,** *23,* 714–717.

Sorensen, K., Brodbeck, U., Rasmussen, A.G., Norgaard–Pedersen, B., *Biochim. Biophys. Acta,* **1987,** *912,* 56–62.

Taylor, P., Lappi, S., *Biochem.,* **1975,** *14,* 1989–1997.

Interactions of Esterases with Soman and Other Chiral Anticholinesterase Organophosphates

L.P.A. de Jong* **and H.P. Benschop**, Prins Maurits Laboratory TNO, P.O. Box 45, Rijswijk, The Netherlands

Our development of a method for the resolution of the four stereoisomers of soman by chiral gas chromatography initiated a number of biochemical and toxicological studies. Stereoselective inhibition of AChE by soman and other chiral organophosphates (OPs) can be understood from a model proposed by Järv for the active site of the enzyme. The model allows the tentative assignment of the absolute configuration of chiral OPs. With the ^{14}C-labelled toxic and nontoxic soman stereoisomers isolated pairwise from ^{14}C-soman, we showed that the toxic stereoisomers of soman are preferentially eliminated by irreversible binding. The nontoxic stereoisomers of soman and of other chiral OPs are rapidly hydrolyzed by phosphorylphosphatases.

Approximately ten years ago Benschop et al. (1981) developed a method for chiral gas chromatography to separate of the four stereoisomers of the nerve agent soman**

$$H_3C-C(CH_3)_2-C^{*}H(CH_3)-O-P^{*}(=O)(CH_3)F$$

in which the chiral C and P atoms are denoted by an asterisk.

In combination with phosphorus-selective detection this method allows a very sensitive assay of the separate isomers (Benschop et al., 1985). Qualitatively, it was shown that only two isomers of soman are powerful inhibitors of AChE and α-chymotrypsin. Moreover, it was found that the two other isomers are preferentially degraded in rat plasma, rat liver homogenate and guinea pig skin. Taking advantage of both types of reactions, Benschop et al. (1984) isolated all four stereoisomers of soman starting from C(+)P(±)- and C(-)P(±)-soman***. The chiral gas chromatographic analysis was indispensable in order to assess the optical purity of the products.

The availability of these techniques initiated a number of investigations on the biochemical and toxicological implications of chirality in soman. In this survey studies will be reported on the interaction of the stereoisomers with (i) the target enzyme, i.e., AChE, (ii) phosphorylphosphatases, which eliminate soman isomers by hydrolysis, and (iii) carboxylesterases and other proteins, which eliminate soman isomers by irreversible binding. Results of similar studies reported on other chiral OPs will also be dealt with.

Interaction with AChE

The epimers of soman having opposite configurations around the central phosphorus atom largely differ in their anticholinesterase activities (Table 1) as expressed by the bimolecular rate constants for the inhibition reaction (k). The configuration around the second centre of asymmetry, the α-carbon of the pinacolyl group, has only a small influence on the inhibitory effects of the soman isomers. Similar results have been reported for other chiral OPs. Some examples are collected in Table 1; see De Jong and Benschop (1989) for a comprehensive survey. In general, high stereoselectivity (ratio of the rate constants for inhibition by enantiomers or epimers > 100) has only been found for compounds of which the more active isomer is a potent inhibitor, i.e., k > 10^6 $M^{-1}min^{-1}$. The limited data reported for OPs in which a chiral centre is introduced into the alkoxy or the leaving group indicate that, as found for soman, asymmetry in one of the substituents at the phosphorus atom leads to minor

** WARNING: in view of their extremely high toxicities, soman and related OPs should be handled only in laboratories where specifically trained medical personnel is available.

*** C stands for the asymmetric pinacolyl carbon and P for the asymmetric phosphorus atom; (+) and (-) are the two configurations around each atom and (±) stands for a mixture of the two configurations in any ratio.

Table 1. Stereoselective inhibition of bovine erythrocyte AChE by stereoisomers of some chiral OPs with the general structure Y(Z)P(O)X, due to asymmetry at the central phosphorus atom or to asymmetry at the carbon atom in Y or X. Rate constants (k) for inhibition by the more active isomer and ratios of the rate constants for enantiomers or epimers (r_a) are presented

Organophosphate[a] Y	Z	X	Chiral centre	k^b ($M^{-1}min^{-1}$)	r_a	Refs[c]
EtO	Me	$S(CH_2)_2N(i\text{-}Pr)_2$	P	$4x10^8$	200	1
i-PrO	Me	$S(CH_2)_2SMe_2$	P	$4x10^7$	340	2
		$S(CH_2)_2NMe_3$	P	$5x10^7$	1200	2
		$OC_6H_4\text{-}4\text{-}NO_2$	P	$1x10^6$	370	2
		F	P	$1x10^7$	4200	3
$Me_3CCH(Me)O$	Me	F	P	$2x10^8$	17500	4
				$3x10^7$	2700	
EtO	Et	SC_6H_5	P	$3x10^6$	9	5
EtCH(Me)O	Et	$S(CH_2)_2NMe_2$	P	$9x10^6$	1400	6
				$3x10^6$	1600	
$Me_3C^*CH(Me)O$	Me	F	C	$2x10^8$	6	4
$Et^*CH(Me)O$	Et	$S(CH_2)_2NMe_2$	C	$9x10^6$	3	6
				$6x10^3$	4	
EtO	EtO	$S^*CH(Me)CH_2NMe_2$	C	$3x10^5$	40	3
		$SCH_2{}^*CH(Me)NMe_2$	C	$3x10^6$	1.2	3

[a] Asymmetry in Y or X is denoted by an asterisk.
[b] At pH 7.5 (ref. 4,6), 7.6 (ref. 5) or 7.7 (refs 1-3) and 25 °C (ref. 1-4) or 30 °C (ref. 5,6).
[c] Ref. 1, Benschop and De Jong, 1988; ref. 2, Benschop et al., 1985; ref. 3, Boter and Van Dijk, 1969; ref. 4, Benschop et al., 1984; ref. 5, Lee et al., 1978; ref. 6, Wustner and Fukuto, 1974.

differences in inhibitory activities of the stereoisomers (see Table 1 for some examples).

In general, the enantioselective anticholinesterase activities are reflected in enantiomeric differences in toxicity of chiral OPs (De Jong and Benschop, 1989), except for profenofos, sulprofos oxon and EtO(Me)P(O)SPr, due to metabolic activation of the less actively inhibiting enantiomer (Wing et al., 1983; Hirashima et al. 1984; Hall et al., 1977). This analogy can be illustrated by the data obtained for the soman stereoisomers (Benschop et al., 1984). The powerfully inhibiting C(-)P(-)- and C(+)P(-)-soman are highly toxic when subcutaneously administered to mice, i.e., LD_{50} values of 38 and 99 µg/kg, respectively, whereas the weakly inhibiting C(±)P(+)-isomers are hardly toxic (LD_{50} > 2000 µg/kg).

Stereoselective inhibition of AChE by OPs can be understood on the basis of the model presented by Järv (1984; Järv et al., 1977) for the active site of the enzyme. In addition to two catalytic sites, i.e., a serine OH group activated by a basic group (B, see Fig. 1) and an acidic group (HA) involved in hydrogen bond formation with the carbonyl oxygen of the substrate or with the phosphoryl oxygen of the inhibitor, Järv proposed two hydrophobic subsites ρ_1 and ρ_2 for binding of the alcohol and acyl moieties of the substrate, respectively. Structure-activity analyses for a series of substrates revealed that ρ_1 can bind alcohol groups up to the size of n-butoxy, whereas the ρ_2 subsite is limited in size, accommodating only acetyl groups. An essential assumption for the model is the difference in binding sites for the leaving groups of substrate and OP inhibitor. This assumption was supported by two arguments. Firstly, the most probable courses of the reactions are an attack of the nucleophilic serine OH group perpendicular to the plane of the ester moiety, but opposite to the leaving group displaced in the OP according to a trigonal-bipyrimidal transition state with axially entering and leaving groups (Fig. 1). The leaving group should interact with a third subsite ρ_3. Secondly, structure-activity analyses for enzyme catalyzed hydrolysis of CH_3COOR and inhibition by RO(Me)P(O)SBu and $(EtO)_2P(O)SR$ show similar dependence of the rate constants on the structure of R when comparing the substrates and the former series of OPs, but a clear different dependence for the substrates and the latter series of OPs.

Since the spatial locations at the AChE active site are fixed for interaction with the phosphoryl oxygen (binding to HA) and with the leaving group, it follows from this model that one of the two other substituents should accommodate to the sterically restricted subsite ρ_2, explaining the preferential interaction of AChE with one of the enantiomers of a chiral OP.

Additional support for the proposed model can be obtained from comparison of the spatial

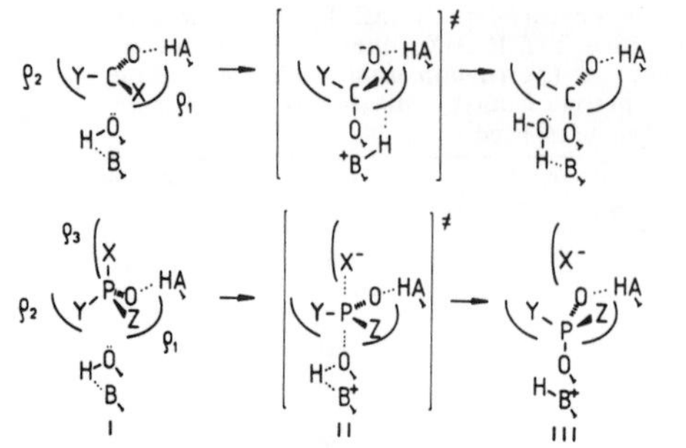

Fig. 1. Schematic representation of hydrolysis of a substrate and of inhibition by an OP at the active site of AChE according to Järv. (Reproduced with permission from Järv, 1984. Copyright 1984 Academic Press)

arrangement of the more active enantiomer for AChE inhibition of chiral OPs for which the absolute configuration has been established. The spatial arrangement of the more active enantiomers of these compounds and of OPs to which absolute configurations were tentatively assigned can be represented by

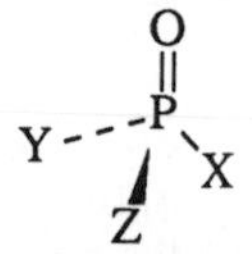

provided that Y is the bulkiest and Z the smallest group of the two substituents, in agreement with the proposed model (De Jong and Benschop, 1989). Moreover, this analysis validates the use of AChE inhibition for tentative assignment of the absolute configuration of the more active enantiomer of chiral OPs.

In this connection the results obtained by Casida and coworkers (Leader and Casida, 1982; Wing et al., 1983; Hirashima et al., 1984) obtained for profenofos and sulprofos oxon are worth mentioning. The investigators showed that these compounds are bioactivated. After activation the S(O)Pr group serves as a leaving group instead of the aryloxy group in the parent compounds (Fig. 2). In accordance with the proposed model, the more active enantiomers of profenofos or sulprofos oxon and of the activated compounds have opposite configurations.

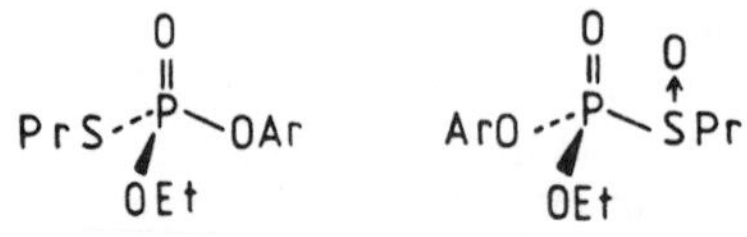

Fig. 2. Spatial arrangement of the more active enantiomer of profenofos (Ar, C_6H_3-2-Cl-4-Br) and sulprofos oxon (Ar, C_6H_4-4-SMe) for inhibition of AChE.

Hydrolysis by Phosphorylphosphatase (A-Esterase)

The procedure for isolation of the single soman stereoisomers and the analytical technique allowing their separate determination were advantageously applied in toxicokinetic studies in rats, guinea pigs, and marmosets after intravenous administration of 2-6 LD_{50} of C(±)P(±)-soman (Benschop and De Jong, 1987; Benschop et al., 1987). From these studies it appeared that the concentrations in blood of the relatively nontoxic C(±)P(+)-isomers decrease within minutes to a very low level, whereas the concentrations of the highly toxic C(±)P(-)-isomers could be followed for periods of 1-4 h.

More insight into the differences in persistence of the C(±)P(+)- and C(±)P(-)-isomers of soman was obtained from studies on the two main pathways for elimination of soman, i.e., hydrolysis catalyzed by phosphorylphosphatases (A-esterases), which are present in plasma and various organs, and irreversible binding to carboxylesterases and other binding sites.

As already mentioned, stereoselective hydrolysis of the soman stereoisomers in rat blood, rat liver homogenate and guinea pig skin was demonstrated in a qualitative manner by using our chiral gas chromatographic method. Quantitative information has been reported by De Bisschop et al. (1985, 1987, 1989) on stereoselective hydrolysis in human serum and in serum of some domestic animals, and by De Jong et al. (1988) on hydrolysis of the single stereoisomers in liver homogenates and plasma of rat, guinea pig, and marmoset. Results obtained for plasma are summarized in Table 2.

Table 2. First-order rate constants[a] ($10^{-3}min^{-1}$) for hydrolysis (pH 7.5, 37 °C) of the soman stereoisomers, catalyzed by diluted plasma of rat, guinea pig, and marmoset

Isomer	Rat	Guinea pig	Marmoset
C(+)P(+)	3400	220	1200
C(-)P(+)	1200	80	310
C(+)P(-)	1.5	0.2	1.2
C(-)P(-)	3.2	0.8	5.5

[a] Calculated for the same enzyme concentration, i.e., hydrolysis in 1 ml 0.6% plasma.

In all these studies it was found that the C(±)P(+)-isomers of soman are much faster hydrolyzed than the C(±)P(-)-isomers. These findings suggest that the rapid disappearance of the C(±)P(+)-isomers from the bloodstream after intravenous administration of C(±)P(±)-soman in rat, guinea pig, and marmoset, is mainly due to elimination by enzymatic hydrolysis.

Enantioselectivity of enzyme catalyzed hydrolysis has been investigated for a limited number of other chiral OPs. As found for soman, the less active enantiomers with respect to AChE inhibition of the phosphofluoridate sarin (i-PrO(Me)P(O)F; Christen and Van den Muysenberg, 1965) and of the p-nitrophenyl esters, EPN oxon ($EtO(C_6H_5)P(O)OC_6H_4$-4-NO_2; Nomeir and Dauterman, 1979), MeO(BuO)P(O)OC_6H_4-4-NO_2 (Dudman et al., 1977), $Et(C_6H_5)$-$P(O)OC_6H_4$-4-NO_2 (Grothusen and Brown, 1986), and $Me(CH_2)_nS(Me)P(O)OC_6H_4$-4-$NO_2$ in which n = 0-4 (Ooms and Boter, 1965), are more rapidly hydrolyzed by phosphorylphosphatases than the enantiomers with higher anticholinesterase activity. In view of these results it should be kept in mind that initial rates of enzymatic hydrolysis for a racemic mixture of a chiral OPs, usually considered as a measure for enzymatic activity, are often not relevant for the rate of detoxification of the compound.

Binding to Carboxylesterases and Other Binding Sites

Information on the relative contribution of hydrolysis and binding to the elimination of the C(±)P(+)- and C(±)P(-)-isomers was obtained from radiometric experiments in which anaesthetized, atropinized and artificially ventilated guinea pigs were challenged with 6 LD_{50} of C(±)P(±)-soman in which a ^{14}C label in the P-CH_3 moiety was present in three stereoisomeric variations: (i) randomly distributed over the four stereoisomers, (ii) stereoselectively in the C(±)P(-)-isomers, and (iii) stereoselectively in the C(±)P(+)-isomers. In the latter two variations intact soman was reconstituted by addition of the equimolar amounts of the nonlabelled pairs of isomers. The labelled and nonlabelled stereoisomeric pairs were isolated from ^{14}C-C(±)P(±)-soman and C(±)P(±)-soman, respectively, as already mentioned. Intact, hydrolyzed and bound soman were separately determined radiometrically 1 h after intoxication by using an extraction procedure, as developed by Harris et al., (1964), and Fleisher and Harris (1965), for work-up of homogenates (Table 3).

The low quantities of residual intact soman found 1 h after intoxication should be expected on the basis of the low blood levels found previously (Benschop and De Jong, 1987) by way of gas chromatographic analysis. The results clearly show that the nontoxic C(±)P(+)-isomers are largely eliminated by hydrolysis but that irreversible binding is the preferential elimination pathway for the toxic C(±)P(-)-isomers. The major part of the C(±)P(-)-isomers but also a minor part of the C(±)P(+)-isomers are eliminated by irreversible binding. This result indicates a high binding capacity of the animal for soman. It has been suggested that carboxylesterases are the main binding sites for soman in guinea pig and rat (Maxwell et al., 1987,1988a). Additional experiments in which we performed in vitro determinations of the quantities of available binding sites reveals that even after administration of 6-8 LD_{50} of the agent the overall occupation of binding sites is still incomplete. Estimations made by Maxwell et al. (1987, 1988b) for the total quantities of carboxylesterase binding sites lead to a similar conclusion for the rat.

Table 3. Quantities[a] (% of total concentrations of labelled compounds found before work-up) of radioactively labelled intact, hydrolyzed and bound soman in anaesthetized, atropinized and artificially ventilated guinea pigs 1 h after iv administration of 6 LD_{50} soman

Challenge	Intact soman	Hydrolyzed soman	Bound soman
^{14}C-C(±)P(±)	0.3±0.1	58±1	40±2
^{14}C-C(±)P(-)/ C(±)P(+)	0.5±0.2	34±3	61±2
^{14}C-C(±)P(+)/ C(±)P(-)	0.3±0.1	75±5	25±4

[a] Mean ± s.d. (n = 5)

It is concluded that differentiation between enantiomers or between various stereoisomers is essential in biochemical and toxicological studies of chiral OPs. A chiral gas chromatographic method in combination with a highly sensitive phosphorus-selective detection turned out to be a powerful tool for such a differentiation.

Acknowledgement

The authors are grateful to Dr. D.E. Lenz, U.S. Army Medical Research Institute for Chemical Defense, Aberdeen Proving Ground, U.S.A., and to Mrs. Isabelle Callebat, Centre d'Etudes du Bouchet, Vert-le-Petit, France, for kind gifts of ^{14}C-soman. This work was supported in part by the U.S. Army Medical Research and Development Command under Grant DAMD17-87-G-7015.

References

Benschop, H.P., De Jong, L.P.A., NTIS report AD-A199 573, 1987.

Benschop, H.P., De Jong, L.P.A., *Acc. Chem. Res.*, **1988**, *21*, 368-374.

Benschop, H.P., Konings, C.A.G., De Jong, L.P.A., *J. Am. Chem. Soc.*, **1981**, *103*, 4260-4262.

Benschop, H.P., Konings, C.A.G., Van Genderen, J., De Jong, L.P.A., *Toxicol. Appl. Pharmacol.*, **1984**, *72*, 61-74.

Benschop, H.P., Bijleveld, E.C., Otto, M.F., Degenhardt, C.E.A.M., Van Helden, H.P.M., De Jong, L.P.A., *Anal. Biochem.*, **1985**, *151*, 242-253.

Benschop, H.P., Bijleveld, E.C., De Jong, L.P.A., Van der Wiel, H.J., Van Helden, H.P.M., *Toxicol. Appl. Pharmacol.*, **1987**, *90*, 490-500.

Boter, H.L., Van Dijk, C., *Biochem. Pharmacol.*, **1969**, *18*, 2403-2407.

Christen, P.J., Van den Muysenberg, J.A.C.M., *Biochim. Biophys. Acta*, **1965**, *110*, 217-220.

De Bisschop, H.C., Mainil, J.G., Willems, J.L., *Biochem. Pharmacol.*, **1985**, *34*, 1895-1900.

De Bisschop, H.C.J.V., De Meerleer, W.A.P., Van Hecke, P.R.J., Willems, J.L., *Biochem. Pharmacol.*, **1987**, *36*, 3579-3585.

De Bisschop, H.C.J.V., Quaeyhaegens, F.J.L., Vanstockem, M.A.H., In: *Enzymes Hydrolysing Organophosphorus Compounds*, Reiner, E., Aldridge, W.N., Hoskin, F.C.G., Eds., Ellis Horwood Ltd., Chichester, England, 1989, pp 98-107.

De Jong, L.P.A., Benschop, H.P., In: *Chemicals in Agriculture*, Ariëns, E.J., Van Rensen, J.J.F., Welling, W., Eds., Elsevier, Amsterdam, 1989, Vol. 1, pp 109-149.

De Jong, L.P.A., Van Dijk, C., Benschop, H.P., *Biochem. Pharmacol.*, **1988**, *37*, 2939-2948.

Dudman, N.P.B., De Jersey, J., Zerner, B., *Biochim. Biophys. Acta*, **1977**, *481*, 127-139.

Fleisher, J.H., Harris, L.W., *Biochem. Pharmacol.*, **1965**, *14*, 641-650.

Grothusen, J.R., Brown, T.M., *Pestic. Biochem. Physiol.*, **1986**, *26*, 100-106.

Hall, C.R., Inch, T.D., Inns, R.H., Muir, A.W., Sellers, D.J., Smith, A.P., *J. Pharm. Pharmacol.*, **1977**, *29*, 574-576.

Harris, L.W., Braswell, L.M., Fleisher, J.H., Cliff, W.J., *Biochem. Pharmacol.*, **1964**, *13*, 1129-1136.

Hirashima, A., Leader, H., Holden, I., Casida, J.E., *J. Agric. Food Chem.*, **1984**, *32*, 1302-1307.

Järv, J., *Bioorg. Chem.*, **1984**, *12*, 259-278.

Järv, J., Aaviksaar, A., Godovikov, N., Lobanov, D., *Biochem. J.*, **1977**, *167*, 823-825.

Leader, H., Casida, J.E., *J. Agric. Food Chem.*, **1982**, *30*, 546-551.

Lee, P.W., Allahyari, R., Fukuto, T.R., *Pestic. Biochem. Physiol.*, **1978**, *8*, 146-157.

Maxwell, D.M., Lenz, D.E., Groff, W.A., Kaminskis, A., Froehlich, H.L., *Toxicol. Appl. Pharmacol.*, **1987**, *88*, 66-76.

Maxwell, D.M., Brecht, K.M., Lenz, D.E., O'Neill, B.L., *J. Pharmacol. Exp. Ther.*, **1988**, *246*, 986-991.

Maxwell, D.M., Vlahacos, C.P., Lenz, D.E., *Toxicol. Lett.*, **1988**, *43*, 175-188.

Nomeir, A.A., Dauterman, W.C., *Biochem. Pharmacol.*, **1979**, *28*, 2407-2408.

Ooms, A.J.J., Boter, H.L., *Biochem. Pharmacol.*, **1965**, *14*, 1839-1846.

Wing, K.D., Glickman, A.H., Casida, J.E., *Science*, **1983**, *219*, 63-65.

Wustner, D.A., Fukuto, T.R., *Pestic. Biochem. Physiol.*, **1974**, *4*, 365-376.

Aryldiazonium Salts as Affinity or Photoaffinity Probes for Labelling Cholinesterases : Towards Enzyme and/or Binding Site Discrimination

L. Ehret-Sabatier, M. Goeldner, C. Hirth, Laboratoire de Chimie Bioorganique, CNRS URA 1386, Université Louis Pasteur, Strasbourg, Faculté de Pharmacie, BP 24, 67401 Illkirch Cedex, France.
B. Rousseau, Service des Molécules Marquées, CEN Saclay, 91191 Gif sur Yvette Cedex, France.

Depending on their structure, aryldiazonium derivatives are either affinity or photoaffinity labels of cholinesterases. A p-butyroxybenzene diazonium derivative is able to selectively inactivate acetylcholinesterase whereas it is a substrate of butyrylcholinesterase. On the other hand, using energy transfer photoactivation, two p-N,N-dialkylbenzene diazonium derivatives are able to discriminate between the active and the regulatory peripheral binding sites of acetylcholinesterase.

Aryldiazonium derivatives were originally described as efficient affinity labels of cholinesterases (Wofsy and Michaeli, 1967) and have also been used as photoactivatable analogs of quaternary ammonium ions (Kieffer *et al.*, 1981). In fact, electron withdrawing groups on the aromatic ring will improve the electrophilic character of the diazonium function, leading to coupling with the nucleophilic residues of the protein active site (affinity labelling). Conversely, electron donating substituents stabilize the diazonium group and prevent this chemical coupling. In this latter case aryldiazonium compounds can be used as photoactivatable probes (photoaffinity labelling).

In this paper, we describe two types of aryldiazonium derivatives substituted in the para position either by a weak or strong electron donating group and used either as affinity or as photoaffinity labels.

Affinity labelling: specific inactivation of acetylcholinesterase

The probe used, p-butyroxybenzene-diazonium fluoroborate 1, is reasonably stable at neutral pH. Compound 1 is hydrolysed both by acetylcholinesterase (AChE) and butyrylcholinesterase (BuChE), giving the corresponding butyrate and 4-diazocyclohexa-2,5-dienone 2. The characteristic chromophores of both 1 and 2 (Fig.1) allowed easy monitoring of the hydrolysis reaction at 350 nm. We could deduce the kinetic parameters of 1, for both enzymes (Table 1):

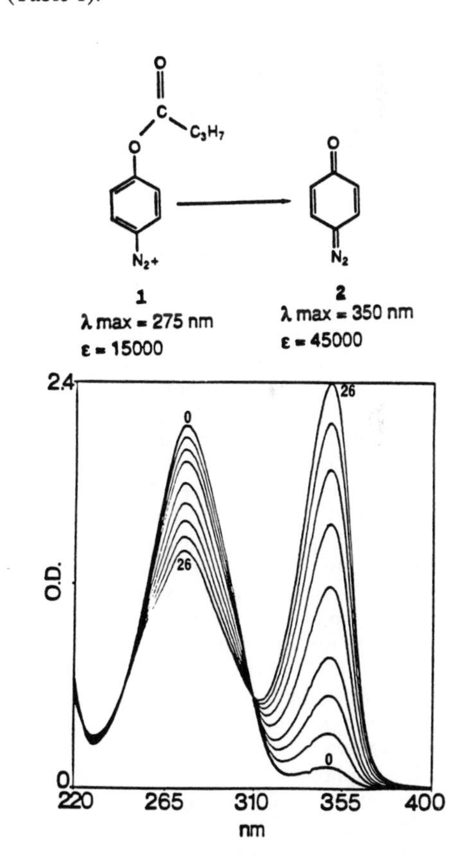

Fig. 1. Enzymatic conversion of compound 1 into compound 2 as a function of time.
BuChE was added at t = 0 to a 1.4 x 10^{-4}M solution of 1 in TrisHCl buffer. UV spectra were taken at times 0, 2, 4, 6, 10, 14, 18, 22, 26 min.

Table 1: Conversion of 1 into 2 by ChEs

	K_m(M)	Relative V_{max}
Torpedo AChE	6 x 10^{-5}	1
Serum BuChE	1.3 x 10^{-4}	2500

0–8412–2008–5/91/0245$06.00/0

The K_m of 1 was slightly better for AChE than for BuChE while BuChE hydrolysed 1 about 2500 times faster than did AChE. This suggests that the diazonium part of the molecule was recognized as a quaternary ammonium by both enzymes and is also in good agreement with the known selectivity of the two enzymes toward choline esters (Silver, 1974, Gnagey *et al.*, 1987).

However, we observed that after a longer incubation in the dark with 1 only AChE was progressively inactivated. Several lines of evidence favor the hypothesis that 1 affinity labels the enzyme.

Firstly, when AChE was incubated in the presence of 1 at pH 7.0, a time-dependent loss of activity was observed which followed a pseudo first-order kinetics. The rate of this inactivation process was also a function of the inhibitor concentration. These experimental results fit well inactivation rate equation deduced from the scheme (I) :

$$E + I \underset{Kd}{\rightleftharpoons} (E - I) \xrightarrow[k]{} E'(\text{inactive})$$

This implies that the inhibitor concentration remained constant during the experiment : actually, less than 10 % of 1 disappeared. From the inactivation experiments performed at pH 7.0, a Kd value of 4.1×10^{-5} M was obtained. This value is close to the previously determined Km of 1. In parallel, we determined the inactivation rate constant ($k = 0.21$ min^{-1}). Under similar incubation conditions we did not observe any inactivation of BuChE in the dark.

Secondly, the enzyme can be selectively protected against this inactivation by adding a specific competitive inhibitor to the incubation medium. AChE is known to have at least two kinds of binding sites : the active site and a peripheral site. Binding of a ligand to the peripheral site completely inactivates the enzyme, whereas occupancy of the active site does not prevent binding of drugs specific for the peripheral site (Taylor and Lappi, 1975). In order to assess which site was labelled by 1, we examined the protective effect of edrophonium which is known to be specific for the active site. Indeed, edrophonium very efficiently prevented the inactivation of AChE by 1 in a concentration dependent manner. From these experiments, it was possible to deduce an apparent affinity constant Kd' of edrophonium for the enzyme of 2.8×10^{-7}M which is close to the published affinity constant for the active site (Taylor and Lappi, 1975). We also studied the protective effect of propidium, a peripheral site specific ligand. An efficient and concentration-dependent protection against inactivation by 1 was observed (not shown), from which an apparent affinity constant of 3.4×10^{-6}M was deduced in agreement with the known affinity of propidium (Taylor *et al.*, 1974). These results strongly suggest that the irreversible inactivation of the enzyme occurred through the binding of 1 to the active site. An alternative to the affinity labelling of AChE by 1 would be a suicide-type inactivation by reaction of residues of the active site with 2 obtained *in situ* during hydroysis of 1. In this view, we checked that compound 2 at a concentration up to 7×10^{-5}M at pH 8.0 was unable to irreversibly inactivate the enzyme.

In summary the aryldiazonium salt 1 is a substrate of both AChE and BuChE with fairly good Michaelis constants for both enzymes but it discriminates efficiently between the two enzymes since it only blocks irreversibly AChE. This phenomenon can be explained by a difference in the cheminal environment of the substrate within the active site: either the positioning of 1 is different in the two active sites or there is a nucleophilic residue in the AChE active site which is lacking in BuChE. Furthermore, the fact that 1 behaves both as a substrate and as an affinity label for AChE indicates that it can react within the active site either by its ester function (1 as a substrate) or by its diazonium group (1 as an affinity label).

The selective inactivation of AChE is remarkable since very few, if any, irreversible ligands are able to discriminate between acetyl- and butyrylChEs. As an example, BW 62c51 was described to block AChE but this inactivation was slowly reversible and not completely selective (Koelle, 1955) ; conversely, iso-OMPA is known to selectively block BuChE. More recently bambutarol was described as an efficient and very selective irreversible inactivator of BuChE (Tunek & Swenson, 1988). Thus compound 1 might also be useful in selectively and irreversibly inactivating AChE.

Photoaffinity labelling: aryldiazonium as topographical probes of cholinesterase binding sites.

A prerequisite for photoaffinity labelling by aryldiazonium ions is the absence of chemical reaction between the target protein and the probe in the dark. Thus, the diazonium function of the probe must be stabilized by strong electron donating substituents such as dialkylamino groups, preferentially located in the para position.

On irradiation, aryldiazonium salts lose nitrogen and give the corresponding arylcation (Ambroz and Kemp, 1979), one of the few reactive species that displays virtually universal reactivity (Angelini *et al.*, 1982, 1984, Kieffer *et al.*, 1986). In addition, a topographical probe must generate reactive species which react with the surrounding active site residues faster than it escapes. We checked this critical assessment using a tetracarboxylate derivative of 18-crown-6 (Behr *et al.*, 1982) as a receptor model. The photochemical reaction of the entrapped diazonium salt did not lead to the formation of the photoproducts formed in the presence of solvent only. Such an exclusive reaction can only be explained by a faster reaction process than the escape of the reactive p-dimethylaminobenzene cation.

Furthermore, the UV spectral characteristics of aryldiazonium salts substituted in the para position by electron donating groups, allowed us to develop an "energy transfer" induced photodecomposition methodology (Goeldner and Hirth, 1980). This technique consists in a selective photoactivation of the protein tryptophan residues (λ_{exc} = 290 nm) which transfer their activation energy to proximal diazonium ion through a radiationless process. Under these conditions a ligand bound to the active site will be preferentially photoactivated as compared to the free ligand. This method improves:

i) the efficiency of labelling since the concentration of the free ligand remains essentially constant throughout the irradiation experiment.

ii) the specificity of labelling by lowering the alkylation yield of residues located outside the active site (Goeldner *et al.*, 1982).

We used two p-N,N-dialkylaminobenzene diazonium salts 3 and 4 which differ by the alkyl chain length. We showed that they are able to discriminate the active from the peripheral binding site. In the dark 3 and 4 are fairly stable, when tested as ligands of AChE and BuChE, they behave as reversible inhibitors of acetylthiocholine hydrolysis (Table 2).

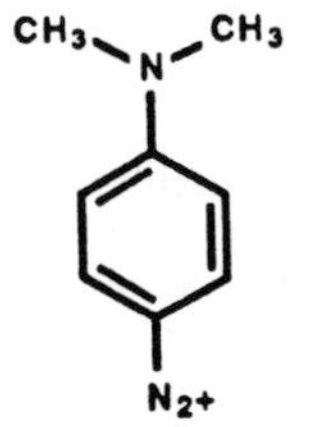

3: $(CH_3)_2N-C_6H_4-N_2^+$ 4: $(C_4H_9)_2N-C_6H_4-N_2^+$

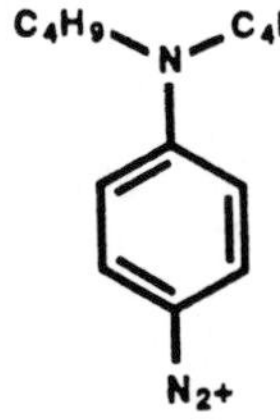

Reversible binding of 3 and 4

These values are consistent with the hypothesis that compound 3 is an acetylcholine analog whereas compound 4 mimics butyrylcholine. In both cases, the diazonium function is recognized as a quaternary ammonium group and the alkyl sustituent as the acyl moiety of the choline ester.

Table 2: Competitive inhibition constants of compounds 3 and 4 deduced from Lineweaver-Burk or Dixon plots

	3	4
AChE	2.6 x 10^{-5}M	7.6 x 10^{-5}M
BuChE	1.7 x 10^{-4}M	3.1 x 10^{-5}M

We checked the ability of compounds 3 and 4 to bind to the AChE peripheral regulatory anionic site by displacement experiments using propidium (Taylor and Lappi, 1975). Compound 4 exhibited a Kd of 3.7 x 10^{-4}M whereas compound 3 was unable to displace bound propidium up to 1.6 x 10^{-2}M. These results indicate that 3 discriminates efficiently between the active and the peripheral site. The behaviour of 4 is more complex suggesting that it might bind to both of the sites.

Irreversible binding of 3 and 4

Under energy transfer irradiation conditions, both 3 and 4 are able to irreversibly block AChE in a time- and concentration-dependent manner. These results were analyzed using kinetic scheme (I) since, under our experimental conditions, the concentration of the free ligand remains practically constant. In this situation, the irreversible rate constant k includes both the photoactivation process and the alkylation reaction.The deduced constants are presented in Table 3.

Table 3: Kinetic constants of irreversible inactivation of AChE by 3 and 4

Compound	Kd(M)	k(min^{-1})
3	5 x 10^{-5}	1.0
4	6 x 10^{-5}	0.2

Clearly, 3 inactivates AChE more efficiently than 4, suggesting differences in the positionning of the two molecules within the active site.

The enzyme was protected against 3 and 4 induced inactivation using $N(Me)_4^+$, Br^- a general ligand for quaternary ammonium sites. On the other hand, edrophonium completely protected AChE against 3 but only partially against 4 while propidium protects the enzyme against 3 and 4 to the same extent as shown in Table 4.

Table 4: Protection of AChE against 3 and 4 inactivation by active site (edrophonium) and peripheral site (propidium) selective ligands

protective agent	inactivation by 3 (10^{-5} M) (% control)	inactivation by 4 (2 x 10^{-5} M) (% control)
none	62	60
propidium 2 x 10^{-6} M	30	26
edrophonium 10^{-6} M	8	27

More complete analysis of the protective effects of edrophonium and propidium was undertaken using various concentrations of both products. As shown in Fig. 2a, edrophonium protection againts 3 is monophasic, indicating that both compounds bind to one class of sites. A Kd' for edrophonium of 4.2 x 10^{-8} M was deduced, which is consistent with its known affinity for the active site. In contrast, protection experiments against 4 resulted in a biphasic plot (2b) indicating the existence of two different alkylated sites. The first site one has an affinity constant for edrophonium of 1.4 x 10^{-7} M suggesting that the alkylation process occured at the active site. The affinity constant of the second one extrapolated to 7 x 10^{-4} M (2c), a value close to the known constant of edrophonium for the peripheral binding site. A unique affinity constant for propidium was obtained by protection experiments against 3 (Kd' = 1.3 x 10^{-6} M) and 4 (Kd' = 4.6 x 10^{-7} M) (not shown). This latter result was expected since propidium blocks the active site upon binding to the peripheral site.

Taken together, these results indicate that 3 irreversibly and exclusively blocks the active site, while 4 is able to inactivate also the peripheral site. To confirm this point, inactivation experiments using various concentrations of 4 in the presence of 2 x 10^{-5}M edrophonium to fully occupy the active site was performed. As before, an efficient alkylation occurred and an affinity constant Kd of 4 for the peripheral site of 1.7 x 10^{-4}M was determined. The kinetic rate constant remained essentially unchanged.

Specificity of the alkylation reaction was studied with radioactive labels 3 and 4. After irradiation, labelled AChE was analysed by PAGE performed under denaturating conditions. The radioactivity profiles (Fig. 3) indicates the labelling of the catalytic subunit (MW 74 KDa). When tetramethylammonium bromide was added as a protective agent to the irradiation medium, incorporation of radioactivity into the protein was dramatically lowered. In fact, 92% and 77% of the total radioactive incorporation of 3 or 4 respectively was abolished. Identical results were obtained using edrophonium as a protective agent against inactivation by 3, confirming an active-site directed alkylation. On the other hand, incorporation of 4 into the enzyme was only partially reduced in the presence of edrophonium, indicating that it binds irreversibly to an another ammonium binding site, probably the peripheral site. Stoichiometry of the incorporation was calculated for each site. A value of 1.1, close to unity, was found for incorporation of 4 into the peripheral site while only 0.55 mol of 3 were inserted *per* mol of enzyme.This latter result can be explained by a partial loss of active sites during preparation of AChE. Labelled AChE was then subjected to proteolysis. Preliminary analysis of the radioactive peptides indicate differences in the pattern of AChE labelled in the active or in the peripheral site. This suggests that both sites involve amino acid residues located in different parts of the primary structure. It will be of particular interest to see if whether or not the small labelled peptide GSXF (X unknown residue) described by Kieffer *et al.* (1986) as being part of the quaternary ammonium binding site of E. electricus AChE is also labelled on the Torpedo enzyme.

Compounds 3 and 4 are also reversible and irreversible inhibitors of BuChE. Using the same methodology as for AChE, we observed a specific radioactive incorporation into the catalytic subunit of BuChE virtually identical to the one observed on the catalytic site of AChE.

Identification of the labelled residues on AChE and BuChE by 3 and 4, which can be compared to those already identified on the acetylcholine binding site of the nicotinic cholinergic receptor (Dennis *et al.*, 1988) will give some interesting information on the way by which proteins are able to complex quaternary ammonium ions.

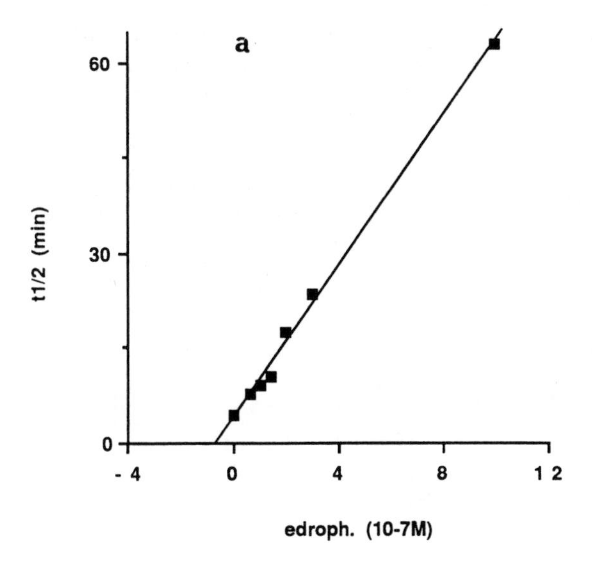

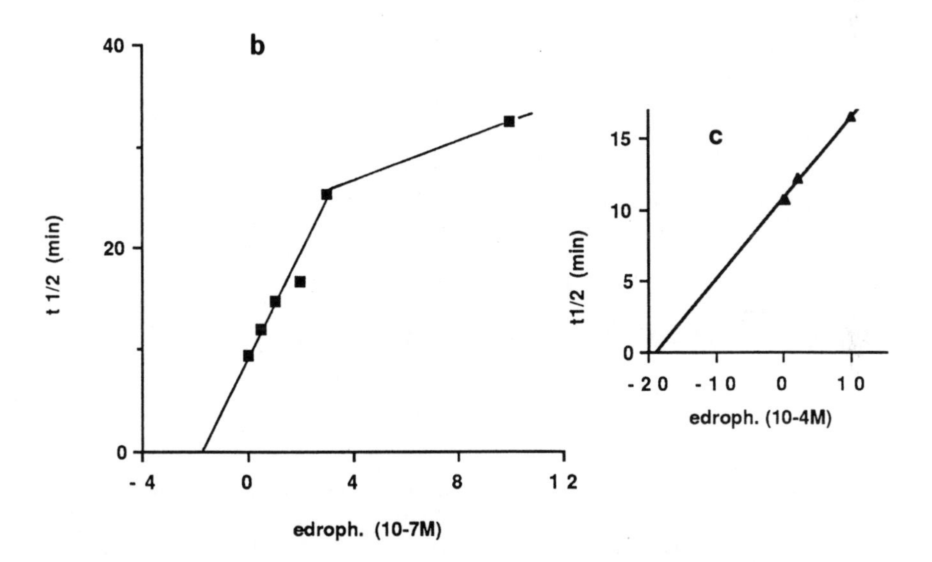

Fig. 2. Photoaffinity labelling of AChE: protection by edrophonium.

Figure 2a shows the protection of inactivation induced by 3 (10^{-5}M). A unique Kd' was deduced.

Figure 2b shows the protection of inactivation induced by 4 (2×10^{-5}M). The first Kd' was calculated using a concentration range of edrophonium from 0 to 3×10^{-7}M. The second constant was extrapolated from plot 2c (10^{-6}M to 10^{-3}M edrophonium).

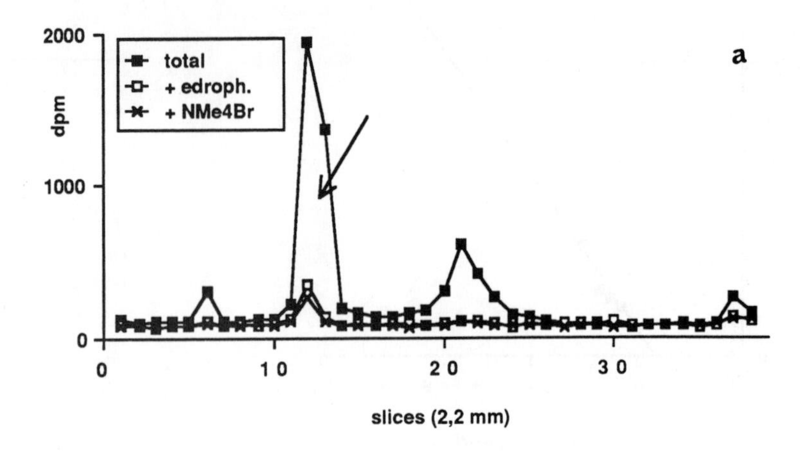

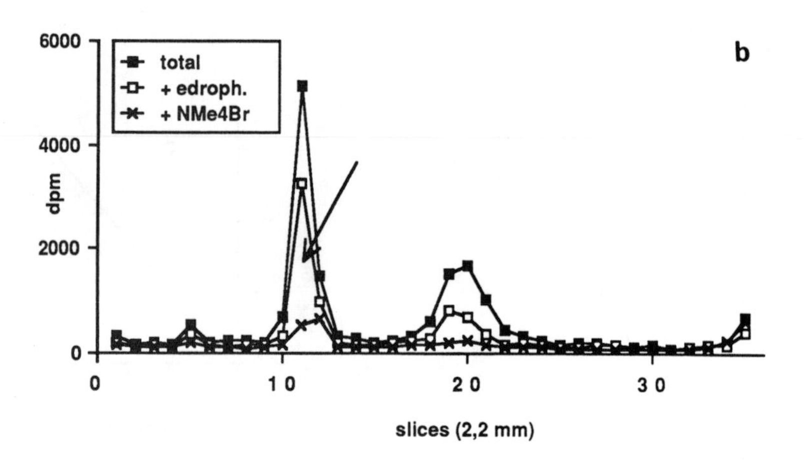

Fig. 3. Analysis of labelled AChE by SDS-PAGE
Figures 3a and 3b show the incorporation of radioactive compounds 3 and 4, respectively, with or without protective agent added during the irradiation. The arrow indicates the catalytic subunit of AChE (MW 74 KDa).

References

Ambroz, H.B., Kemp, T.J. *Chem. Soc. Rev.* **1979**, 8, 353.
Angelini, G., Fornarini, S., Speranza, M. *J. Am. Chem. Soc.* **1982**, 104, 4473-4480.
Angelini, G., Sparapani, C., Speranza, M. *Tetrahedron* **1984**, 40, 4865-4871 and 4873-4884.
Behr, J.P., Lehn, J.M., Vierling, P. *Helv. Chim. Acta* **1982**, 65, 1853-1857.
Dennis, M., Giraudat, J., Kotzyba-Hibert, F., Goeldner, M., Hirth, C., Chang, J.Y., Lazure, C., Chrétien,M., Changeux, J.P. *Biochemistry* **1988**, 27, 2346-2357.
Gnagey, A.L., Forte, M., Rosenberry, T.L. *J. Biol. Chem.* **1987**, 262, 13290-13298.
Goeldner, M., Hirth, C. *Proc. Natl. Acad. Sci. USA* **1980**, 77, 6439-6442.
Goeldner, M., Hirth, C., Kieffer, B., Ourisson, G. *TIBS* **1982**, 7, 310-312.
Kieffer, B., Goeldner M., Hirth, C., *J.C.S. Chem. Comm.* **1981**, 398-399.
Kieffer, B., Goeldner, M., Hirth, C., Aebersold, R., Chang, J.Y. *FEBS Lett.* **1986**, 202, 91-96.
Koelle, G.B. J. *Pharmacol. Exp. Ther.* **1955**, 114, 167-184.
Taylor, P., Lwebuga-Mukasa, J., Lappi, S., Rademacher, J. *Mol. Pharmacol.* **1974**, 10, 703-708.
Silver, A. *Biology of Cholinesterases*, A. Neuberger Ed., Frontiers of Biology, North Holland Co., Amsterdam, **1974**, 36.
Taylor, P., Lappi, S. *Biochemistry* **1975**, 14, 1989-1997.
Tunek, A., Swenson, L.A. *Drug. Metab. and Dispos.* **1988**, 16, 759-764.
Wofsy, L., Michaeli, D. *Proc. Natl. Acad. Sci. USA* **1967**, 58, 2296-2298.

The Chemical Mechanism of Acetylcholinesterase Reactions. Biological Catalysis at the Speed Limit

Daniel M. Quinn,* Alton N. Pryor, Trevor Selwood and Bong-Ho Lee, The University of Iowa, Department of Chemistry, Iowa City, Iowa 52242, U.S.A.
Scott A. Acheson, Glaxo Research Laboratories, 5 Moore Drive, Research Triangle Park, North Carolina 27709, U.S.A.
Paul N. Barlow, Oxford University, Biochemistry Department, South Parks Road, Oxford OX1 3QU, England

The mechanism of acetylcholinesterase (AChE) catalysis is marked by high catalytic power and transition state plasticity. For physiological choline substrates V/K (acylation) is rate limited by encounter, but for less reactive choline esters, aryl esters and anilides V/K is rate limited by both induced-fit and consequent chemical transition states. In contrast, the maximal velocity V is always rate limited by chemical steps. Though an Asp-His-Ser triad likely resides in the AChE active site, evidence for charge-relay transition state stabilization is lacking.

Acetylcholinesterase (AChE) is one of nature's most elegant catalysts. For the physiological substrate acetylcholine (ACh) the bimolecular rate constant k_{cat}/K_m (i.e. V/K divided by $[E]_t$) is $>10^9$ M^{-1} s^{-1} and is likely rate limited by diffusion or binding (Nolte et al, 1980). This reaction thus proceeds at the "speed limit", a hallmark of an evolutionarily perfect enzyme (Knowles & Albery, 1977). Delineation of the origins of this amazing catalytic power is a challenge for the mechanistic biochemist.

Until recently the AChE mechanism was more anecdote than enzymology. However, developments in molecular biology have facilitated determinations of protein sequences via their respective cDNA sequences. Regions of the primary sequence of AChE show extensive amino acid analogy to corresponding regions of various esterases, as shown in Fig. 1. The sequence around S200 of the *Torpedo californica* AChE active site (MacPhee-Quigley et al., 1985) is similar to those that contain the active site residues S198 and S194, respectively, of butyrylcholinesterase (Lockridge et al., 1987) and cholesterol esterase (Kissel et al., 1989). In addition, regions that contain conserved histidine and aspartate residues are compared in Fig. 1. Therefore, the AChE active site likely contains an Asp-His-Ser triad like that of the serine proteases (Stroud, 1974; Blow, 1976; Kraut, 1977; Polgar, 1987).

In this paper solvent isotope effect and pH-rate experiments are described that illuminate mechanistic features and transition state structures of AChE reactions. The experiments probe the operation of the active site triad and reveal a rich mechanistic diversity for AChE function.

Results and Discussion

Though the physiological role of AChE is the hydrolytic destruction of the neurotransmitter acetylcholine (ACh) in the peripheral and central nervous systems, *in vitro* the enzyme exhibits a broad substrate specificity (Froede & Wilson, 1971; Rosenberry, 1975a; Quinn, 1987). This feature has been exploited to investigate an array of AChE substrates whose reacti-

0–8412–2008–5/91/0252$06.00/0

Active Site Sequences

CEase	I-T-I-F-G-E-S^{194}-A-G-A-A-S-V
AChE	V-T-I-F-G-E-S^{200}-A-G-G-A-S-V
BuChE	V-T-L-F-G-E-S^{198}-A-G-A-A-S-V

Sequences Containing Glu/Asp

CEase	E^{78}-D^{79}-C-L-Y-L-N-I-W-V-P
AChE	E^{92}-D^{93}-C-L-Y-L-N-I-W-V-P
BuChE	E^{90}-D^{91}-C-L-Y-L-N-V-W-I-P

Sequences Containing His

CEase	W-M-G-A-D-H^{435}-A-D-D-L-Q-Y-V-F-G-K-P
AChE	W-M-G-V-I-H^{440}-G-Y-E-I-E-F-V-F-G-L-P
BuChE	W-M-G-V-M-H^{438}-G-Y-E-I-E-F-V-F-G-L-P

Fig. 1. Sequence comparisons for serine esterases.

vities span a range of 10^5. These substrates fall into three classes: anilides, aryl esters and choline esters.

In 1975 Rosenberry suggested the following kinetic mechanism for AChE catalysis (Rosenberry, 1975a, 1975b):

$$E \underset{k_2}{\overset{k_1 A}{\rightleftharpoons}} EA_1 \underset{k_4}{\overset{k_3}{\rightleftharpoons}} EA_2 \xrightarrow[\downarrow P]{k_5} F \xrightarrow[\downarrow Q]{k_7} E$$

A is the substrate, P and Q are products, F is the acylenzyme intermediate (Froede & Wilson, 1984), and EA_1 and EA_2 are noncovalent complexes that are interconverted by an induced-fit conformational change. The steady-state kinetic parameters for this mechanism are given by the following equations:

$$V/K = \frac{k_1 k_3 k_5 [E]_t}{k_2(k_4 + k_5) + k_3 k_5} \qquad (1)$$

$$V = \frac{k_3 k_5 k_7 [E]_t}{k_3 k_5 + k_7(k_3 + k_4 + k_5)} \qquad (2)$$

As Eq. 1 shows, V/K monitors only microscopic steps in the acylation stage of catalysis. For highly reactive substrates such as ACh, phenyl acetate (Quinn, 1987) and o-nitrophenyl acetate (Acheson *et al*., 1987a), k_1 probably contributes to rate limitation of V/K. However, for less reactive anilides (Quinn & Swanson, 1984; Acheson *et al*.,

1987b) and esters (e.g. p-nitrophenyl acetate; Hogg *et al*., 1980) binding does not contribute to rate determination, and Eq. 1 reduces to Eq. 3:

$$V/K = \frac{k_1 k_3 k_5 [E]_t}{k_2(k_4 + k_5)} \qquad (3)$$

Proton transfer elements of the acylation transition state of AChE-catalyzed hydrolysis of anilides have been characterized by the proton inventory technique (Venkatasubban & Schowen, 1985; Schowen & Schowen, 1982), which consists of a series of rate measurements in equivalently buffered mixtures of H_2O and D_2O. Assume that the chemical transition state of the k_5 step is stabilized by a single proton transfer. In this case, the appropriate expression for the dependence of k_5 on n, the atom fraction of deuterium in the mixed isotopic buffers, is given by Eq. 4:

$$k_{5,n} = k_{5,0}(1 - n + n\phi_5^T) \qquad (4)$$

In this equation ϕ_5^T is the isotopic fractionation factor of the proton that is transfered in the transition state. Combination of equations 3 and 4 gives the following equation for the proton inventory of V/K (cf. Acheson *et al*., 1987b, for a detailed derivation):

$$\frac{V/K_n}{V/K_1} = \frac{(1+C\phi_5^T)(1-n+n\phi_5^T)}{\phi_5^T + C\phi_5^T(1-n+n\phi_5^T)} \qquad (5)$$

$C = k_5/k_4$ is the commitment to catalysis, and is related to the fractional contributions to rate determination of the serial k_3 and k_5 steps:

$$f_5 = 1-f_3 = (1+C)^{-1} \qquad (6)$$

The intrinsic isotope effect generated by the transition state proton transfer of the k_5 step is ${}^{D_2O}k_5 = 1/\phi_5^T$. Eq. 5 predicts a nonlinear, upward-bulging dependence of V/K on n for C>0 and ϕ_5^T <1, but reduces to a linear function of n when C=0.

Fig. 2 shows an upward-bulging dependence of V/K on n for the *Electrophorus electricus* AChE-catalyzed hydrolysis of o-nitroacetanilide (Acheson *et al*., 1987b). The observed isotope effect for this reaction is small, ${}^{D_2O}V/K = 1.55$. Least squares analysis of the data in Fig. 2 gives C = 1.4 and ϕ_5^T = 0.40. Therefore, the proton transfer gives an isotope effect ${}^{D_2O}k_5 = 2.5$ and the k_5 step is 42% rate determining. A similar accounting of acylation reaction dynamics obtains for human erythrocyte AChE-catalyzed hydrolysis o-nitrochloroacetanilide (Acheson & Quinn, 1990), and for *E. electricus* AChE-catalyzed hydrolyses of o-nitroformanilide, o-nitrochloroacetanilide (Acheson *et al*., 1987b), and p-nitrophenyl acetate (Hogg *et al*., 1980). These results provide quantitative support for Rosenberry's induced-fit model of AChE catalytic function.

In the preceding paragraph substrates whose reactivity is 10^3 to 10^5-fold less than that of ACh were discussed. For a similarly unreactive choline ester, butyrylthiocholine (BuTCh), the isotope effect is ${}^{D_2O}V/K = 1.87$ and the proton inventory is linear. Therefore, C=0 and the chemical transition state is solely rate limiting. This result indicates that for choline substrates the induced-fit step is kinetically insignificant, which suggests that induced-fit is accelerated by the interaction of a quaternary amino group of the substrate with active site carboxylate sidechains. Work by Berman and Decker (1986) also suggests that active site conformation responds to substrate structure. They inhibited AChE from *Torpedo californica* with 3,3-dimethylbutyl and ß-(trimethylamino)ethyl methylphosphonofluoridates. The complexes that are formed are analogs of tetrahedral intermediates in the acylation stage of catalysis, and thus should elicit functionally relevant active site conformations:

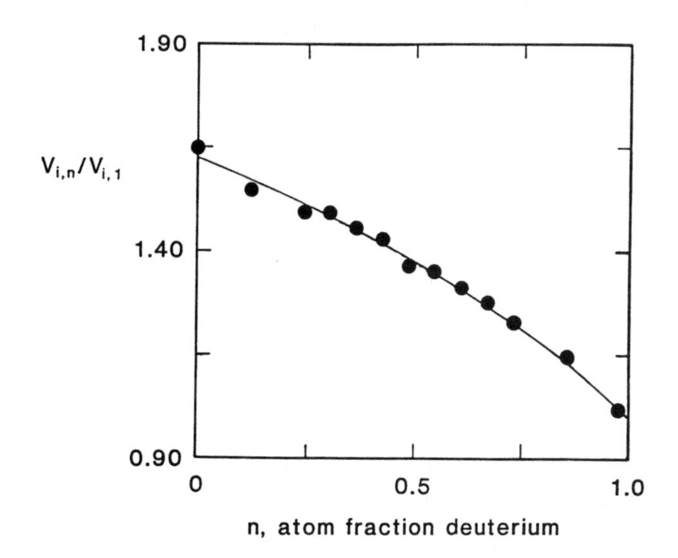

Fig. 2. Proton inventory for V/K of E. electricus AChE-catalyzed hydrolysis of o-nitroacetanilide.

Reprinted from Acheson et al. (1987b). Copyright 1987.

```
    Me            Me
    |             |
 Me-X-CH2CH2O-P=O          X = C, N+
    |             |
    Me            O
                   \AChE
```

When X=C phosphonylated AChE binds the bisquaternary amino ligand decidium about as well as does the native enzyme (K_d = 54 nM and 21 nM, respectively). However, when $X=N^+$ decidium binds ~100-fold more weakly to the phosphonylated enzyme. These and other data discussed by Berman and Decker suggest that when $X=N^+$ an active site conformation is induced that involves ion pairing between the quaternary amino group of the phosphonyl ligand and an aspartate or glutamate side chain of the enzyme. When X=C this conformational change does not occur and the enzymic anion is free to interact with reversible quaternary amino active site ligands.

The pH-rate profile in Fig. 3 for V/K of Electrophorus electricus AChE-catalyzed hydrolysis of benzoylcholine strongly supports the situation in the active site of glutamate and/or aspartate sidechains. The fit in Fig. 3 is to a model in which V/K depends on the basic forms of two enzyme groups that have the same pK_a:

$$V/K = \frac{V/K_{lim}\, K_a^2}{[H^+]^2 + K_a[H^+] + K_a^2} \qquad (7)$$

The fit gives pK_a = 4.7 for the two sidechains on AChE, a value consistent with the involvement of carboxylate residues in catalysis. The data cannot be fit to a model in which V/K depends on the ionization of a single active site amino acid.

The proton inventory probe has been used to determine whether the Asp-His-Ser triad of AChE functions as a concerted proton transfer catalyst or as a one

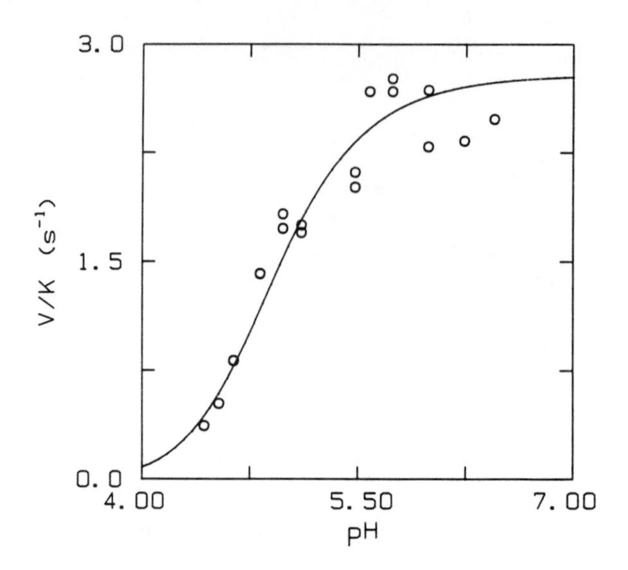

Fig. 3. pH-Rate profile for V/K of E. electricus AChE-catalyzed hydrolysis of benzoylcholine. The sigmoidal line is a least-squares fit of the data to Eq. 7 of the text.

proton, simple general acid-base catalyst. Concerted proton transfers are manifested in a proton inventory plot that is nonlinear and downward-bulging, whereas single proton transfer gives a plot that is linear (Schowen & Schowen, 1982; Venkatasubban & Schowen, 1985). Linear proton inventories have been observed for the following E. electricus AChE-catalyzed reactions: V/K of BuTCh hydrolysis; V of phenyl chloroacetate, o-nitrophenyl acetate (Acheson et al., 1987a), phenyl acetate (Kovach et al., 1986), acetylthiocholine and BuTCh hydrolyses. In addition, the proton inventory for V of human erythrocyte AChE-catalyzed hydrolysis of acetylthiocholine is linear. Therefore, the rule for AChE is that single proton transfers stabilize chemical transition states, and there is no need to invoke the operation of a charge-relay system like that of the serine proteases (Stroud, 1974; Blow, 1976; Kraut, 1977; Polgar, 1987).

Acknowledgements

The work described herein was supported by NIH grant NS21334. We thank Terry Rosenberry for kindly providing purified human erythrocyte AChE.

References

Acheson, S.A.; Dedopoulou, D.; Quinn, D.M. J. Am. Chem. Soc. **1987a**, 109, 239-245.

Acheson, S.A.; Barlow, P.N.; Lee, G.C.; Swanson, M.L.; Quinn, D.M. J. Am. Chem. Soc. **1987b**, 109, 246-252.

Acheson, S.A.; Quinn, D.M. Biochim. Biophys. Acta. **1990**, in press.

Berman, H.A.; Decker, M.M. Biochim. Biophys. Acta **1986**, 872, 125-133.

Blow, D.M. Acc. Chem. Res. **1976**, 9, 145-152.

Froede, H.C.; Wilson, I.B. In The Enzymes, 3rd ed.; Boyer, P.D., Ed.; Academic Press: New York, 1971, Vol. 5; pp 122-152.

Froede, H.C.; Wilson, I.B. J. Biol. Chem. **1984**, 259, 11010-11013.

Hogg, J.L.; Elrod, J.P.; Schowen, R.L. J. Am. Chem. Soc. **1980**, 102, 2082-2086.

Kissel, J.A.; Fontaine, R.N.; Turck, C.W.; Brockman, H.L.; Hui, D.Y. Biochim. Biophys. Acta **1989**, 1005, 177-182.

Knowles, J.R.; Albery, W.J. Acc. Chem. Res. **1977**, 10, 105-111.

Kovach, I.M.; Larson, M.; Schowen, R.L. J. Am. Chem. Soc. **1986**, 108, 3054-3056.

Kraut, J. Ann. Rev. Biochem. **1977**, 46, 331-358.

Lockridge, O.; Bartels, C.F.; Vaughan, T.A.; Wong, C.K.; Norton, S.E.; Johnson, L.L. J. Biol. Chem. **1987**, 262, 549-557.

Nolte, H.J.; Rosenberry, T.L.; Neumann, E. Biochemistry **1980**, 19, 3705-3711.

Polgar, L. In Hydrolytic Enzymes, Neuberger, A., Brocklehurst, K., Eds.; New Comprehensive Biochemistry; Elsevier: Amsterdam, 1987, Vol. 16; pp 159-200.

Quinn, D.M. Chem. Rev. **1987**, 87, 955-979.

Quinn, D.M.; Swanson, M.L. J. Am. Chem. Soc. **1984**, 106, 1883-1884.

Rosenberry, T.L. Adv. Enzymol. Relat. Areas Mol. Biol. **1975a**, 43, 103-218.

Rosenberry, T.L. Proc. Natl. Acad. Sci. U.S.A. **1975b**, 72, 3834-3838.

Schowen, K.B.; Schowen, R.L. Methods Enzymol. **1982**, 87, 551-606.

Schumacher, M.; Camp, S.; Maulet, Y.; Newton, M.; MacPhee-Quigley, K.; Taylor, S.S.; Friedmann, T.; Taylor, P. Nature **1986**, 319, 407-409.

Stroud, R.M. Sci. Am. **1974**, 231, 74-88.

Venkatasubban, K.S.; Schowen, R.L. CRC Crit. Rev. Biochem. **1985**, 17, 1-44.

Rapid Postmortem Decrease in Ectocellular Acetylcholinesterase in the Rat Striatum

Guy Testylier, Patrick Gourmelon, Didier Clarençon, Marc Fatome and Jacques Viret.
Centre de Recherches du Service de Santé des Armées. Grenoble - BP 87
38 702 La Tronche-cedex (France)

It is generally assumed that rapid dissection and cooling of the tissue within few minutes of decapitation avoids dramatic changes of the neurological system. However, little is known about early post-mortem changes in brain enzymes. It is due to the necessity to measure the enzyme activity before death, which implies to use a reliable "in vivo" method for direct and *in situ* assay. Push-push cannula method allows "in vivo" enzyme assays but only the released enzymes can be monitored as shown with acetylcholinesterase (AChE) (Taylor 1989).

"In vivo" spectrophotometric method (Testylier 1987) allows direct, local and long-term measurements of AChE activity in brain tissue of live animals. We investigated the effects of death on the stiatal ectocellular AChE activity in anaesthetized rats. Surprisingly, we observed a rapid post-mortem decrease in AChE activity by 10 minutes after death, which reached about 55% after bleeding or about 35% after intoxication by pentobarbital (Fig.1). The AChE activity levels remained stable in the following hours.

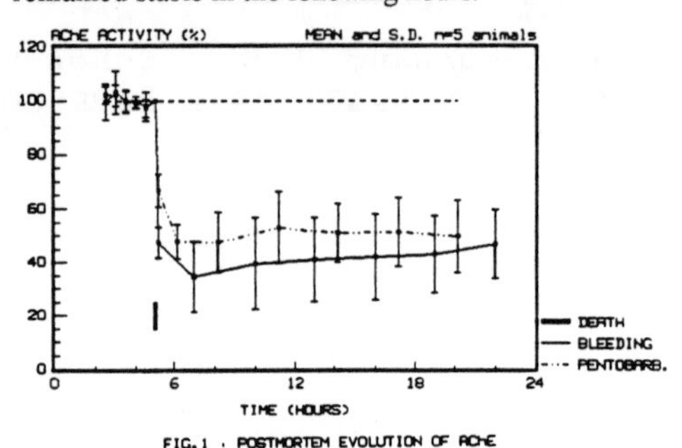

FIG.1 . POSTMORTEM EVOLUTION OF ACHE

The fall in the AChE activity in a such short time after death cannot be explained by optical modifications of the measuring site (as controlled by injections of calibrated dye solutions) or by variations of pH or cerebral temperature which were experimentally checked, or by spontaneous proteolysis of the enzyme. It is tempting to speculate that this post-mortem phenomenon is secondary to the decreasing in the abundance of the ectocellular enzyme seen by the optical probe because of the death interruption of the soluble secreted enzyme exocytosis. But, this hypothesis implies that a major part of the ectocellular enzyme would be in a soluble fraction form (about 35-55 % of the whole compartment) when only ten percent of brain AChE is soluble. Moreover, to explain a such rapid decreasing, we have to imagine a rapid removal of the soluble AChE from the recording site. However, Kreutzberg (1974) estimated the velocity of the exogenous AChE into brain at few millimeters per hour. This velocity which would decrease after the death in account of the interruption of the cerebral blood flow, seems us too low for explaining the post-mortem AChE activity fall. Exactly how this drastic and rapid decreasing in enzyme activity can occur after death awaits eludication.

References

Kreutzberg, G.W., Kaiya, H., *Histochemistry*, **1974**, *42*, 233-237.

Taylor, S.J., Haggblad, J., Greenfield, S.A., *Neurochem. Int.*, **1989**, *15* (2), 199-205.

Testylier, G., Gourmelon, P., *Proc. Natl. Acad. Sci. USA*, **1987**, *84*, 8145-8149.

"In Vivo" Monitoring of the Return of Striatal Ectocellular Acetylcholinesterase Activity in Rat and Primate After Inhibition by Methylphosphonothioate

Guy Testylier, Patrick Gourmelon, Didier Clarençon, Jean Claude Mestries, Marc Fatome and Jacques Viret.
Centre de Recherches du Service de Santé des Armées. Grenoble - BP 87
38 702 La Tronche-cedex (France)

It is now well established that recovery of the acetylcholinesterase (AChE) activity after inhibition by organophosphorus compounds is biphasic and characterized by an initial phase with a rapid, but incomplete, recovery of the enzyme activity, and a slow final phase lasting several days, as show in "ex vivo" experiments in crude brain homogenates (Davison, 1953, Goossens, 1984). The origin of these two phasis is not still elucidated.

We presents here the recovery of AChE activity after inhibition by the methylphosphonothioate (MPT) in "in vivo" experiments. AChE activity was followed using microspectrophotometry technique which allows long-term measurements of the ectocellular enzyme activity directly inside the brain, in unanesthetized animals (Testylier, 1987).

The injection of methylphosphonothioate (MPT) (13 microg/kg) in chronically implanted rat induced in 30 minutes a fast decreasing in about 60 - 65% in the ectocellular AChE activity. This enzyme inhibition was followed by a regular and constant phase of reactivation with a return to the basal level in about 16 to 20 hours (Fig. 1).

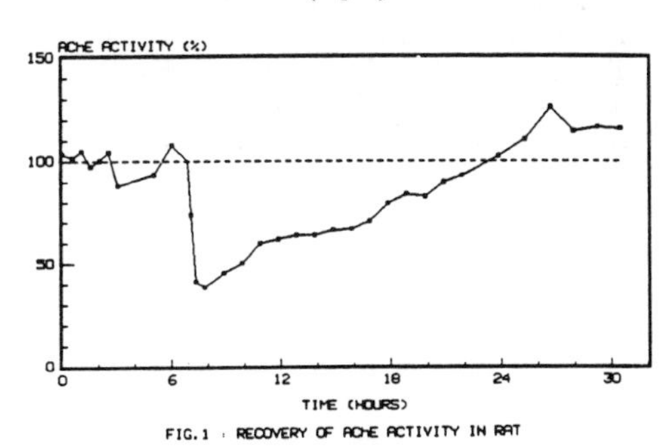

FIG. 1 : RECOVERY OF ACHE ACTIVITY IN RAT

Recently, we adapted this method to the primate (*Cynomolgus* monkey). After AChE inhibition by 33 % in striatum by 30 minutes, we observed a complete reactivation phase lasting 8 hours and followed by an overshoot of 30 % (Fig.2).

Thus, the "in vivo" spectrophotometric method revealed a single, complete and rapid recovery phase of the enzyme activity of the ectocellular compartment after inhibition by an organophosphorus compound. These data obtained in "in vivo" experiments could explain the two phase AChE recovery observed in "ex vivo" experiments in which the total AChE activity is assayed (ectocellular and intracellular enzyme). Indeed, the initial and incomplete recovery phase obtained "ex vivo" could correspond to the reactivation of the ectocellular compartment of the enzyme and the final slow phase of several days would reflect the recovery of the intracellular enzyme pool.

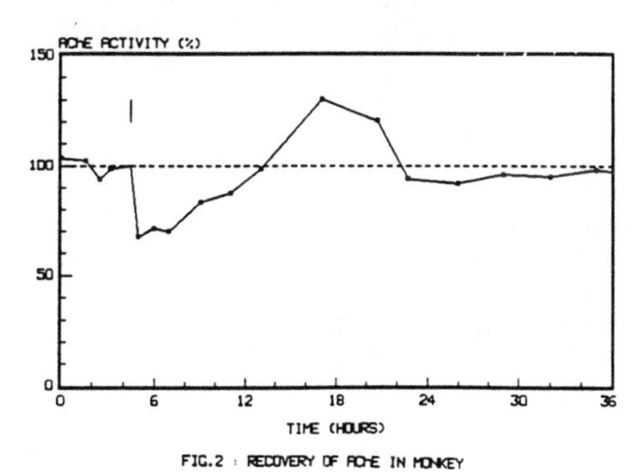

FIG. 2 : RECOVERY OF ACHE IN MONKEY

References

Davison, A.N., *Biochem. J.*, **1953**, *54*, 583-590.
Goossens, P.,Viret, J., Leterrier F., *Biochem. Biophys. Res. Commun.*, **1984**, *123* (1) 71-77.
Testylier, G., Gourmelon, P., *Proc. Natl. Acad. Sci. USA*, **1987**, *84*, 8145-8149.

Selective Inhibition of Brain Acetylcholinesterase After Chronic Administration of Phenylmethylsulfonyl Fluoride

Kenneth A. Skau*, Div. Pharmacology & Med. Chem., University of Cincinnati College of Pharmacy, Cincinnati, OH 45267.

Phenylmethylsulfonyl fluoride (PMSF) has been shown to inhibit acetylcholinesterase (AChE) with selectivity for the brain enzyme after a single injection (Moss, et al. 1985). This tissue selectivity is believed to be related to a selective inhibition of membrane associated AChE (Skau & Shipley, 1988). These studies have now been extended to examine the effects of chronic PMSF on AChE in various tissues.

Adult male rats were treated with PMSF by gavage as 85 mg/kg for 3 days followed by 50 mg/kg for 25 days. The animals were sacrificed and the carcasses perfused with ice cold saline to remove residual blood. Selected tissues were removed, homogenized in high ionic strength buffer containing Triton X-100 and protease inhibitors. After centrifugation the supernatants were assayed for AChE activity. The molecular forms of AChE were separated on linear 5-20% sucrose density gradients.

Chronic PMSF treatment, even at the reduced dosage, showed greater effects on brain AChE when compared to acute drug treatment (85 vs 71% inhibition). Peripheral organ AChE was inhibited about the same or less after chronic PMSF. Examination of the molecular forms confirmed that membrane-associated forms, such as asymmetric and the globular tetramer, were most severely affected.

It appears that the combination of a high fraction of membrane-bound AChE and slow regeneration of the enzyme in brain results in a greater overall inhibition after chronic PMSF treatment.

References

Moss, D.E., Rodriguez, L.A., Selim, S., Ellett, S.O., Devine, J.V., Steger, R.W.; In Senile Dementia of the Alzheimer Type; Hutton, J.T., Kenny, A.D. Ed.; Alan R. Liss, Inc. NY, NY, 1985;337-350.

Skau, K., Shipley, M.; The Pharmacologist 1988, 30, A81.

0–8412–2008–5/91/0260$06.00/0

Quantitative autoradiographic analysis of neuroreceptors in rat brain after acute or chronic soman intoxication

C. Bouchaud, P. Mailly, A. Chollat-Namy, D. Vergé, Département de Cytologie Université P. & M. Curie, CNRS UA 1199. 75252 Paris Cedex 05 (France).
I. S. Delamanche, Centre d'Etudes du Bouchet BP3, 91710 Vert le Petit (France).

Soman (pinacoloxymethylphosphoryl fluoride) is a potent irreversible inhibitor of cholinesterases. Soman intoxication induces long-term behavioral, neurological and biochemical disorders, which cannot be all attributed to hypercholinergy. Various neurotransmitter changes have been demonstrated in the brain after the treatment; the aim of this study was to analyze possible perturbations of various cerebral receptors.

Muscarinic, dopaminergic, GABAergic and serotonergic binding sites in the rat brain have been studied by quantitative autoradiography, either after chronic administration of repetitive sublethal doses or after acute administration of a single LD_{50} dose of soman. Cholinesterase inhibition was detected by histochemistry after autoradiography.

A marked decrease (-30%) in $[^3H]$QNB specific binding to muscarinic receptors was observed in chronically treated rats, as shown by densitometric analysis of autoradiograms. In contrast, GABAergic sites labelled with 5 nM $[^3H]$Muscimol, dopaminergic D_2 receptors labelled with 1 nM $[^3H]$N-n-propylnorapomorphine (NPA), 5-HT_{1A} sites labelled with $[^3H]$8-hydroxyl-2-(di-n-propylamino)tetralin (8-OH-DPAT) and 5-$HT_{1\text{-non-}A}$ sites labelled with 2 nM $[^3H]$5-hydroxytryptamine (5-HT) in the presence of 0.1 mM 8-OH-DPAT were not affected by the repeated exposure to sublethal doses of soman.

Six hours after the acute injection of soman, no change could be observed in the specific labelling of muscarinic and serotonergic receptors in convulsing animals. In contrast, specific binding of radioligands to the GABAergic and particularly dopaminergic receptors was significantly enhanced in various areas of the brain such as in the thalamic, medial geniculate and caudate nuclei.

These results indicate that muscarinic receptors are down-regulated after repeated sublethal administration of soman. This down-regulation could be due to a desensitization induced by increased acetylcholine levels after acetylcholinesterase inhibition.

The observed increases in GABA and DA receptor densities after acute soman treatment suggest that the GABA and the DA systems counteract the enhanced cholinergic activity induced by this organophosphorous compound.

Supported by Direction des Recherches et Etudes Techniques (DRET) grant N°89/039.

Reference

Fosbraey P., Wetherell J.R. & French M.C. *J. Neurochem.* **1990**, 54, 72-79.

0–8412–2008–5/91/0261$06.00/0

Physostigmine and Organophosphorus Action on the CA1 Rabbit Hippocampal Cells discharge during Ontogenesis

Jacques NIO, Patrick BRETON, Centre d'Etude du Bouchet, BP N°3 91710 Vert-le-Petit, France.

Electrophysiological investigations were carried out on rabbit brain hippocampus from 6 to 28 days old, CA1 pyramidal cells discharge were recorded after iontophoresis or intracarotidian drugs injection. Muscarinic receptor (AChR) - binding method - and acetylcholinesterase (AChE) - KOELLE method - cartography in hippocampus were realised in the same time. Iontophoretic investigation show a distinctly difference between the two drugs effect. This difference is age dependant and more pronounced in adult rabbit with a strong activation of cell discharge after physostigmine injection but an inhibition of most cell discharge under paraoxon influence. Studies of AChR binding and AChE localization indicate a rather similar evolution in CA1 field.
Data suggest that the paraoxon toxic effect involve a neuronal mechanism not only procholinergic. Further investigations are required in order to precise this mechanism.

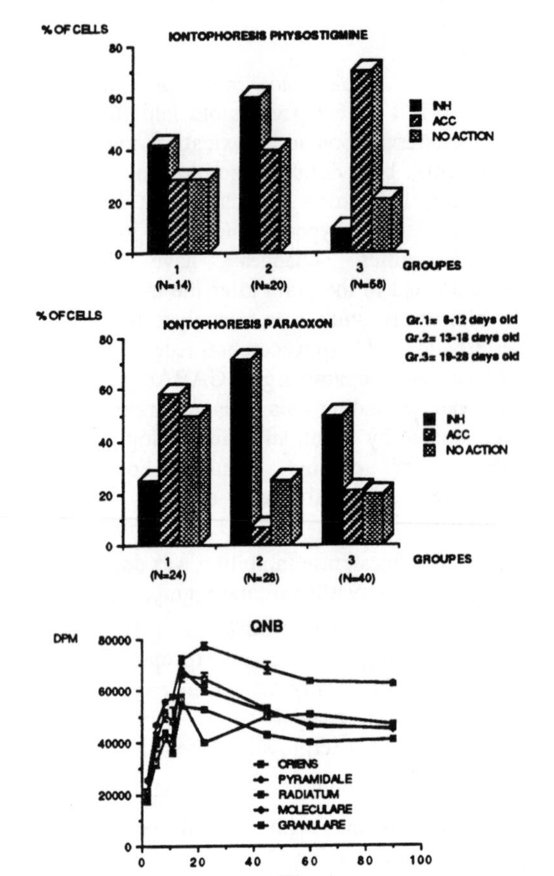

References

David J. Spencer et al, Br. Res., **1986,** 380, 59-68.
Koelle, G.B. et al, Proc. Soc. Exp. Biol. Med., **1949**, 70, 617-622.

Blockade of Synaptic Transmission in Sympathetic Ganglia Treated with Anticholinesterase Drugs

Nera Ya. Lukomskaya, Marina V. Samoilova, Vadim Yu. Bolshakov,
Sechenov Institute of Evolutionary Physiology and Biochemistry, Academy of Sciences of the USSR, 44 M.Thorez pr., Leningrad, 194223, U.S.S.R.

It is well known that AChE inhibitors are able to depress ganglionic transmission (Lukomskaya et al., 1980; Yarowsky et al., 1984). The mechanisms of this effect are not yet clear.

The experiments were made on the isolated rabbit superior cervical ganglion and on IX-X ganglia of frog sympathetic trunk using the intracellular recording. Superfusion of repetitively (0.4-1 Hz) stimulated ganglia with the organophosphorus compound armin (1 μM) or neostigmine (3 μM) induced sustained depolarization of the frog (4.7±0.4 mV, n=41) and rabbit (10.0±5.0 mV, n=48) ganglionic neurons, accompanied by partial or complete blockade of the synaptic transmission. These effects depended on the stimulation rate and EPSPs quantal content. The effects of armin were abolished by the AChE reactivator - TMB-4 (10 μM). Hence, accumulation of ACh leads to prolonged postsynaptic depolarization, which is the main cause of transmission failure. Both muscarinic and nicotinic receptors are involved in the depolarization (Bolshakov, Lukomskaya, 1988; Lukomskaya et al., 1988).

AChE inhibitors as well as oxotremorine (1 μM) decreased EPSPs quantal content to 47% and presynaptic facilitation. Both presynaptic effects were prevented by atropine (3 μM). The negative feedback is suggested to diminish the ACh quantal release via activation of the presynaptic muscarinic receptors by accumulated ACh, and thereby to attenuate the postsynaptic effects of AChE inhibition.

References

Bolshakov, V.Yu., Lukomskaya, N.Ya., *Neirofiziologia*, **1988**, *20*, 227-234 (in Russian).

Lukomskaya, N.Ya. et al., *Gen.Pharmacol.*, **1980**, *11*, 83-87.

Lukomskaya, N.Ya. et al., *Zhurnal Evol.Biokhim.Fiziol.*, **1988**, *24*, 668-678 (in Russian).

Yarowsky, P. et al., *Cell. Mol. Neurobiol.*, **1984**, *4*, 351-356.

The Alkylating Derivate of Hexadecamethonium Protects Muscle Synaptic Acetylcholinesterase Against Inhibition

Victor M. Grigoriev, Lev G. Magazanik, Natalya K. Prokhorenko, Oleg E. Sherstobitov, Vladimir A. Snetkov, Sechenov Institute of Evolutionary Physiology and Biochemistry, Academy of Sciences of the USSR, 44 M.Thorez pr., Leningrad, 194223, U.S.S.R.

One of the most potent alkylating agents blocking nicotinic cholinoreceptors is derivate of hexadecamethonium (hexadecamethylene-bis-methylchloroethylamine, C-16a) (Michelson, Shelkovnikov, 1980). In our experiments treatment of the frog *m.sartorius - n.ishiadicus* preparations with 1 μM of C-16a led to irreversible decrease of the amplitude (*A*) and the decay time constant (τ) of the end-plate currents (EPC) without change of EPC quantal content (experiment/control ratio was 0.99±0.05, n=4). Channel blocking action was also not revealed.

The decrease of sensitivity of the postsynaptic membrane to acetylcholine was not a single effect of C-16a observed. As a result of C-16a pretreatment effects of the acetylcholinesterase (AChE) inhibitors (prostigmine, armine) became much less prominent. In control, prostigmine (6.6 μM) increased ***A*** to 132±8% and τ to 390±30% (n=6) in Ringer solution containing 0.9 mM Ca^{2+} and 7 mM Mg^{2+}. After about 4-fold reduction of ***A*** caused by C-16a (1 μM) prostigmine increased ***A*** to 121±4% and τ to 129±5% (n=8) only This effect was not due to a reduction of receptor density. Indeed, after the decrease of ***A*** by α-bungarotoxin to the same extent prostigmine increased ***A*** to 167±4% and τ to 302±25%.

C-16a (10 μM) did not change catalytic activity of AChE in perineural segments of frog *m. cutaneous pectoris* measured by Ellman method but decreased the sensitivity to the organophosphorus inhibitor armine from k_{II} = 2.60 ± 0.1•10⁵ $M^{-1}min^{-1}$ to k_{II} = 1.21 ± 0.05•10⁵ $M^{-1}min^{-1}$ (n=5).

The results of biochemical studies showed that C-16a indeed reduces the action of AChE inhibitors. Thus the attenuation of their effects by C-16a observed in electrophysiological experiments may be at least partly explained by the protection of AChE.

References

Michelson, M.Ya., Shelkovnikov, S.A., *Gen.Pharmacol.*, **1980**, *11*, 75-82.

0–8412–2008–5/91/0264$06.00/0

Changes in Time Course of Free Transmitter and Endplate Current Caused by Acetylcholinesterase Inhibition at the Neuromuscular Junction: Computer Simulation

Serge A. Slobodov, Vladimir A. Snetkov, Lev G. Magazanik, Sechenov Institute of Evolutionary Physiology and Biochemistry, Academy of Sciences of the USSR, 44 M.Thorez pr., Leningrad, 194223, U.S.S.R.

We used a new approach to modeling of postsynaptic events: single quantal response (EPC) in synaptic zone was simulated by total tracing of stochastic behavior of free transmitter molecules (T) motion, binding to receptors (R) and dissociation. Different kinetic schemes of channel activation and desensitization (DS) were tested. Under normal conditions the number of T in synaptic zone fell close to zero during rising phase of EPC. As a result, EPC decay time constant is about of mean open time of single channel. When acetylcholinesterase (AChE) was absent, decay of T number became biphasic, the "tail" being even more slow than EPC decay. The results obtained show that (i) without AChE EPC decay reflects relatively slow diffusional removal of T; (ii) R in single-bound state (TR) and open channels are in local equilibrium with T; (iii) EPC decay rate depends strongly upon the ratio of free and bound transmitter.

The treatments inducing DS of R may shorten EPC decay after AChE inhibition without prominent decrease of amplitude (Giniatullin et al., 1989; Magazanik et al., 1990). We proposed that some DS-induced R states may be involved in this shortening. The simulation shows that any entity which is possible to bound T for the time longer than EPC duration accelerates decay by reduction of the ratio of free and bound transmitter like AChE does (see iii). The most probable candidate is DS-state of R itself, which has a much higher affinity to T than resting one. Another possibility is that the DS promotes the formation of the relatively long-lived TR state served as possible "traps" for T during EPC.

References

Giniatullin R.A., Khamitov Kh.S., Khazipov R., Magazanik L.G., Nikolsky E.E., Snetkov V.A., Vyskocil F., *J.Physiol.*, **1989**, *412*, 113-122.

Magazanik,L.G., Snetkov,V.A., Giniatullin R.A., Khazipov R.N., *Neurosci. Let.*, **1990** (in press).

0–8412–2008–5/91/0265$06.00/0

Desensitization of AChR as a Result of AChE Inhibition in the Neuromuscular Junction

Natalya N. Potapyeva, Vladimir A. Snetkov, Lev G. Magazanik,
Sechenov Institute of Evolutionary Physiology and Biochemistry, Academy of Sciences of the USSR, 44 M.Thorez pr., Leningrad, 194223, U.S.S.R.

Evoked and spontaneous end-plate currents (EPCs and MEPCs) were recorded in the frog *m.sartorius-n.ischiadicus* preparations by convenient two-microelectrode voltage clamp. The effects of AChE inhibition by prostigmine (6.6 μM) or armine (1 μM) were studied up to 9 hours. At 20°C the inhibition of AChE led to increase of decay time constants (τ) of EPCs and MEPCs up to 404±49% (n=7) and 443±49% (n=4) at 15-30 min. EPCs and MEPCs amplitudes increased up to 131±19% (n=7) and 133±21% (n=4). However, the maximal prolongation was followed by slow decay of τ, and after 2 hours it was equal 305±63% (n=6) for EPCs and 301±75% (n=3) for MEPCs only. In most experiments the shortening of EPCs or MEPCs was accompanied by deviation of decay from single exponential. Two exponentials with time constants τ ≈1.7 ms and τ ≈6.4 ms and relative amplitudes 0.2 and 0.8 fitted the decays satisfactory.

It was proposed that the desensitization of synaptic receptors may lead to shortening of end-plate currents after AChE inhibition (Magazanik et al., 1990). Indeed, the decrease of temperature to 11°C prevented shortening completely, and maximal values of τ and amplitude were stable during several hours. To test the influence of desensitization on the effects of AChE inhibition we used the preparations pretreated with 10 μM of carbacholine (Fiekers, Neel, Parsons, 1987). The application of prostigmine after 15 min of carbacholine washout caused lengthening of MEPCs decay about 2-fold only.

As a possible source of endogenous acetylcholine the non-quantal transmitter release from motor nerve ending may be proposed (Katz, Miledi, 1975).

References

Fiekers, J.F., Neel, D.S., Parsons, R.L., *J.Physiol.*, **1987**, *391*, 109-124.

Katz, B., Miledi, R., *Proc. Roy. Soc. B.*, **1975**, *192*, 27-38.

Magazanik, L.G., Snetkov, V.A., Giniatullin R.A., Khazipov R.N., *Neurosci.Let.*, **1990** (in press).

0–8412–2008–5/91/0266$06.00/0

Anticholinesterase and Neuromuscular Blocking Activity of Ranitidine and Nizatidine on the Toad Rectus Abdominis Muscle; Comparison with Physostigmine

Georgios Kounenis, Department of Pharmacology, Veterinary Faculty, Aristotelian University of Thessaloniki, 540 06 Thessaloniki, Greece

In previous studies it was shown that the H_2-receptor antagonists ranitidine (Kounenis *et al.*, 1986) and nizatidine (Kounenis *et al.*, 1988) augment the acetylcholine-induced contractions on the isolated guinea pig ileum.

The present study was designed to examine the effect of ranitidine, nizatidine and physostigmine on the acetylcholine and carbachol-induced contractions of the toad rectus abdominis muscle.

Materials and Methods

Rectus abdominis muscles (3 cm long) from toad were mounted in 20 ml organ baths containing Frong Ringer solution.The solution in the organ baths was bubbled with 95% O_2 - 5% CO_2 gas and maintained at a temperature of 20° C. The responses of the preparations were recorded on a physiograph recorder via isotonic myograph transducers.

Results

In the presence of ranitidine, nizatidine (3.2 x 10^{-4} and 10^{-3} M),and physostigmine (3.2 x 10^{-5} M), the acetylcholine concentration - response curves were shifted to the left. On the other hand, at higher concentrations of ranitidine nizatidine (3.2 x 10^{-3} M), and physostigmine (10^{-4} and 3.2 x 10^{-4} M), there was a non-parallel rightward shift of the acetylcholine concentration-responce curves. Also, in the presence of ranitidine, nizatidine (10^{-4} and 10^{-3} M), and physostigmine (3.2 x 10^{-5} and 10^{-4} M), the carbachol concentration-response curves were shifted to the right in a concentration-dependent manner with a marked decline in the maximum response.

Discussion

At low concentrations, ranitidine, nizatidine, like physostigmine augmented the acetylcholine-induced contractions and this augmentation can be attributed to their anticholinesterase activity, since in our previous studies it was shown that ranitidine and nizatidine (Kounenis *et al.*, 1986, 1988) possess an anticholinesterase activity. On the other hand, at higher concentrations, the above compounds prevented the action of acetylcholine. These findings provide evidence that at higher concentrations, they possess a neuromuscular blocking activity. It was also found, that these compounds prevented the responses to carbachol. It should be noted that none of the concentrations of the compounds used, potentiated the tissue responses to carbachol, whereas potentiation occured when acetylcholine was used as an agonist. This was expected, since carbachol is highly resistant to hydrolysis by acetylcholinesterase, while acetylcholine undergoes hydrolysis readily.

References

Kounenis, G., Koutsoviti-Papadopoulou, M., Elezoglou, V., *J. Pharmacobio-Dyn.*, **1986**, 9, 941-945.

Kounenis, G., Voutsas, D., Koutsoviti-Papadopoulou, M., Elezoglou, V., *J. Pharmacobio-Dyn.*, **1988**, 11, 767-711.

0–8412–2008–5/91/0267$06.00/0

Effect of Acute and Trained Exercise on the Rate of Decarbamylation of ChE in RBC and Tissues of Rat after Physostigmine Administration

S.M. Somani and S.N. Dube, Department of Pharmacology, Southern Illinois University, School of Medicine, P.O. Box 19230, Springfield, IL 62794-9230 U.S.A.

Cholinesterase inhibitors are likely to be used as a protective agent against oranophosphate intoxication in the fields when a person is engaged in strenuous work (a review by Somani & Dube 1989). This presentation addresses the question whether the pharmacodynamics of Phy (rate of decarbamylation of cholinesterase enzyme) alters due to acute and trained treadmill exercise in RBC and various tissues of rat.

Male Sprague-Dawley rats (initial weight 160-200 g) were divided into 6 groups: Gr. I. - Sedentary Control (SC), Saline Administration; Gr. II. -Acute Exercise (80% VO_2 max) (AE); Gr. III. - Endurance Trained + Acute Exercise (80% VO_2 max) (ET); Gr. IV. - Phy (70 μg/kg, I.M.) (Phy); Gr. V. - Acute Exercise (80% VO_2 max) + Phy (70 μg/kg, I.M.) (AE + Phy); Gr. VI. - Endurance Trained + Acute Exercise (80% VO_2 max) + Phy (70 μg/kg, I.M.)

Groups	Times(min)	RBC	Brain	Heart	Diaphragm	Muscle
II. AE	2-30	114-95	91-97	90-96	97-95	99-105
III.ET	5-60	66-76	96-92	85-75	85-78	89-79
IV. Phy	2-60	72-79	85-78	86-89	90-81	86-83
	$K_d(min^{-1})$	.0207	.0139	.0185	.0103	.0126
V. AE	2-60	78-93	99-100	79-101	93-102	81-80
+ Phy	$K_d(min^{-1})$	.0239	.0252		.0396	.0083
VI. ET	2-60	70-71	75-63	71-73	81-68	81-76
+ Phy	$K_d(min^{-1})$		.0092	.0082	.0083	.0116

(ET + Phy). Rats were sacrificed at various time points soon after the exercise or administration of Phy. RBC and tissues were analyzed for ChE enzyme activity by radiometric method. The results are shown in table as range of percent of control ChE activity at times and the rate of decarbamylation (K_d) of ChE in RBC and tissues and the halftime of ChE recovery. The rate of decarbamylation and the halftime of ChE recovery was determined as per Somani & Khalique (1987).

Acute exercise produced transient increase in RBC ChE activity without significantly affecting brain, heart, diaphragm and thigh muscle. Endurance training slightly decreased ChE activity in RBC, brain, heart, diaphragm and thigh muscle. Acute exercise + Phy showed increased ChE activity in RBC and tissues as compared to Phy alone at all time points studied. Endurance training + Phy further decreased ChE activity in RBC and tissues as compared to Phy alone and endurance training alone. Acute exercise + Phy enhanced the regeneration of ChE enzyme in RBC, brain and diaphragm i.e. decreased the half time of ChE recovery in RBC (29 min), brain (27.5 Min) and diaphragm (17.5 min) compared to Phy alone 33.5, 50 and 67.5 min respectively. Endurance training + Phy lessened the recovery of ChE enzyme in brain, heart and diaphragm i.e. increased the half time of ChE recovery in brain (75 min), heart (85 min) and diaphragm (84 min) compared to Phy alone. The results indicate that acute exercise and endurance training have opposite effect on the rate of recovery of ChE after Phy administration. (Supported by U.S. Army Contract No. DAMD 17-88-C-8024).

References

Somani, S.M. and Dube, S.N., *Intl. J. Cli. Pharmacol. Ther. & Tox.* **1989**, 27:307-387.

Somani, S.M. and Khalique, A., *Drug Metab. Dispos.* **1987**, 15:627-633.

0–8412–2008–5/91/0268$06.00/0

The Molecular Forms of Acetylcholinesterase in Relation to Organophosphate Toxicity

R.W. Busker* **and J.J. Zijlstra**; Medical Biological Laboratory T.N.O., P.O. Box 45, 2280 AA Rijswijk, The Netherlands.

A hardly understood aspect of organophosphate toxicity is the lack of correlation between total AChE activity, as it is usually determined, and function (neuromuscular transmission (NMT), in situ ACh hydrolysis, behaviour and survival). If only one of the molecular forms of AChE correlates with function, this functional form must be preferentially formed, protected or inhibited. If not, total AChE would correlate just as well. Thus the effect of the AChE inhibitor soman on the ratio of molecular forms, and their contribution to cholinergic function was studied.

Methods

Following i.v. injection of rats (♂; ±200 g) with soman, they were sacrificed; part of the diaphragm was homogenized and applied to a sucrose gradient, followed by AChE-assay. The other part was used to determine NMT. The contribution of spontaneous recovery to NMT was corrected for by addition of an excess of soman in vitro.

Results and Discussion

With increasing sedimentation, G_1-G_4-A_{12} were seen. At 0.5 h after injection with soman, AChE was inhibited, but the ratio in which the isoforms occur had not changed (Table 1). However, at 3 h after injection G_1 was much less inhibited than the other forms. This is not caused by differential reactivity or by selective protection of G_1 by its intracellular location, since this difference is neither seen at 0.5 h, nor after incubation with soman in vitro. Probably de novo synthesis of AChE, starting with G_1 building blocks, leads to the relative increase of G_1.

To assess the contribution of G_1 to function, we determined NMT, either 0.5 or 3 h after injection with 42 or 60 μg/kg soman. These dose-time combinations were chosen to obtain similar total AChE levels, but different ratios of molecular forms. The NMT of 3 h animals was significantly lower than after 0.5 h or after incubation with soman in vitro, although the mean total AChE activity was in all cases the same (Table 1).

G_1, although enzymatically as active as other forms per subunit, has a less than proportional contribution in the maintenance of NMT, may be because it is less available for extracellular ACh. In view of these preliminary results, it seems probable that one of the other forms (A_{12} ?) is a better measure of function than total AChE activity.

Table 1. Influence of exposure-time to soman on the composition of AChE (calculated as the fraction of AChE which is in a particular form) and on NMT in rat diaphragm; (*: $p<0.005$; •: $p<0.001$; t test)

time after soman	%NMT	%G_1	%G_4	%A_{12}	% total AChE
untreated control	100	32±2	31±1	37±3	100
0.5 h; 42 μg/kg (n=7)	59±14*	33±1*	31±1	36±2*	12±3
3 h; 60 μg/kg (n=15)	36±13	53±11	23±7	24±7	14±8
15 nM in vitro (n=8)	57±10•	34±3•	29±2	37±4•	15±4

Does Acetylcholine Induce the Synthesis of Inhibited Acetylcholinesterase in Fish Brain ?

Probodh Ghosh[1], **Samir Bhattacharya**[2], **Shelley Bhattacharya**[1], Environmental Toxicology Laboratory[1], Endocrinology Laboratory[2], Department of Zoology, Visva-Bharati University, Santiniketan-731 235, India.

ACh accumulates in the synaptic regions due to anticholinesterase effects of organophosphates and carbamates (Simeon, 1974; Loskowski and Dettbarn, 1975). Pesticide treated fish have a higher level of brain ACh than the untreated one leading to faster recovery of AChE (Jash, 1982). Exogenously administered ACh caused a dose-dependent increase in AChE activity (Jash *et al*, 1982). Thus a pertinent question was addressed – how this stimulation is effected ? To answer this we designed two experiments. In the first *in vivo* experiment *Channa punctatus,* a freshwater murrel, received intramuscular injections of Lomustine, an inhibitor of protein synthesis which crosses the blood-brain barrier, and brain AChE activity measured (Ellman *et al*, 1961).

In the second *in vitro* experiments, ^{14}C–leucine (0.5 μCi) incorporation was monitored in short term fish brain slice cultures (Deb *et al*, 1985) treated with Actinomycin–D (3.5×10^{-6}M) followed by ACh addition (1 μmole).

The first group of brain slice with ^{14}C–leucine was kept as control and incubated for 3 hr. The second group of brain slice with ACh and ^{14}C–leucine added simultaneously was incubated for 3 hr. In the third set the brain slice was incubated for 3 hr with Actinomycin–D and ^{14}C–leucine added together. The fourth set of brain slice was incubated with Actinomycin–D and ^{14}C–leucine for 1 hr. The fifth set of brain slice was preincubated with Actinomycin–D for 1 hr at the end of which ACh and ^{14}C–leucine were added and the incubation carried further for 2 hr. In the last set the brain slice was pretreated with ACh for 30 min then Actinomycin–D and ^{14}C–leucine were added together and incubation continued for 2.5 hr.

Lomustine inhibited brain AChE activity by 22% at 48hr and 35% at 72hr of post injection. Exogenous administration of ACh (1.8 mg/fish/day) reversed the inhibition caused by Lomustine 18% above the control level.

In vitro experiments showed that ^{14}C–leucine incorporation into the TCA precipitable protein significantly increased when the brain slice was incubated with ACh. Contrastingly, pretreatment with Actinomycin–D inhibited the incorporation by 27% which was revoked by 2hr of ACh incubation, the amount of ^{14}C–leucine incorporation into protein was more than that of the control incubation. A 30 min preincubation with ACh effectively counteracted the inhibitory effect of Actinomycin–D.

It was unequivocally noted that ACh was able to stimulate both AChE activity and ^{14}C–leucine incorporation to protein in the absence or presence of Actinomycin–D, irrespective of the time of addition of ACh, whether added 1 hr after the addition of Actinomycin–D or 30 min earlier to it. Enhancement of AChE activity can not possibly be due to direct activation of the inhibited enzyme because excess of ACh combines with the alcohol binding site of the acylated enzyme, retarding deacylation (Dixon and Webb, 1975). Simeon (1974) have suggested that *in vivo* ACh accumulation provides protection against circulating anticholinesterase compounds which resulted in a faster rate of spontaneous reactivation. The present data provide evidence that the recovery of AChE activity by ACh is not due to stimulation of the catalytic activity of the enzyme but due to its synthesis.

References

Deb, S., Jamaluddin, Md., Bhattacharya, S., Bhadra, R., Datta, A.G., *Gen. Comp. Endocrinol*;, 1985, *57*, 491-497.

0–8412–2008–5/91/0270$06.00/0

Dixon, M., Webb, E.C., "*Enzymes*", *Longman Group Ltd.*, London, 1979.

Ellman, G.L., Courtney, K.D., Andres, Jr. V., Featherstone, R.M., *Biochem. Pharmacol.*, 1961, *7*, 88-95.

Jash, N.B., *Doctoral dissertation*, Visva-Bharati University, Santiniketan, 1982.

Jash, N.B., Chatterjee, S., Bhattacharya, S., *Comp. Physiol. Ecol.*, 1982, *7*, 56-58.

Loskowski, M.B., Dettbarn, W.D., *J. Pharmac. Expl. Ther.*, 1975, *194*, 351-361.

Simeon, V., *Arhiv Za Higijenu Rada I Toksikologiju*, 1974, *25*, 51-56.

Actions of Anticholinesterases on Airway Smooth Muscle

M. Adler, D. H. Moore and M. G. Filbert U. S. Army Medical Research Institute of Chemical Defense, Aberdeen Proving Ground, MD 21010-5425

Exposure to organophosphorus (OP) cholinesterase (ChE) inhibitors results in excess stimulation of nicotinic and muscarinic acetylcholine (ACh) receptors. The predominant signs of toxicity are due to persistence of ACh and are mediated by muscarinic receptors. Such persistence is due to continuous release of ACh coupled with its slow diffusion.

To elucidate the mechanism(s) underlying the toxicity of the (OP) anti-ChE agents, we examined the actions of soman, sarin, VX, paraoxon and diisopropylfluoro-phosphate (DFP) on isolated canine tracheal smooth muscle. Contractions to electric field stimulation (EFS) were examined prior to and during exposure to the above OPs. The baths were enclosed in a fume hood to minimize vapor hazard. Methodological details are similar to Adler et al. (1987).

All of the OPs were found to increase the amplitude and prolong the decay of contractions elicited by EFS and to potentiate the tensions due to bath applied ACh (Table 1).

This sequence paralleled the ability of the OPs to inhibit ChE. A similar relationship was also observed for the carbamate inhibitors neostigmine, physostigmine and pyridostigmine. Exposure to the anti-ChE agents also caused large sustained contractures. The latency and rise-time of the contractures varied with the inhibitor but the maximal tensions were similar for all agents. The OP-induced alterations were completely antagonized by anti-muscarinic compounds, with scopolamine being the most potent member. Relaxations produced by the beta-adrenergic agonists were generally transient and relaxations produced by the organic calcium blockers nifedipine and verapamil were incomplete. With the exception of soman, the actions of the OPs were rapidly reversed by incubation with the reactivator HI-6 (10-100 μM).

Table 1. Relationship between muscle tension and ChE activity[a]

Inhibitor	Onset[b] nM	ChE activity (% Control)
VX	0.3	27
Soman	0.7	23
Sarin	2.1	35
Paraoxon	10	31
DFP	30	28

[a]Determined by radiometric method of Siakotos, A.N., Filbert, M.G., Hester, R. Biochem. Med., **1969**, 3,1-12. [b]Concentration producing ≤ 25% increase in EFS amplitude.

References

Adler, M., Filbert, M.G., Moore, D., Reutter, S. In Cellular and Molecular Basis of Cholinergic Function, Dowdall, M., Hawthorne, J.N. eds. Ellis-Horwood, U.K., **1987**, pp. 582-597.

Cholinesterases from Invertebrates : A Comparison of Kinetic Constants with Different Thiocholine Esters as Substrates

Giovanni B. Principato *, Sandro Contenti , Vincenzo Talesa, Carla Mangiabene, Elvio Giovannini, Gabriella Rosi, Department of Experimental Medicine, Division of Cellular and Molecular Biology, University of Perugia, Via del Giochetto, 06100 Perugia, Italy
Rita Pascolini, Institute of Comparative Anatomy, University of Perugia, Via Elce di Sotto, 06100 Perugia, Italy

In Vertebrata, Cholinesterases can be divided in two subclasses, Acetylcholinesterase (AChE, EC 3.1.1.7) and less-specific Cholinesterase (ChE, EC 3.1.1.8).

In Invertebrata, Cholinesterases are represented by enzyme(s) with intermediate charasteristics in between vertebrate's AChE and ChE. It is possible that these two subclasses have been originated by a common precursor esterase in the course of the evolution. The specificity for choline esters could have been originated by the generation of a negative charge into the active site. Information on this possible ancestral protein can come from the study of invertebrate ChEs, whose properties are typical of the species and vary within an apparent continuous range of specificity pattern.

Comparison of Vmax/Km and Km values

The ChEs from *Hirudo medicinalis* (Principato et al., 1983), *Palaemonetes varians* and *Murex brandaris* (Talesa et al., 1990) on the basis of Vmax/Km ratios, hydrolyze propionylthiocholine (PTC) better than the acetic- (ATC) or butyric-thiocholine (BTC) esters which are hydrolyzed at a similar rate.

The ChEs from *Parascaris equorum* (Aisa et al., 1982), *Helix pomatia* and *Asterias bispinosa* (Principato et al., 1984), *Squilla mantis* (Principato et al., 1988), and *Allolobophora caliginosa* (Principato et al., 1989) hydrolyze BTC as a poor substrate in respect to ATC and PTC.

The role of electric charge and hydrophobic forces in substrate binding emerges by Km values. Similar Km for ATC, PTC and BTC, that is a major role of electrostatic interaction, are showed by *Murex*, *Eisenia*, *Parascaris*, *Asterias*, and *Palaemonetes* ChEs. The opposite situation is reported for *Helix* ChE, that is similar Km among these substrates.

Intermediate behaviours are also reported: Km decreasing from ATC to BTC by *Hirudo* ChE, that is a role for hydrophobic forces; Km increasing from ATC to BTC by *Allolobophora* and *Squilla* ChEs, that is a role for steric hindrance.

It is possible that the ChEs from Invertebrata might represent an interesting intermediate evolutionary level between more evolved forms from Vertebrata and a possible ancestral esterase.

References

Aisa, E., Principato, G.B., Biagioni, M., Giovannini E., *Comp.Biochem.Physiol.*, **1982**, 71C,119-122.

Principato, G.B., Rosi, G., Biagioni, M., Giovannini, E., *Comp.Biochem.Physiol.*, **1983**, 75C,185-192.

Principato, G.B., Rosi, G., Bocchini, V., Giovannini, E., *Comp.Biochem.Physiol.*, **1984**, 77B,211-219.

Principato, G.B., Talesa, V., Giovannini, E., Pascolini, R., Rosi, G., *Comp. Biochem. Physiol.*, **1988**, 90C,413-416.

Principato, G.B., Contenti, S., Talesa, V., Mangiabene, C., Pascolini, R., Rosi, G., *Comp.Biochem.Physiol.*, **1989**, 94C, 23-27.

Talesa, V., Contenti, S., Mangiabene, C., Pascolini, R., Rosi, G., Principato, G.B., *Comp.Biochem.Physiol.*, **1990**, in press.

Physicochemical and Crystallographic Studies on the Stability and Structure of Aged and Nonaged Organophosphoryl Conjugates of Chymotrypsin

Ching-Tang Su, Nitza Steinberg, Israel Silman*, Department of Neurobiology, The Weizmann Institute of Science, Rehovot 76100, Israel
Michal Harel, Joel Sussman, Department of Structural Chemistry, The Weizmann Institute of Science, Rehovot 76100, Israel
Jacob Grunwald, Yacov Ashani, Israel Institute for Biological Research, Ness Ziona 70450, Israel

Many serine hydrolases, such as acetylcholinesterase (AChE) and chymotrypsin (Cht), are inhibited by organophosphorus (OP) esters which form a stoichiometric (1:1) covalent conjugate with the active site serine. Such conjugates can often be reactivated by nucleophilic reagents. However, they may undergo a process called 'aging', which involves detachment of an alkyl group from the bound OP moiety with concomitant generation of a negatively charged oxygen. This 'aged' conjugate is totally resistant to nucleophilic reactivation. A homologous pair of nonaged and aged conjugates, diethylphosphoryl-Cht (DEP-Cht, nonaged) and monoethylphosphoryl-Cht (MEP-Cht, aged), were utilized in a comparative study aimed at finding a physicochemical basis for this resistance. Transverse urea gradient electrophoresis showed that at three different temperatures the urea concentration at the mid-point of transition, $[U]_{0.5}$, was greater for MEP-Cht, than for DEP-Cht. Monitoring of the circular dichroism (CD) spectra of MEP- and DEP-Cht as a function of the concentration of the protein denaturant, guanidine HCl, also showed that the aged enzyme is significantly more stable than the nonaged enzyme. The data obtained by these two techniques suggest that ΔG_f is increased by ca. 2-5 Kcal/mole. It was postulated that this stabilization originates in an interaction of the negative charge of the oxygen in the aged OP moiety with the imidazole ring of His57. This assumption was substantiated by comparison, by X-ray crystallography, of the nonaged conjugate, DEP-Cht and the aged conjugate, monoisopropyl-Cht (MIP-Cht). The X-ray crystallography data show a striking resemblance of the two structures, the main difference being in the positioning of the OP group bound at the active site. The negatively charged oxygen of the MIP group in the aged enzyme appears to be involved in an H-bond to His57$N^{\varepsilon 2}$ (2.64 Å), whereas in the nonaged enzyme, the corresponding noncharged oxygen of the DEP group is 3.58 Å from this nitrogen. Thus, both the physicochemical and X-ray data provide a physical basis for the difference in stability of the aged and nonaged conjugates and for the resistance to reactivation of the aged enzyme by nucleophilic reagents.

References

Creighton, T. E., *J. Mol. Biol.*, **1979**, *129*, 235-264.

Grunwald, J., Segall, Y., Shirin, E., Waysbort, D., Steinberg, N., Silman, I., Ashani, Y., *Biochem. Pharmacol.*, **1989**, *38*, 3157-3168.

Masson, P., Goasdoue, J. L., *Biochim. Biophys. Acta*, **1986**, *869*, 304-313.

Steinberg, N., van der Drift, A. C. M., Grunwald, J., Segall, Y., Shirin, E., Haas, E., Ashani, Y., Silman, I., *Biochemistry*, **1989**, *28*, 1248-1253.

0–8412–2008–5/91/0274$06.00/0

Stability of native and organophosphate–inhibited butyrylcholinesterase under high pressure

Patrick Masson, Cécile Cléry, Didier Huchet
Centre de recherches du Service de Santé des Armées, Unité de Biochimie, B.P.87, 38702, La Tronche, France.

The tetrameric form (G_4) of human plasma butyrylcholinesterase (BChE) was used as a model enzyme to investigate the molecular basis of organophosphate–aged cholinesterase resistance to reactivators.

The enzyme was phosphorylated by four organophosphates (OP) : DFP, soman, 1–pyrenebutylphosphorodichloridate (PBPDC) and 1–pyrenebutylethylphosphorofluoridate (PBEPF). The PBEPF–inhibited enzyme could be reactivated by a pyridinium oxime (2–PAM). On the other hand, due to the aging process, the other phosphonylated forms were irreversibly inhibited.

The combined effects of high pressure and a protein denaturing agent (propylene carbonate, PC) upon the stability of native and phosphonylated BChE were investigated. The study was carried out in phosphate and in Tris buffers, pH7, containing PC at different concentrations varying from 0 to 2M. We operated in a thermostated high–presure vessel at 20°C in the pressure range 10^{-3} to 3.5kbar. The pressure and PC perturbing effects were analyzed : a). by polyacrylamide capillary gel electrophoresis at the desired hydrostatic pressure using a high–pressure electrophoresis apparatus designed for denaturation studies (Balny, 1989 ; Masson, 1990) ; b). by slab gel electrophoresis and kinetic measurements after pressure release.

All the G_4 species, native and conjugated, appeared more stable in phosphate than in Tris. There was no noticeable pressure–induced dissociation of G_4 to observe. However, when operating in phosphate, several pressure–induced isomeric states were observed. This isomerization process was found to be enhanced by PC. In Tris, the combined effects of pressure and PC induced aggregation of the various enzyme species. Heavy water used as solvent in place of water did not change the threshold of appearance of these effects.

Several ligands were used for stabilization of the enzyme structure (Payne, 1989). The rank order for ligand stabilization was : 2–PAM > Edrophonium. On the other hand, Procainamide and Decamethonium acted as destabilizers.

The native, reactivable OP–inhibited and aged forms showed differences in pressure/denaturant sensitivity. In particular, in phosphate, the PBPDC–aged enzyme was more sensitive to pressure than the native and reactivable phosphonylated ones. Conversely, the DFP–and soman–aged species did not generate isomeric states below 3.5kbar. In Tris, the pressure–induced aggregation of aged G_4 occured at higher pressure than the native G_4. Soman–aged enzyme was found to be the more stable species . This confirms previous urea–denaturation results (Masson, 1986) and fits Raman spectroscopic data we have on secondary structure of OP– aged ChE (Aslanian, in preparation). Thus, the non reactivability of DFP–and soman–aged BChE may be interpreted as the result of aging–induced stability/conformation changes.

References

Balny, C., Masson, P., Travers, F., *High Pressure Res.*, **1989**, *2*, 1–28.
Masson, P., Goasdoué, J.C., *Biochim. Biophys.Acta*,**1986**, *869*, 304–313.
Masson, P., Arciero, D., Hooper, Balny,C.,*Electro–phoresis.*, **1990**, *11*, 128–133.
Payne, C.S., Saeed, M., Wolfe, A.D., *Biochim. Biophys. Acta*, **1989**, *999*, 46–51.

Electrostatic Effect on Cholinesterase Reaction Kinetics

T.Kesvatera, Laboratory of Bioorganic Chemistry, Institute of Chemical Physics and Biophysics of the Estonian Academy of Sciences, P.O. Box 670, Tallinn 200026, Estonia

The electrostatic effect on the ligand binding of cholinesterases (EC 3.1.1.7 and EC 3.1.1.8) was monitored by the salt effect studies. Salt effect data were described as (Record, 1976; Kesvatera, 1990a):

$$pK_d = pK_d^o - \psi \log[M^+] + \Delta\kappa c,$$

where K_d^o is the nonelectrostatic component of the observed dissociation constant of the enzyme-ligand complex, $\Delta\kappa c$ is the binding contribution due to the salting effect (Kesvatera, 1988, 1990b) and $\psi \log[M^+]$ is the electrostatic term (Record, 1976), originally introduced for the analysis of salt effects on the protein-nucleic acid complex formation, proceeding from the counterion condensation theory of polyelectrolytes (Manning, 1969). The electrostatic effect on the interaction of acetylcholinesterase with 15 cationic ligands was uniformly characterized by the ψ=0.49±0.03, which is determined by the charge density of the polyampholyte molecule of the enzyme. Similar value of ψ was observed for butyrylcholinesterase (Kesvatera, 1990a). Polyelectrolyte field effects of the enzyme molecule totally account for the electrostatic contribution ($pK_d - pK_d^o - \Delta\kappa c$) that can be observed also in the QSAR data for cholinesterases (Järv, 1984). Accordingly, the presence of specific anionic site at the catalytic center of these enzymes is not required.

References

Järv, J., *Bioorg. Chem.*, **1984**, *12*, 259-278.

Kesvatera, T., Aaviksaar, A., Peenema, E., Järv, J., *Bioorg. Chem.*, **1988**, *16*, 429-439.

Kesvatera, T., Langel, A., Järv, J., *Bioorg. Chem.*, **1990a**, *18* , 13-18.

Kesvatera, T., *Biochim. Biophys. Acta*, **1990b**, *1039*, 21-24.

Manning, G.S., *J. Chem. Phys.*, **1969**, *51*, 924-938.

Record, M.T. Jr., Lohman, T.M., De Haseth, P., *J. Mol. Biol.*, **1976**, *107*, 145-158.

Anticholinesterase Potency of Hostaquick, Pirimor, Croneton and Methomyl for Aphids and Entomophages

I.N.Sazonova, K.V.Novozhilov, E.V.Nikanorova, E.K.Shvets, All-Union Institute of Plant Protection, 3 Podbelskogo St., Puskin-Leningrad, USSR 188620

Anticholinesterase potency of insecticides Hostaquick, Pirimor, Croneton and Methomyl was investigated. The different anticholinesterase potency of insecticides for aphids and their entomophages was established.

The cholinesterase sensitivity of aphids and their entomophages to Hostaquick, Pirimor, Croneton, Methomyl was investigated. The catalytic activity of the acetylcholinesterase (AChE) was determinated by Ellman method(1), the anticholinesterase potency of the insecticides (k_a) - by Yakovlev method (2), the toxicity (LC_{50}) - by Popov one(3).

The sensitivity of AChE of peach aphid *Myzus persicae Sulz* to Pirimor, Hostaquick, Croneton, Methomyl has been within the limits of $k_a = 10^4-10^5\ M^{-1}*min^{-1}$. AChE of aphids are 1-3 order of magnitude more sensitive to Hostaquick and Pirimor as compared to entomofages *Crysopa carnea Steph.* and *Cocinella septempunctata L.* For assessing the selectivity of anticholinesterase compounds the anticholinesterase coefficient of the $k_{ache} = k_a$ aphids / k_a entomophageshas has been suggested by analogy with coefficient of selectivity , CS , (LC_{50} entomophages / LC_{50} insects).

References

1. Ellman G.L. et al. *Pharmacol.*, **1961**, v.7, p.88-95.

2. Yakovlev V. The kinetics of enzymic catalysis, M.:Nauka, **1965**, 248p.

3. Popov P.V. *Chimia v selskom khozyaistve*, **1965**, n.10, p.72-79.

The effect of Conformation Isomerism and Hydrophobicity of Organophosphates on their Interaction with Cholinesterases of Different Origin

Tamara I.Vasiljeva, All'Union Institute of Plant Protection, Podbelskogo av., 3, Pushkin-8, Leningrad, 189620, USSR
Ljudmila I.Kugusheva, Victor I.Rozengart, Institute of Evolutionary Physiology and Biochemistry Academy of Sciences of the USSR, Thorez pr., 44, Leningrad, 194223, USSR

Derivatives of lupinine (1) and of epilupinine (2) (Dalimov et al.,1978) of the general structure $(RO)_2P(O)SCH_2-$[quinolizidine] with $R=C_2H_5-C_5H_{11}$ were studied. Firstly, distinct dependency of their action from conformation of the heterocycle was established. Derivatives 1 were shown seemed to be more potent inhibitors for arthropods ChE. For example, *Schizaphis gramina Rond.* ChE was 50 times more sensitive to derivatives 1 (R=C2H5) as compared to derivatives 2. Some of studied compounds possess a unique ability to inhibit *Musca domestica L.* ChE ($k_a=10^8$). Derivatives 2 seemed to be more potent inhibitors for mammalian ChE. This fact testified to deferences between the environment of anion center in ChEs of arthropoda and mammals. Secondly, the dependence of anticholinesterase action and toxicity on R-lenght was established. Lenghtening of R increased the inhibition of mammalian enzyme in derivatives 1. When changing from C_2H_5 to C_5H_{11} for AChE of human erythrocytes, k_a increased by 27-fold, whereas sensitivity of arthropoda ChE was reduced by 15-120 fold. It shows different structures of hydrophobic areas of ChE in arthropoda and mammals. The toxicity decreases for all objects studied when R is lenghtening. The conformity between toxicity and anticholinesterase action is observed for arthropoda. Unlike it the toxicity decreases for mammals but the anticholinesterase activity increases. Thus, changing conformation of heterocycle and structures of the molecule it is possible to succeed in increasing the selective action of organophosphates.

Reference

Dalimov, D.N., Abduvakhabov,A.A., Aslanov, Kh.A., Godovikov,N.N., *Izv.AN SSSR, ser.khim.*, **1978**, *2*, 480-486.

0–8412–2008–5/91/0278$06.00/0

Conformation-Activity Relationships of the Ligands of Active Sites of Cholinesterases

Boris S. Zhorov, Pavlov Institute of Physiology of the Academy of Sciences of the USSR, Nab. Makarova, 6, Leningrad B-034, U.S.S.R.
Evgeny V. Rozengart, Natalya N. Shestakova, Sechenov Institute of Evolutionary Physiology and Biochemistry of the Academy of Sciences of the USSR, pr. Thoreza, 44, Leningrad K-223, U.S.S.R.

With the goal of mapping of anionic point in butyrylcholinesterase active site, all equilibrium conformations of 19 substrates, $MeCOOCH_2CH_2NR^1R^2R^3$ (R - alkyl or cycloalkyl substituents), were calculated by molecular mechanics method. It was suggested that the anionic point is located in a cavity of the enzyme molecule. Dimensions and shape of this cavity were chosen to provide satisfactory correlation between the total population of those substrate conformers which fit the cavity and the rate of cholinesterase hydrolysis of the substrates. Basing on this results, the mechanism of nonproductive sorbtion was explained (Rozengart, Zhorov, 1989).

To determine productive conformation of acetylcholinesterase substrates, all equilibrium conformations of 23 acetylcholine derivatives were calculated. In a series of 9 flexible substrates $MeCO\text{-}O\text{-}CHR^1\text{-}CHR^2NMe_3$ (R^1, R^2 = H or Me), a correlation was found between population of completely extended all-trans(*ttt*)conformer and the rate of enzymic hydrolysis of the substrates. In the subset of semirigid compounds, substrates have while non-active compounds do not have low-energy conformers which are spatially compatible with *ttt*-conformer of acetylcholine. Activity of substrates $MeCOO(CH_2)_nNMe_3$ (n=1,2,3,4) correlate with population of those their conformers which are compatible with *ttt*-conformer of acetylcholine. The regions of essential volume and excluded volume in the active center of enzyme are suggested. The location of these regions is determined relative to productive *ttt*-conformer of acetylcholine (Shestakova et al., 1989).

References

Rozengart, E.V., Zhorov, B.S. *J.Evoluts.Biochim.Fiziol.* **1989**, *25*, 189-194.

Shestakova, N.N., Rozengart, E.V., Khovanskikh, A.E., Zhorov, B.S., Govyrin, V.A. *Bioorgan. Khimiya* **1989**, *15*, 335-344.

0–8412–2008–5/91/0279$06.00/0

Inhibitory Effect of Paraquat on Acetylcholinesterase Activity

Yasuo Seto* and **Toshiaki shinohara**, National research Institute of Police Science, 6, Sanban-cho, Chiyoda-ku, Tokyo 102, Japan

A nonselective contact herbicide, paraquat (1,1'-dimethyl-4,4'-bipyridinium, PQ), possesses high toxicity for animals, and the biochemical mechanism of its toxicity is explained mainly in terms of the peroxidation of membrane lipids (Autor 1977). Here, inhibitory effect of PQ on human erythrocyte and electric eel acetylcholinesterase (AChE) activities was investigated.

Paraquat Binding Site

The degree of AChE inhibition was not altered by changing the time of preincubation with PQ prior to substrate addition. PQ gave effective protection against the irreversible inhibition by an anionic site inhibitor, dibenamine, but not by esteratic site inhibitors, dichlorvos and methanesulfonylchloride (Seto and Shinohara 1987). Therefore, it is concluded that PQ functions as a reversible inhibitor binding to AChE anionic site.

Structure-Activity relationship

Inhibition powers (I_{50}) and Hill coefficients of reversible AChE inhibitors were compared each other (Seto and Shinohara 1988). PQ showed strong inhibition for AChE and high inhibition selectivity (IS) compared to inhibition for human serum butyrylcholinesterase. Negative correlations were observed between I_{50} and IS among PQ derivatives, monoquaternary ammonium compounds and anticholinergic drugs, respectively. Similar to d-tubocurarine, PQ showed negative cooperativity for AChE inhibition. However, diquat showed weak and competitive inhibition for AChE. Therefore, it is concluded that the strong and specific inhibition with negative cooperativity for AChE is derived from its dialkyl 4,4'-bipyridinium structure.

References

Autor,A.P., Biochemical mechanism of paraquat toxicity, Academic Press, New York, 1977

Seto,Y., Shinohara,T., Agric.Biol. Chem., **1987**, 51, 2131.

Seto,Y., Shinohara,T., Arch.Toxicol. **1988**, 62, 37.

0–8412–2008–5/91/0280$06.00/0

Salt Effect on Multivalent Ligand Binding by Acetylcholinesterase

V. Tõugu, T. Kesvatera, Laboratory of Bioorganic Chemistry, Institute of Chemical Physics and Biophysics of the Estonian Academy of Sciences, P.O. Box 670, Tallinn 200026, Estonia

The influence of inorganic salts on the affinity of native and chemically modified acetylcholinesterases (EC 3.1.1.7) for cationic substrates and multivalent ligands was studied. The data were analyzed according to the equations (Record, 1976):

$$pK_d = pK_d^o - Z_L \psi_{+z} \log[M^{+z}], \quad (1)$$

$$\psi_{+z} = (z - 1 + \psi_+)/z^2, \quad (2)$$

where ψ_+, determined by the charge density of the polyelectrolyte, is the fraction of condensed monovalent counterions per one negative charge of polyion (Manning, 1969), Z_L is the charge of the ligand, z is the charge number and $[M^{+z}]$ is the concentration of salt cation. The observed $Z_L\psi_{+z}$ values for the complex formation between acetylcholinesterase and hexamethonium (Z_L=+2) or gallamine (Z_L=+3) were in quantitative agreement with the ψ_+ = 0.5 (Kesvatera, 1989) for the binding of univalent cationic ligands. Chemical modification of lysine amino groups with pyromellitic dianhydride resulted in an increased value of ψ_+, equal to 0.6. Eqn. 1 failed to describe the salt effect data for the enzyme, where a fraction of carboxylic groups was modified by tris(hydroxymethyl)aminomethane, apparently due to the low negative charge density of the obtained enzyme derivative. These results show that the influence of salts on the electrostatic contribution to the binding of cationic ligands by anionic acetylcholinesterase can be described proceeding from the counterion condensation theory of Manning by using only one empirical parameter ψ_+.

References

Kesvatera, T., *Proc. Estonian Acad. Sci. Chem.*, **1989**, *38*, 216-218.

Manning, G.S., *J. Chem. Phys.*, **1969**, *51*, 924-938.

Record, M.T. Jr., Lohman, T.M., De Haseth, P., *J. Mol. Biol.*, **1976**, *107*, 145-158.

Similarity in Pocket Selectivities of Acetylcholinesterase and Cyclodextrins in Reactions with Substituted Phenylacetates. Modelling of the Spatial Structure of the Acetylcholinesterase Active Site

Peep Palumaa and **Kalev Kask,** Laboratory of Bioorganic Chemistry, Tartu University, Jakobi 2, 202400 Tartu, Estonia

Pocket selectivity of acetylcholinesterase (AChE) has been studied by making use of a series of *meta-* and *para*-substituted phenylacetates with alkyl (-CH_3, -$C(CH_3)_3$) and halogen (-Cl, -Br, -I) substituents. As *meta* isomers from the selected series are chemically almost equireactive with the corresponding *para* isomers (Hansch, C. and Leo, A., 1979), the *meta-para* specificity of the enzyme must originate from the spatial structure of the active site.

It has been found that the AChE-catalyzed hydrolysis of *meta* isomers is always faster than that of *para* isomers and the difference in AChE reactivity toward *meta* and *para* isomers depends on the volume of the substituent. Temperature dependences indicate that *meta*-isomers have considerably lower enthalpy and slightly lower entropy of activation at the acylation step as compared with the corresponding *para* isomers.

The specificity of AChE towards substituted phenylacetates has been compared to the specificity of cyclodextrins (CDs) towards a similar series of substrates (Komijama, M. and Bender, M. L., 1978; Van Etten, R. L., *et al.*, 1967). A surprisingly good agreement between both the kinetic and thermodynamic behaviour of AChE and β-CD has been found, allowing us to use β-CD as a biomimetic model for AChE. Since the unusual specificity of CDs towards phenylacetates lie in the spatial structure of the CD's cavity (Komijama, M. and Hirai, H., 1980), a model of the AChE active site has been created on the basis of the spatial structure of the CDs cavities.

The model represents a cylindrical rigid cavity with a diameter of appr. 0.75 nm and a nucleophile, located appr. 0.2 nm away from the edge of cavity in an almost parallel orientation to the axis of the cavity.

The proposed model allows to explain the *meta*-specificity of AChE towards different derivatives of substituted phenols (acetates, carbamates, phosphates, sulphonates etc.) with orientation effects leading to the acceleration of the reactions of *meta*-substituted isomers in the comparision with *para*-isomers via proximity effects.

References

Hansch, C.; Leo, A. Substituents Constants for Correlation Analysis in Chemistry and Biology; J. Wiley and Sons Inc.: New York, 1979.

Komiyama, M.; Bender, M.L. J. Am. Chem. Soc. **1978**, 100, 4576-4579.

Komiyama, M.; Hirai, H. Chem. Lett. **1980,** 1471-1474.

Van Etten, R.L.; Sebastian, J.F.; Clowes, G.A.; Bender, M. J. Am. Chem. Soc. **1967**, 89, 3242-3253.

Cholinesterases of Spring Grain Aphid. Activation Effect of some Tetraalkylammonium Ions and Alcohols

Eugene B. Maizel, Alexander P. Brestkin, Serge N. Moralev, Alexander E. Khovanskikh, I.M.Sechenov's Institute of Evolutionary Physiology and Biochemistry Academy of Sciences of the USSR, Thorez pr. 44, 194223 Leningrad

Irina N. Sazonova, Kapiton V. Novozhilov, All-Uninon Research Institute of Plant Protection, Podbelskogo av. 3, Pushkin-6, 188620 Leningrad

The acetylcholinesterase (AChE) and butyrylcholinesterase (BuChE) of aphids were found by us (Brestkin et al.,1985,1986) to differ substantially from the same enzymes of mammals. The unique feature of aphid AChE is its sensitivity to the inhibiting effect of SH-reagents (Novozhilov et al., 1989).

The experiments with reversible inhibitors have shown a more interesting peculiarity of AChE of spring grain aphid *Schizaphis gramina Rond.* The catalytic activity of this enzyme rises up to 2 times in presence of low concentrations (up to 5- 20 mM) of choline, tetramethyl-, tetraethyl- and tetrapropylammonium ions, and is inhibited by their higher concentrations. BuChE of the aphid is sensitive neither to inhibiting nor to activating effect of those ions. It is only inhibited by tetrabutylammonium and some bisammonium ions.

Aliphatic alcohols from methanol to butanols at high concentrations (from 0.5 to 5 M) reversible inhibit both AChE and BuChE of the aphid, but lower concentrations of the same alcohols, with the exception for methanol, activate hydrolysis of substrate by aphid AChE, but not by aphid BuChE.

Activating effect of tetraalkylammonium ions and aliphatic alcohols was not displayed in aphid AChE interaction with diisopropyl fluorophosphate.

It is supposed that the activation of the acetylcholine enzymatic hydrolysis by these effectors is the result of acceleration of enzyme deacetylation stage. On this basis a scheme of interaction between enzyme-substrate system and the effectors was suggested and the corresponding equation describing the variations of aphid AChE catalytic activity was derived.

References

Brestkin, A.P., Maizel,E.B., Moralev, S.N.,Novozhilov, K.V.,Sazonova, I.N., *Insect Biochem.*,**1985**,15,309-314.

Brestkin, A.P., Khovanskikh, A.E., Maizel, E.B., Moralev, S.N., Novozhilov, K.V., Sazonova, I.N., Abduvakhabov, A.A., Godovikov, N.N.,Kabachnik, M.I., Khaskin, B.A., Mastryukova, T.A., Shipov, A.E.,*Insect Biochem.*,**1986**,16,701-707.

Novozhilov, K.V., Brestkin, A.P., Khovanskikh, A.E., Maizel, E.B.,Moralev, S.N., Nikanorova, E.V.,Sazonova, I.N., ***Insect Biochem***,**1989**,19,15-18.

Alkylation of the Anionic Centre of Acetylcholinesterase with a Cationic Label Accelerates the Acylation Step of Ester Substrate Hydrolysis

Armin Sepp, Laboratory of Bioorganic Chemistry, Tartu University, Tartu 202400, Estonia, USSR

Catalytic properties of DPA-labelled acetylcholinesterase are similar to those of the native enzyme reversible complex with tetraalkylammonium ions (Purdie, 1968). QSAR of the modified enzyme in its reaction with non-ionic acetic acid esters $CH_3C(O)OR$ (R=Me,Et,n-Pr,n-Bu,n-Pe,n-Hex,n-Hep,$C_2H_4CH(CH_3)_2$,-$C_2H_4(CH_3)_3$,-$CH_2CH=CH_2$,-CH_2C-CH,-C_2H_4Cl, Phe) has been studied in this work in order to understand the origin of the effects that can cast light to the functional role of the anionic centre.

The DPA modified enzyme displayed Michaelis-Menten kinetics. Both K_m and k_{cat} were estimated for all substrates. The dependencies of K_m and k_{II} on the leaving group properties were found to coincide quantitatively. k_{cat} had the same value $\approx 10000\ s^{-1}$ for all substrates. The results of the QSAR analysis are in agreement with the three-step reaction scheme where $k_2 >> k_3$:

$$E + S \overset{K_s}{====} ES \overset{k_2}{\underset{P_1}{-------}} EA \overset{k_3}{------} E + P_2$$

It was concluded that the affected interaction between substrate and enzyme is localized to the acetyl group of the ester molecule because the intensities of the leaving group hydrophobic and electronegative interactions were found to be the same as in the case of native enzyme (Järv, 1976).

It is proposed that the increased reactivity of DPA-acetylcholinesterase is caused by conformational change in the esteratic centre that adapts its spatial structure to the (partially) tetrahedral transition state of the substrate ester group in the acylation step (Quinn, 1987).

References

Järv, J.; Kesvatera, T.; Aaviksaar, A. *Eur. J. Biochem.* **1976**, *67*, 315-322.
Purdie, J.E. *Biochim. Biophys. Acta* **1968**, *185*, 122-133.
Quinn, D.M. *Chem. Rev.* **1987**, 87-114.

Characterization of Peripheral Anionic Site Peptides of AChE by Photoaffinity Labeling with Monoazidopropidium (MAP)

G. Amitai, Division of Chemistry, IIBR, Ness Ziona 70450, Israel.
P. Taylor, Department of Pharmacology, M-036, UCSD. La Jolla, CA. 92093, USA.

Certain ligands (e.g. propidium, and gallamine) which bind to the peripheral anionic site (PAS) of AChE modulate the enzyme catalytic activity by an allosteric mechanism (Taylor and Lappi, 1975). The exact location of the PAS in the enzyme primary structure is unknown but it appears physically separated from the active center. Delineation of the PAS topography is important since site-directed mutagenesis in this region may help to assign specific amino acids involved in allosteric regulation of catalysis.

The fluorescent photolabile ligand monoazidopropidium (MAP) was prepared by modification of methods described by Graves et al. (1977). MAP inhibits Torpedo (T.) AChE in the dark with an IC50 of 10 nM. Upon UV irradiation MAP binds covalently to 11S T. AChE. Reversible PAS ligands (propidium and decamethonium) protect 50-80% of the MAP sites. [^{3}H]MAP was prepared to allow for monitoring the protein fragments by fluorescence or by radioactivity. Covalent photolabeling was demonstrated by the monomer band in the autoradiogram of the SDS gel of [^{3}H]MAP- AChE conjugate. Two samples of AChE conjugates were labeled in parallel: a totally bound (TOT) and a nonspecifically bound MAP (NSP) with 10 fold excess of propidium.

Both conjugates were denatured in guanidine-HCl, reduced by DTT, alkylated with IAA and then digested with trypsin. The tryptic peptides were purified by gel filtration on Sephadex G-50. The fraction which contained the highest radioactivity was injected to HPLC (C-4 reverse phase). The HPLC chromatograms showed several labeled peptides in the TOT sample. One of the labeled peptides in the TOT sample was isolated and has the following sequence: NLNCNLNSDEELIH. This sequence corresponds to residues 251-264 of the T. AChE (Schumacher et al. 1986) which resides in loop B of AChE (MacPhee-Quigley et al. 1986). The presence of one Asp residue adjacent to two consecutive Glu residues (259-261) is consistent with a negatively charged PAS which is remote from the catalytic site of AChE.

References

Graves, D.E., Yielding L.W., Watkins, C.L., and Yielding K.L. Biochim. Biophys. Acta, **1977**, 479, 98-104.

MacPhee-Quigley, K., Vedvick, T.S., Taylor, P., and Taylor, S.S. J. Biol. Chem., **1986**, 261, 13565-13570.

Schumacher, M., Camp, S., Maulet, Y., Newton, M., MacPhee-Quigley, K., Taylor, S.S., Friedman, T. and Taylor, P. Nature, **1986**, 319, 407-409.

Taylor, P. and Lappi, S. Biochemistry, **1975**, 14, 1989-1997.

Interaction of Reversible and Irreversible Inhibitors with a Peripheral Site on Acetylcholinesterase

Alain Friboulet*, Laboratoire de Technologie Enzymatique; URA 523 du CNRS; UTC, 60206 Compiègne Cedex France.

Palmer Taylor, Department of Pharmacology, M036, University of California, San Diego, CA 92093, USA.

François Rieger, Unité de Biologie et Pathologie Neuromusculaires, INSERM U153, 17 rue du Fer-à-Moulin, 75005 Paris, France.

The interaction of organophosphates (OP) with acetylcholinesterase (AChE) is generally studied in terms of classical irreversible active site-directed inhibition. We have previously demonstrated that mouse skeletal muscle and *Torpedo californica* AChE inhibitions by an OP compound, 0-ethyl S-[2-(diisopropylamino)-ethyl] methylphosphonothioate (MPT), follow second-order irreversible kinetics at low inhibitor concentrations. However, at high concentrations, MPT modifies its own inhibitory properties; the reaction deviates from classical kinetics (Friboulet et al., 1986, 1990). This behavior was explained by a sequential mechanism in which the phosphorylation of the active site serine by MPT induces a conformational change of AChE. In this induced state, the reversible binding of a second MPT molecule to a peripheral site induces the dephosphorylation of the active center. Ligands which are known to interact with an anionic peripheral site of AChE (propidium, gallamine, d-tubocurarine) compete with the binding of MPT to the peripheral site.

Back-titration curves of propidium-AChE complex by MPT are biphasic. First, an increase in fluorescence occurs, without modification of the propidium dissociation constant. At higher concentrations, MPT displaces propidium from its peripheral site. According to the proposed model, the increase in fluorescence is explained by a conformational change of AChE concomitant to its phosphorylation, followed by the competition of MPT with propidium binding.

The study of the influence of different OP and alkyl sulfonate compounds on the propidium-AChE complex fluorescence allows a better characterization of chemical and steric interactions involved both in the conformational modifications of the enzyme and in the relationships between the propidium pheripheral site and the pheripheral locus of OP association.

References

Friboulet, A., Goudou, D., Rieger, F., *Neurochem. Int.*, **1986**, *9*, 323-328.

Friboulet, A., Rieger, F., Goudou, D., Amitai, G., Taylor, P., *Biochemistry*, **1990**, *29*, 914-920.

Determination of the Rate Constants of the Phosphonylation of Purified Serum Cholinesterase from Man, Dog and Pig by the four Isomers of Soman

H.C.J.V. De Bisschop, Royal Military Academy, Brussels, Belgium
C.W. Michiels and L.B.C.Vlaminck, Technical Division of the Army, Vilvoorde, Belgium
S.O.Vansteenkiste, Laboratory for Organic Chemistry, State University of Ghent, Belgium

The nerve agent soman (1,2,2-trimethylpropyl methylphosphonofluoridate) has two asymmetric centers and, therefore, consists of four stereo-isomers. As a result of soman poisoning butyrylcholinesterase (serum cholinesterase, ChE, EC 3.1.1.8.) is inhibited by the formation of a covalent bond between the enzyme and soman (phosphonylation). This communication reports on the stereoselectivity of this reaction.

The four isomers are isolated on a sub-microgram-scale according to a method modified from Benschop et al. (1984). BuChE is 400-600 fold purified from human, canine and porcine serum in essentially three steps: ion exchange chromatography (on DEAE sepharose-6B), affinity chromatography (with N,N-dimethyl ethylenediamine ligand, bound to activated CH-sepharose 4B) and ultrafiltration. The BuChE concentration is expressed as a concentration of soman binding sites (De Bisschop et al.,1987).

Free soman isomer is assayed in a reaction mixture and second-order rate constants are determined (De Bisschop et al., 1987).

Table 1 shows the results. BuChE from the three sera is highly selective towards the C(+)P(-)-isomer. The P(+)-isomers show less affinity towards BuChE of human and canine serum. The overall stereoselectivity is less marked for the reaction with canine BuChE.

References

Benschop, H.P.,Konings, CAG,Van Genderen, J. and De Jong, L.P.A. *Toxicol. Appl. Pharmacol.* **72**, *61-74, 1984.*

De Bisschop, H.C.J.V., De Meerleer W.A.P. and Willems, J.L. *Biochem. Pharmacol.,* **36**, *3587-3591, 1987.*

Table 1: Second-order phosphonylation rate constants (10^5.mol^{-1}.l .s^{-1}) of serum BuChE and soman isomers at pH 7.4 and 25°C (standard deviation in brackets)

Species	C(+)P(-)	C(-)P(-)	C(-)P(+)	C(+)P(+)
man	6.5 (1.5)	0.60 (0.27)	0.055 (0.018)	0.015 (0.060)
dog	1.2 (0.3)	0.11 (0.04)	0.067 (0.023)	0.045 (0.020)
pig	8.8 (1.7)	1.20 (0.50)		

Human Serum Esterases: Differentiation of EDTA-insensitive Enzymes

Zoran Radić *, Elizabeta Pavković and Elsa Reiner, Institute for Medical Research and Occupational Health, University of Zagreb, P.O. Box 291, 41001 Zagreb, Croatia, Yugoslavia

It has been shown earlier that beta-naphthyl acetate (BNA), phenyl acetate (PA) and paraoxon are each hydrolysed by two enzymes in human sera: one enzyme inhibited by EDTA and the other insensitive to EDTA (Reiner et al., 1989a and 1989b).The question was asked whether the EDTA-insensitive enzyme(s) are paraoxonase (EC 3.1.8.1) , arylesterase (EC 3.1.1.2), cholinesterase (EC 3.1.1.8) or carboxylesterase (EC 3.1.1.1). The hydrolysis of paraoxon is by definition due to paraoxonase.

Experiments were done in Tris/HCl buffer (0.1 M, pH 7.4) at 37 °C. The substrate concentrations were 1.0 mM BNA, 5.0 mM PA, 1.0 mM PTCh (propionylthiocholine) and 1.0 mM ATCh (acetylthiocholine). EDTA was 1.0 mM.

The hydrolysis of BNA, PA and PTCh was measured in serum samples from 82 individuals: BNA and PA in presence of EDTA, and PTCh in absence of EDTA. The correlation coefficients between activities of BNA vs PTCh, PA vs PTCh and BNA vs PA were 0.50, 0.52 and 0.72, respectively. The correlation coefficient between activities of PTCh vs ATCh was 0.95 (measured in serum samples from 237 individuals).

The EDTA-insensitive hydrolysis of PA and BNA was tested for inhibition by tabun (550 nM), iso-OMPA (11 μM), BNPP (110 μM) and eserine (110 nM). Experiments were done on serum samples from six individuals. The inhibition by tabun, iso-OMPA and eserine was time-dependent thus excluding paraoxonase and arylesterase as possible enzymes. The time course of inhibition by all three inhibitors deviated slightly from linearity. The rate constant of inhibition by eserine or iso-OMPA was identical, and for tabun similar, to the rate constant of inhibition measured with PTCh/ATCh as substrate. The carboxylesterase inhibitor BNPP did not inhibit the hydrolysis of BNA and PA.

The results concerning progressive inhibition indicate that the EDTA-insensitive hydrolysis of BNA and PA is primarily due to cholinesterase. However, cholinesterase cannot be the only enzyme involved in hydrolysis because the time course of inhibition deviates from linearity and because the correlation coefficients between activities of BNA, PA and PTCh are much lower than one expects for a single enzyme (cf. ATCh vs PTCh).

References

Reiner,E.,Radić,Z.,Pavković,E.,Simeon V.,*Biol.Chem.Hoppe-Seyler*,**1989**a, *370*, 987-988.

*Enzymes Hydrolysing Organophosphorus Compounds;*Reiner, E., Aldridge,W.N., Hoskin, F.C.G., Eds.; Ellis Horwood Ltd: Chichester, UK, 1989b.

0–8412–2008–5/91/0288$06.00/0

The Structure-Anticholinesterase Activity Relationships for New Carbamoyloximes. QSAR Study

G.A. Shataeva, V.B. Sokolov, V.I. Fetisov, Y. Y. Ivanov, T.A. Epishina, N.V. Kovaleva, I.V. Martynov, Institute of Physiologically Active Substances, Academy of Sciences USSR, I42432 Chernogolovka, Moskow region, USSR

A series of new earlier unknown O-carbamoyl alkylchloroformoximes have been prepared and examined for anticholinesterase activity, alkaline hydrolysis, and toxicity to white mice. Antichlolinesterase activity of the compounds depended both on the structure of the O-carbamoyl alkylchloroformoximes, and the origin of the enzyme.

The interest in carbamoyl and phosphorylated oximes have been arisen because of their high anticholinesterase activity. At the present the oxime carbamated have been studied sufficiently well (Fukuto, 1969; Jones, 1972; Magee, 1982; Mrlina, 1980 a, b, 1989). These investigations considered the changes of substituents in the alcohol (hydroximidate) moiety only. The search for the optimal insecticide among carbamate esters of hydroximino compounds led us to synthesize new unknown 0-carbamoyl alkylchloroformoximes with general formula :

$R_1R_2NC(O)ON=C(Cl)R_3$, WHERE R_1=H, CH_3, C_2H_5, C_6H_5, 2-$CH_3C_6H_4$, 3-$CH_3C_6H_4$, 4-$CH_3C_6H_4$, 3,4-Cl-C_6H_3, 3-Cl,4-F-C_6H_3, 4-Br-C_6H_4; R_2=H, CH_3,C_2H_5; R_3=CH_3, C_2H_5, C_3H_7, i-C_3H_7 (changes both in oxime moiety and carbamate substitution).

The interaction of these compounds with human erythrocytes acetylcholinesterase (AChEh), from the housefly (AChEf) Musca domestica L. and horse serum butyrylcholinesterase (BuChE), the alkaline hydrolysis and acute toxicity to mice (LD_{50}, per os) were studied. The great majority of carbamoyloximes exhibited the highest specificity with respect to BuChE (bimolecular constant of inhibition rate $k_{II} = 9.8x10^3$-$1.8x10^7 M^{-1}$ x min^{-1}). The values of the k_{II} for AChEh and AChEf ranged in $1.1x10^2$-$1.0x10^6$M-1 x min^{-1} and 21.0-$6.5x10^4 M^{-1}$ x min^{-1} respectively. The second order constant of alkaline hydrolysis varied from $1x10^2$ to $1.2x10^5 M^{-1}$ x min^{-1}. The toxicity of carbamoyloximes with alyphatic substituents (LD_{50}=32-530mg/kg) correlated with anti-AChEh activity, where no aromatic oximes were not highly toxic (LD_{50}=500 mg/kg) although they had high anti-AChE activity.

Only fairly good correlation between ability for hydrolysis (K_{OH}) and summary hydrophobicity ($\Sigma\pi$) on the one hand and anit-AChEh and anti-BuChE activity (R=0.603, and 0.5 respectively, for 18 compounds) on the other hand was established. More detailed investigation of "structure activity" relationships in series, where R_1, R_2 and R_3 change, led us to equation which bind steric and electronic properties of substituent R_3 with E_S, σ Taft's, K_{OH}, K_{II} AChE, K_{II}, and BuChE. The series of correlations was found for the case, were R_2, R_3=const and R_1 changed; so, with R_3=CH_3, R_2=H

$$lgK_{OH}=4.311 + *1.495\sigma \text{ (meta)} - *1.769\sigma \text{ (para)}$$
n=5, r=0.990, S=0.147, F=50.3

$$lgK_{II}, AChEh=8.862 + 1.180\Sigma\pi - 1.2371\ lgK_{OH}$$
n=5, r=0,766, S=0.521, F=1.4

$$lgK_{II}, BuChE=1.201 + 0.599\Sigma\pi + 1.431\ lgK_{OH}$$
n=5,r=0.614, S=0.357, F=0.6

References

Fukuto T.R., Metcalf R.L., Jones R.L., Myers R.O., *J.Agr.Food Chem.*,**1969**,*17*,923-930.

Jones R.L., Fukuto T.R., Metcalf R.L., *J.Econ.Entomol.*,**1972**,*65*,28-32.

Mrlina G., Calmon J.-P., *J.Agric.Food. Chem.*,**1980**,*28*,605-609.

Mrlina G., Calmon J.-P.,ibid.,**1980**,*28*,673-675.

Vilarem G., Mrlina G., Calmon J.-P., ibid,**1989**,*37*,173-177.

Magee T.A., In *Insecticide Mode Action*; Coates, J.R., Ed; Academie Press, New York: London,**1982**,p.3,pp.71-100.

0–8412–2008–5/91/0289$06.00/0

Phosphorylated Alkylchloroformoximes as Inhibitors of Mammalian Cholin- and Carboxylesterases. Structure – Activity Relationships

Galina F.Makhaeva, Victoria L.Yankovskaya, Vladimir V. Malygin, Vladimir B.Sokolov, Ivan V.Martynov, Institute of Physiologically Active Substances USSR Academy of Sciences, Chernogolovka, 142432, USSR

The second order rate constants of inhibition of human erythrocyte acetylcholinesterase (AChE), horse serum butyrylcholinesterase (BuChE) and pig liver carboxylesterase (CE) by series of organophosphorus pesticides – O-phosphorylated alkylchloroformoximes I-III were determined

$(R_1O)_2P(O)ON{=}C(R_2)Cl$ (I)

$R_1O(CH_3)P(O)ON{=}C(R_2)Cl$ (II)

$(R_1S)_2P(O)ON{=}C(F)Cl$ (III)

R_1 = Me, Et, Pr, Bu, i-Bu, Am; R_2 = Me, Et, Pr, i-Pr, Bu, CH_2Cl, $CHCl_2$, F (42 compounds)

Phosphates I,III were more inhibitory against BuChE and CE, than against AChE, were as antienzymatic activity of phosphonates II was comparable for all three esterases. With multiple regression analysis in groups I-II, were R_2=const, it was shown that hydrophobic interactions are most important in BuChE and CE inhibition. Substituent R_2 (CH_3, CH_2Cl, $CHCl_2$ or F) determins the kind of dependence of antiAChE activity on alkoxy groups hydrophobicity (Makhaeva, 1986; Shataeva, 1988). The regression analysis of inhibition data with steric, hydrophobic and electronic properties of R_1 and R_2 substituents demonstrated that AChE, BuChE and CE inhibition rate constants correlate strongly with electronegative properties of R_2 substituents (Swain and Lapton F field constants) and suggest the occurence of either charge transfer or polar interactions of the oxime fragment in enzyme active site. AntiAChE activity was found to be also correlated with steric properties of R_1 substituents (Charton E_s^v) and Verloop sterimol constants of R_2. AntiBuChE and antiCE activities were dominantly function of molar refractivity of of R_1 substituents. It is possible to suggest that a high correlation with MR reflects, as it was recently shown for chymotrypsin inhibition (Gupta,1987),"a second type of hydrophobic bonding" in which groups with their surrounding flickering clusters of water are held together in solution without desolvatation playing the major role.

References

Makhaeva G.F., Yankovskaya V.L., Fetisov V.I., Sokolov V.B., Ivanov A.N., Malygin V.V., Martynov I.V. *In Molecular Basis of Neural Function*, Tuček S., Štípec S., Šťastny F., Křivánek J., Eds.; ISSN *0862-0385;* Eur. Soc. Neurochem. : Prague, **1986**, p.159.

Shataeva G.A, Makhaeva G.F., Yankovskaya V.L., Sokolov V.B., Ivanov A.N, Martynov I.V. , *J. Evol. Biochem. Phisiol.* (Russ), **1988**, *24*, p.791-796.

Gupta S.P. , *Chem. Rev.*, **1987**, *87*, p.1183-1253

Some Peculiarities of Inhibition of Cholinesterase Activity by Organophosphorous Compounds

Agabekyan, R.S., Lubimov, V.S., Institute of Physiologically Active Substances, Academy of Science of USSR, Chernogolovka, Moscow Region, 142432, USSR.

The results of kinetic experiments on human erythrocyte acetylcholinesterase and horse serum butyrylcholinesterase inactivation by organophosphorous inhibitors (OPI) of different structure were analyzed and computer simulation of kinetic curves was made. It was found that the process of irreversible inhibition of cholinesterase activity by the OPI containing hydrophobic substituents can be described as two-phase, one including an inactivating step owing to conformational changes of the enzymes.

For explanation of the concentration dependence of the bimolecular inhibition constants (Aldridge and Reiner, 1972) and (Brestkin et al., 1971) introduce a notion of "anomalous enzyme-inhibitor complex" is assumed, which does not transform into phosphorylated enzyme. However this mechanism does not principally differ from the one proposed by (Main, 1964). We analyzed the data from steady-state kinetic experiments for more than 200 OPI which have different groups at the phosphorous atom and in the leaving part of the inhibitor molecule. The detailed measurements of the kinetic parameters of the acetylcholine hydrolysis by the enzyme which was preincubated with the OPI, provide basis to support that the irreversible inhibition proceeds as a two-phase process with the inhibitor having a large hydrophobic group. The more hydrophobic a substituent, the more obvious is the stage of conformational change of the enzyme. On the basis of these facts, some kinetic schemes were proposed for investigating this process. The computer simulation of kinetic curves offers the possibility to determine the fast and slow stages of a reaction and to investigate the dependence of effective kinetics of inhibition and of pre-exponential factors from the inhibitor concentration. We had the good agreement between the theoretical and experimental concentration dependence of the above parameters. This work provides the basis to propose the more general scheme of inhibition of chloinesterase activity by the OPI. This scheme includes both irreversible phosphorylation of esterase sites of cholinesterases and inactivation of enzymes by conformational changes of enzyme-inhibitor complex due to the interaction with the second molecule of the inhibitor.

$$E + I \underset{}{\overset{K_i}{\rightleftharpoons}} EI \xrightarrow{k_p} E_p$$

$$EI \xrightarrow{k_{in}} EII$$

References

Aldridge W.N., Reiner E.,Enzyme Inhibitors as Substrates; In *Interaction of Esterases with Esters of Organophosphorous and Carbamic Acids;* Frontiers of Biology, North-Holland, Amsterdam,**1972**,26.

Brestkin A.P., Brik L.I., Volkova R.I., Godovikov N.N., Cabachnik M.I., Kardanov N.A.,*Doklady AN SSSR (russ)*, **1971**,*200*,103-106.

Main A.R., *Science*,**1964**,*144*,3621,992-993.

Kinetic Anomalies in Cholinesterase Inactivation by Some S-Benzhydrylic Esters of Monothioacids of Phosphorous

R.S.Agabekyan, V.V.Kuusk, O.V.Antonova, Institute of Physiologically Active Substances, Academy of Science of USSR, Chernogolovka, Moscow Region, 142432, USSR

The temperature dependencies for processes of human erythrocyte acetylcholinesterase and horse serum butyrylcholinesterase inactivation by reversible inhibitor (S-benzhydryl(diisopropyl)thiophosphinate, I) and irreversible one (S-benzhydryl(diphenyl)thiophosphate, II) have been investigated with steady-state kinetic methods. Vant Hoff's plots for the temperature dependencies of the enzyme-inhibitor-I complex dissociation constants were found to have fracture at nearly 20 C In addition the Arhenius plots for temperature dependencies of bimolecular rate constants k_{II} ($k_{II}=k_{in}/K_i$) of cholinesterase inactivation by inhibitor-II have fracture at the same temperature . The fact that for the different enzymes and for reversible and irreversible inhibitors the fracture of plots were revealed at the same temperature, as well literature evidence (Kochnev I.N. et al., 1986, Zhukovsky A.P. et al.,1986) for temperature dependence of IR-spectra of water and water-protein systems allowed us to propose that the kinetic anomaly mentioned above is connected with the water-induced conformational changes of the enzyme molecules.

References

Kochnev I.N., Vinnitchenko M.B., Smirnova L.V., in: *State of water in various physico-chemical conditions, Molecular physics and biophysics of water systems Issue 6*; Leningrad, SU, **1986**, p.42-52

Zhukovsky A.P., Khaloimov A.I., Rovnov N.V., Raev A.N., in: *State of water in various physico-chemical conditions, Molecular physics and biophysics of water system, Issue 6;* Leningrad, SU, **1986**, p.105-114

The Fluorocontaining Derivatives of α-Aminoalkylphosphonates a New Type of Cholinesterase Inhibitors

Kuusk V.V., Agabekyan R.S., Morozova I.V., Kovaleva N.V., Rasdolsky A.N., Sokolov V.B., Aksinenko A.Y., Fetisov V.I. and Martynov I.V., Institute of Physiologically Active Substances, Academy of Science of USSR, Chernogolovka, Moscow Region, 142432, USSR

A series of O,O-diethyl-1-(N-α-hydrohexafluoroisobutyryl) aminoalkylphosphonates (APh) has been synthesized and their interaction with human erythrocyte acetylcholinesterase (AChE) and with horse serum butyrylcholinesterase (BuChE) has been studied. It was found that the most of APhs inactivated the both ChEs irreversibly. On the basis of QSAR models of APhs-ChE binding it was concluded that 1) hydrophobic interactions play an dominant role in APh-AChE binding; 2) on the surface of ChE there two APh binding sites.

Eight APhs were synthesized.

$$(EtO)_2P(O)-C(R^1)(R^2)-N(R^3)-C(O)-CH(CF_3)_2$$

$R^1 = H, CH_3$
$R^2 = CH_3, C_2H_5, n\text{-}C_4H_9, i\text{-}C_4H_9, C_6H_5$
$R^3 = H, CH_3$

It was shown that only one APh ($R = C_6H_5$) inactivate ChE reversibly whereas its binding constant may be described within QSAR models for all eight APhs. Nonlinear binding constant dependence on hydrophobic parameters of substituent (π) (Fig.1) indicate that there are two binding sites on the surface of ChE (Kubinyi, 1976).

For BuChE there is no dependence on π. For this reason in QSAR models we use additional steric parameters (E_s) which were calculated by modified method (Galkin et al., 1981) with consideration of two variants of reaction center position (P-atom and C-atom in -NR -CO- group). Conformations of APhs were obtained from MM2 calculations.

References

Galkin V.I., Cherkasov R.A., *Reaczionnaya sposobnost organicheskich soedineny* (russ), SU,**1981**,*18*,111-132.

Kubinyi H., *Arzneimittel-Forsch.*,**1986**,*26*,1991-1997.

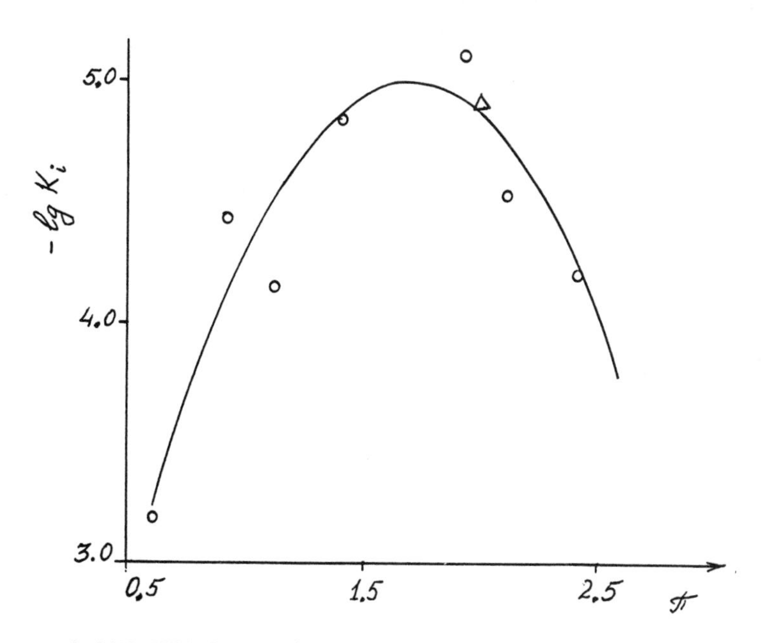

Multiple QSAR Models for Organophosphorous Inhibitors of Cholinesterase

Raevsky O.A., Chistyakov V.V., Agabekyan R.S., Sapegin A.M., Zefirov N.S.,
Institute of Physiologically Active Substances, Academy of Sciences of USSR, Chernogolovka, Moscow Region, 142432,USSR

Results of computer modeling of relationships between structure and anticholinesterase activity of organophosphorus compounds are presented. Compounds were grouped in clusters using an original method developed by the authors, and for most of clusters, quantitative relationships were obtained.

A large number of organophosphorous inhibitors of cholinesterase were analyzed for possible quantitative relationships between structure descriptors and kinetic constants (of their logarithms) using an original discrete regression model method developed by the authors (Raevsky, 1986, Sapegin, 1987). We used experimental data (Agabekyan, 1977, 1990) and literature data on kinetic constants of inhibition of human erythrocyte acetylcholinesterase (EC 3.1.1.7) and horse serum butyrylcholinesterase (EC 3.1.1.8) for 240 phosphoryl and thiophosphoryl compounds. Of these, 147 were irreversible and 93 were reversible inhibitors. To describe of their structure, 43 topological, informational, physicochemical descriptors such as molar refractivity, partition coefficients, molecular volume, etc., were used. Despite such detailed description, the discrimination between reversible and irreversible inhibitors was not possible without the consideration of energy factors. Inclusion of donor-acceptor factors in descriptions allowed grouping of compounds into structurally conjugated clusters using various procedures of cluster and discriminant analysis. At the stage of quantitative analysis, multiple regression techniques were used to obtain statistically verified equations for nearly all clusters of the model.

Results

1. The important role of donor-acceptor factors was revealed, while commonly-used physicochemical factors were shown to be less influential.
2. A trial connecting this fact with mechanism of biological action suggested that clusters were formed according to specificity of action.
3. Results of investigations not only disclosed some peculiarities in structure and action but also gave new criteria for search strategies based on the type of activity.

References

Agabekyan R.S., Berchamov M.C., Kabachnik M.J., *Izvestia AN SSSR ser.chim.* (russ),**1977**,*8*,1868-1872.
Agabekyan R.S., Pegova Z.K., Suynduykova V.K., Vilchevskaya V.D., Godovikov N.N., *Metallorg.Chim.* (russ),**1990**,*3*,275-280.
Raevsky O.A., in *QSAR in Drug Design*, Hadzi D., Ed.; Proceedings of the Sixth European Symposium on QSAR, Yugoslavia, **1986**, Vol.10,pp31-36
Sapegin A.M., Razdolsky A.S., Chistyakov V.V., Raevsky O.A., *Chim.Pharm.J.* (russ),**1987**, *21*,1341.

Stereospecificity of the active centers in different cholinesterases

Galina M. Grigorjeva, Tatyana I.Krasnova, Sechenov Institute of Evolutionary Physiology and Biochemistry of the Academy of Sciences of the USSR, pr. Thoreza, 44, Leningrad, 194223, U.S.S.R.

Stereospecificity of cholinesterase (ChE) from nervous ganglia of cockroach *Periplaneta americana,* acetylcholinesterase (AChE) of bovine erythrocytes and butyrylcholinesterase (BuChE) of horse serum was studied using stereoisomers of organophosphorus inhibitors (OPI) with asymmetric phosphorus atom, $iC_3H_7O(CH_3)P^*(O)S(CH_2)SC_2H_5$ (1) and $iC_3H_7O(CH_3)P^*(O)S(CH_2)_2CH_3$ (2). Kinetics of the enzyme inhibition (25°C, 0.05 M phosphate buffer, pH 7.5) was analyzed according to the biphasic scheme of reaction (Main, 1964):

$$E+I \xrightarrow{K_a} EI \xrightarrow{k_p} EI'$$

It was shown that much greater selectivity in the action of (-)isomers of both OPI as compared to (+)isomers on insect ChE and AChE was observed. No marked differences were discerned between isomers in their binding to BuChE. Selectivity of the action of (-)isomers on enzyme was displayed mainly in the affinity constant K_a, the effect being the most pronounced for the insect ChE, particularly with (-)isomer of (1); the phosphorylation rate constants k_p were similar. In the case of AChE, the differences in the K_a values of the isomers of (1) and (2) were much lower comparing to those of insect ChE; there was found also the clear-cut distinction between the k_p values. Thus, the stereospecificity of the AChE active center reveals not only in the EI complex formation but also in the phosphorylation stage.

References

Main A.R. *Science* **1964,** *144,* 992-993

Onchidal: A Naturally Occurring Irreversible Inhibitor of Acetylcholinesterase With a Novel Mechanism of Action

Stewart N. Abramson*, Zoran Radic*, Denise Manker#, D. John Faulkner#, and Palmer Taylor*. Department of Pharmacology* and Scripps Institution of Oceanography#, University of California-San Diego, La Jolla, California, 92093, USA.

Onchidal is the major lipid-soluble component of the defensive secretion of the mollusc *Onchidella binneyi*, and it has been proposed as the compound responsible for the chemical protection of *Onchidella*. In support of this hypothesis, onchidal was found in several different species of *Onchidella* and it was found toxic to fish. Since onchidal is an acetate ester similar to acetylcholine, its ability to interact with cholinergic proteins such as nicotinic acetylcholine receptors and acetylcholinesterase was investigated. Onchidal did not inhibit the binding of (^{125}I)α-bungarotoxin to nicotinic acetylcholine receptors. However, onchidal produced a progressive, time dependent, inhibition of the activity of acetylcholinesterase. This inhibition of acetycholinesterase was irreversible, in that enzyme activity could not be recovered even after a 10,000-fold dilution of onchidal. Kinetic analysis revealed that the apparent affinity of onchidal for the initial reversible binding to acetylcholinesterase (K_d) was approximately 300 μM, and the apparent rate constant for the subsequent irreversible inhibition of enzyme activity (k_{inact}) was approximately 0.1 min^{-1}. A sensitive assay for acetate revealed that onchidal can be utilized as a substrate by acetylcholinesterase. Approximately 3250 moles of onchidal were hydrolyzed by acetylcholinesterase per mole of enzyme that was irreversibly inhibited. The calculated k_{cat} for onchidal was therefore 325 min^{-1}. Irreversible inhibition apparently resulted from either onchidal itself or from a reactive intermediate in the enzyme catalyzed hydrolysis of onchidal, rather than from the final hydrolysis products. Irreversible inhibition of enzyme activity was prevented or reduced in magnitude by coincubation with reversible agents that either sterically block (edro-phonium and decamethonium) or allosterically modify (propidium) the acetylcholine binding site. Since enzyme activity inhibited by onchidal could not be regenerated by oxime reactivators, the mechanism of irreversible inhibition does not appear to involve acylation of the active-site serine. Since onchidal contains a potentially reactive α,β-unsaturated aldehyde, irreversible inhibition of acetylcholinesterase may result from formation of a novel covalent bond between the toxin and the enzyme. Thus onchidal may represent a new class of anticholinesterase insecticides and it may be useful in the identification of amino acids that contribute to the binding and hydrolysis of acetylcholine.

References

Ireland, C., Faulkner, D. J. *Bioorganic Chemistry*, **1987**, 7, 125-131.

Abramson, S.N., Radic, Z., Manker, D., Faulkner, D.J., Taylor, P., *Mol. Pharmacol.*, **1989**, 36, 349-354.

0–8412–2008–5/91/0296$06.00/0

Detection of Acetylcholine Using Enzyme Sensor : Application for the Determination of Organophosphorus and Carbamates Compounds

Jean-Louis Marty*, **Nathalie Mionetto** and **Regis Rouillon**, Groupe d'Etudes et de Recherches Appliquées Pluridisciplinaires, U.R.A. C.N.R.S. 461,Université de Perpignan, Chemin de la Passio Vella, 66025 Perpignan, Cedex, France.

Many works have been published on the determination of acetylcholine. Generally the enzyme was immobilized covalently and the activity determined by spectrophotometric, potentiometric or amperometric methods. We developed a two-enzymes system based on amperometric detection of acetylcholine using physical entrapment of acetylcholine esterase and choline oxidase in polyvinylalcohol bearing styrylpyridinium groups (PVA-SbQ).

Characteristics of enzyme sensor

The acetylcholine sensor has to be used between pH 7 and 9. The molar concentration of the buffer must be above 0.025 M. The detection limit is 2 10^{-8} M. A linear relationship between current and acetylcholine can be obtained up to 10^{-3} M depending on the ratio choline oxidase / acetylcholinesterase. After one year in a dry state at 4°C there is little or no loss of activity.

Detection of insecticides

The measurement of carbamates and organophosphorus insecticides is a problem attracting considerable attention. They act as inhibitors of cholinesterases. Through calibration procedure, by measuring the decrease of the current, the concentration of insecticides can be determined. The activity of the immobilized acetylcholine esterase decreases in a time dependent manner and the rate of inhibition is dependent on the insecticide concentration. We can detect 1 nM paraoxon. Based on their inhibition capacity, the compounds studied can be arranged in a series : paraoxon> carbofuran>aldicarb>monocrotophos >parathion methyl>phosphamidon. The sensitivity of the system is dependent on the amount of immobilized acetylcholinesterase. The lowest detection is achieved with the membrane which has the lowest immobilized acetylcholinesterase. We are testing different kinds of cholinesterases; human , bovine, equine and electric eel acetyl and butyryl choline esterases to optimize the response of the sensor.

The enzyme sensor proves to be a good tool for analysis. It is inexpensive and can be used by unskilled operators.

0–8412–2008–5/91/0297$06.00/0

Electrochemical Biosensor for the Microdetection of Choline, Acetylcholine and Choline esterase Activity

R.M.Morelis, A. Heijbel, C.Duret and P.R.Coulet. Laboratoire de Génie enzymatique, UMR 106 CNRS - Université Lyon 1. 43 Bd du 11 novembre 1918. 69622 Villeurbanne Cédex. France.

Biosensors are chemical sensors which incorporate a biological element in a sensing layer associated to a transducer giving an electrical signal depending on the concentration of a target analyte. Enzyme electrodes are a kind of biosensor incorporating an enzymatic membrane and an electrochemical transducer. Such an enzyme electrode was prepared for the determination of choline and acetylcholine by co-immobilizing choline oxidase and acetylcholine esterase on a chemically preactivated polyamide membrane. Detection of enzymatically generated H_2O_2 was performed by amperometric detection . The amperometric transducer consisted of a platinum anode at a potential fixed at + 650 mV *vs.* an Ag / AgCl reference and was plugged into a polarograph type PRGE. The enzymatic membrane was maintained in close contact with the platinum tip by a screw cap.

The minimal concentration of choline or acetylcholine which could be determined was as low as $5x10^{-8}$ M. A linear relationship between the current increase and their concentration could be obtained up to $2x10^{-5}$ M. The biosensor sensitivity for choline and acetyl choline was 2.6 mA.M $^{-1}$ when 2 mg of enzymatic preparation were used for immobilization on the membrane. After 8h. of continuous use, the decrease of electrode sensitivity was no more than 5%; after an overnight storage in 0.1M phosphate buffer / 0.1M KCl (pH 8) at 4°C, the probe still kept 70% of its initial sensitivity. The long term stability of the immobilized enzyme was examined after different periods of storage at -20°C in 0.1M phosphate buffer / 0.1M KCl / 1% BSA / 10% glycerol. The sensitivity of sensors was not affected by a 3-months storage (Morelis & Coulet - *Anal.Chim.Acta* <u>231</u>, (1990), 27-32).

To improve the efficiency of the choline sensor, choline oxidase was purified prior coupling. The enzymatic solution was chromatographied through a column of Sephadex G 200 equilibrated with 0.1 M phosphate buffer (pH 8). 2 ml fractions were collected and assayed for choline oxidase activity by spectrophotometry and protein amount by the measure of absorbance at 280 nm or by the technique of Bradford. The collected fractions (from 108 to 128) were pooled, lyophylized and stored at -20°C. The electrode sensitivity was then 2.7 mA. M $^{-1}$ with 0.127 mg of enzymatic preparation immobilized on the membrane and this sensitivity was still 2.2 mA. M $^{-1}$ after 8 days of continuous use.

Furthermore, with such a choline sensor it is possible to determine esterase activities . For this purpose, calibration curves were obtained using acetylcholine esterase and butyryl choline esterase as enzymes and acetyl choline and butyrylcholine as substrates respectively. The limit of detection was 3 μU.ml $^{-1}$ buyrylcholine esterase activity and the detection range was linear up to 3mU.ml^{-1}. One of our goals is to detect pesticides at a very low level. Preliminary results obtained in our group have shown that for ethyl-parathion, concentrations as low as 10^{-10} M could be detected.

Propidium, a Peripheral Anionic Site Binding Ligand, Retards Aging and Increases Reactivatability of Soman Inhibited AChE in Rat Diaphragm In Vitro

Z. Grubič*, Institute of Pathophysiology, School of Medicine, 61105 Ljubljana, Yugoslavia

Soman-AChE complex ages very fast; reaction rate is especially high at human AChE (half times for rat and human AChEs are 20 and 2 min respectively), which makes efficient oxime therapy almost impossible at soman poisoning. Searching for a good aging effector seems therefore a rational approach to the problem of reactivation of AChE-soman complex. Most of the compounds, which retard aging, belong to the group of bisquaternary compounds, which bind to an allosteric site of AChE; allosteric interaction with phosphonylated active site seems therefore the most probable mechanism of retardation of aging. As reduction of aging rate improves recovery of AChE activity only if it proceeds simultaneously with or prior to reactivation, aging effector should not interact with binding of reactivator to the active site; it should bind exclusively to the peripheral anionic site of AChE and should not interfere with the reactivation process.Propidium, which binds only at the site, peripheral to the active center of the enzyme (Taylor and Lappi, 1975) therefore fulfills at least first of the two conditions. The aim of this work was to test propidium as aging effector and an agent, which might improve reactivatability of AChE-soman complex. Both the effects on aging and on reactivatability were tested in pieces of rat diaphragm in vitro, using procedures described by Hallek and Szinicz (1988) and Grubič and Tomažič (1989).

At low concentration (1 uM), propidium reduces the rate of aging as additional 15% of AChE could be recovered in muscle samples in our conditions. At 0.5 mM concentration this effect could not be observed. Reactivatability of AChE-soman complex was enhanced in the endplate region of rat diaphragm but not in the endplate-free region. At low ionic strenght, percent of reactivatable AChE almost doubles. At 0.5 mM propidium, which did not reduce the aging rate, reactivatability could also not be increased. The observed effects of propidium are more like the effects of decamethonium and differ from the effects of suxamethonium (Hallek and Szinicz, 1988).

References

Grubič, Z., Tomažič A., *Arch. Toxicol.*, **1989,** *63*, 68-71.

Hallek, M., Szinicz, L., *Biochem. Pharmacol.*, **1988,** *37*, 819-825.

Taylor, P., Lappi, S., *Biochemistry*, **1975,** *14*, 1989-1997.

0–8412–2008–5/91/0299$06.00/0

Pretreatment with Atropine Retards Aging and Improves Reactivatabilty of Soman-inhibited AChE in Rat and Human Skeletal Muscle Preparations In Vitro

Z. Grubič,* Institute of Pathophysiology, School of Medicine, 61105 Ljubljana, Yugoslavia
A. Šerko, Institute of Legal Medicine, School of Medicine, 61105 Ljubljana, Yugoslavia

A part of the beneficial action of atropine, given at soman poisoning, might be due to its action on phosphonylated AChE. Retardation of aging of human red cell AChE, phosphonylated by soman, was first observed by Kuhnen et al.(1985) and later confirmed by Van Dongen et al.(1987). Atropine has no effect on reactivation process itself (Van Dongen et al.,1987), however, reduced rate of aging is a mechanism which improves reactivatability of phosphonylated AChE and hence percent activity recovered by subsequent action of oxime reactivator. The aim of the present work was to test in pieces of skeletal muscle the effects of atropine on the rate of aging of AChE-soman complex and the influence of atropine on reactivatability of soman inhibited AChE. As the enzyme species also determines aging and reactivatability of AChE soman complex (Grubič and Tomažič, 1989), both effects were tested in rat and human muscle. The effects of atropine on the rate of aging process were determined by the modified procedure of Hallek and Szinicz (1988);the effect on reactivatability of AChE-soman complex was tested in modified experiments described before (Grubič and Tomažič, 1989). Rat hemidiaphragms (endplate and andplate-free regions) were used as rat and mm. intercostales ext., isolated no later than 5 h after death, as human muscle samples.

Atropine reduces the rate of aging of AChE-soman complex in both rat and human skeletal muscle but only at 0.5 mM concentration, while at 1 uM concentration no such effect could be observed. In absolute terms there is practically no difference between endplate region of rat diphragm and human intercostal muscle,however taking into account that in the former 10% difference means an increase from 38% to 48% while in the latter 9% difference an increase from 2% to 11%, the effect appears relatively better expressed in human muscle.

References

Grubič, Z., Tomažič A., *Arch. Toxicol.*, **1989**, *63*, 68-71.

Hallek, M., Szinicz, L., *Biochem. Pharmacol.*, **1988**, *37*, 819-825.

Kuhnen, H., Schrichten, A., Schoene, K., *Arzneim.-Forsch/Drug Res.*, **1985**, *35*, 1454-1456.

Van Dongen, C.J., Elskamp, R.M., De Jong, L.P.A., *Biochem. Pharmacol.*, **1987**, *36*, 1167-1169.

Mechanism of Action of the Protecting Effect of Combinations of Antidotes in Intoxications with Anticholinesterases

Ingrid Nordgren*, Bo Karlén, Monica Kimland, Lena Palmér and Bo Holmstedt, Department of Toxicology, Karolinska Institutet, Box 60400, S-104 01 Stockholm, Sweden

Diazepam in combination with atropine and an oxime is of great value as prophylaxis and treatment of organophosphorus poisoning. Since the central cholinergic system is involved in organophosphorus poisoning, this prompted us to study the effect of different combinations of these antidotes on the acetylcholine (ACh) synthesizing system in mouse brain in vivo.

Methods

ACh and choline (Ch) were analyzed using gas chromatography - mass spectrometry and deuterated internal standards. Turnover of ACh was studied by following the incorporation of Ch into ACh after an i.v. injection of $[^2H_6]$Ch. The fractional rate constant (K_a) of ACh synthesis was calculated from the specific activities of $[^2H_6]$ACh and $[^2H_6]$Ch 20 and 50 sec after the $[^2H_6]$Ch injection. The mice were killed by focussed microwave irradiation on the head. The antidotes we have studied so far are diazepam, l-hyoscyamine (the active antipode of atropine), and the oxime HI-6 ([[[(4-aminocarbonyl) pyridino] methoxy] methyl]-2-[(hydroxyimino)methyl]-pyridinium dichloride).

Results

l-Hyoscyamine (2mg/kg i.v.) decreases the endogenous level of ACh, presumably due to an increased ACh output. When diazepam (1 mg/kg i.v.) is added to the injection solution, the endogenous ACh levels are normalized. Diazepam separately does not, or very slightly, affect (increase) the ACh level.

l-Hyoscyamine increases K_a and the synthesis rate of ACh, while diazepam has the opposite effect. Interestingly, the combination of diazepam and l-hyoscyamine decreases ACh synthesis to the same extent as diazepam alone. The increased synthesis rate of ACh after l-hyoscyamine is probably an unfavourable feature of a protecting agent. These results may therefore, at least partly, explain the advantage of combining atropine with diazepam in cases of nerve gas poisoning.

HI-6 (25 mg/kg i.v.) induces a marginally increasing effect on the ACh level, and a slight increasing effect on ACh synthesis rate.

When all three treatments are combined in the same injection the result is a slight decrease of the ACh synthesis rate.

The aim of these experiments has been to collect background data on the protective effect of this three-drug combination on soman poisoning and the correlation with ACh dynamics. In mice administered soman 150 µg/kg s.c. (0.5xLD_{50}) after pretreatment with the combination of these three antidotes the most marked effect is a 50% increase of ACh levels. However, K_a is in the same range as in animals administered diazepam.The animals show no cholinergic symptoms. Animals administered the same dose of soman with no antidotal pretreatment suffer from severe tremor and salivation, and in addition to elevated ACh levels K_a is very much lowered.

0–8412–2008–5/91/0301$06.00/0

Protection By Butyrylcholinesterase Against Soman Poisoning

D.E. Lenz, C.A. Broomfield, D.M. Maxwell, R.P. Solana, A.V. Finger and C.L. Woodard, U.S. Army Medical Research Institute of Chemical Defense, Aberdeen Proving Ground, MD, USA 21010-5425
S. McMaster, United States Environmental Protection Agency, Washington, DC, USA 20460.

We have studied the properties of the naturally occurring enzyme, butyrylcholinesterase (BuChE) as a biological scavenger for toxic organophosphorus (OP) compounds in guinea pigs and rhesus monkeys. The monkeys were trained to perform a cognitive task so to evaluate any debilitating side effects from the enzyme or the OP after protection.

Guinea Pigs (300-350g; male) were given 4000 I.U. of equine BuChE (Sigma Chemical) or 1000 I.U. human BuChE (gift of Dr. O. Lockridge) iv 1 hr before receiving soman. The median lethal dose (LD_{50}) for soman in guinea pigs was 28 ug/kg (controls) and 45.6 ug/kg and 56.1 ug/kg in animals receiving human and equine BuChE respectively.

Equine BuChE (25,000 I.U.; 460 nmol) was administered iv to rhesus monkeys which had been trained to perform a Serial Probe Recognition task. Soman or sarin was administered iv one hour later at the doses indicated in the Table. To study the pharmacokinetics of BuChE, blood (1.5 ml) was drawn at selected time intervals from control animals and from animals that received soman or sarin and the BuChE activity was determined.

Serum from monkeys treated with BuChE was tested for the presence of antibodies against equine BuChE by standard ELISA. After three injections of BuChE the serum tested positive for anti-BuChE antibodies.

BuChE provided protection against a lethal dose of soman or sarin in two mammalian species. The enzyme produced no behavioral side effects, but since the enzyme source was not from the same species, antibodies developed after multiple injections. This in turn altered the pharmacokinetics of the enzyme in circulation (Fig. 1,2). In rhesus monkeys, the pharmacokinetics of BuChE were such that protection should be possible for up to 13-14 hr after enzyme administration (Fig. 1). Our results indicate that the in vivo properties of BuChE, (1) decreased OP toxicity (Table), (2) elimination $t_{1/2}$ of 620 hr (Fig. 1), and (3) lack of behavioral side effects, are requisite for a biological scavenger to protect against the physiological and behavioral toxicity of OP poisons (See also Maxwell et al. and Wolfe, et al., These proceedings).

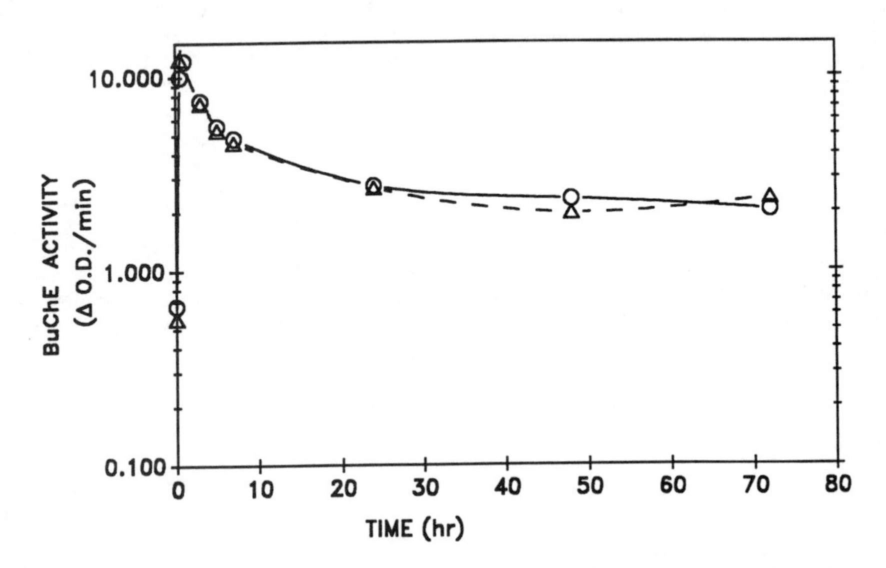

Fig 1. Pharmacokinetic profile of BuChE in Rhesus Monkeys. Monkeys given 25,000 I.U. of equine-BuChE, iv. Pharmacokinetic constants were estimated using PCNONLIN (two compartment model): $k_{el} = 0.0011\ hr^{-1}$; half time = 620 hr; $C_{max} = 12.5$ (O.D./min).

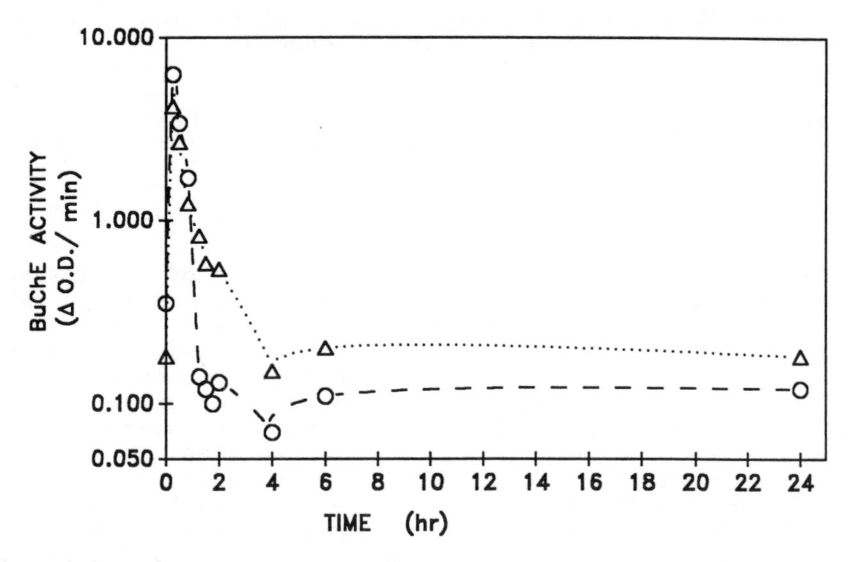

Fig 2. Comparison of pharmacokinetics of BuChE in Rhesus Monkeys with (O) or without (Δ) soman (460 nmol) given one hr after BuChE. Antibodies for BuChE were detected in both animals. Pharmacokinetic constants were estimated using PNONLIN (one compartment model): $k_{el} = 2.0\ hr^{-1}$; half time = 0.35 hr; $C_{max} = 6.6$ (O.D./min).

Table 1

BuChE Protection Against Soman or Sarin in Rhesus Monkeys

SPECIES	OP	ENZYME	CHALLENGE
Rhesus Monkey	Soman	460 nmol	340 nmol (4 LD_{50})
Rhesus Monkey	Soman	460 nmol	253 nmol (3 LD_{50})
Rhesus Monkey	Sarin	460 nmol	280 nmol (1 LD_{50})
Rhesus Monkey	Soman	460 nmol	140 nmol (1.6 LD_{50})

Human Serum Cholinesterase Phenotypes: Interaction with Some Organophosphorus, Carbamate and Oxime Compounds

Vera Simeon*, Mira Škrinjarić-Špoljar, Andjelka Buntić, Institute for Medical Research and Occupational Health, University of Zagreb, P.O. Box 291, 41001 Zagreb, Croatia, Yugoslavia

The interaction of human serum cholinesterase phenotypes, usual and atypical, with the organophosphates (paraoxon, phosphostigmine, sarin, soman, tabun and VX), a dimethylcarbamate (Ro 02-0683) and two oximes (HI-6 and PAM-2) was studied. The time course of progressive inhibition by the organophosphates and the carbamate, the reversible inhibition by the oximes as well as the reactivation of the phosphylated and carbamylated enzymes were followed by measuring the residual activity with acetylthiocholine or propionylthiocholine as substrates. The charged progressive inhibitors (VX, phosphostigmine and Ro 02-0683) inhibited faster (22 to 165 times) the usuaL than atypical serum cholinesterase. No difference in the inhibition rate constants of usual and atypical cholinesterases was found when the inhibitors were uncharged (paraoxon, sarin, soman, tabun). The charged mono-oximes, HI-6 and PAM-2 were stronger inhibitors (two times) of the usual than of the atypical cholinesterase. This was shown in the reversible inhibition of the substrate hydrolysis as well as in the protection of the enzymes by HI-6 against progressive inhibition by paraoxon, soman and tabun. The usual and atypical serum cholinesterases inhibited either by paraoxon, sarin or VX did not reactivate spontaneously; the rates of spontaneous reactivation of dimethylcarbamylated enzymes were found to be the same. The usual cholinesterase inhibited by paraoxon, sarin or VX was more refractive to the reactivation by the oximes than the atypical enzyme. The results confirmed the difference between the usual and atypical serum cholinesterases in the affinity towards charged compounds reported by other authors (Kalow and Davies, 1958; Lockridge and La Du, 1978; Valentino et al., 1981).

References

Kalow, W., Davies, R.O., *Biochem. Pharmacol.*, **1958**, *1*, 183-192.

Lockridge, O., La Du, B.N., *J. Biol. Chem.*, **1978**, *253*, 361-366.

Valentino, R.J., Lockridge, O., Eckerson, H.W., La Du, B.N., *Biochem. Pharmacol.*, **1981**, *30*, 1643-1649.

0-8412-2008-5/91/0304$06.00/0

Pyrydinium Oximes: Reaction with Cholinesterase Substrates

Mira Škrinjarić-Špoljar*, Zoran Radić, Vera Simeon, Institute for Medical Research and Occupational Health, University of Zagreb, P.O. Box 291, 41001 Zagreb, Croatia, Yugoslavia

Oximes are used as antidotes in intoxication with organophosphorus compounds. The reaction of oximes with cholinesterase substrates is therefore of special interest. All oximes tested so far were known to react with acetylthiocholine (Bergmann and Govrin, 1973; Simeon et al., 1981; Matsubara and Harikashi, 1985; Škrinjarić-Špoljar et al., 1988).

This communication presents results obtained on reactions of pyrydinium monooximes PAM-2 and HI-6 and dioximes LüH6 and TMB-4 with acetylcholine (AcCh) and acetylthiocholine (AcTCh). Reactions were studied in 0.1 M phosphate buffer pH 7.4 at 37 °C and followed spectrophotometrically by measuring the decrease in oxime concentration at its maximum absorbance. AcCh did not react with any of the tested oximes while AcTCh reacted with all four oximes. Reactions with AcTCh were also measured by the release of thiocholine (TCh) with the thiol reagent DTNB. In the experiments with the AcTCh concentration over the oxime concentration the rate constants for the initial reaction were between 17 and 25 L mol^{-1} min^{-1} and the total amount of produced TCh was higher than the initial oxime concentration: about 1.4-fold for monooximes and 4-fold for dioximes. In the experiments with the oxime concentration over the AcTCh concentration the rate constants were between 34 and 36 L mol^{-1} min^{-1} and the total amount of TCh was equal to the initial amount of AcTCh. The obtained results show that in dioxime both oxime groups react with AcTCh and indicate on the occurence of side reactions producing TCh.

References

Bergmann, F., Govrin, H., *Biochemie,* **1973**, *55,* 515-520.

Simeon, V., Radić, Z., Reiner, E., *Croat. Chem. Acta,* **1981**, *54,* 473-480.

Matsubara, T., Harikashi, I., *Jap. J. Pharmacol.*, **1985**, *39,* 336-340.

Škrinjarić-Špoljar, M., Simeon, V., Reiner, E., Krauthacker, B., *Acta Pharm. Jugosl.*, **1988**, *38,* 101-109.

Esterases as Organophosphate Scavengers

A.D. Wolfe, Y. Ashani and B.P. Doctor. Walter Reed Army Institute of Research, Washington, D.C., 20307-5100.

D.M. Maxwell, U.S. Army Medical Research Institute of Chemical Defense, Aberdeen Proving Ground, MD., 21010-5425.

Organophosphorus (OP) chemical warfare agents react rapidly and irreversibly with neural acetylcholinesterase (AChE, EC 3.1.1.7), thereby causing convulsions, paralysis, and death. These effects have proven difficult to prevent, despite the use of drug formulations which generally contain a cholinesterase (ChE) reactivator, an antimuscarinic, and an anticonvulsant. It has, therefore, proven desirable to evaluate alternative pretreatment modalities, i.e., enzymes which react rapidly with OPs to remove them from circulation or destroy them before they reach their target. The properties which these enzymes should possess include a high reaction rate with and specificity towards OPs, a long circulatory half-life, be symptom-free upon administration, and provide symptom-free protection without performance decrement. The candidates for such a pretreatment modality were fetal bovine serum AChE (FBS-AChE; Ralston, 1985), human butyrylcholinesterase (BuChE, EC 3.1.1.8), and carboxylesterases (CaEs, EC 3.1.1.1). Prior to use as an OP scavenger, diisopropylphosphonofluoridate, 7-(methylethoxyphosphinyloxy)-1-methyl quinolinium iodide (MEPQ; Levy and Ashani, 1986), and pinacolyl methylphosphonofluoridate (soman) were shown to titrate FBS-AChE in vitro and/or in vivo with bimolecular rate constants typical of AChE-OP interaction. (Wolfe, 1987; Raveh, 1989). Mice administered FBS-AChE were protected stoichiometrically from multiple median lethal doses of ethyl-S-2-diisopropylaminoethylmethylphosphonothiolate (VX), MEPQ, and soman. BuChE also protected mice stoichiometrically from soman and MEPQ (Ashani et al, 1990). Both enzymes reduced OP toxicity, maintained viability, prevented overt OP induced side effects, and possessed circulatory half-lives in excess of 20 hours. CaEs act as endogenous OP scavengers and are preferentially inhibited by cresylbenzodioxaphosphorin oxide (CBDP; Boskovic, 1979; Clement, 1984; Maxwell, 1987). Use of CBDP and FBS-AChE jointly protected mice stoichiometrically from 6.5 median lethal doses of soman (Wolfe et al, 1988). Non-human primates administered FBS-AChE were protected against soman and MEPQ without performance decrement. Thus ChEs have been found to be safe, effective single pretreatment agents which maintain neurological integrity against highly toxic OPs.

References

Ashani, Y., Shapira, S., Levy, D., Wolfe, A.D., Doctor, B.P., and Raveh, L. Biochem. Pharm., **1990,** in press.

Boskovic, B. Arch. Toxicol., **1979,** 42, 207-216.

Clement, J.G. Biochem. Pharm., **1984,** 33, 3807-3811.

Levy, D. and Ashani, Y. Biochem. Pharm. **1986,** 35, 1079-1085.

Maxwell, D.M., Brecht, K.M., and O'Neill, B.L. Toxicol. Lttrs., **1987,** 39, 35-42.

Ralston, J.S., Rush, R.S., Doctor, B.P. and Wolfe, A.D. J. Biol. Chem., **1985,** 260, 4312-4318.

Raveh, L., Ashani, Y., Levy, D., De La Hoz, D., Wolfe, A.D., and Doctor, B. P. Biochem. Pharm., **1989,** 38, 529-534.

Wolfe, A.D., Maxwell, D.M., Doctor, B. P., and Y. Ashani. FASEB J. 2, A791 **1988.**

Wolfe, A.D., Rush, R.S., Doctor, B.P., Koplovitz, I. and Jones, D. Fund. Appl. Toxicol., **1987,** 9, 266-270.

0–8412–2008–5/91/0306$06.00/0

Hydrolysis of Nerve Agents by Plasma Enzymes

C. A. Broomfield and K. W Ford

U. S. Army Medical Research Institute of Chemical Defense,
Aberdeen Proving Ground, Maryland 21010-5425

Although much progress has been made in protection against the nerve agents, it is apparent that we are approaching the limiting efficacy for pharmacological intervention with the drugs now in development. Any significant increase of protection without debilitating side effects must come as a result of new approaches. One such approach is the prophylactic administration of scavengers capable of detoxifying nerve agents before they are able to reach their critical targets. The feasibility of the use of certain enzymes administered exogenously has been demonstrated (Maxwell, et al, Lenz, et al, Wolfe, et al, These proceedings). While those scavengers offer some protection against the nerve agents, including soman, they have in common the disadvantage that they are high molecular weight and react stoichiometrically 1:1 with the OP toxin. Thus, a half gram of the scavenger must be kept active in the blood stream for each mg of OP to be neutralized. It would be preferable to use an effective hydrolytic catalyst so that a fairly small amount of protein could destroy larger amounts of nerve agent. We recently became aware of significant progress in studies of a class of OPA hydrolases, referred to as paraoxonases. One group at the University of Michigan (La Du and Novais, 1989) has classified two populations of human serum paraoxonase as type "A" or "B" and has purified the enzyme from each population type as well as a heterozygous "AB" type. Dr. LaDu generously provided samples of each of his enzymes and we studied the kinetics of hydrolysis of sarin, soman and tabun in 1 M NaCl and 1 mM $CaCl_2$. Unfortunately, the hydrolysis of tabun by all of the enzymes was too slow to measure accurately and therefore kinetic constants could not be obtained. The results are summarized in Table 1. Since the spontaneous hydrolysis rate of soman is very fast at pH 8.5 the constants reported here were measured at pH 7.4. Most of the OPA hydrolases studied previously have shown a decided preference for the P(+) isomers of soman. Furthermore, soman appears not to be completely hydrolyzed, indicating that one or more of the stereoisomers might be resistant to hydrolysis by these enzymes. Therefore, we felt that it was important to determine if any of these enzymes have a chiral preference. This was done by an enzyme inhibition assay described previously (Ray, et al, 1988). None of the enzymes tested displayed a preference for the nontoxic isomers of soman (Table 1). This experiment has been repeated several times. Soman is definitely detoxified by these enzymes. These results suggest that enzymes such as these might be used as protection against OP toxins.

Literature Cited

La Du, B.N. and Novais, J. In *Enzymes Hydrolyzing Organophosphorus Compounds* Reiner, E., Aldridge, W.N. and Hoskin, F.C.G, Eds. Ellis Horwood, Chichester, England 1989.

Ray, R., Boucher, L.J., Broomfield, C.A., Lenz, D.E. *Biochem Biophys Acta*, **1988**, 967, 373-381.

TABLE 1. Hydrolysis of Sarin and Soman by OPA Hydrolases

ENZYME	SARIN		SOMAN		
	K_M (mM)	V_{max}[1]	K_M (mM)	V_{max}[1]	P(-)[2]
Human A	0.21	69	0.41	82	+
Human B	0.31	21	0.25	31	+
Human AB	0.15	24	0.39	36	+

[1] Units for V_{max} = micromoles/minute/milligram protein
[2] + indicates loss of AChE inhibiting activity by soman

Plasma cholinesterase-variation among greenhouse workers, spraying organophosphate- and carbamate insecticides

Flemming Lander, MD, **Kaj Hinke**, MD. Department of Occupational Medicine Odense University Hospital, Denmark

The aim of the study is to describe the ChE-variation in greenhouse workers, spraying with organophosphate- (OP) and carbamate insecticides.

MATERIAL AND METHODS.
The study was designed as a cross-sectional study, measuring the ChE-activity twice among greenhouse workers in the spraying season (August and September, 1988). The baseline blood samples were drawn January 1989. Fifty nurserygardens were selected including 126 subjects.

RESULTS.
The study shows that the changes in the mean ChE-activity in September, but not in August, was significantly lower compared to the baseline value (p=0.001). The decline was only significant for those spraying OP-insecticides.

The change in the ChE-activity declines according to increasing grades of personal protections, but only significantly at time of first sample (table 1).

DISCUSSION.
The present study showed that the wearing of protective clothing when spraying, seems to offer the workers some protection against dermal uptake. In the intervals between application, the greenhouse workers described, had substantial upper extremity dermal contact with flowers contaminated with residues of insecticides. The significant reduction in ChE activity at the time of second measurement are probably caused by uptake of OP-insecticides from these sources, because no dose-effect could be detected.

TABLE 1

The change in the mean ChE-activity ($ChE_{january}$ - $ChE_{in\ season}$) in greenhouse workers spraying with organophophate insecticides more than once a month according to personal protective clothing

	In season					
	1. sample			2. sample		
Personal protection	N	ChE	SD	N	ChE	SD
No protective clothing	13	0,49	0,54*	13	0,35	0,86
Only a rubber apron	6	0,17	0,67	6	0,47	0,67
Whole body clothing	25	-0,08	0,61	26	0,15	0,63
	44	0,12	0,64	45	0,25	0,70

* F = 3,76 p = 0,03 r = 0,39

0–8412–2008–5/91/0308$06.00/0

PHARMACOLOGICAL UTILIZATION OF ANTICHOLINESTERASE AGENTS: NEUROPATHOLOGY OF CHOLINERGIC SYSTEMS

Cholinesterase Molecular Forms in Neurological Diseases

Zoltan Rakonczay, Central Research Laboratory, Albert Szent-Györgyi Medical University, Szeged, Hungary

The significance of changes in cholinesterases (ChE) seen in many pathological states is still unknown (for review see Rakonczay, 1988). Considerable progress in both acetylcholinesterase (AChE) and butyrylcholinesterase (BuChE) research has been made and several authors report changes in both enzyme activity and distribution of molecular forms in different human diseases. In this paper, based on the results of our own and other laboratories, we summarize recent knowledge on the alteration of ChE molecular forms in neurological diseases.

Alzheimer's Disease (AD):

Different molecular forms of AChE in postmortem cortical tissue from AD (Atack et al., 1983; Fishman et al., 1986) and aged matched controls have recently been measured. Selective loss of G_4 (10S), a detergent solubilized form of the enzyme, has been demonstrated. In the diseased cortex the ratio of G_4/G_1 forms was markedly decreased. In addition, 342 and 406% elevation were noted in asymmetric A_8 and A_{12} AChE (Younkin et al., 1986), however, they represent only minor forms. It is interesting to note that during the development of human fetal brain, a relative increase in the specific activity of membrane-associated G_4 is present (Zakut et al., 1985).

We measured the level of AChE in lumbar cerebrospinal fluid (CSF), including its molecular forms, to investigate whether or not the loss of G_4 in the central nervous system could be detected peripherally. The CSF AChE activity of normal controls increased continuously as a function of age. Five different molecular forms (G_1, G_2, G_4, A_8 and A_{12}) were detected in the CSF by density gradient centrifugation. No significant differences were found in the AChE activities of the lumbar CSF from subjects with different severities of AD and age-matched controls (Table 1). The distribution of CSF AChE molecular forms did not change during aging in normal subjects and AD patients (Table 2). However, recently a statistically significant decline in CSF AChE was found with advancing AD (Elble et al., 1989).

Table 1
CSF AChE Activity in Alzheimer Disease and Amyotrophic Lateral Sclerosis

Disease	Ages (y)	AChE nmol/min/ml	n
Control	71.0 ± 6.2	13.4 ± 3.4	(6)
AD1	72.2 ± 5.9	12.3 ± 1.8	(8)
AD2	73.8 ± 6.2	10.9 ± 3.2	(4)
AD3	74.3 ± 7.4	13.0 ± 0.4	(3)
ALS		11.7 ± 2.8	(11)

AD1 = mild; AD2 = medium; AD3 = severe

Pick's Disease:

Postmortem brains from cases with Pick's disease did not change significantly in AChE activity (Yates et al., 1980a).

Parkinson's Disease (PD):

The molecular forms of AChE and BuChE were studied in the frontal cortex (grey and white matter), postmortem and in CSF of demented patients with PD and compared to controls and non-demented parkinsonians (Perry et al., 1985; Ruberg et al., 1986). In frontal cortex, AChE activity decreased significantly in both demented and non-demented parkinsonian subjects compared to controls. The G_4 form of the enzyme was lower in demented parkinsonians

0–8412–2008–5/91/0310$06.00/0

Table 2

Distribution of CSF AChE Molecular Forms in Alzheimer Disease and Amyotrophic Lateral Sclerosis

		AChE Molecular Forms (%)				
Disease	n	G_1	G_2	G_4	A_8	A_{12}
Control	n=6	15.3 ± 5.9	36.9 ± 13.6	51.0 ± 14.8	0.89 ± 1.3	2.5 ± 2.2
AD1	n=6	15.1 ± 5.3	27.6 ± 16.9	52.8 ± 13.7	6.1 ± 5.5	4.8 ± 5.1
AD2	n=4	15.4 ± 3.2	27.8 ± 10.1	52.0 ± 7.0	2.9 ± 2.1	1.5 ± 1.0
AD3	n=2	9.5	32.0	46.9	7.6	2.9
ALS	n=10	13.9 ± 4.6	15.7 ± 5.3^b	64.7 ± 8.2	5.8 ± 2.5^c	4.7 ± 1.3^a

[a] $p < 0.05$; [b] $p < 0.01$; [c] $p < 0.002$; Student's t-test
AD1 = mild; AD2 = medium; AD3 = severe

than in the non-demented subjects. No decreases in AChE or BuChE activity were observed in the CSF of patients studied (Ruberg et al., 1986).

Huntington's Disease (HD):

The AChE level was normal in several brain areas sampled from postmortem cases (McGeer and McGeer, 1976).

Down's Syndrome (DS):

In parallel with choline acetyltransferase, AChE activity was markedly reduced in seven brain areas from DS cases (Yates et al., 1980b).

Neural Tube Defects (NTD):

Neural tube defects have been associated with a large increase of AChE activity in the amniotic fluid of abnormal fetus which probably originated from the developing CNS. In polyacrylamide gel electrophoresis a faster moving AChE band appears which is associated with the increase of G_4 molecular form separated by density gradient centrifugation (Rakonczay, 1988).

Brain Tumors:

High levels of ChE activity have been detected in human primary brain tumors of different tissue origin, including glioblastomas and meningiomas. In meningiomas, only the G_1 (4S) form was observed whereas both gliomas and normal forebrain samples the major form is G_4 (Razon et al., 1984).

Olivopontocerebellar Atrophy (OPCA):

Mean AChE activity was significantly reduced in cerebral (-51% to -65%) and cerebellar (-47%) cortex with less severe change (-37%) in hippocampus (Kish et al., 1988).

Guillain-Barré Syndrome (GBS):

Similar to the electrophoretic pattern of AChE in NTD, Guibaud and coworkers (1982) found two bands in polyacrylamide gel from CSF samples of GBS subjects.

Hirschsprung's Disease:

In Hirschsprung's disease an accumulation of AChE can be seen with histochemistry in the nerve fibers of the aganglionic megacolon. In parallel with this phenomenon, a large increase of the membrane-bound G_4 molecular form can be detected in the affected bowel (Rakonczay and Németh, 1984). Therefore, this form is probably associated with axons.

Paroxysmal Nocturnal Hemoglobinuria (PNH):

In PNH, a chronic hemolytic disorder, the AChE (G_2 form) is missing in one population of erythrocytes (Brimijoin et al., 1986). The AChE deficit in abnormal erythrocyte may depend on deficiency of glycan-inositol-phospholipid-anchor proteins which are required to bind the enzyme to the membrane (Ueda et al., 1990).

Muscle Diseases:

The complete disappearance of heavy and medium forms of AChE was seen in 60% of myopathic non-dystrophic patients (Poiana et al., 1986). This pattern was never observed in biopsies from neurogenic and dystrophic patients or from controls.

Motor Neurone Degenerative Disease:

In muscular dystrophy biopsies (Fernandez et al., 1986) or CSF samples from amyotrophic lateral sclerosis both globular (G_2) and asymmetric (A_8 and A_{12}) forms are severely affected (Table 2). We were unable to detect changes in total CSF AChE activity (Table 1).

Conclusion:

Molecular forms of cholinesterase exhibit a tissue-specific distribution. Each form is presumable functionally adapted to different cellular locations. Changes in ChE activity in different diseases almost always reflects changes in the distribution of molecular forms (Table 3). In AD and PD the ratio G_4/G_1 is decreased. This illustrates the loss of presynaptic AChE in brain areas subsequent to degeneration of nerve cell bodies in distant areas. The deficiency in cholinergic neurons may not be directly reflected in the CSF in AD, PD or ALS. In Hirschsprung's disease the high G_4/G_1 ratio could simply be explained by the increased number of proliferating presynaptic fibers. These results support the idea that presynaptic and postsynaptic compartments may be characterized by different AChE molecular form patterns. In NTD, AChE (mostly G_4) is secreted by CNS into the CSF from the open lesion which gradually accumulates in the amniotic fluid. Human primary brain tumors, especially in some gliomas, exhibit exceptionally high levels of cholinesterases. It is not clear whether differences between gliomas and meningiomas are regulated at the post-translational level, or different mRNA species are responsible for the formation of G_4 and G_2 forms. We do not know whether the above mentioned differences are the cause or the consequence of the pathological states but they may reflect impairments in ChE biosynthesis or degradation at the subcellular level.

Although in most of the cases cholinesterase cannot serve as a tool for diagnostic purposes there are few exceptions. AChE activity measurement and separation of molecular forms either by density gradient ultracentrifugation or electrophoresis together with alpha-fetoprotein assay could be used to screen amniotic fluids from neural tube defects. In Hirschsprung's disease they would be a valuable tool for diagnosis or to determine the length of the aganglionic part of the affected intestine. Fluorescence-activated cell sorting and monoclonal antibodies against AChE could help to separate erythrocyte populations and assess the clinical status of patients.

Finally, it would be interesting to know whether there are any changes in distribution of cholinesterase molecular forms in diseases such as Pick's HD, OPCA, GBS or Down's syndrome. More information is needed in order to assess changes in cholinesterases in different diseases which are essential components of cholinergic synapses.

The author is especially indebted to Professor E. Giacobini for helpful comments and discussion. He is also grateful to Mrs. Diana Smith for typing this manuscript. Supported by the Scientific Research Council Ministry of Health, Hungary (06/4-20/457).

Table 3

Cholinesterase Activity and Molecular Forms in Different Diseases

Disease	enzyme activity AChE	BuChE	molecular forms AChE	BuChE
Alzheimer[1,2,3]	↓ (-10 to -60%)	?	↓ G_4, ↑ A_8, ↑ A_{12}	?
Pick[4]	no change	no change	?	?
Parkinson[2,5]				
demented	↓ (-50%)	no change	↓ G_4	?
non-demented	↓ (-42%)	↓ (-42%)	no change	↓ G_4
Huntington[6]	no change	no change	?	?
Down[7]	↓ (-43 to -82%)	?	?	?
Neural tube defect[8]				
(anencephal)	↑ (+817)	?	↑ G_4	?
Brain tumors[9]				
glioma	↑ (+114 to +479)	↑(+177 to +1505)	no change	?
meningioma	no change	no change	↓ G_4	no change
Olivopontocerebellar atrophy[10]	↓ (-37 to -65%)	?	?	?
Guillain-Barré syndrome[11]	↑ CSF	?	?	?
Hirschsprung[12]				
(intestine)	↑ (+640%)	↓ (-40%)	↑ G_4	no change
Myopathy[13]				
non-dystrophic	no change	no change	↓ G_1, ↓ G_4, ↓ A_{12}	?
neurgenic	no change	no change	no change	?
dystrophic	no change	no change	no change	?
Amyotrophic lateral sclerosis[14]				
muscle	↓ (-65%)	?	↓ G_1, ↓ G_4, ↓ A_{12}	?
CSF	no change	?	↓ G_2, ↓ A_8, ↓ A_{12}	?
Paroxysmal nocturnal hemoglobinuria[15]	↓ (-23 to -35%)	-	↓ G_2	-

References: [1]Atack et al. (1983); [2]Perry et al. (1985); [3]Younkin et al. (1986); [4]Yates et al. (1980a); [5]Ruberg et al. 1986); [6]McGeer and McGeer (1976); [7]Yates et al. (1980b); [8]Rakonczay (1988); [9]Razon et al. (1984); [10]Kish et al. 1988; [11]Guibaud et al. (1982); [12]Rakonczay and Németh (1984); [13]Poiana et al. (1986); [14]Fernandez et al. (1986); [15]Brimijoin et al. (1986)

References

Atack, J.R., Perry, E.K., Bonham, J.R., Perry, R.H., Tomlinson, B.E., Blessed, G., Fairbairn, A., *Neurosci. Lett.*, **1983**, 40, 199-204.

Brimijoin, S., Hammond, P.I., Petitt, R.M., *Mayo Clin. Proc.*, **1986**, 61, 522-529.

Elble, R., Giacobini, E., Higgins, C., *Neurobiol. Aging*, **1989**, 10, 45-50.

Fernandez, H.L., Stiles, J.R., Donoso, J.A., *Muscle Nerve*, **1986**, 9, 399-406.

Fishman, E.B., Siek, G.C., MacCallum, R.D., Bird, E.D., Volicer, L., Marquis, J.K., *Ann. Neurol.*, **1986**, 19, 246-252.

Guibaud, S., Simplot, A., Mercatello, A., *Lancet*, **1982**, 2, 1456.

Kish, S.J., Schut, L., Simmons, J. Gilbert, J., Chang, L., Rebbetoy, M., *J. Neurol. Neurosurg. Psychiat.*, **1988**, 51, 544-548.

McGeer, P.L., McGeer, E.G., *J. Neurochem.*, **1976**, 26, 65-76.

Perry, E.K., Curtis, M., Dick, D.J., Candy, J.M., Atack, J.R., Bloxham, C.A., Blessed, G., Fairbarn, A., Tomlinson, B.E., Perry, R.H., *J. Neurol. Neurosurg. Psychiat.*, **1985**, 48, 413-421.

Poiana, G., Leone, F., Longstaff, A., Scarsella, G., Biagioni, S., *Neurochem. Int.*, **1986**, 9, 239-245.

Rakonczay, Z., *Prog. Neurobiol.*, **1988**, 31, 311-330.

Rakonczay, Z., Németh, P., *J. Neurochem.*, **1984**, 43, 1194-1196.

Razon, N., Soreq, H., Roth, E., Bartal, A., Silman, I., *Exp. Neurol.*, **1984**, 84, 681-695.

Ruberg, M., Rieger, F., Villageois, A., Bonnet, A.M., Agid, Y., *Brain Res.*, **1986**, 362, 83-91.

Ueda, E., Kinoshita, T., Terasawa, T., Shichishima, T., Yawata, Y., Inoue, K., Kitani, T., *Blood*, **1990**, 75, 762-769.

Yates, C.M., Simpson, J., Maloney, A.F.J., Gordon, A., *J. Neurol. Sci.*, **1980a**, 48, 257-263.

Yates, C.M., Simpson, J., Maloney, A.F.J., Gordon, A., Reid, A.H., *Lancet*, **1980b**, 2, 979.

Younkin, S.G., Goodridge, B., Katz, J., Lockett, G., Nafzinger, D., Usiak, M.F., Younkin, L.H., *Fed. Proc.*, **1986**, 45, 2982-2988.

Zakut, H., Matzkel, A., Schejter, E., Avni, A., Soreq, H., *J. Neurochem.*, **1985**, 45, 382-389.

Aspects of Acetylcholinesterase in Human Neurodegenerative Disorders, Particularly Alzheimer's Disease

John R. Atack, Merck Sharp & Dohme, Neuroscience Research Centre, Terlings Park, Eastwick Road, Harlow, Essex CM20 2QR, United Kingdom

Cerebrospinal fluid (CSF) acetylcholinesterase (AChE) activity has been studied as a possible diagnostic index of the loss of cortical cholinergic innervation in Alzheimer's disease (AD). However, reductions in activity in AD are only modest (16%), presumably because a number of regions of the CNS not involved in AD also contribute to CSF AChE. Consequently, CSF AChE activity seems to be of limited diagnostic use in AD. Inhibition of AChE by physostigmine has been used as a treatment strategy for AD. Limited success in these trials may be related to poor pharmacokinetic properties of physostigmine. Novel physostigmine analogues with improved pharmacokinetic and pharmacodynamic properties may prove more beneficial in the treatment of AD.

Alzheimer's disease (AD) is the leading cause of senile dementia accounting for over 50% of all dementia cases. It has been estimated that about 2 million people suffer from AD in the United States alone, at an annual cost for institutional care of over $25 billion (Katzman, 1986). There is, however, no reliable premortem diagnostic test for AD (Hollander et al, 1986); clinical diagnosis of probable AD is made by eliminating all other possible causes of dementia and definitive diagnosis can only be made on the basis of abundant senile plaques and neurofibrillary tangles observed during microscopic examination of the postmortem brain. Furthermore, there is currently no effective treatment to reverse the symptoms of AD (Bagne et al, 1986).

Considerable effort has been devoted to the characterisation of the neurochemical changes that occur in postmortem brain samples of AD patients. The most reliable deficits observed to date are seen in the cholinergic projection systems from the basal forebrain to the cortex (Coyle et al, 1983; Candy et al, 1986). More specifically, there is a degeneration of the cholinergic projections originating in the nucleus basalis of Meynert (NbM) and innervating the neocortex. It should be noted, however, that not all central cholinergic systems degenerate in AD. For example, striatal cholinergic neurons are relatively spared by the disease process. In view of the cortical cholinergic deficit in AD, acetylcholinesterase (AChE) has aroused considerable interest as a possible antemortem diagnostic tool in AD cerebrospinal fluid (CSF) and as a target for therapeutic enhancement. In the present paper, the selective loss of the G_4 form of cortical AChE will be reviewed and the use of CSF AChE as a diagnostic marker and anticholinesterases as a therapy for AD discussed.

Molecular Forms of AChE in AD

A number of reports have described a loss of the G_4 form of AChE in AD (Fig. 1a: Atack et al, 1983; Fishman et al, 1986; Younkin et al, 1986). A similar loss of cortical G_4 AChE is also seen in demented Parkinson's disease patients (Perry et al, 1985). The G_4 form of AChE is the most abundant molecular species of AChE within the CNS (Atack et al, 1986), and its functional importance is further emphasised by observations that there is a selective increase in the abundance of this form during development (Muller et al, 1985; Zakut et al, 1985; Perry et al, 1986).

The loss of the cortical G_4 form of AChE in AD is similar to the loss of G_4 AChE from rat hippocampus following lesions of the cholinergic input from septal neurons (Fig. 1b) and in neocortex following lesions of the rat substantia innominata (Bisso et al, 1988). These results are consistent with the G_4 form of AChE being relatively selective for presynaptic cholinergic nerve terminals (Marquis and Fishman, 1985).

In contrast to the cortex, the distribution of molecular forms in the NbM in AD is similar to that seen in normal brain tissue (Fig. 2). This suggests that the assembly of the G_4 form of AChE in the remaining NbM neurons is relatively normal - it is only the reduced number of these neurons (which presumably transport the G_4 form to the neocortex) that results in the loss of cortical G_4 AChE.

CSF AChE: An Antemortem Marker in AD?

A number of studies examining AChE activity in CSF have been reported, yet the data has been inconsistent (Giacobini, 1986).

0–8412–2008–5/91/0315$06.00/0

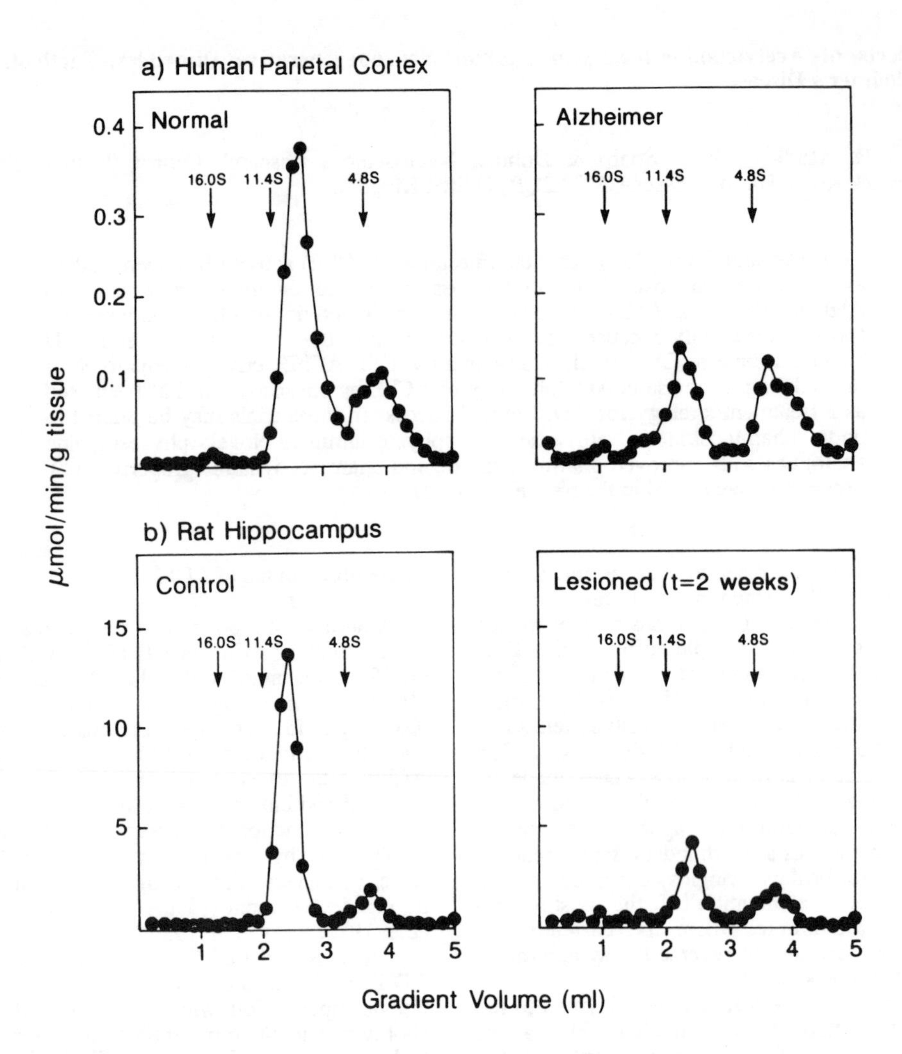

Fig. 1. Distribution of AChE molecular forms in a) normal and Alzheimer's disease (AD) parietal cortex and b) hippocampus of sham-operated and fimbria-fornix lesioned rats. The selective loss of the G4 form of AChE in AD is similar to that seen in the hippocampus of rats with lesions of the fimbria-fornix (which carries cholinergic fibres from cell bodies in the septal nuclei to the hippocampus). These results suggest that the G4 form of AChE in the CNS may be preferentially associated with presynaptic cholinergic nerve terminals and loss of the G4 form in AD is consistent with a degeneration of cortical presynaptic nerve terminals. (Adapted from Atack et al, 1983, 1987.)

However, studies of CSF are complicated the fact that a number of parameters influence CSF measurements. For example, drugs, time of day of lumbar puncture, diet, environment (inpatient or outpatient) and fraction of CSF assayed can all influence lumbar CSF neurochemical levels (Atack, 1989). Furthermore, the selection of appropriate age-matched, control subjects (preferably healthy normal volunteers rather than patient controls) is crucial. However, it is not always possible to select control subjects and consistently sample lumbar CSF and this probably, at least to a certain extent, accounts for some of the inconsistencies in the literature.

Fortunately, we have been able to obtain lumbar CSF from age-matched healthy normal volunteers and patients with probable AD (diagnosed according to established guide-lines; McKhann et al, 1984) under strictly controlled conditions (e.g. lumbar punctures performed with subjects drug-free for at least 2 weeks and following overnight bedrest and a controlled

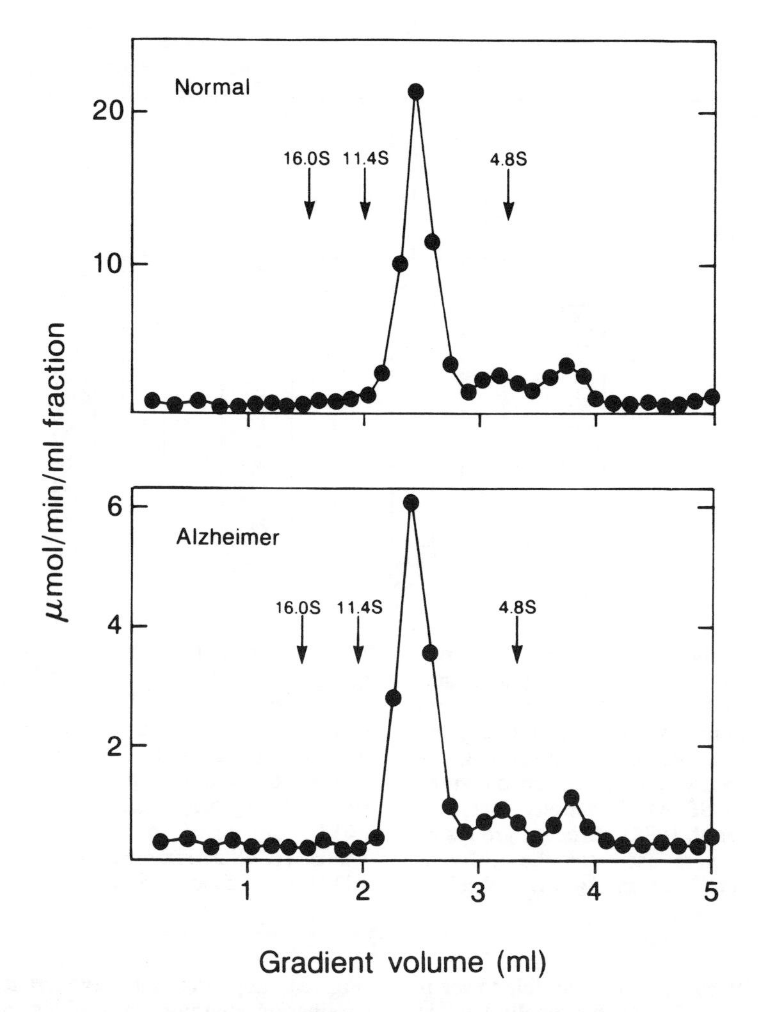

Fig. 2. Distribution of AChE molecular forms in the nucleus basalis of Meynert in normal and Alzheimer's disease (AD) brain tissue. Micropunches of tissue were obtained (Perry et al, 1984) and homogenised and subjected to density gradient centrifugation as described elsewhere (Atack et al, 1986). Although total AChE activity was lower in AD compared to normal tissue, the distribution of AChE molecular forms was similar. The presence of the G_1, G_2 and G_4 forms is noteworthy since it is consistent with the sequential assembly of the G_4 form from the monomer and dimer prior to anterograde transport to the cerebral cortex. These results indicate that in AD, despite loss of cholinergic cell bodies in this region, the remaining cell bodies have relatively normal metabolism of AChE molecular forms.

diet for the preceding 3 days). Under these conditions, we were able to observe a significant, albeit relatively modest, 16% decrement in lumbar CSF AChE activity in AD (Fig. 3). However, there was no correlation between the severity of dementia and CSF AChE activity, nor was there any tendency for CSF AChE activity to decrease with progression of the disease (Atack et al, 1988).

With respect to the longitudinal study of AChE, the period over which CSF AChE activity was measured (18 ± 7 months) was relatively short in comparison to the duration of AD (up to 10-15 years). Consequently, alterations in CSF AChE activity over this relatively short period might be too small to be reliably detected. We therefore measured CSF AChE activity in young and old Down's syndrome (DS) subjects since virtually all old DS subjects (older than 35-40 years) acquire the neuropathological changes characteristic of AD. Therefore, by comparing CSF AChE activity in

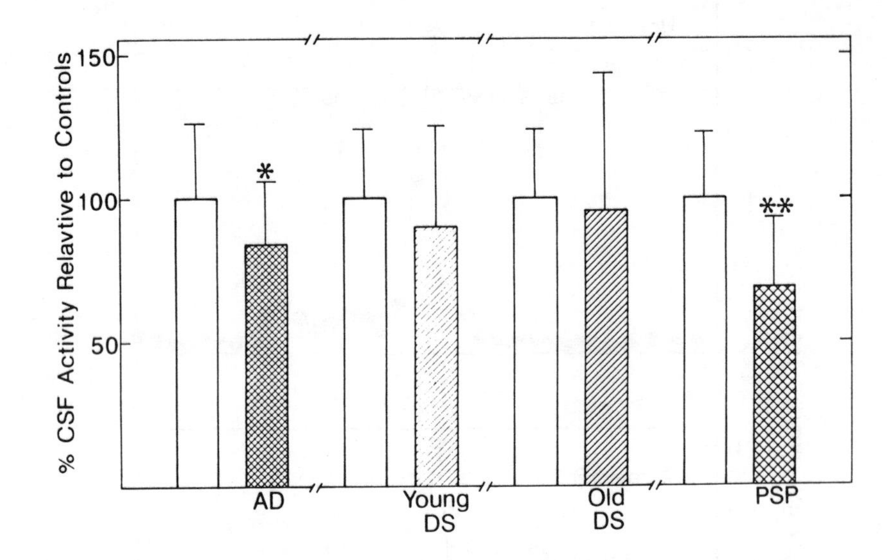

Fig. 3. AChE activity in lumbar CSF from Alzheimer's disease (AD) patients, young and old Down's syndrome (DS) subjects and individuals with progressive supranuclear palsy (PSP). Values are expressed as a percentage (± SD) of activity observed in age-matched control subjects. (Since CSF AChE activity increases with age (Atack et al, 1988), different groups of control subjects were used for the different patient populations.) Mean (± SD) absolute AChE activities (nmol/min/ml) for each patient group (with values for age-matched controls in parentheses) were AD, 18.0 ± 4.8, n=37 (control, 21.5 ± 5.6, n=13); young DS, 15.0 ± 5.8, n=11 (control, 16.7 ± 4.0, n=17); old DS, 19.4 ± 9.6, n=18 (control, 20.3 ± 4.8, n=13); and PSP, 14.7 ± 5.1, n=11 (control, 21.3 ± 4.8, n=18). Compared to their respective control values, AChE activity was significantly lower in AD (*, $p<0.05$) and PSP (**, $p<0.002$).

young and old DS subjects it is possible to see if the appearance of AD-type neuropathology is accompanied by reduced CSF AChE activity (in which case CSF AChE activity should be lower in old relative to young DS subjects). As can be seen from Fig. 3, young DS and old DS subjects had CSF AChE activities that did not differ significantly from age-matched controls, suggesting that the appearance of AD-type neuropathology does not, *per se*, result in reduced CSF AChE activity.

To further examine the extent to which lumbar CSF AChE activity reflects degeneration of central cholinergic neurons, we measured CSF AChE activity in patients with progressive supranuclear palsy (PSP), in which there is a degeneration of the brainstem pedunculopontine tegmental cholinergic neurons - a region which projects to many areas of the extrapyramidal system. In these patients, there was a significant reduction of 31% in CSF AChE activity relative to age-matched controls (Fig. 3), implying that degeneration of these cholinergic neurons in PSP is reflected by reduced CSF AChE activity. Therefore, it is reasonable to assume that a number of cholinergic neuronal systems within the brain contribute to AChE activity measured within lumbar CSF. Indeed, AChE has been shown to be secreted *in vivo* from a number of regions of the brain, for example the caudate nucleus, cerebellum, hippocampus, hypothalamus and substantia nigra (Greenfield, 1985; Appleyard and Smith, 1985; De Sarno et al, 1987; Appleyard et al, 1988; Taylor et al, 1989).

It would therefore appear that in AD, the modest decrement in lumbar CSF AChE activity may be due to the fact that reduced secretion of cortical AChE into CSF is masked by relatively normal secretion from areas of the brain and spinal cord not involved in the pathogenesis of AD. Consequently, lumbar CSF AChE activity does not reliably reflect the loss of cortical AChE in AD.

Anticholinesterases in the Treatment of AD

In an attempt to enhance the residual cortical innervation to the cortex in AD (in a manner analogous to the enhancement of the dopaminergic nigrostriatal pathway in Parkinson's disease), three approaches have been used. First, dietary choline or lecithin has been used in an attempt to increase choline concentrations in the brain and thereby increase the synthesis of acetylcholine. Second, agonists, for example arecoline, have been employed to directly stimulate muscarinic receptors, the numbers of which remain unchanged in AD. Third, and most relevant to the present discussion, anticholinesterases have been administered to inhibit AChE and thereby prolong the actions of the residual cortical cholinergic innervation.

With regards to this latter approach, the most widely used anticholinesterase has been physostigmine, although more recently tetrahydroaminoacridine (THA) has also received considerable attention (Summers et al, 1986). Physostigmine has proved particularly attractive for the treatment of AD since it has been in use clinically since 1877, when it was first used in the treatment of glaucoma (Karczmar, 1970). Furthermore, in normal subjects, physostigmine administration results in a reversal of the dementia-like symptoms induced by scopolamine (Drachman, 1977).

The treatment of AD patients with physostigmine has, unfortunately, proved largely disappointing, with little, if any, benefit being seen (Becker and Giacobini, 1988). However, the underlying hypothesis, namely inhibition of AChE in AD can prolong the action of residual acetylcholine and thereby improve aspects of cognitive and memory function, remains to be proven. Thus, very little is known of the pharmacokinetics and pharmacodynamics of physostigmine administration in humans and it has not previously been shown whether physostigmine penetrates the blood-brain barrier and produces inhibition of central AChE. Consequently, lack of clinical effects could be due to an insufficient degree of central inhibition of AChE.

In order to address this issue, we have recently measured CSF AChE activity before and after physostigmine administration in PSP as an index of the degree to which physostigmine enters the CNS and inhibits brain (and presumably CSF) AChE activity (Litvan et al, 1989). We found that using a conventional oral dosing schedule (0.5 to 2.0 mg/dose, 6 times a day) we could not observe significant inhibition of CSF AChE activity, indicating that using this type of regime, there was negligible penetration of physostigmine into the CNS. This failure to produce significant inhibition of central AChE activity may be related to the short plasma half life of this compound in humans (about 30 min; Gibson et al, 1985; Sharpless and Thal, 1985; Whelpton and Hurst, 1985).

Consequently, it may be preferable to administer physostigmine as a continual i.v. infusion or as a controlled-release formulation (Thal et al, 1989) in order to maintain sustained plasma physostigmine levels. Indeed, using a continual i.v. dosing regime in AD patients (Giuffra et al, 1990), we were able to achieve a significant (21%) inhibition of CSF AChE activity (Atack and Giuffra, unpublished observations), although it should be noted that significant side-effects were also observed. It would therefore appear that physostigmine does indeed penetrate into the CNS and produce inhibition of AChE. However, the doses required produced significant side-effects.

A different approach to the use of physostigmine in the treatment of AD has been the development of novel physostigmine analogues that may possess better pharmacokinetic and pharmacodynamic properties than physostigmine itself. For example, we have synthesized a number of analogues of physostigmine with substituents at either the carbamoyl side chain or at the N(1) position (Atack et al, 1989). We found that a variety of substitutions can be made at both positions whilst retaining potency versus human brain AChE (Fig. 4). It remains to be determined whether these substituents confer improved pharmacokinetic and pharmacodynamic prop-erties *in vivo*. Nevertheless, a similar approach has been used by De Sarno and colleagues who have described a heptyl substitution on the carbamoyl side group which results in a prolonged half-life of the compound (De Sarno et al, 1989).

Future Directions

With regards to the role of AChE in the pathogenesis of AD, it would be interesting to determine whether the degeneration of a number of subcortical nuclei (NbM, locus coeruleus, raphe nucleus) is related in any way to the fact that they all contain AChE (Smith and Cuello, 1984). In addition, AChE has been reported to possess proteolytic activity (Small, 1989) and is associated with senile plaques (Perry et al, 1980; Struble et al, 1982). Since abnormal proteolysis of the amyloid precursor protein results in the deposition of amyloid in senile plaques (Sisodia et al, 1990), an intriguing possibility is that AChE may be involved in some way in the abnormal processing of the amyloid precursor protein.

With respect to the use of anticholinesterases, the development of novel physo-

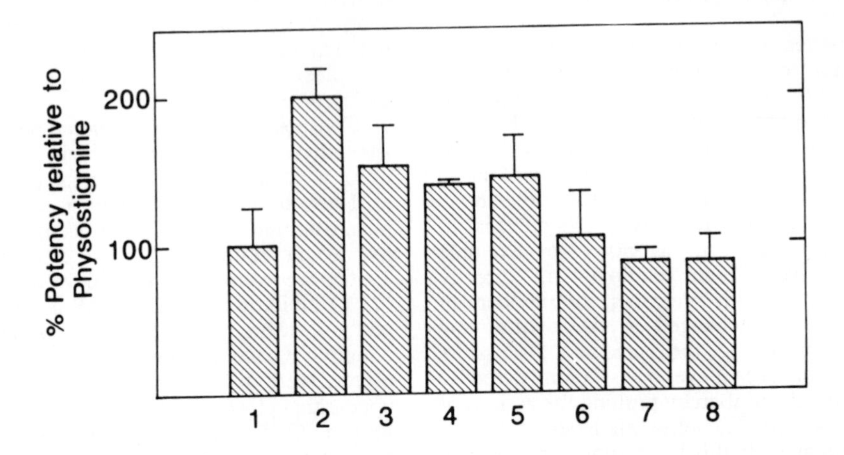

Fig. 4. Potency of a number of novel physostigmine analogues against human cerebral cortex AChE. For each compound, the IC_{50} was determined and expressed as a percentage of the IC_{50} observed for physostigmine. 1=Physostigmine; 2=octylcarbamoyl eseroline; 3=butylcarbamoyl eseroline; 4=N(1)-nor-physostigmine; 5=benzylcarbamoyl eseroline; 6=N(1)-allyl physostigmine; 7=N-phenylcarbamoyl eseroline; and 8=physostigmine methosulphate. Values shown are mean (± SD) of 3-5 separate determinations. The main point to note is that potency can be retained despite changing the structure quite considerably. It is therefore possible that modifications to the structure of physostigmine that confer more favourable pharmacokinetic and pharmacodynamic properties can be made whilst retaining potency. (Adapted from Atack et al, 1989.)

stigmine-based compounds with improved pharmacokinetic and pharmacodynamic properties is encouraging, although it will be a while before such compounds become available for testing in the clinic.

References

Appleyard, M.E., Smith, A.D., *Neurosci. Lett.*, **1985,** *Suppl. 21*, S48.

Appleyard, M.E., Vercher, J.-L., Greenfield, S.A., *Neuroscience*, **1988,** *25*, 133-138.

Atack, J.R., In *Biological Markers of Alzheimer's Disease*; Boller, F., Katzman, R., Rascol, A., Signoret, J.-L., Christen, Y., Eds.; Springer-Verlag: Berlin, 1989; pp 1-16.

Atack, J.R., Perry, E.K., Bonham, J.R., Perry, R.H., Tomlinson, B.E., Blessed, G., Fairbairn, A., *Neurosci. Lett.*, **1983,** *40*, 199-204.

Atack, J.R., Perry, E.K., Bonham, J.R., Candy, J.M., Perry, R.H., *J. Neurochem.*, **1986,** *47*, 263-277.

Atack, J.R., Perry, E.K., Bonham, J.R., Boakes, R., Candy, J.M., *Neurosci. Lett.*, **1987,** *79*, 179-184.

Atack, J.R., May, C., Kaye, J.A., Kay, A.D., Rapoport, S.I., *Ann. Neurol.*, **1988,** *23*, 161-167.

Atack, J.R., Yu, Q.-S., Soncrant, T.T., Brossi, A., Rapoport, S.I., *J. Pharmacol. Exp. Therap.*, **1989,** *249*, 194-202.

Bagne, C.A., Pomara, N., Crook, T., Gershon, S., In Treatment Development Strategies for Alzheimer's Disease; Crook, T., Bartus, R., Ferris, S., Gershon, S., Eds.; Mark Powley: Madison, CT, 1986; pp 585-638.

Becker, R.E., Giacobini, E., *Drug Dev. Res.*, **1988,** *12*, 163-195.

Bisso, G.M., Diana, G., Fortuna, S., Meneguz, A., Michalek, H., *Brain Res.*, **1988,** *449*, 391-394.

Candy, J.M., Perry, E.K., Perry, R.H., Court, J.A., Oakley, A.E., Edwardson, J.A., In *Aging of the Brain and Alzheimer's Disease*; Swaab, D.F., Fliers, E., Mirmiran, M., Van Gool, W.A., Van Haaren, F., Eds.; (Progress in Brain Research, vol.70) Elsevier: Amsterdam, 1986; pp 105-130.

Coyle, J.T., Price, D.L., DeLong, M.R., *Science*, **1983**, *219*, 1184-1190.
De Sarno, P., Giacobini, E., Downen, M., *J. Neurosci. Res.*, **1987**, *18*, 578-590.
De Sarno, P., Pomponi, M., Giacobini, E., Tang, X.C., Williams, E., *Neurochem. Res.*, **1989**, *14*, 971-977.
Drachman, D.A., *Neurology*, **1977**, *27*, 783-790.
Ellman, G.L., Courtney, K.D., Andres, V. Jr., Featherstone, R.M., *Biochem. Pharmacol.*, **1961**, *7*, 88-95.
Fishman, E.B., Siek, G.C., MacCallum, R.D., Bird, E.D., Volicer, L., Marquis J.K., *Ann. Neurol.*, **1986**, *19*, 246-252.
Geneser, F.A., *J. Comp. Neurol.*, **1986**, *254*, 352-368.
Giacobini, E., *Neurobiol. Aging*, **1986**, *7*, 392-396.
Gibson, M., Moore, T., Smith, C.M., Whelpton, R., *Lancet*, **1985**, *I*, 695-696.
Greenfield, S.A., *Neurochem. Int.*, **1985**, *7*, 887-901.
Giuffra, M., Mouradian, M.M., Bammert, J., Claus, J.J., Mohr, E., Ownby, J., Chase, T.N., *Neurology*, **1990**, *40(Suppl.1)*, 229.
Hollander, E., Mohs, R.C., Davis, K.L., *Neurobiol. Aging*, **1986**, *7*, 367-387.
Karczmar, A.G., In *Anticholinesterase agents*; Karczmar, A.G., Ed.; (International Encyclopedia of Pharmacology and Therapeutics, section 13, vol.1) Pergamon Press: Oxford, 1970; pp 1-44.
Katzman, R., *N. Engl. J. Med.*, **1986**, *314*, 964-973.
Litvan, I., Gomez, C., Atack, J.R., Gillespie, M., Kask, A.M., Mouradian, M.M., Chase, T.N., *Ann. Neurol.*, **1989**, *26*, 404-407.
Marquis, J.K., Fishman E.B., *Trends Pharm. Sci.*, **1985**, *6*, 387-388.
McKhann, G., Drachman, D., Folstein, M., Katzman, R., Price, D., Stadlan E.M., *Neurology*, **1984**, *34*, 939-944.
Muller, F., Dumez, Y., Massoulie, J., *Brain Res.*, **1985**, *331*, 295-302.
Perry, R.H., Blessed, G., Perry, E.K., Tomlinson, B.E., *Age Ageing*, **1980**, *9*, 9-16.
Perry, R.H., Candy, J.M., Perry, E.K., Thompson, J., Oakley, A.E., *J. Anat.*, **1984**, *138*, 713-732.
Perry, E.K., Curtis, M., Dick, D.J., Candy, J.M., Atack, J.R., Bloxham, C.A., Blessed, G., Fairbairn, A., Tomlinson, B.E., Perry, R.H., *J. Neurol. Neurosurg. Psychiat.*, **1985**, *48*, 413-421.
Perry, E.K., Smith, C.J., Atack, J.R., Candy, J.M., Johnson, M., Perry, R.H., *J. Neurochem.*, **1986**, *47*, 1262-1269.
Sharpless, N.S., Thal, L.J., *Lancet*, **1985**, *I*, 1397-1398
Sisodia, S.S., Koo, E.H., Beyreuther, K., Unterbeck, A., Price, D.L., *Science*, **1990**, *248*, 492-495.
Small, D.H., *Neuroscience*, **1989**, *29*, 241-249.
Smith, A.D., Cuello, A.C., *Lancet*, **1984**, *I*, 513.
Struble, R.G., Cork, L.C., Whitehouse, P.J., Price, D.L., *Science*, **1982**, *216*, 413-415.
Summers, W.K., Majovski, L.V., Marsh, G.M., Tachiki, K., Kling, A., *N. Engl. J. Med.*, **1986**, *315*, 1241-1245.
Taylor, S.J., Haggblad, J., Greenfield, S.A., *Neurochem. Int.*, **1989**, *15*, 199-205.
Thal, L.J., Lasker, B., Sharpless, N.S., Bobotas, G., Schor, J.M., Nigalye, A., *Arch. Neurol.*, **1989**, *46*, 13.
Whelpton, R., Hurst, P., *N. Engl. J. Med.*, **1985**, *313*, 1293-1294.
Younkin, S.G., Goodridge, B., Katz, J., Lockett, G., Nafziger, D., Usiak, M.F., Younkin, L.H., *Fed. Proc.*, **1986**, *45*, 2982-2988.
Zakut, H., Matzkel, A., Schejter, E., Avni, A., Soreq, H., *J. Neurochem.*, **1985**, *45*, 382-389.

New Cholinesterase Inhibitors for Treatment of Alzheimer Disease

Ezio Giacobini, Department of Pharmacology, Southern Illinois University School of Medicine, Springfield, IL 62794-9230 USA

New Cholinesterase Inhibitors for Treatment of Alzheimer Disease

The finding of a severely damaged and underactive cholinergic system in the brain of Alzheimer disease (AD) patients has led to clinical trials of cholinomimetics including cholinesterase inhibitors (ChEI) (for review see Becker and Giacobini, 1988). Based on available experimental and clinical information (Giacobini and Becker, 1989), an ideal ChEI suitable for symptomatic treatment of memory and cognitive impairment should satisfy the following requirements: a) produce a long-term acetylcholinesterase (AChE) inhibition in brain with a steady-state of increased cortical acetylcholine (ACh); b) not inhibit ACh synthesis or release in nerve endings; c) and produce only mild side effects at therapeutic doses. Such requirements have not been met by the ChEI used so far (Becker and Giacobini, 1988). We have studied a number of new ChEI in our laboratory (Giacobini and Becker, 1989). Based on our experimental results in animals we have proposed two new ChEI for experimental therapy of AD, heptyl-physostigmine (heptyl-Phy) (Fig. 1), a physostigmine (Phy) derivative, and metrifonate (MTF) (Fig. 2), a slow release formulation (Hallak and Giacobini, 1987, 1989; DeSarno et al., 1989; Tang et al., 1989).

Following the results of Summers et al. (1986) with tetrahydroaminoacridine (THA) and the results obtained in animal trials with MTF (Hallak and Giacobini, 1987, 1989), the clinical effects of THA, MTF and Phy have been compared at the Regional Alzheimer Disease Assistance Center on AD patients (Table 1).

Of the three clinical approaches with Phy (oral, i.v. and i.c.v.), followed by us (Table 1, Elble et al., 1988; Giacobini et al., 1988) the intracerebroventricular (i.c.v.) administration seems to be the most effective. Oral and single dose i.v. administrations show only modest and short-lasting improvement in the neuropsychological tests performance (Elble et al., 1988). In addition, side effects such as nausea and dizziness, which limit their use, are common. The low efficacy of the oral and i.v. route is probably related to the short-lasting effect of the drug in the brain and the appearance of toxic effects at doses which show only low (20-25%) butyrylcholinesterase (BuChE) inhibition in plasma. The i.c.v. route produces practically no BuChE inhibition in plasma and no peripheral side effects (Giacobini et al., 1988). On the other hand, the i.c.v. route produces a high percent inhibition (90%) of AChE in CSF and a demonstrable increase of ACh in the CSF. This effect lasts for at least three hours with single doses and is not accompanied by side effects (Giacobini et al., 1988).

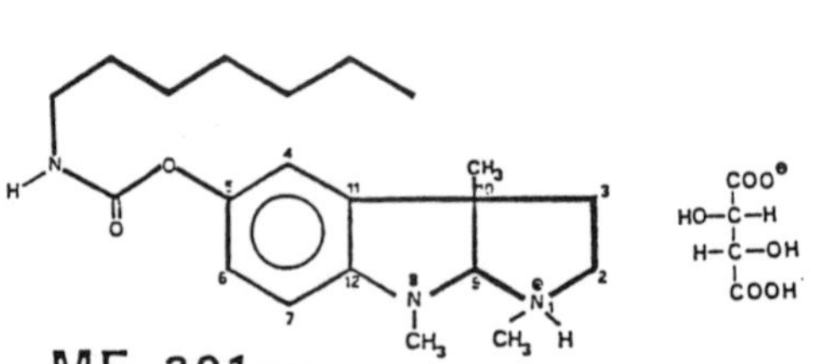

MF 201 tartrate

Fig. 1. Chemical structure of heptyl-physostigmine [MF-201, heptastigmine (Mediolanum, Milan, Italy)].

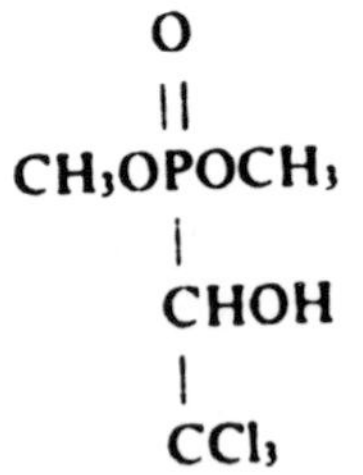

$CH_3OP(O)OCH_3$ – CHOH – CCl_3

METRIFONATE

0,0-dimethyl-(1-hydroxy-2,2,2-trichloroethyl)-phosphonate

Fig. 2. Chemical structure of metrifonate.

0–8412–2008–5/91/0322$06.00/0

Table 1

Comparison of Clinical Trials of Four Cholinesterase Inhibitors on Alzheimer Patients (AD) and Normal Volunteers (V)

Drug	Route	Dose (mg/kg)	% Plasma BuChE Inhib (60 min)	% CSF AChE Inhib.	Side ** Effects	n	Subjects
Phy	oral	.06	24	--	+ +	12	AD
	i.v.	.01-.02	12-25	--	+ + +	20	AD
	i.c.v.	.001	5	85 (60 min)	0	3	AD
Heptyl-Phy	oral	.6	40	--	(+)	10	V
THA	oral	.4-3	25-30	--	+ + + *	18	AD
MTF	oral	5	80	37-47 (24 hrs)	(+)	20	AD

Phy = physostigmine
Heptyl-Phy = heptyl-physostigmine
THA = tetrahydroaminoacridine
MTF = metrifonate

n = number of patients
i.v. = intravenous
i.c.v. = intracerebral ventricular

SIDE EFFECTS: moderate-severe + + +
minimal-moderate + +
none-minimal +

* hepatotoxicity
** nausea, vomiting, dizziness

a. Metrifonate

Metrifonate (Fig. 2) is an organophosphorus ChEI with a duration of inhibition of brain ChE four times longer than Phy (Hallak and Giacobini, 1987). In contrast to Phy it is not a directly acting inhibitor of ChE but requires non-enzymatic metabolism to form the active compound, dichlorvos (2,2-dichlorovinyl dimethyl phosphate, Nordgren et al., 1978). Our results (Hallak and Giacobini, 1987) and that of Nordgren et al. (1978) show that the maximal concentration of the active drug which reaches the brain is only around 2%. This suggests a considerable blood-brain barrier for the compound or a rapid metabolism which does not allow a high concentration to build up in brain. This is a safety mechanism that, together with the slow reversibility of inhibition, explains the long-lasting effect and the minimal side effect level. Differences in levels and metabolism in brain between MTF and Phy explain the difference in time of maximal inhibition and rate of enzyme activity recovery seen between the two drugs in the rat (Hallak and Giacobini, 1987). When given acutely i.m., Phy and MTF show significant differences. First, ACh levels reach peak values in brain faster and are higher with Phy (500 μg/kg) than with MTF (80 mg/kg), even if ChE inhibition is four times longer with MTF (Hallak and Giacobini, 1987, 1989). In humans, a dose of 10 mg/kg of oral MTF produces an 80% inhibition of plasma ChE, which endures for 2-3 days without inducing significant side effects (Nordgren et al., 1980). In AD patients, over 80% inhibition of plasma and red blood cell (RBC) ChE was achieved with only minor side effects (Becker et al., 1990).

Becker et al. (1990) performed a first study of a multiple dose trial of MTF conducted over a prolonged period of time in humans. They administered MTF to 20 patients who met NINCDS-ADRDA criteria for probable AD. Patients were given, under open conditions, single oral doses of MTF, 2.5, 5, 7.5 and 15 mg/kg/week. A statistically significant improvement in the Alzheimer Disease Assessment Scale (ADAS) scores was observed with the 5 mg/kg/week dose. Maximal improvement on the ADAS was associated with a mean 55.9% (±12.6% standard deviation) activity level of RBC AChE. Over 80% inhibition of plasma and RBC ChE was achieved with only minor side effects (Table 1). Cholinesterase inhibition in the CSF of two patients was 37% and 47.5%, 24 hrs after a second dose of 5 mg/kg/week of MTF separated by 7 days from the first dose.

b. Analogues of Physostigmine

Several authors (Oliverio et al., 1986; Pavone et al., 1986; Brufani et al., 1987; Castellano et al., 1987) demonstrated the effect of a number of lipophilic derivatives of Phy characterized by reduced toxicity and reduced activity in the peripheral nervous system. The inhibitory effect (in vitro - versus in vivo) of these drugs on AChE has been reported (Marta and Pomponi, 1987, 1988; DeSarno et al., 1989). One of these lipophilic derivatives, the heptyl-Phy (C8) - derivative (Fig. 1), produced an increase of ACh levels in brain and behavioral modifications suggesting a possible therapeutic use (Brufani et al., 1987; DeSarno et al., 1989). The toxicity (LD_{50}) of heptyl-Phy (35 mg/kg) is about sixty times lower than that of Phy (0.6 mg/kg) (Marta et al., 1988). At doses ranging between 0.1 mg/kg and 3.0 mg/kg, heptyl-Phy does not affect spontaneous activity in mice (Marta et al., 1988). At higher doses (3 mg/kg) its administration results in a depressant effect and in an inhibition of the activity-stimulating effect induced by scopolamine (1 mg/kg). Passive avoidance (step-through latencies) of mice injected post-trial with different doses of heptyl-Phy is significantly modified (Oliverio et al., 1986). Longer step-through latencies are present 24 and 48 hrs following the post-trial injection of heptyl-Phy, indicating that the drug acts on consolidation mechanisms, by facilitating memory.

In human volunteers, 40 mg of a single oral dose of heptyl-Phy produced 40% plasma BuChE inhibition at 1 hr and 17.7% at 6 hrs (Table 1). Red blood cell AChE was 46% and 30% inhibited at the same time points (Unni et al., 1990). No side effects were recorded. The results suggest that heptyl-Phy produces different duration of effects on AChE and BuChE. This could explain the low level of peripheral side effects mainly mediated by BuChE inhibition seen with this compound, as compared to Phy and THA.

c. The Huperzines

Pharmacologists at the Chinese Academy of Science and Shanghai School of Medicine, isolated a new series of alkaloids from the plant Huperzia serrata, which they called Huperzine A and B (HUP-A, HUP-B) (Liu et al., 1988; Tang et al., 1986; Wang et al., 1986) (Fig. 3).

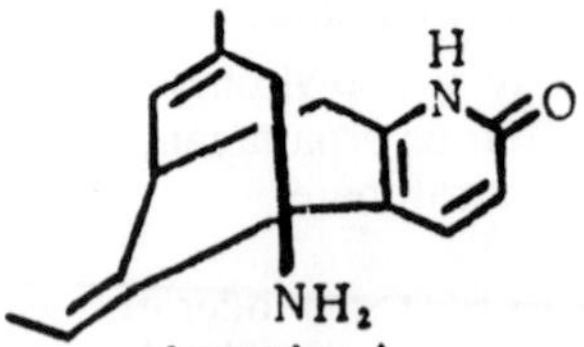

H
N
O
NH_2
huperzine A
(5 R, 9 R, 11 E)-5-amino-11-ethylidene-5, 6, 9, 10-tetrahydro-7-methyl-5,9-methanocycloocteno [b]pyridin-2 (1 H)-one

Fig. 3. Chemical structure of Huperzine A.

A recent study of Tang et al. (1989) shows that HUP-A can produce a long-term inhibition of AChE activity in brain (up to 360 min) and increases the ACh levels up to 40% at 60 min (Fig. 4). There is considerable regional variation in the degree of ACh elevation after HUP-A with maximal values seen in frontal (125%) and parietal (105%) cortex and smaller increases (22-65%) in other brain regions. Huperzine A at concentrations of 10^{-6} to 10^{-4}M does not significantly alter the electrically evoked release of ^{3}H-ACh from cortical slices. With the exception of the highest concentrations (6 x 10^{-4}M) the displacement effect of HUP-A for cholinergic ligands is stronger for ^{3}H-(-)-nicotine than for ^{3}H-quinuclidinylbenzylate (QNB). A parallel autoradiographic study in the mouse shows that 60 min after i.v. injection (183

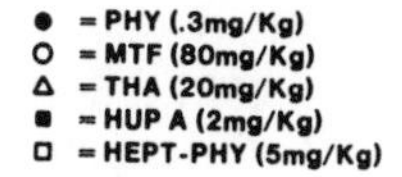

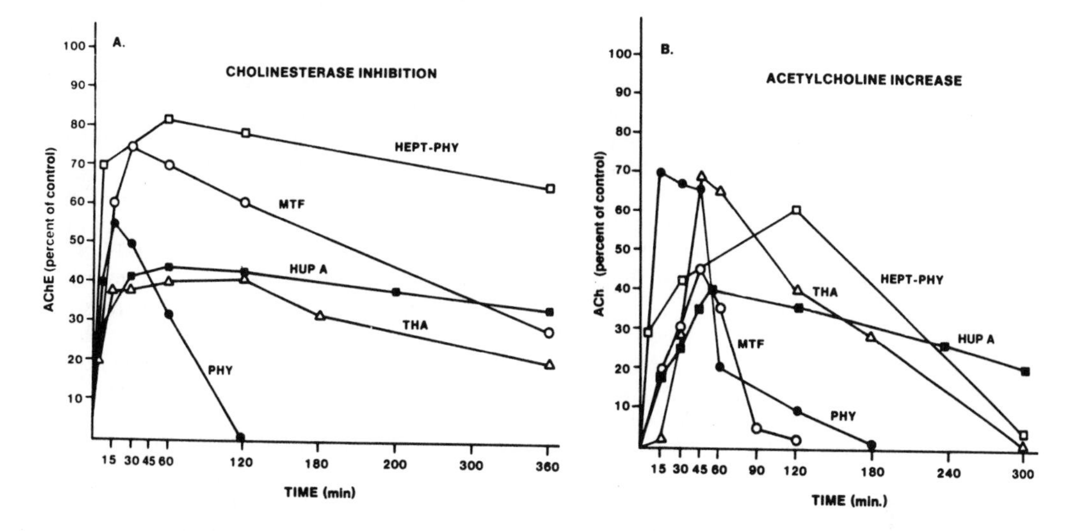

Fig. 4. Comparison of the effect of five ChEI on percent AChE inhibition (A) and ACh increase (B) in whole rat brain following a single dose of the drug administered i.m. [• = physostigmine (Phy); o = metrifonate (MTF); Δ = tetrahydro-aminoacridine (THA); ■ = Huperzine A (HUP-A); □ = heptyl-physostigmine (heptyl-Phy)]. (Adapted from Giacobini et al., 1988).

μg/kg) the drug is present in all brain regions, but it is particularly concentrated in certain areas such as frontoparietal cortex, nucleus accumbens, hippocampal, and striatal cortex.

d. Comparison of Cholinesterase Inhibitor Effects and General Conclusions

Figure 4 compares the effect of five ChEI on AChE inhibition and ACh levels in rat brain following i.m. administration of one single dose of the drug. The dosage has been selected according to the criteria of assuring a maximal increase in ACh levels in brain with high ChE activity inhibition at a zero mortality level. Figure 4A shows that with MTF and heptyl-Phy it is possible to reach peak levels of 70-80% AChE inhibition within 30-60 min. from the administration. Huperzine A and THA produce lower peak levels of inhibition (49%); however, this is maintained for almost 6 hours with HUP-A and 3 hours with THA. Physostigmine produces a rapid (15 min) peak of 55% inhibition but this effect is over within 2 hours.

The effect on brain ACh levels is also markedly different for the different drugs (Fig. 4B). High ACh increases (70%) are obtained following both Phy and THA administration; however, - particularly with Phy, this effect is short-lasting (1 hr). Heptyl-physostigmine shows the most prolonged effect on brain ACh levels with a peak at 2 hrs. The effect of MTF instead is rather short-lasting (1.5 hrs)

These results reveal major differences in biochemical effects on rat brain including extent and duration of ChE inhibition, inhibition of ACh release and increase in levels of ACh. Side effects are also markedly different in time of appearance, duration and severity. These findings suggest significant differences in mechanisms of action of various ChE inhibitors with potential clinical implications for a future symptomatic therapy of AD patients.

References

Becker R., Giacobini, E., *Drug Devel. Res.* **1988**, 12, 163-195.

Becker R.E., Colliver J., Elble R., Feldman E., Giacobini E., Kumar V., Markwell S., Moriearty P., Parks R., Shillcutt S.D., Unni L., Vicari S., Womack C., Zec R.F., *Drug Devel. Res.* **1990**, (In Press).

Brufani M, Castellani C., Marta M., Oliverio A., Pagella P.G., Pavone F., Pomponi M., *Pharmacol. Biochem. & Behavior* **1987**, 26, 625-629,

Castellano C., Oliverio A., Jacopino C., Pavone F., Brufani M., Marta M., Pomponi M., In *Bulgarian Academy of Sciences, Trends in the Pharmacology of Neurotransmission*; House of the Bulgarian Academy of Sciences: Sofia, Bulgaria, **1987**; pp. 57-64.

DeSarno P., Giacobini E., *J. Neurosci. Res.* **1989**, 22, 194-200.

DeSarno P., Pomponi M., Giacobini E., Tang X.C., Williams, E., *Neurochemical Res.* **1989**, 14(10), 971-977.

Elble R., Giacobini E., Becker R., Zec R., Vicari S., Womack C., Williams E., Higgins C., In *Current Research in Alzheimer Therapy*, Giacobini, E., Becker, R., Eds.; Taylor and Francis: New York, NY, **1988**; pp. 123-140.

Giacobini E., Becker, R., In *Familial Alzheimer's Disease - Molecular Genetics and Clinical Perspectives*, Miner, G.D., Richter, R.W., Blass, J.P., Valentine, J.L., Winters-Miner, L.A., Eds.; Marcel Dekker: New York, NY, **1989**; pp. 223-268.

Giacobini E., Becker R., McIlhany M., Kumar, V., In *Current Research in Alzheimer Therapy*, Giacobini, E., Becker, R., Eds.; Taylor and Francis: New York, NY, **1988**; pp. 113-122.

Hallak M.E., Giacobini E., *Neuropharmacology* **1987**, 26(6), 521-530.

Hallak M., Giacobini E., *Neuropharmacology*, **1989**, 28(3), 199-206.

Liu J.S., Yu C.M., Zhou Y.Z., *Can. J. Chem.* **1988**, 64, 837.

Marta M., Pomponi M., *Acta Med. Rom.* **1987**, 25, 433-437.

Marta M., Pomponi M., *Biomed. Biochem. Acta* **1988**, 47, 285-288.

Marta M, Castellani C., Oliverio A., Pavone F., Pagella P.G., Brufani M., Pomponi M., *Life Sci.* **1988**, 43, 1921-1928.

Nordgren I., Bergstrom M., Holmstedt B., Sandoz M., *Arch. Toxicol.* **1978**, 41, 31-41.

Nordgren I., Holmstedt B., Bengtsson E., Finkel Y., *Am. J. Trop. Med. Hyg.* **1980**, 29(3), 426-430.

Oliverio A., Castellano C., Iacopino C., Pavone F., Brufani M., Marta M., Pomponi M., In *Modulation of Central and Peripheral Transmitter Function, Fidia Research Series, Symposia in Neuroscience III*; Biggio, G., Spano, P.F., Toffano, G., Gessa, G.L., Eds.; Liviana Press: Padova, Italy, **1986**, pp. 305-309.

Pavone F., Castellano C., Oliverio A., Brufani M., Marta M., Pomponi M., *2nd Congresso della Societ Italiana di Neuroscienze,* **1986**, pp. 424.

Summers W.K., Majorski L.V., Marsh G.M., Tachiki K., Kling A., *New Eng. J. Med.* **1986**, 315(20), 1241-1245.

Tang X.C., Han Y.F., Chen X.P., Zhu X.D., *Acta Pharmacologica Sinica* **1986**, 7(6), 507-511.

Tang X.C., DeSarno P., Sugaya K., Giacobini E., *J. Neurosci. Res.* **1989**, 24, 276-285.

Unni, L., Becker R.E., Hutt V, Bruno P., *Intl. Contr. Pharmacology* **1990**, (In Press).

Wang Y.E., Yue D.X., Tang X.C., *Acta Pharmacologica Sinica* **1986**, 7(2), 110.

THA - a Potential Drug in the Treatment of Alzheimer's Disease

A. Nordberg, Department of Pharmacology, Uppsala University, Box 591, S-751 24 Uppsala, Sweden

1,2,3,4-tetrahydro-9-aminoacridine (THA) is a moderately long-acting acetylcholinesterase (AChE) inhibitor which easily penetrates to the brain after systemic administration. It inhibits plasma cholineserase (ChE) and tissue AChE with similar affinity and with somewhat higher affintity for buturyl cholinesterase (BuChE). Except for being a AChE inhibitor THA can interact with various transmittorsystems and receptors Clinical trials with THA are now performed in Alzheimer patients and it is possible that the multiple effects of THA in brain will favour the therapeutic effect of the drug in Alzheimer´s disease

Introduction

The neurodegenerative disorder Alzheimer's disease, senile dementia of Alzheimer type (AD/SDAT) is the most common form of dementia (40-60 %). It is characterized by a progressive loss of intellectual capacity and personality. An early clinical diagnosis of the disease is hampered by the lack of diagnostic markers and can finally be verified at autopsy by histopathological examination. Neurochemical studies in autopsy AD/SDAT brains reveal deficits in multiple transmitter systems. The cholinergic system seem to be most consistently affected. The decrease in choline acetyltransferase (ChAT) activity and acetylcholine (ACh) synthesis correlates with cognitive impairement (Perry et al. 1978; Francis et al. 1985).
Great efforts have been paid in finding therapeutic agents which reveal the symtoms of the disease and/or slow down the progress of the disease. For the cholinergic system the precursor, receptor and enzyme inactivating strategies have been tried with modest result except for the acetylcholinesterase (AChE) inhibitorts (Thal et al. 1988; 1989).The interest for the AChE inhibitors as a terapeutic tool in AD/SDAT came in focus when Summers and coworkers in 1986 reported some positive effects with the AChE inhibitor tetrahydroaminoacridine (THA) in Alzheimer patients. The aim of this chapter is to summarize our knowledge about THA focussing on the underlying neurochemical mechanisms for its effect in neurodegenerative disorders such as AD/SDAT.

Tetrahydroaminoacridine

1,2,3,4-tetrahydro-9-aminoacridine (THA) was already in 1945 described as an AChE inhibitor (Albert et al. 1945). THA was then used therapeutically as a drug potentiating morphine analgesia, antagonizing morphine induced respiratory depression (Shaw et al. 1953), enhancing neuromuscular blockade caused by suxamethonium (Gordh, Wåhlin 1961; Hunter 1965) or as an antagonist to tubocurarine chloride (Hunter 1965), barbiturates (Gordh 1962) or antidepressants (Summers et al. 1980).
The cholinergic hypothesis of memory function and the cholinomimetic effect obtained via AChE inhibition provide a rationale for using AChE inhibitors in memory disorders and dementia. Physostigmine has shown some modest improvement of memory in animals (Bartus et al 1983) and humans (Davies et al. 1978) as well as AD/SDAT patients (Thal et al. 1989)
Physostigmine has a short half life in plasma and a narrow therapeutic window. The advantage of THA would therefore be the longer duration of the AChE inhibiting effect due to its longer elimination half life. THA inhibits plasma cholinesterase (ChE) and tissue AChE in vitro in concentrations roughly comparable to physostigmine. The affinity for buturyl cholinesterase (BuChE) is greater than for AChE (Sherman & Messamore, 1988). In contrast to physostigmine which interacts with the catalytic site of AChE, THA is supposed to interact with a hydrophobic binding adjacent to the catalytic site causing an inhibition via an allosteric interaction of the AChE molecule (Steinberg et al., 1975; Sherman & Messamore, 1988). A rapid distribution of THA to the brain has been observed in mice following i.p. administration with brain concentrations exceeding plasma levels 8-12-fold (Nielsen et al., 1989). In vivo whole body autoradiography in rats reveals a regional localisation in brain of THA in areas such as cortex, hippocampus, striatum which consistently does not correlate with the distribution of AChE. THA which shows an

0–8412–2008–5/91/0327$06.00/0

variable biovailability is partly metabolised to 1-hydroxy-THA (Nielsen et al., 1989; Forsyth et al. 1989; Hartvig et al., 1990) which has a weak inhibitory effect on AChE (Shutske et al., 1988).

Effect of THA on Neurotransmitter Release

THA can induce multiple effects in brain. Since THA has a structural similarity to 4-aminopyridine it blocks K^+ and Na^+ channels in the synaptic membranes (Drukarch et al. 1987; Freeman et al. 1988; Halliwell & Grove, 1989, Elinder et al., 1989) THA may thus by prolongating the action potential increase the release of transmitters. An increased release of noradrenaline (NA), dopamine (DA) and serotonin (5-HT) by THA has been observed (Drukarch et al.,1988; Robinson et al., 1989) in rat brain. THA also inhibits monoaminoxidase (MAO) (Adem et al., 1989). and depresses the potassium induced release of GABA in a similar manner as 4-aminopyridine (Belleroche & Gardiner,1988). THA decreases also the potassium induced cortical ACh release (Nilsson et al.,1987). This effect is probably induced via AChE inhibition which enhances the content of ACh in the synaptic cleft and via feed-back mechanism through muscarinic presynaptic autoreceptors slows down the ACh release. 4-aminopyridine enhances the release of ACh (Buyukuysal & Wurtman,1989) probably by blocking presynaptic K^+ channels. THA may also act via this mechanism but at higher concentrations than it inhibits AChE (Harvey & Rand,1988).

Modulation of ACh release in AD/SDAT Brain Tissue by THA

An in vitro method which allows measurement of synthesis and release of ACh from human cortical brain slices has been devloped in our laboratory (Nilsson et al.,1986; 1987) In AD/SDAT cortical tissue the release of ACh is markedly diminished compared to control tissue (Nilsson et al., 1986,). In the presence of physostigmine and THA the release of ACh is restored to control level (Nilsson et al. 1986; 1987).This effect is opposite to the effect observed in normal human brain where THA depresses the release of ACh. The latter effect can be explained by a muscarinic feed back mechanism. The efffect of AChE on ACh release in AD/SDAT tissue is abolished in the presence of nicotinic (mecamylamine, tubocurarine) and muscarinic receptor antagonists (atropine) (Nordberg et al.,1988; 1989b). These observations indicate that both muscarinic and nicotinic receptors may directly/indirectly be involved in the mechanism of action of THA in AD/SDAT brains. Possible mechanisms by which THA might facilitate ACh release in AD/SDAT cortex are illustrated in Fig.1. Of interest in this connection is the reduced number of nicotinic receptors and change in proportion of nicotinic receptor subtypes measured in autopsy cortical AD/SDAT tissue (Nordberg & Winblad,1986; Nordberg et al.,1988; 1989a). The observation of a facilatory effect of THA on ACh release in AD/SDAT brain tissue illustrates the importance of using diseased brain tissue instead of rat brain or normal human brain tissue for studying the effects and underlying mechanisms of putative drugs in Alzheimer´s disease.

Interaction of THA with Cholinergic and Ohter Neuroreceptors in Brain

THA was reported by Nilsson et al. (1987) to bind to both muscarinic and nicotinic binding sites in human autopsy brain tissue. This has been confirmed in rat (Flynn & Mash, 1989; Hunter et al., 1989; Nielsen et al.,1989), rabbit (Pearce & Potter,1988, Potter et al.,1989), guinea pig (Freeman et al.,1988) and human (Perry et al.,1988) brain. The interaction of physostigmine with the cholinergic binding sites in brain is negligible (Nilsson et al.,1987, Perry et al.,1988). THA shows a lower affinity for the nicotinic receptors than for the muscarinic receptors in normal cortical human tissue (Nilsson et al., 1987). Possible the affinity might be different in AD/SDAT tissue where the proportion of nicotinic subtypes are changed (Nordberg et al., 1988). Subchronic treatment with THA to rats causes an increase in number of nicotinic binding sites in cortical regions (Nilsson-Håkanssonet al., 1990). THA shows no selectivity between muscarinic M_1 and M_2 receptors (Hunter et al.; Potter et al, 1989; Flynn & Mash, 1989) but can act through allosteric sites and stabilize antagonist conformation (Potter et al., 1989; Flynn & Mash,1989). An interactions via muscarinic receptors might be relvant to the action of THA at higher concentrations.

THA also interacts with the phencyclidine receptor binding sites in rat brain with mixed agonist-antagonist properties (Albin et al., 1988) and has also been suggested to attenuate NMDA receptor-mediated neurotoxicity (Davenport et al., 1988) but also to potentiate neurotoxicity (Zhu et al., 1988). These effects might thus have a positive or a negative effect on the progress of the AD/SDAT disease. Considerable higher doses are needed for these effects than for AChE inhibition. THA has also been suggested to

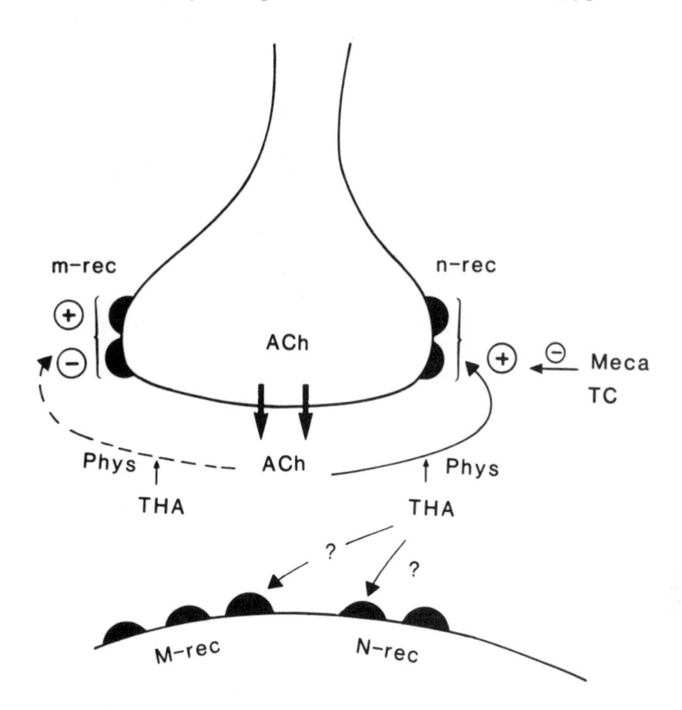

Fig 1. Cholinergic nerve terminal and putative mechanisms by which THA modulates ACh release in AD/SDAT cerebral cortical tissue. Abbrev. ACh=acetylcholine; M-rec= muscarinic receptor; N-rec= nicotinic receptor; Phys= physostigmine; THA= tetrahydroaminoacridine; Meca= mecamylamine; TC= tubocurarine

interact with adenosine receptors (Freeman et al., 1988) and on histamine neurons (Reiner & McGeer, 1988).

Clinical Trials with THA

THA has in animal studies improved memory function both in adult and old animals as well as in those where memory have been impaired by drug treatment (scopolamine) or by experimental lesions (Bartus & Dean, 1988; Fitten et al., 1988 a,b; Flood 1988; Shutske et al., 1988; Nielsen et al., 1989). Several human clinical trials with THA have been performed in AD/SDAT patients (Summers et al., 1981; Kay et al., 1982; Summers et al., 1986; Gauthier et al, 1986; Fitten et al., 1988a, 1990; Nybäck et al., 1988; Chatellier et al., 1990) and further studies are ongoing. Since an asymptomatic transaminitis induced by THA is common the patients must be carefully followed during the treatment period. Studies using positron emission tomography (PET) to visualize nicotinic receptors in the brain of AD/SDAT patients (Nordberg et al. 1990) might be a technique to further investigate in vivo whether THA treatment in Alzheimer patients improve the cholinergic activity in brain. The hypothesis raised by Mesulam and coworkers (1987) that physostigmine and THA by inhibiting plaque and tangle-bound cholinesterases might arrest the pathological process of AD/SDAT is an intersting theory. Whether THA is a useful and safe drug for the treatment of Alzheimer´s disease has to be further proven. THA might be "the first generation" of a new drug profile where selective analogues of THA might prove to have beneficial effects (Shutske et al., 1989; Adem et al., 1990). Other cholinesterase inhibitors such as metrifonate (Nordgren and Holmstedt, 1988), huperzine (Tang et al., 1988), galanthamine (Sweeney et al., 1989), physostigmine deratives (Brufani et al., 1988) also await further exploration.

References

Adem, A., Jossan, S.S., Oreland, L., Neurosci. Lett, **1989**, 107. 313-317.

Adem, A., Mohammed,A., Nordberg, A., Winblad, B., In: Nagatsu,T., Fisher,

A.,Yoshida, M., eds., Alzheimer's and Parkinson's diseases 11: Basic and Therapeutic Strategies, Plenum Press, New York, **1990**, in press.
Albert, A.,Gledhill, W., J. Soc. Chem. Ind., **1945**, 64, 169.
Albin, R..L. Young, A.B., Penney, J.B., Neurosci. Lett., **1988**, 88, 303-307.
Bartus, R.T., Dean, R.L., In: Giacobini,E., Becker, R.,eds.,Current Research in AlzheimerTherapy, Cholinesterase inhibitors, Taylor and Francis, New York, **1988**,.pp.179-190
Bartus, R.T., Dean, R.L., Beer, B., Psychopharmacol., **1983**, 19, 168-184.
Brufani, M., Castellano, C., Martz, M., Murroni, F., Oliverio, A., Pagella, P.G., Pavone, F., Pomponi, M., Rugarli, P.L., In: Giacobini, E., Becker, R., eds., Current Research in Alzheimer Therapy, Cholinesterase inhibitors, Taylor and Francis, New York, **1988**, pp.343-352.
Buyukuysal, R.L., Wurtman, R.J., Brain Res.,**1989**, 482, 371-375.
deBelleroche, J., Gardiner, I.M., Br. J. Pharmacol. **1988**, 94, 1017-1019.
Chatelllier, G., Lacombles, L., Br. Med.J., **1990**, 300, 495-499.
Davenport, C.J, Monyer, H., Choi, D.W., Eur.J Pharmacol., **1988**, 154, 73-78.
Davis, K.L., Mohs, R.C., Tinklenberg, J., Pfefferbaum, A., Hollister, L.E., Kopell, B.S., Science, **1978**, 201, 272-274.
Drukarch, B., Leysen, J.E., Stoof, J.C., Life Sci., **1988**, 42, 1011-1017.
Elinder, F., Mohammed, A.K., Winblad,B., Århem, P., Eur. J Pharmacol., **1989**, 164, 599-602.
Fitten, L.J., Perryman, K., Gross, P., in: Giacobini, E., Becker, R., eds., Current Researchin Alzheimer Therapy, Cholinesterase inhibitors, Taylor and Francis, New York, **1988**, pp. 211-216.
Fitten, L.J., Perryman, K.M., Gross, P.L., Fine, H., Cummins,J., Marshal,C. Am. J. Psychiatry, **1990**, 147, 239-242.
Fitten, L.J., Perryman, K., Tackiki, K., Kling, A., Neurobiol.Aging, **1988**, 9, 221 -224.
Flood, J.F., J. Gerontology: Biol. Sci., **1988**, 43, B54-56.
Flynn, D.D., Mash, D.C., J.Pharmacol. Exp. Ther., **1989**, 250, 573-581.
Forsyth, D.R., Wilcock, G.K., Morgan, R.A., Truman, C.A., Ford, J., Roberts,C.J.C.,Clin. Pharmacol. Ther., **1989**, 46, 634-641.
Francis, P.T., Palmer, A.M., Sims, N.R., Bowen, D.M., Davison, A.N., Esisri, M.N., Neary, M.D., Snowdon, J.S., Wilcock, G.K., New. Engl. J. Med.,**1985**, 313, 7 -11.
Freeman, S.E., Lau, W-M., Szilagi, M., Eur, J. Pharmacol., **1988**, 154, 59-65.
Gauthier, S., Masson, H., Gauthier, L. et al., In:Giacobini, E., Becker, R., eds., Current Research in Alzheimer Therapy, Cholinesterase inhibitors, Taylor and Francis, New York, **1988**, pp. 237-245.
Gordh, T., Nord. Med. (swe), **1962**, 68, 132.
Gordh, T., Wåhlin, Å., Acta Anaesthiol. Scand.,**1961**, 5, 51-55.
Halliwell, J.V., Grove, E.A., Eur. J. Pharmacol., **1989**, 163, 369-372.
Harvey, A.L., Rowan, E.G., In: Giacobini, E., Becker, R., eds., Current Research in Alzheimer Therapy, Cholinesterase Inhibitors, Taylor and Francis, New York, **1988**, pp. 191-209.
Hartvig, P., Askmark, H., Aquilonius, S.M., Wiklund, L., Lindström, B., Eur. Clin. Pharmacol., **1990**, 38, 259-263.
Hunter, A.R., Br. J. Anesth., **1965**, 37, 505-513.
Hunter, AJ., Murray, T.K., Jones, J.A., Cross, A.J., Green, A.R., Br. J. Pharamcol., **1989**, 98, 79-86.
Kay, W.H., Stiaram, N., Weingartner, H., Ebert, M.H., Smallberg, S. Gillin, J.C., Psychiatry, **1982**, 17, 275-280.
McNally, W., Roth, M., Young, R., Brockbrader, H., Chang, T., Pharmaceut. Res., **1989**, 6, 924-930.
Mesulam, M.M., Gesula, C., Moran, M.A., Ann. Neurol., **1987**, 22, 683-691.
Nielsen, J.A., Mena, E.E., Williams, I.H., Nocerini, M.R., Liston, D., Eur. J. Pharmacol., **1989**, 173, 53-64.
Nilsson,L., Adem,A, Hardy, J., Winblad, B., Nordberg, A., J. Neural Transm. **1987**, 70, 357-368.
Nilsson, L., Nordberg, A., Hardy, J., Wester., Winblad, B., J. Neural Transm., **1986**, 67, 275-285.
Nilsson-Håkansson, L., Lai, Z., Nordberg, A., Eur. J. Pharmacol., **1990**, submitted.
Nordberg, A., Adem, A., Hardy, J., Winblad, B., Neurosci. Lett., **1988**, 86, 317 -321.
Nordberg, A., Hartvig, P., Lilja, A., Viitanen,.M., Amberla, K., Lundqvist, H.,Andersson, Y., Ulin, J., Winblad, B., Långström, B., J.Neural Transm, **1990**, in press.
Nordberg, A., Nilsson, L., Adem, A., Hardy, J., Winblad, B., In: Giacobini, E., Becker,R., eds., Current Research in Alzheimer Theraphy, Cholinesterase Inhibitors, Taylor and Francis, New York, **1988**, pp. 247-257.
Nordberg, A., Nilsson-Håkansson, L., Adem, A., Hardy, J., Alafuzoff, I., Lai., Z.,Herrera-Marschitz, M., Winblad, B., Progr. Brain Res., **1989a**, 79, 353-362.
Nordberg, A., Nilsson-Håkansson, L., Adem, A.,Lai, Z., Winblad, B., In: Iqbal, K., Wisniewski, H.M., Winblad, B., eds.,

Alzheimer's disease and Related Disorders, Alan Liss, New York, **1989b**, vol. 317; pp. 1169-1178.
Nordberg, A., Winblad, B., Neurosci. Lett., **1986**, 72, 115-119.
Nordgren, I., Holmstedt, B., In: Giacobini, E., Becker, R., eds., Current Research in Alzheimer Therapy, Cholinesterase Inhibitors, Taylor and Francis, New York, **1988**, pp. 281-288.
Nybäck, H., Nyman, H., Öhman, G., Nordgren, I., Lindström, B., In: Giacobini, E., Becker, R., eds., Current Research in Alzheimer Therapy, Cholinesterase Inhibitors, Taylor and Francis, New York, **1988**, pp. 231-236.
Pearce, B.P., Potter, L.T., Neurosci. Lett., **1988**, 88, 281-285.
Perry, E.K., Smith, C.J., Court, J.A., Bonham, J.R., Rodway, M., Atack, J.R., Neurosci. Lett., **1988**, 9, 211-216.
Perry, E.K., Tomlinson, B.E., Blessed, G., Bergman, K, Gibson, P.H., Perry, R.H., Br. Med. J., **1978**, 2, 1456-1459.
Potter, L.T., Ferrendelli, C.A., Hanchett, H.E., Hollifield, M.A., Lorenzi, M.V., Molec. Pharmacol., **1989**, 35, 652-660.
Reiner, P.B., McGeer, E.G., Eur. J. Pharmacol., **1988**, 155, 265-270.
Robinson, T.N., De Souza, R.J., Cross, A.J., Green, A.R., Br. J. Pharmacol., **1989**, 98, 1127-1136.
Rogawski, M.A., Eur. J. Pharmacol., **1987**, 142, 169-172.
Shaw, F.H., Bently, G.A., Australian J. Expthl. Biol. Med. Sci., **1953**, 31, 573 -576.
Sherman, K.A., Messamore, E., In: Giacobini, E., Becker, R., eds., Current Research in Alzheimer Therapy, Cholinesterase Inhibitors, Taylor and Francis, New York, **1988**, pp. 73-86.
Shutske, G.M., Pierrat, F.A., Cornfeldt, M.L., Szewczak, M.R., Huger, F.P., Bores, G.M., Haroutunian, V., Davis, K.L., J. Med. Chem., **1988**, 31, 1279-1282.
Shutske, G.M., Pierrat, F.A., Kapples, K.J., Cornfeldt, M.L., Szewczak, M.R., Huger, F.P., Bores, G.M., Haroutunian, V., Davis, K.L., J. Med. Chem, **1989,** 32, 1805-1813.
Summers, W.K., Kaufman, K.R., Altman, F., Fisher, J.M., Clin. Toxicol., **1980**, 16, 269-281.
Summers, W.K., Majorski, L.V., Marsh, G.M., Tachiki, K., Kling, A., New Engl. J. Med., **1986**, 315, 1241-1245.
Summers, W.K., Viesselman, J.O., Marsh, G.M., Candelora, K., Biol. Psychiatry, **1981**, 16, 145-153.
Sweeney, J.E., Puttfarchen, P.S., Coyle, J.T., Pharmacol. Biochem. Behav., **1989**, 34, 129-137.
Tang, X.C., Zhu, X.D., Lu, W.H., In: Giacobini, E., Becker, R., eds., Current Research in Alzheimer Therapy, Cholinesterase Inhibitors, Taylor and Francis, New York, **1988**, pp. 289-293.
Thal, L.J., Lasker, B.R., Masur, D.M., Blau, A.D., Knapp, S., In: Giacobini, E., Becker, R., eds., Current Research in Alzheimer Therapy, Cholinesterase Inhibitors, Taylor and Francis, New York, **1988**, pp. 103-111.
Thal, L.J., Masur, D.M., Blau, A.D., Fuld, P.A., Klauber, M.R., JAGS, **1989**, 37, 42 - 48.
Zhu, S.G., McGeer, E.G., Singh, E.A., McGeer, P.L. Neurosci. Lett.,**1988**, 95, 252 -256.

Experimental Autoimmunity to Neural Acetylcholinesterase

S. Brimijoin*, P. Hammond M. Balm and V. A. Lennon, Departments of Pharmacology, Neurology and Immunology, Mayo Clinic, Rochester Minnesota 55905

Systemically administered monoclonal antibodies to acetylcholinesterase gained access to peripheral cholinergic synapses in adult rats, and one antibody also accumulated to a biologically significant extent in the brain. In parasympathetic and motor nerves, the complexation of enzyme with IgG did not produce major functional impairment. In sympathetic ganglia, however, presynaptic cholinergic terminals were permanently destroyed. Selective, preganglionic immunosympathectomy with cholinesterase antibodies provides a new investigative tool for autonomic physiology and disease.

Because acetylcholinesterase is a synaptic protein largely exposed to extracellular fluid (McIsaac and Koelle, 1959), it is a potential target of autoimmunity. With the aid of monoclonal antibodies to rat brain acetylcholinesterase (AChE), we have created a model autoimmune disorder of cholinergic nerves.

One antibody with a surprising tendency to accumulate in the central nervous system has the potential to cause widespread abnormalities (Brimijoin et al., 1990). Most of our antibodies do not have access to AChE epitopes in the central nervous system. When injected systemically, these antibodies induce a unique disorder of preganglionic sympathetic nerves but have minimal apparent effects on other cholinergic systems.

Antibodies

The murine IgG antibodies ZR1, 2, 3, 4, 5, and 6, raised against the AChE of rat brain (Rakonczay and Brimijoin, 1986), lacked intrinsic inhibitory activity against AChE. Antibodies were injected intravenously in various combinations in doses totaling 0.2 to 1.5 mg of IgG, into adult male rats. The biologic half-life of the injected material was approximately 30 days but murine IgG could not be detected in the serum beyond 2 months.

Distribution of Immune Complexes

Intravenously injected antibodies were distributed more widely than expected. Most of the IgG was confined to the bloodstream, where it caused immediate and nearly total clearance of plasma AChE without affecting butyrylcholinesterase or hematocrit. However, measurements of immune complexes showed that antibody also had *in vivo* access to the AChE of sympathetic ganglia (Fig. 1) and endplate regions of skeletal muscle (not shown). Studies of the sedimentation of AChE in extracts of diaphragm indicated that asymmetric forms of the enzyme were depleted by 75% and that surviving asymmetric forms were completely complexed with mouse IgG. Tetrameric AChE was partly depleted and complexed, but monomeric and dimeric enzyme were not affected. This pattern was predicted from the established cellular location of AChE forms in muscle (Younkin et al., 1982, Fernandez et al., 1984). Thus, the smaller molecular forms, mainly intracellular, were immunologically sequestered. The larger forms, predominantly anchored in plasma membranes and basal lamina, were accessible to antibody.

Most of the antibodies failed to interact with brain AChE *in vivo*, but they avidly bound the AChE of brain extracts and brain slices. A striking exception was antibody ZR1, which

0–8412–2008–5/91/0332$06.00/0

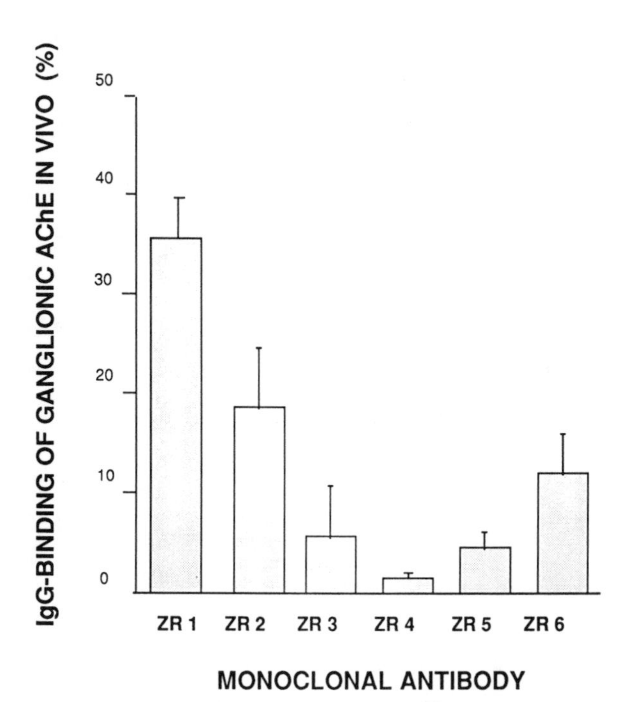

Fig. 1. Accumulation of AChE antibody in sympathetic ganglia. Superior cervical ganglia were removed 3 days after injection of the indicated monoclonal antibodies (0.25 mg each) into the tail veins of adult Sprague-Dawley rats (four to five per group). Antibody-bound enzyme was determined by immunoadsorption with insolubilized rabbit anti-mouse IgG, followed by spectrophotometric AChE assay.

accumulated gradually in the brain and spinal cord, reaching levels that approximated the local concentration of the enzyme. The time course (Fig. 2) and the specificity of this accumulation (Table 1) indicated that it was not a postmortem artifact.

Three days after injection of ZR1 antibody (0.25 mg, i.v.) the IgG content of saline-perfused cerebral cortex was 103 ± 30 ng/g wet weight (about 2% of the serum IgG level). The cortical AChE activity at that time was 30% less than the control value, and 63% of the residual AChE was complexed with IgG. Similar figures were recorded for other regions of the central nervous system.

Passage of measurable amounts of antibody into the brain, though surprising, is consistent with known properties of the blood-brain and blood - cerebrospinal fluid barriers. These barriers maintain protein gradients on the order of 5000-fold, but they are not absolute (Cserr et al., 1988). Our calculations indicate that high affinity antibody circulating in micromolar concentrations might well reach concentrations in CSF and in the extracellular fluid of the brain in the range of the antigen-antibody dissociation constant (Brimijoin et al., 1990). Because AChE is a trace protein, even in brain, a major portion of the available enzyme could be complexed with immunoglobulin.

Clinical Observations

Although the foregoing results showed that autoimmune involvement of brain AChE is theoretically possible, immunological attack on peripheral synapses is more likely. To document the clinical characteristics of experimental autoimmunity restricted to peripheral AChE, we injected rats with antibodies that did not interact with brain AChE *in vivo*. Motor abnormalities in these rats were minimal on clinical inspection (detailed studies of neuromuscular physiology are currently in progress). Signs typical of anticholinesterase

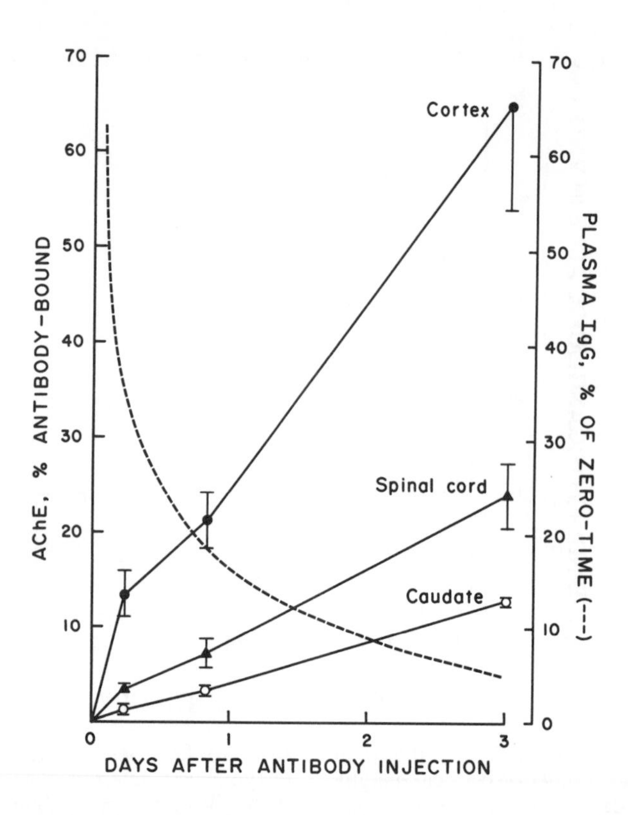

Fig. 2. Rate of appearance of AChE-antibody complexes in CNS. Tissues sampled after i.v. injection of an equal part mixture of ZR antibodies (total dose, 2 mg IgG). Antibody-bound AChE activity shown as percentage of total activity in frontal cortex, striatum, and lumbar spinal cord. Each group comprised four to five rats. Dashed line = double exponential curve fitted to plasma concentrations of radioiodinated ZR1. (Reprinted from Brimijoin et al, 1990, with permission from Raven Press).

intoxication, such as salivation, diarrhea, vomiting, profuse respiratory secretion, and muscle fasciculation, were also conspicuously lacking. However, a marked drooping of the eyelids (ptosis) appeared regularly within 2 hours after injection of a mixture of any three AChE antibodies. The ptosis lasted as long as the animals were observed (up to 15 months). Ptosis was never seen in control animals treated with equivalent doses of normal mouse IgG.

Physiological -Biochemical Characterization

Other observations suggested that ptosis reflected a permanent immunologic lesion in the autonomic nervous system. These observations, to be described elsewhere in depth, included postural syncope and chronic, atropine-resistant hypotension and bradycardia. Since autonomic dysfunction occurred without accumulation of antibody in the brain, the syndrome was presumed to be purely peripheral in origin. Evidently, the postganglionic sympathoadrenal pressor system was intact, because the rats exhibiting autonomic abnormalities gave normal pressor responses to tyramine (a norepinephrine-releaser) and dimethylphenylpiperazinium (an agonist at nicotinic receptors of ganglia). The parasympathetic system was also functional, since heart rate was reduced in normal fashion by electrical stimulation of the vagus nerve.

Direct physiological evidence of sympathetic impairment was obtained as follows. When tested under pentobarbital anesthesia, none of the rats with ptosis showed eyelid contraction in response to electrical stimulation of preganglionic sympathetic fibers to the

Table 1. Differential antibody-access to cerebral AChE

Monoclonal Antibody	Binding[1] after Exposure *in vitro*[2] (n = 4 to 7)	*in vivo*[3] (n = 3 to 4)
ZR1	45 ± 6	74 ± 11
ZR2	90 ± 3	3 ± 0.7
ZR3	90 ± 3	0.5 ± 0.3
ZR4	90 ± 5	0.3 ± 0.1
ZR5	85 ± 4	0.2 ± 0.1
ZR6	88 ± 3	0.3 ± 0.1

[1] Percentage of cortical AChE complexed with IgG.

[2] Brain extracts mixed with 5 x 10^{-7} M IgG for 3 hr

[3] Three days after i.v. injection of 0.2 mg MAb IgG

superior cervical ganglion. The same animals responded in a normal or exaggerated way to direct electrical stimulation of the ganglion itself. This result localized the immunologic lesion to the preganglionic neurons of the sympathetic system.

Biochemically, the ganglia of antibody-treated rats were found to contain normal dopamine-β-hydroxylase activity, an adrenergic marker. This information confirmed the physiological evidence for sparing of post-ganglionic neurons. On the other hand, the ganglia were grossly deficient in choline acetyltransferase, a marker of cholinergic cytoplasm. The loss of enzyme activity in superior cervical ganglia was virtually complete. Reductions of 80-90% were also seen in the stellate, thoracic, lumbar and coeliac ganglia, as well as in the adrenal gland. These data were supported by ultrastructural evidence that synaptic terminals disappeared from the affected ganglia (unpublished data). We concluded that AChE antibodies selectively destroy the presynaptic cholinergic fibers of the sympathetic nervous system.

Mechanism of Immunological Lesion

Preliminary results indicate that preganglionic immunosympathectomy involves complement. Thus, depletion of hemolytic complement by pretreatment with cobra venom factor completely prevents ptosis and the loss of choline acetyltransferase from sympathetic ganglia. We suspect that the immunologic lesion is triggered by the appearance of AChE-IgG complexes on the surface of cholinergic terminals, leading to complement-mediated lysis of terminal membranes. The major unresolved question concerns the selectivity for preganglionic sympathetic neurons with apparent sparing of parasympathetic and motor neurons. It is important to determine whether this selectivity reflects variations in the molecular architecture of AChE, in the relation of the enzyme to the plasma membrane, or in the ability of cells to recover from complement-mediated membrane damage.

Conclusion

AChE-antibodies induce global and selective destruction of preganglionic sympathetic fibers in the adult rat. This unexpected reaction provides a novel tool for physiological studies of the autonomic nervous system. Problems for which this tool might be useful include 1) elucidating the sympathoadrenal role in hypertension, 2) defining neurotrophic effects that depend on sympathetic impulse traffic, and 3) clarifying "wiring diagrams" in the peripheral autonomic nervous system. There is also potential relevance to acquired dysautonomias and other neurological diseases in man.

(Supported by grants NS 18170 and NS 15057 from NINCDS and by the William K. Warren Foundation).

References

Brimijoin, S.; Balm, M.; Hammond, P.; Lennon, V.A. *J. Neurochem.* **1990,** 236-241.

Cserr, H. F.; DePasquale, M. J.; Knopf, P. *Soc. Neurosci. Abstr.* **1988**, 14, 617.

Fernandez, H. L.; Inestrosa, N. C.; Stiles, J. R. *Neurochem. Res.* **1984**, 9, 1211-1230.

McIsaac, R. J.; Koelle, G. B. *J. Pharmacol. Exp. Ther.* **1959**, 126, 9-20.

Younkin, S. G.; Rosenstein, C.; Collins, P.L.; Rosenberry, T.L. *J. Biol. Chem..* **1982**, 257, 13630-13637.

Anaesthesia and Abnormal Plasma Cholinesterase

Frank Samsøe Jensen, Department of Anaesthesia, Glostrup Hospital, DK-2600 Glostrup, Denmark
Jørgen Viby-Mogensen, Department of Anaesthesia, Rigshospitalet, DK-2100 Copenhagen, Denmark

Succinylcholine is usually hydrolysed in plasma in a few minutes by plasma cholinesterase. However, if the cholinesterase activity is depressed for some other reason a clinically significant prolonged recovery may be seen in phenotypically normal as well as in heterozygous patients. Some old and new aspects of the clinical significance of plasma cholinesterase is discussed.

The depolarizing neuromuscular blocking agent succinylcholine is used in anaesthesia to facilitate tracheal intubation. After intravenous injection, it is usually hydrolysed in plasma by pChe in a few minutes, and only a fraction of the injected drug actually reaches the neuromuscular junction. If the pChe activity is decreased, a larger proportion of succinylcholine reaches the neuromuscular end plate and the patient may show a prolonged neuromuscular blockade (Whittaker 1986, Viby-Mogensen 1985, Pantuck & Pantuck 1987).

Six to seven percent of patients in most surgical populations have an abnormal pChe activity, but practical and economical considerations make it impossible to test this activity preoperatively in all patients.

About 65% of all cases of prolonged neuromuscular blockade following succinylcholine is due to genetic factors. It is therefore possible by examination of both patients who have reacted abnormally to succinylcholine and their relatives to identify those persons at risk of developing prolonged neuromuscular blockade.

Since 1972 when Danish Cholinesterase Research Unit was set up in Copenhagen more than 6000 patients in approximately 2000 families have been recorded. To this unit blood samples from the whole country are forwarded and tested for abnormal pChe activity.

If the pChe studies reveal biochemical grounds for a prolonged response to succinylcholine other members of the family will be offered the possibility of being investigated.

The center permits us to make prospective controlled clinical studies of the significance of pChe for the reaction to different drugs used in clinical medicine.

The normal enzyme

Plasma cholinesterase is a glycoprotein synthesized in the liver. It has two active sites, one negatively charged binding to positively charged quaternary ammonium groups at the substrate, and an esteratic site breaking ester bands.

Succinylcholine has two positively charged quaternary ammonium groups, and it is to these groups that the anionic site of the pChe binds. The esteratic site of the pChe then breaks the ester bonds of the drug. In this way succinylcholine is first hydrolysed to the monocholine ester of succinic acid and choline (Muensch et al. 1976). This happens in a few minutes. Later succinic acid and choline are formed from the ester of succinic acid. Due to this rapid hydrolysis only 5-10% of the injected dose actually reaches the neuromuscular end plates.

A decrease of pChe activity causes a reduction in the rate of hydrolysis of succinylcholine, which again means an increased amount of succinylcholine reaches the receptor sites.

0–8412–2008–5/91/0336$06.00/0

The inherited variants

There are many causative factors for low pChe activity but most important is the genetic variants. Several allelic genes at locus E_1 at chromosome 3 control the synthesis of the variants of the enzyme (Yang et al. 1984, Soreq et al. 1987). The clinically most important variants are: the usual, the atypical, the fluoride resistant, and the silent. Combination of those genes can give any one of nine abnormal phenotypes and one normal.

Plasma cholinesterase activity and reaction to succinylcholine

When discussing the correlation between phenotypes and recovery of neuromuscular function following succinylcholine the phenotypes can conveniently be divided into four main groups, i.e. those homozygous for the usual gene, those heterozygous for the usual and one of the abnormal genes, those heterozygous for two abnormal genes, and finally those homozygous for two abnormal genes.

The average duration of apnoea in patients with normal phenotype is about 5 min, and time to sufficient respiration is about 10 min (Viby-Mogensen 1980).

There is a good correlation between pChe activity and the duration of the neuromuscular blockade in patients with normal phenotypes and time to first evoked response is practically independent of the enzyme activity.

Only when the cholinesterase activity is severely depressed will the time to first evoked response be significantly prolonged.

The neuromuscular function is monitored by evaluating the muscular response to stimulation of a peripheral motor nerve using the so-called train-of-four (TOF) stimulation. In this type of nerve stimulation, four stimuli are given at intervals of 0.5 sec (2 Hz). When used continuously the train of stimuli is normally repeated every 10th to 12th sec. Each stimulus in the train will cause the muscle to contract, and the amplitude of the fourth response in relation to the first gives the TOF ratio. In the control response, i.e. before any relaxant has been given, all four responses are the same: the TOF ratio is 1.0 (Ali et al. 1970).

During a partial non-depolarizing block the ratio is reduced (fade) and inversely proportional to the degree of blockade. During a partial depolarizing block (succinylcholine) there is no fade in the TOF response, i.e. the TOF ratio is around 1.0. Fade in the TOF response following injection of succinylcholine signifies the development of a so-called phase II block (see later).

Prolonged administration of succinylcholine, either as repeated injections or infusions, alters the reaction to nerve stimulation. The original depolarizing block, or phase I block, changes to a non-depolarizing like block or a phase II block, characterized by fade in the TOF response and an abnormally prolonged block. There are large individual variations as to how easily normal patients develop a phase II block. Some patients develop this block after relatively small doses, whereas other might receive large doses of succinylcholine and not show a phase II block. In general, the lower the pChe activity the more easily the patients will develop a phase II block.

Genotypically abnormal patients, especially those homozygous for two abnormal genes, show an increased prolonged duration of the neuromuscular block and time to sufficient respiration (Viby-Mogensen 1981a, 1981b).

Mivacurium

Until recently the patients' plasma cholinesterase activity has only been of importance when using succinylcholine. However, a new non-depolarizing agent, mivacurium, has been introduced, which has been designed to undergo hydrolysis by pChe.

In vitro the hydrolysis rate of mivacurium by purified human pChe is about 70-80% of the rate of hydrolysis of succinylcholine (Savarese et al. 1988).

The significance of different genotypes for the neuromuscular blocking effect of mivacurium is presently studied by the Danish Cholinesterase Research Unit.

In phenotypically normal patients, with a wide range of pChe activity a good correlation between the pChe activity and the duration of action of mivacurium was found (Østergaard 1989).

An increase in the duration of action of mivacurium was also found in patients heterozygous for the usual atypical gene. However, in no patient a very prolonged neuromuscular blockade was found (Østergaard 1990).

Bambuterol

Bambuterol is a prodrug formed by linking two carbamate groups to terbutaline at the phenolic hydroxyl groups. This chemical manipulation has resulted in a drug with 1) a more complete absorption, 2) a low first pass metabolism, 3) a high affinity for lung tissue, and 4) a prolonged effect. Being an inert prodrug, bambuterol has no β_2 agonist activity itself, but is broken down to terbutaline by a combination of oxidative metabolism and esterase-catalyzed hydrolysis by pChe. However, as the carbamates are strong inhibitors of pChe bambuterol is slowing down its own hydrolysis by inhibition of pChe (Fischer et al. 1988).

Two studies have recently been performed investigating if a bambuterol-induced depression of pChe may cause a clinically significant prolongation of the effect of succinylcholine. Patients with normal phenotype and patients heterozygous for one abnormal gene were studied (Bang et al. 1990a, 1990b).

A pronounced inhibition of pChe activity was found in all patients. At two hours, the median inhibition of pChe activity was 91%, and the neuromuscular recovery following succinylcholine (1 mg kg^{-1}) was significantly prolonged up to 3-4 times, and in some patients with very low pChe activity a phase II block developed, which resulted in a clinically very significant prolonged neuromuscular blockade.

In the heterozygous patients a very pronounced depression of pChe was found, which resulted in the same clinically significant prolonged neuromuscular blockade.

References

Ali, H.H., Utting, J.E., and Gray, C. Br. J. Anaesth. **1970**, 42, 967-977.

Bang, U., Viby-Mogensen, J., Wirén, J.E., Skovgaard, L.T. Acta Anaesth. Scand. **1990a** (submitted for publication).

Bang, U., Viby-Mogensen, J., Wirén, J.E. Acta Anaesth. Scand. **1990b** (submitted for publication).

Fischer, D.M., Caldwell, J.E., Sharma, M., Wirén, J.E. Anesthesiology **1988**, 69, 757-759.

Muensch, H., Goedde, H.W., Yoshida, A. Eur. J. Biochem. **1976**, 70, 217-223.

Pantuck, E.J., Pantuck, C.B. In Muscle Relaxants. Side effects and a rational approach to selection; Azar, I., Ed.; M. Dekker, Inc.: New York, 1987, p. 205-229.

Savarese, J.J., Ali, H.H., Basta, S.J., Embree, P.B, Scott, R.P.F., Sunder, N., Weakly, J.N., Wastila, W.B., El-Sayad, H.A. Anesthesiology **1988**, 68, 723-732.

Soreq, H., Zamir, R., Zevin-Sonkin, D., Zakut, H. Human Genet. **1987**, 77, 325-328.

Whittaker, M. Cholinesterase; Monographs in Human Genetics; Karger: Exeter, 1986, vol. 11, pp. 1-125.

Viby-Mogensen, J. Anesthesiology **1980**, 53, 517-520.

Viby-Mogensen, J. Anesthesiology **1981a**, 55, 231-235.

Viby-Mogensen, J. Anesthesiology **1981b**, 55, 429-434.

Viby-Mogensen, J. Lectures in Anaesthesiology. Blackwell Scientific Publications, Oxford, 1985, pp.63-73.

Yang, F., Lum, J.B. McGill, J.R. Moore, C.M., Naylor, S.L., van Bragt, P.H. Baldwin, W.D., Bowman, B.H. Proc. Natl. Acad. Sci. USA **1984**, 81, 2752-2756.

Østergaard, D., Jensen, F.S., Jensen, E., Viby-Mogensen, J. Acta Anaesth. Scand. **1989** (suppl. 91), 33, 164.

Østergaard, D., Jensen, F.S., Jensen,E., Viby-Mogensen, J. Anesthesiology **1990** (submitted for publication).

Neurotrophic Actions of Glycyl-l-Glutamine

George B. Koelle, Department of Pharmacology, Medical School, University of Pennsylvania, Philadelphia, PA 19104-6084, U.S.A.

A few years ago we initiated an investigation of a putative neurotrophic factor (NF) for the maintenance of acetylcholinesterase (AChE) in the ganglion cells of the cat superior cervical ganglion (SCG) (Koelle and Ruch, 1983). The study was instigated by the electron microscopic finding that in the normal cat SCG, AChE is present at both pre- and postsynaptic locations; following preganglionic denervation the enzyme disappears from both sites (Davis and Koelle, 1978, 1981). Morphological considerations seemed to eliminate the possibilty that AChE is normally released from the presynaptic terminals and migrates to postsynaptic locations. A more likely possibility was that the preganglionic terminals secrete a NF that is essential for the synthesis of AChE by the ganglion cells. The working hypothesis was tested as follows:

Twenty-four hours after bilateral preganglionic denervation of the SCG, cats were reanesthetized and set up for intraarterial infusion of the SCG via a common carotial artery with an aqueous extract of cat brain and spinal cord (CNS). Following twenty-four hours' continuous infusion, the SCG were excised, homogenized, and assayed for AChE and protein contents. It was found that this treatment maintained the AChE content at approximately 85% of normal (the level at 24 hours post-denervation) in comparison with similarly treated saline (0.9% NaCl)-infused controls, in which the level had fallen to approximately 50%.

Infusions of extracts of liver and skeletal muscle were ineffective, as were acetycholine and nerve growth factor, 2.5S and 7S; cyclic AMP and gut-extracts showed borderline activity. Further investigation showed the active NF to be a small (MW<1000), heat-stable peptide (Koelle et al., 1984, 1985, 1986).

A report by Haynes and Smith (1985), that glycyl-l-glutamine (GlyGln) maintains the AChE of cultured embryonic rat and chick skeletal muscle, led us to test this compound, since it met the criteria for the NF noted above. GlyGln was found to be highly effective in the cat preparation, but only following prior incubation with heat-inactivated plasma. It was postulated that combination with some component of plasma enables GlyGln, a highly polar compound, to penetrate the ganglion cell membrane (Koelle et al., 1987).

In-vitro studies conducted in the laboratory of J. Massoulié indicated that GlyGln acts at an early stage of AChE synthesis or activation, prior to the aggregation of the initial G_1 monomer to higher polymers (G_4, A_{12}) (Koelle et al, 1988). It has not been determined whether GlyGln, which is the terminal dipeptide of ß-endorphin and is present in brain (Parish et al, 1983), is identical with the endogenous NF of the CNS studied originally.

In reviewing the foregoing results it was proposed that GlyGln may enhance the transcription of the gene for AChE to the corresponding mRNA, which would then be translated by the ribonsomes of the rough endoplasmic reticulum to synthesize monomeric G_1 AChE (Koelle, 1988). This mechanism of action is analogous to that proposed for triiodothyronine (Baxter et al., 1979; Ramsden and Hoffenberg, 1983), to which GlyGln bears a rough resemblance in size and structure.

In collaboration with J.J. O'Neill we have extended the study to another enzyme of the cholinergic system, choline acetyltransferase (ChAcTr). To our surprise GlyGln was found to be just as effective in maintaining ChAcTr as AChE (Fig.1). Since the former enzyme, in contrast with the latter, is present in the SCG only in the preganglionic fibers and their terminals (Banister and Scrase, 1950; Hebb and Waites, 1956; Kuhar, 1976; Hallanger and Wainer, 1988), the latest finding suggests that GlyGln opposes the degeneration of transected axons (Koelle et al, 1989).

Current efforts in collaboration with M.S. Han have been directed to examination of the latter possibility. Rats are anesthetized, and both sciatic nerves are transected (Sawyer, 1946). The renal artery is exposed and a catheter, extending to the aorta, is inserted. To the catheter is attached on Alzet miniosmopump (2MLI) containing various concentrations of GlyGln. The pump is embedded intraperitoneally and the wound sutured (Koelle and Han, 1989). Two to four days

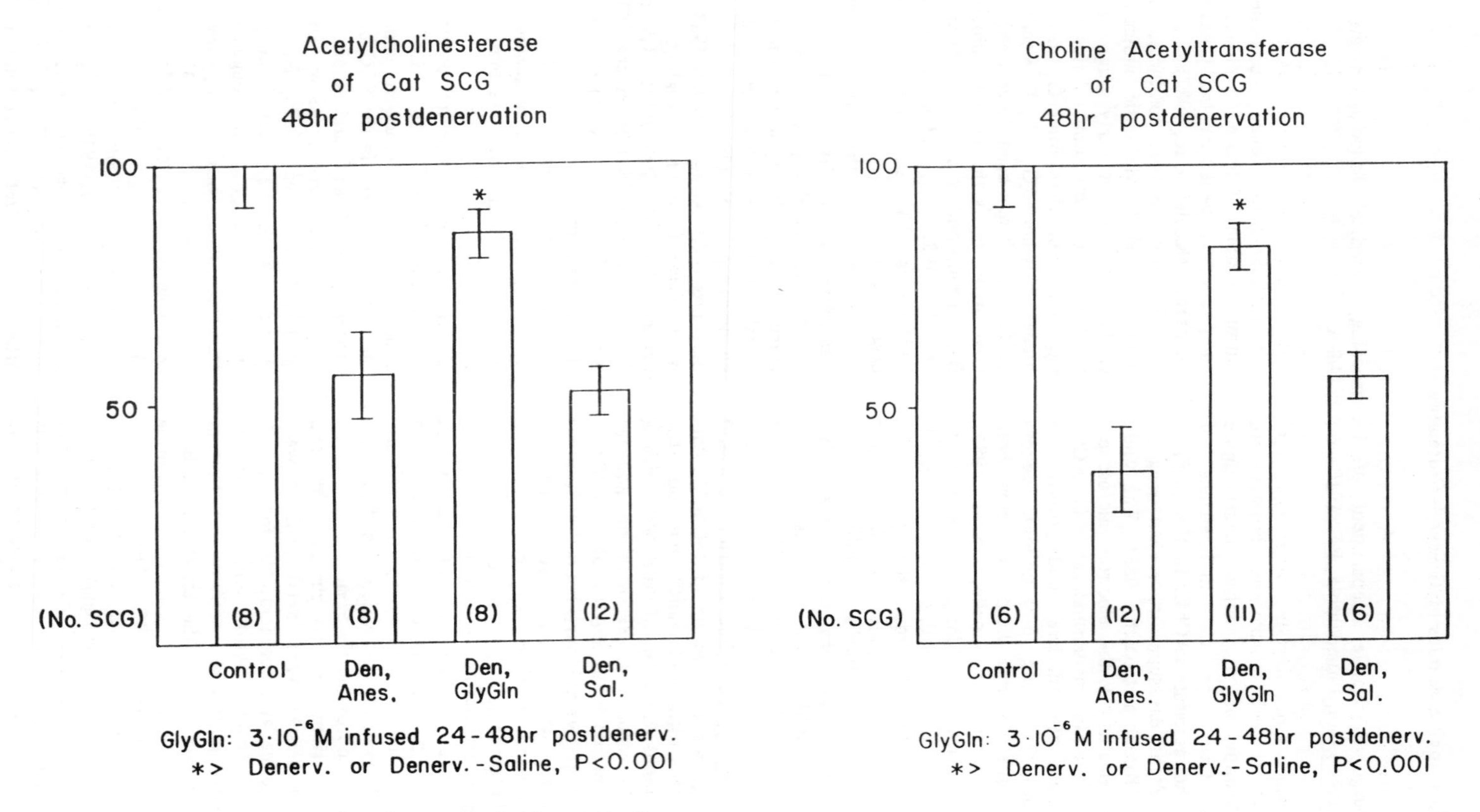

Fig.1. Acetylcholinesterase (left) and choline acetyltransferase (right) contents of cat SCG in comparison with controls (100%); 48 hours postdenervation following maintenance of anesthesia (Den, Anes.); 24 hours infusion of 3.0 μm GlyGln (Den, GlyGln); or 0.9% NaCl solution (Den, Sal.).

later the rats are re-anesthetized and the sciatic nerves are excised, homogenized in AChE- extraction medium, and centrifuged, and the supernatents assayed for AChE and protein. While no definite results have been obtained to date, this study is being continued with GlyGln alone and in combination with other agents reported to enhance the survival of injured axons (Hall et. al., 1984)

The current investigation has been supported by Research Grant NS 00282-36, National Institute of Neurological Disorders and Stroke, National Institutes of Health, U.S.A., and by generous contributions from The Barra Foundation, Foundation for Vascular-Hypertension Research, Philadelphia, and Hoffman-LaRoche, Inc.

References

Banister, J.,Scrase, M.J., *J.Physiol.(London)*, **1950**, *11*, 437-444.

Baxter, J.D., Eberhardt, N.L., Apriletti, J.W., Johnson, L.K., Ivarie, R.D., Schachter, B.S., Morris, J.A., Seeburg, P.H., Goodman, H.M., Latham D.R., Polanksy, J.R., Martial, J.A., *Recent Prog. Horm. Res.*, **1979**, *35*,97-153.

Davis, R., Koelle, G.B., *J.Cell Biol.*, **1978**, *78*, 785-809.

Davis, R., Koelle, G.B., *J.Cell Biol.*, **1981**, *88*, 581-590.

Haines, L.W., Smith, M.E., *Biochem. Soc. Trans.*, **1985**, *13*, 174-175.

Hall, E.D., Wolf, D.L., Braughler, J.M., *J.Neurosurg.*, **1984**, *61*, 124-130.

Hallanger, A.E., Wainer, B.H., *J.Comp. Neurol.*, **1988**, *278*, 486-497.

Hebb, C.O., Waites, G.M.H., *J.Physiol. (London)*, **1956**, *132*, 667-671.

Koelle, G.B., *Trends Pharmacol. Sci.*, **1988**, *9*, 318-321.

Koelle, G.B., Han, M.S., *Proc. Natl. Acad. Sci. USA*, **1989**, *86*,4331-4333.

Koelle, G.B., Massoulié, J.,Eugène, D., Melone, M.A.B., *Proc. Natl. Acad. Sci. U.S.A.*, **1988**, *85*, 1686-1690.

Koelle, G.B., O'Neill, J.J., Thampi, N.S., Han, M.S., Caccese, R., *Proc. Natl. Acad. Sci. USA.*, **1989**, *86*, 10153-10155.

Koelle, G.B., Ruch, G.A., *Proc. Natl. Acad. Sci. USA*, **1983**, *80*, 3106-3110.

Koelle, G.B., Sanville, U.J., Richard, K.K., Williams, J.E., *Proc. Natl. Acad. Sci. USA*, **1984**, *81*, 6539-6542.

Koelle, G.B., Sanville, U.J., Thampi, N.G., Wall, S.J., *Proc. Natl. Acad. Sci. U.S.A.*, **1986**, *83*, 2751-2754.

Koelle, G.B., Sanville, U.J., Thampi, N.G., *Proc. Natl. Acad. Sci. U.S.A.*, **1987**, *84*, 6944-6947.

Koelle, G.B., Sanville, U.J., Wall, S.J., *Proc. Natl. Acad. Sci. USA.*, **1985**, *82*, 5213-5217.

Kuhar, M.J. In *Biology of Cholinergic Function*; Goldberg, A.M., Hanin, I., Eds.; Raven Press, New York, NY, 1976, pp. 3-27.

Parish, D.C., Smyth, D.G., Normanton, J.R., Wolstencroft, J.H., *Nature (London)*, **1983**, *306*, 267-269.

Sawyer, C.H., *Amer. J. Physiol.*, **1946**, *146*, 246-253.

Effect of Cholinesterase Inhibitor and Exercise on Cholineacetyltransferase and Acetylcholinesterase Activities in Rat Brain Regions

S.R. Babu, S.P. Arneríc, S.N. Dube and S.M. Somani, Department of Pharmacology, Southern IL University School of Medicine, Springfield, IL 62794-9230

This presentation addresses the question whether subacute administration of Phy and subacute exercise or the combination of the two elicits any changes in the biosynthetic, cholineacetyltransferase (ChAT), and degradative, acetylcholinesterase (AChE), enzymes for acetylcholine.

Male Sprague-Dawley rats (150-175 g) were divided into five groups: Gr. I, sedentary control; Gr. II, subacute treadmill exercise for two weeks; using an incremental exercise program; Gr. III, subacute Phy (70 μg/kg, i.m. twice daily) for two weeks; Gr. IV, subacute Phy and single bout of acute exercise 100% VO_2 max; Gr. V, subacute Phy + subacute exercise (Fig.). The rats were sacrificed 24 hr after the last dose of Phy and/or the exercise. ChAT and AChE activities in different brain regions were determined by the method of Fonnum (1975). See Fig.

The ChAT activity decreased significantly ($p < 0.05$) in corpus striatum to 88%, 68%, 88% of control in Gr. III, IV and V respectively. AChE activity was depressed in all groups but was significantly depressed in Gr. IV up to 24 hr. ChAT activity decreased in cerebral cortex in all groups but was significantly ($p < 0.05$) decreased in Gr. V (79% of control) indicating that only combination of subacute Phy + subacute exercise affected this region. Phy alone or exercise alone did not affect - significantly - ChAT and AChE activities in cerebral cortex. AChE activity in cerebral cortex has recovered to control level in all groups within 24 hr. ChAT activity significantly decreased ($p < 0.05$) in brainstem to 73%, 78%, 82%, 81% of control in Gr. II, III, IV and V respectively. AChE activity in brain stem was depressed in all groups but significantly ($p < 0.05$) in Gr. II (81% of control) and Gr. III (82% of control). The ChAT activity decreased significantly in hippocampus ($p < 0.05$) to 72% and 73% of control in Gr. IV and V respectively. However, Gr. II and III also showed decreased ChAT activity. AChE activity recovered to control level in all groups by 24 hr.

These results suggest that the Phy or physical exercise or combination of two stressors depress ChAT and/or AChE activities in a regionally selective pattern that depends on the type and interaction of these two stressors. It is unclear whether the susceptibility of the brain regions to the different stressors is a function of the level of ongoing cholinergic neurotransmission, or the dynamic responsiveness of these brain regions involved with motor control, autonomic functions and cognition to adapt to stress. (Supported by U.S. Army DAMD-17-88-C-8024.)

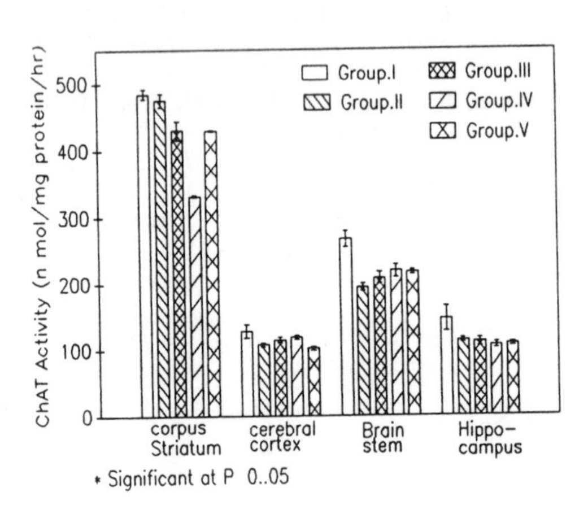

References

Fonnum, F., *J. Neurochem.*, **1975**, *Vol 24*, pp. 407-409.

Inhibition of Rat Acetyl- and Butyrylcholinesterase Activity by Tacrine and Related Aminoacridines

R.D. Schwarz*, C.J. Spencer, R.E. Davis, W.H. Moos, M.R. Pavia, and A.J. Thomas Parke-Davis Pharmaceutical Research Division, Warner-Lambert Co., Ann Arbor, MI 48105 USA.

In 1986, Summers et al. reported a significant improvement in Alzheimer's patients with the acetylcholinesterase (AcChE) inhibitor tacrine (tetrahydroaminoacridine; THA). The present in vitro study examined tacrine and related aminoacridines on their ability to inhibit rat AcChE and butyrylcholinesterase (BuChE) activity.

AcChE and BuChE were measured spectrophotometrically according to the method of Ellman et al. (1961). Tacrine was found to potently inhibit both AcChE and BuChE activity with IC50 values of 0.20uM and 0.075uM respectively. Addition of hydroxyl groups at the 1, 2, and 4- positions of tacrine reduced actvity in both assays, with IC50 values for inhibiting AcChE being 0.78uM, 4.47uM, and 0.86uM and for BuChE 0.42uM, 4.56uM, and 0.14uM respectively. 1-Keto-tacrine (AcChE = 2.25uM; BuChE=10.40uM) and 9-benzyl-1-OH-tacrine (AcChE= 8.38 uM; BuChE=1.97uM) were also weaker than tacrine. Unsaturation in the acyclic ring did not affect AcChE activity as much as BuChE activity with 1,2-dehydrotacrine and 9-aminoacridine being as potent as tacrine on AcChE (0.24uM and 0.37uM) but weaker on BuChE activity (0.13uM and 0.35uM). Groups on the heterocyclic N also reduced activity as examplified by N-methyl-tacrine (AcChE = 7.61uM; BuChE=2.52uM and 9-amino-acridine-N-oxide (AcChE = 4.98uM; BuChE=3.67uM). Modifications on the 9-amino position also proved to reduce activity. 9-(Amino-n-butyl)-tacrine was slightly reduced in its ability to inhibit AcChE (0.96uM) and BuChE (0.57uM) activity compared tacrine, but 9-(acetylamino)-tacrine was substantially weaker (AcChE=19.0uM; BuChE = 7.10 uM). For comparison, results of reference standards were: physostigmine (IC50 for AcChE=0.14uM; BuChE=0.49uM), BW 284C51 (AcChE = 0.0095uM; BuChE = 166uM), and iso-OMPA (AcChE = >100uM; BuChE = 1.15uM).

These in vitro results demonstrate that tacrine is more potent in its ability to inhibit both AcChE and BuChE activity than other aminoacridines tested.

References

Ellman, G.L., Courtney, D., Andres, V., and Featherston, R.M., Biochem. Pharmacol., **1961**, 7, 88-95.

Summers, W.K., Majovski, L.V., Marsh, G., Tachiki, K., and Kling, A., N. Engl. J. Med., **1986**, 315, 1241-1248.

0–8412–2008–5/91/0343$06.00/0

Inhibition of Human Erythrocyte DSAChE by Chlorpromazine and Trifluoperazine is Independent from Drug Effects on Membrane Environment

A. Spinedi, L. Pacini and P. Luly, Department of Biology, University of Rome 'Tor Vergata', Via Orazio Raimondo 00173 Rome (Italy).

A role of membrane lipids in mediating the effects of a number of amphiphilic molecules on the activity of detergent-soluble acetylcholinesterase (DSAChE) has been proposed. Interest in verifying this hypothesis resides in the fact that membrane anchoring of DSAChE is not provided by a hydrophobic sequence of the aminoacidic chain but by covalently linked phosphatidylinositol (Low, 1989). In this framework we investigated the mechanism by which chlorpromazine (CPZ) and trifluoperazine (TFP), two amphiphilic drugs with membrane perturbing properties, affect the activity of human erythrocyte DSAChE. Although human erythrocyte DSAChE is largely resistant to phosphatidylinositol-specific phospholipase C (PIPLC) hydrolysis, owing to covalent modification of the inositol ring (Roberts et al., 1988), a small fraction of the enzyme can be solubilized by this treatment (Futerman et al.,1985): on this fraction and on the native enzyme a parallel study was carried out. CPZ and TFP inhibit native DSAChE activity in a micromolar concentration range (the IC_{50} being 120 μM and 70 μM for CPZ and TFP respectively), within which they have also been demonstrated to affect the physical state of erythrocyte membranes (Minetti & Di Stasi, 1987). Membrane solubilization with 1% Triton X-100 and enzymatic activity measurement in presence of the same detergent concentration result in a complete loss of CPZ and TFP inhibitory potencies.Although these observations might suggest a role of the membrane lipid environment in mediating DSAChE inhibition , we also observed that: i) CPZ and TFP retain their inhibitory activity when assayed on the fraction of DSAChE which is solubilized from erythrocytes upon treatment with PIPLC from B. thuringiensis ; ii) also in the case of the PIPLC - solubilized enzyme the presence of 1% Triton X-100 abolishes the inhibitory effect of CPZ and TFP. We have recently suggested (Spinedi et al., 1989) that the amphiphilic drugs imipramine and lidocaine inhibit human erythrocyte DSAChE activity by direct molecular interaction, although at the same concentrations at which they induce enhancement of 'membrane fluidity'. Data concerning DSAChE inhibition by CPZ and TFP strongly support this view, thus indicating that amphiphile membrane perturbation and DSAChE inhibition may be two unrelated phenomena.

References

Futerman, A.H., Low, M.G., Michaelson, D.M., Silman, I., *J. Neurochem.*, **1985**, 45, 1487-1494.
Low, M., *Biochim. Biophys. Acta,* **1989**, 988, 427-454.
Minetti, M., Di Stasi, A.M.M., *Biochemistry*, **1987**, 26, 8133-8137.
Roberts, W.L., Myher, J.J., Kuksis, A., Low, M.G., Rosenberry, T.L., *J. Biol. Chem.*, **1988**, 263,18766-18775.
Spinedi,A., Pacini, L., Luly, P., *Biochem. J.*, **1989**, 261, 569-573.

0–8412–2008–5/91/0344$06.00/0

Bambuterol, a Selective Inhibitor of Human Plasma Butyryl Cholinesterase

M. Sharma, Department of Anesthesia, University of California, San Francisco, CA 94143-310, USA
Leif-Å. Svensson, Research & Development Department, AB DRACO, Box 34, S-22100 Lund, Sweden

Bambuterol is a **bis**-N,N-dimethylcarbamate prodrug of the bronchodilator terbutaline (Olsson, 1984). As a carbamate, bambuterol is an inhibitor of plasma butyryl cholinesterase (EC 3.1.1.8), (BCHE). **In vitro** studies with human blood have shown that bambuterol is a dramatically weaker inhibitor of acetyl cholinesterase (EC 3.1.1.7), (ACHE), IC_{50}= 41μM, compared to BCHE, IC_{50}= 17 nM. **In vivo,** in man, a dose related inhibition has been documented (Svensson, 1988), but **in vivo**-evidence in man on the extent of ACHE inhibition is lacking. However, studies in animals have shown no toxicological effects resulting from inhibition of ACHE, even when the plasma concentration of bambuterol was above the IC_{50}-value for ACHE.

In the present study we have examined the inhibition of BCHE and erythrocyte ACHE (RBC-ACHE) in six human patients receiving either a high, 30 mg, oral dose of bambuterol, or placebo. Blood samples were withdrawn prior to and 10-12 hours following administration of drug.

Determination of ACHE and BCHE activity.

Each heparinized blood sample, 5 mL, was centrifuged (10 000x g, 10 min.) and the plasma phase separated. The erythrocyte pellet was washed 3 times with a 4-fold volume of normal saline by centrifugation and finally was reconstituted to the original volume with normal saline. The plasma phase and the reconstituted erythrocytes were immediately analyzed for BCHE and ACHE activities, respectively. Both plasma BCHE and RBC-ACHE activities were measured by an adaptation of the radioisotope assay method of Calvey **et al.** (Calvey, 1976).

Results and Discussion.

Blood withdrawn prior to bambuterol administration was assayed to establish pre-bambuterol levels of plasma BCHE and RBC-ACHE activities. Normal human plasma BCHE activity values under our assay conditions ranged from 0.4-1.3 units (n=10) and pre-drug treatment values for patients were within this range. Similarly, pre-drug treatment RBC-ACHE values were within the normal range of 0.8-1.6 units. The time interval between the administration of bambuterol and the first measurement of enzyme activities ranged from 10-12.5 hours. It is obvious that bambuterol causes a significant inhibition; mean 74.0%, a decrease from 0.89+/- 0.22 units to 0.23+/- 0.05 units of plasma BCHE activity within 10-12 hour of bambuterol administration. There was no corresponding decrease in RBC-ACHE activity; mean control value 1.31+/-0.19, and post drug value 1.30+/-0.19.

The selective inhibition of human plasma BCHE activity by bambuterol in effect contributes strongly to the slow **in vivo** conversion of the prodrug to terbutaline. However, also oxidative metabolism of bambuterol at the carbamate methyl groups takes place and contributes to the slow bioconversion of prodrug to active drug. Intermediate metabolites formed in the oxidative process are still carbamate prodrugs of terbutaline, and hence, also inhibitors of plasma BCHE, although less so than bambuterol itself.

In conclusion, the results of this effort clearly document that **in vivo** in humans, bambuterol selectively inactivates BCHE but not ACHE. This is what could be expected, since maximal plasma concentration obtained with 30 mg bambuterol in humans rarely exceeds 25 nmol/L, which is of the same order as the IC_{50}-value for BCHE, but 2000-fold below the IC_{50}-value for ACHE.

References

Olsson, T., Svensson, L-Å., *Pharm. Res.,* **1984,** *1,* 19-23.
Svensson, L.-Å., Tunek, A., *Drug Metab. Rev.,* **1988,** *19,* 165-194
Calvey, T.N., Williams, N.E., Muir, K.T., Barber, H.E., *Clin. Pharmacol. Ther.,* **1976,** *19,* 813

0–8412–2008–5/91/0345$06.00/0

Pharmacology of Tacrine: Comparisons with Other Cholinesterase Inhibitors

Robert E. Davis*, **Linda L. Coughenour, David Dudley, Thomas A. Pugsley, Roy D. Schwarz, Walter H. Moos**, Parke-Davis Pharmaceutical Research Division, Warner-Lambert Co., Ann Arbor, MI, 48105, USA.

Alzheimer's disease is characterized by significant neuronal pathology in discrete brain regions. Loss of forebrain cholinergic neurons accompanied by decreased neocortical choline acetyltransferase activity is consistently seen in brains of demented subjects (Whitehouse, 1981; Davies, 1976) and may be responsible in part for the cognitive dysfunction associated with this disease (Perry, 1978). One approach to the treatment of the cognitive symptoms of Alzheimer's disease is to increase the synaptic availability of acetylcholine (ACh) by preventing its breakdown through inhibition of acetylcholinesterase (AChE). Marked improvement in AD patients has been reported after treatment with the cholinesterase inhibitor Tacrine (Summers, 1986). Because of this, the pharmacologic properties of Tacrine were compared to a series of known cholinesterase inhibitors to elucidate possible mechanisms of action leading to clinical effectiveness.

Tacrine is equipotent (IC_{50}=220 nM) as an acetylcholinesterase inhibitor with physostigmine, miotine and RA-10 but more potent than velnacrine (1-hydroxy-tacrine) and suronacrine. However, Tacrine is substantially more potent (IC_{50}=60 nM) than any of these other agents in inhibiting butyrylcholinesterase activity. Unlike other ChE inhibitors Tacrine binds with μM affinity to numerous noncholinergic receptor and channel sites. At muscarinic receptors it exhibits a binding profile similar to weak muscarinic antagonists; a property not shared by physostigmine and structurally related ChE inhibitors. Consistent with action as a muscarinic antagonist at high concentrations Tacrine reverses arecoline-induced decreases in ACh release and carbachol-elicited increases in PI turnover.

Unlike simple aminopyridines K^+ channel blockers, Tacrine does not increase nonstimulated release of ACh.

However, Tacrine does decrease the presynaptic release of ACh, inhibit high affinity choline uptake and increase the turnover of catecholamines. Like other cholinomimetic ChE inhibitors in rats, Tacrine increases gastrointestinal motility, reduces scopolamine-induced and spontaneous swimming activity, decreases body temperature, and decreases the amount of high-voltage, low-frequency activity in the cortical EEG. This latter effect is also seen in rhesus monkeys.

Tacrine also improves the ability of cognitively impaired mice and basal forebrain lesioned rats to find a hidden platform in a water-maze and modestly improves delayed match-to-sample performance of aged-rhesus monkeys. In addition, tacrine does not increase local cortical blood flow in a manner similar to physostigmine.

Thus, Tacrine is a cholinomimetic agent at reasonable concentrations and cognitively active doses and much of its pharmacologic activity resides in its ability to enhance cholinergic function.

References

Whitehouse, P.J., Price, D.L., Clark, A.W., Coyle, J.T., DeLong, M.R., *Ann. Neurol.* **1981**, 10, 122-126.

Davies P., Maloney A.J.F., *Lancet*, **1976**, 2, 1403.

Perry E.K., Tomlinson, B.E., Blessed, G., Bergmann, K., Gibson, P.H., Perry, R.H., *Br. Med. J.*, **1978**, 2, 1457-1459.

Summers, W.K., Majovski, L.V., Marsh, G.M., Tachiki, K., Kling, A., *New Engl. J*,. Med., **1986**, 315, 1241-1245.

Acetylcholinesterase (ACHE) And Substance P (SP) Levels In The Plasma And Pericardial Fluid Of Patients With Angina Pectoris

JR Kambam, W Merrill, W Parris, R Naukam, R Alhadad, BVR Sastry. Department of Anesthesiology, Vanderbilt University Medical Center, Nashville,Tennessee 37232-2125

Previous data indicate that SP- containing cardiac afferents participate in the transmission of cardiac pain and ACHE hydrolyzes SP. If SP is involved in the transmission of cardiac pain and is released into the pericardial fluid and possibly plasma, then the peptide levels should be altered in patients with angina. Therefore we measured SP and ACHE in the pericardial fluid (PF) and plasma.

Methods & Results:

Nine patients (no angina group) undergoing mitral valve and/or aortic valve replacement surgery, and 10 patients (angina group) undergoing coronary artery graft surgery were included in the study. Before the pericardiotomy was performed, PF was collected with an iv cannula. Simultaneously 3 ml of blood was collected via an arterial cannula. ACHE activity in the plasma and PF was determined using a colorimetric method. A radioimmuno assay was used in the determination of SP levels. There is a two fold increase in SP levels and three fold decrease in ACHE in the pericardial fluid of patients with angina when compared with no angina. In conclusion we demonstrated that SP and ACHE are present in the pericardial fluid and the levels of SP, however, are significantly higher and the ACHE activity is lower in patients with angina pectoris when compared to the patients without angina. Increased levels of SP may be responsible for some of the symptoms including angina in these patients.

References:

Chubb, I.W., et al. *Neuroscience* **5:2065-72,1980**

Reinecke, M., et al. *Neuroscience Lett* **20:265-9,1980**

Weihe, E., et al. *Neuroscience Lett* **26:283-8,1981**

Table: SP levels (fmols) and ACHE activities (units)

	PERICARDIAL FLUID		PLASMA	
	SP	ACHE	SP	ACHE
ANGINA GROUP	1.7±.2	0.06±.02	0.5±.3	0.29±.15
NO ANGINA GROUP	0.9±.1	0.16±.02	0.5±.1	0.26±.08
P	<0.05	<0.05	>0.05	>0.05

Amniotic Fluid Acetylcholinesterase in the Antenatal Diagnosis of Open Neural Tube Defects and Abdominal Wall Defects

A.G. Rasmussen Loft, B. Nørgaard-Pedersen, Department of Clinical Biochemistry, Statens Seruminstitut, DK-2300 Copenhagen, Denmark
K. Nanchahal, H.S. Cuckle, N.J. Wald, Department of Environmental and Preventive Medicine, The Medical College of St. Bartholomew's Hospital, Charterhouse Square, London EC1M 6BQ
M. Hulten, P. Leedham, Regional Cytogenetics Laboratory, East Birmingham Hospital, Burdesley Green, East Birmingham B9 5ST

Two methods for amniotic fluid acetylcholinesterase (AChE) determination are compared: a monoclonal antibody enzyme antigen immunoassay - EAIA, and polyacrylamide gel electrophoresis - PAGE (details published elsewhere; Loft et al., 1990).

Samples from 5689 singleton pregnancies were analysed. PAGE yielded detection rates of 97% for anencephaly (n=36), 99% for open spina bifida (n=77) and 94% for abdominal wall defects (n=17); the false positive rate was 0.24%. EAIA yielded similar results; using appropriate cut-off values to allow for differences in AChE levels in blood stained and clear samples, a similar false positive rate of 0.22% was associated with detection rates of 97%, 95% and 71% respectively, for the three types of defect.

This study demonstrates that either PAGE or EAIA are similarly effective, the former having a slightly higher detection rate, the latter being more objective and simpler to perform.

References

Loft, A.G.R., Nanchahal, K., Cuckle, H.S., Wald, N.J., Hulten, M., Leedham, P., Nørgaard-Pedersen, B, Prenat Diagn, 1990, in press.

0-8412-2008-5/91/0348$06.00/0

NONCHOLINERGIC ROLES OF CHOLINESTERASES

Expression and Possible Functions of Cholinesterases During Chicken Neurogenesis

Paul G. Layer, Max-Planck-Institut für Entwicklungsbiologie, Spemannstraße 35/IV, D-7400 Tübingen, FRG

Along with final celldivision, butyrylcholinesterase (BChE) is transiently elevated on the ventricular side of the chicken neural tube. Shortly after BChE activation, cells will start to express acetylcholinesterase (AChE) and then extend axons. AChE as a postmitotic marker is useful to map patterns of primary brain development. Before spinal motor axons start to grow, cholinesterases are found at their origins, at their targets and along their pathways. *In vitro* experiments show that BChE can regulate cell proliferation and AChE synthesis, and that both cholinesterases can modulate axon growth.

I. After final cell division of neuroblasts, first BChE is transiently elevated shortly before AChE is expressed.

In addition to the classical localization of AChE at the cholinergic synapse (Silver, 1974; Massoulié and Bon, 1982), AChE is expressed in single neuroblasts located on the surface of the young neural tube (Fig. 1; Layer, 1983; Puelles et al., 1987; Layer et al., 1988a; Weikert et al., 1990; Layer and Alber, 1990). As development progresses, more and more cells will synthesize AChE, and as a consequence a confluent layer of AChE-positive cells builds up that finally covers the entire brain. BChE is also expressed during these early stages, although to a much lesser extent than AChE. In the sections shown in Fig. 1, we have applied ATC as substrate; therefore both AChE and BChE are revealed. The diffuse staining detectable near AChE-positive cells is due to BChE activity. By comparing the expression of cholinesterases with thymidine-pulse experiments, we found that it takes roughly 10 hours from the last cell division until the same cell will begin with its AChE-synthesis (Layer & Sporns, 1987; Vollmer & Layer, 1987). We have shown quantitatively for various brain parts, that the general decrease of cell proliferation is shortly followed by a transient expression of BChE activity, before cells really start to produce AChE. Following AChE-expression, these young postmitotic cells are going to extend the first axons (Weikert et al., 1990), as is shown by a double staining technique using AChE and the neurite-specific antibody G4 (Rathjen et al. 1987). It is predominantly the small molecular forms of both cholinesterases that are expressed during neurogenesis (Layer et al., 1987).

Since cholinesterases are expressed at a significant point of neuronal differentiation (scheme Fig. 1, lower), several questions are raised, that will be briefly discussed below.

II. Does BChE regulate cell proliferation and differentiation in vitro?

To test this question, we have added a highly purified BChE preparation from horse serum (Sigma, further purified,

0–8412–2008–5/91/0350$06.00/0

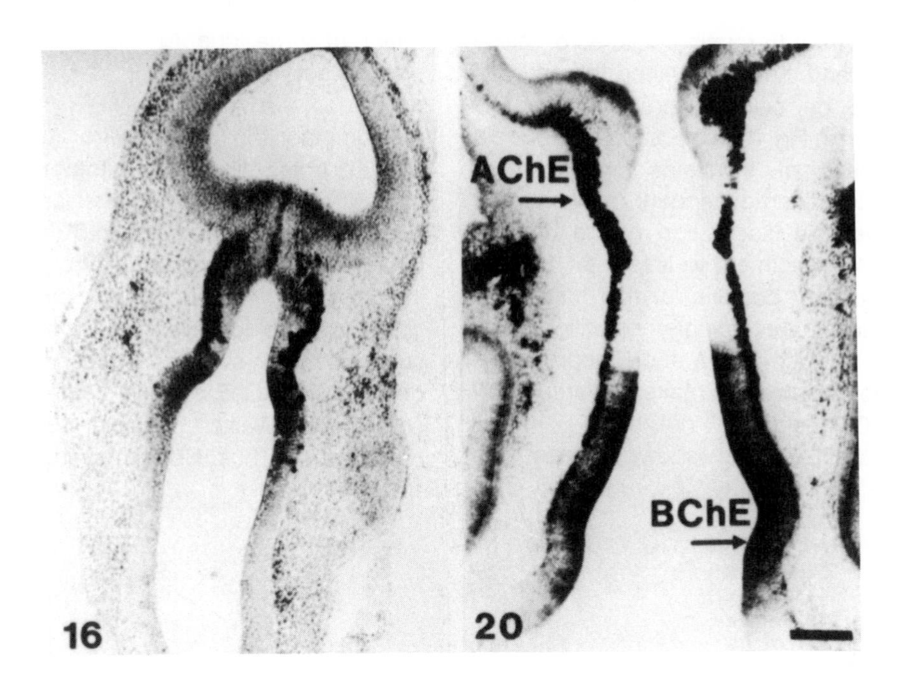

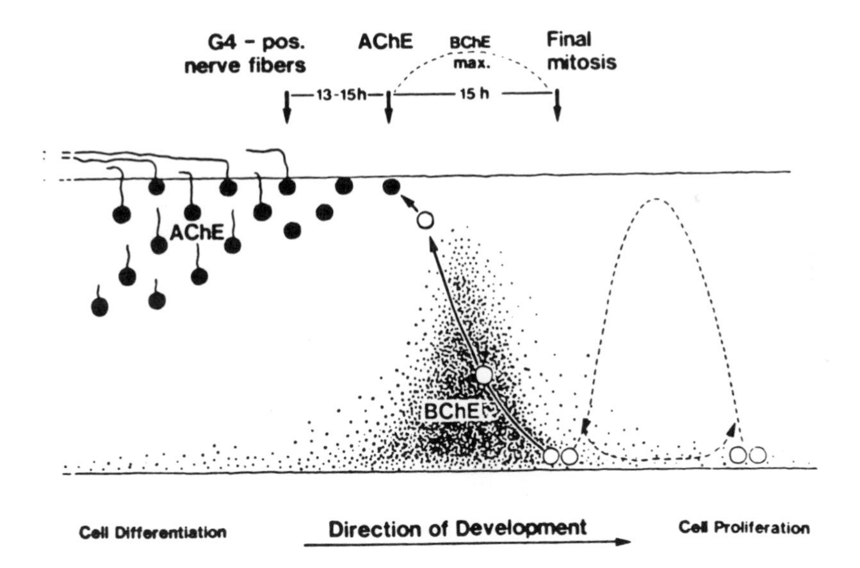

Fig. 1, upper. Expression of AChE and BChE in close vicinity to each other at early stages of neurogenesis (stages HH 16, left and HH20, right; method Karnovsky & Roots, 1964). **Lower.** Scheme showing the sequence of activation of BChE, of AChE and of neurite growth, all shortly following after final proliferation.

250 U/mg) to rotating cultures of retinal cells, and then determined the effect of BChE on cell proliferation rates. As shown in Fig. 2, BChE causes a 2-3-fold increase of thymidine uptake in a concentration-dependent manner. After irreversible inactivation of the enzyme by DFP-treatment to less than 1%, the stimulatory potential of this material is reduced. Accordingly, specific inhibitors, e.g. iso-OMPA and Ethopropazine inhibit thymidine-uptake, whereas BW 284C51 showed no effect on cell proliferation (not shown). Thus, the effect may be due to the esterolytic enzyme activity of BChE.

In summary, in this *in-vitro* system BChE acts like a mitogen, or alternatively it speeds up the last cell cycle to drive neuroblasts into the differentiation compartment. The latter interpretation is supported by our finding that blocking BChE activity in such *in-vitro* systems leads to a decrease of AChE synthesis within the cells (Layer, Weikert, Willbold, 1990; in preparation). Thus, BChE may interfere

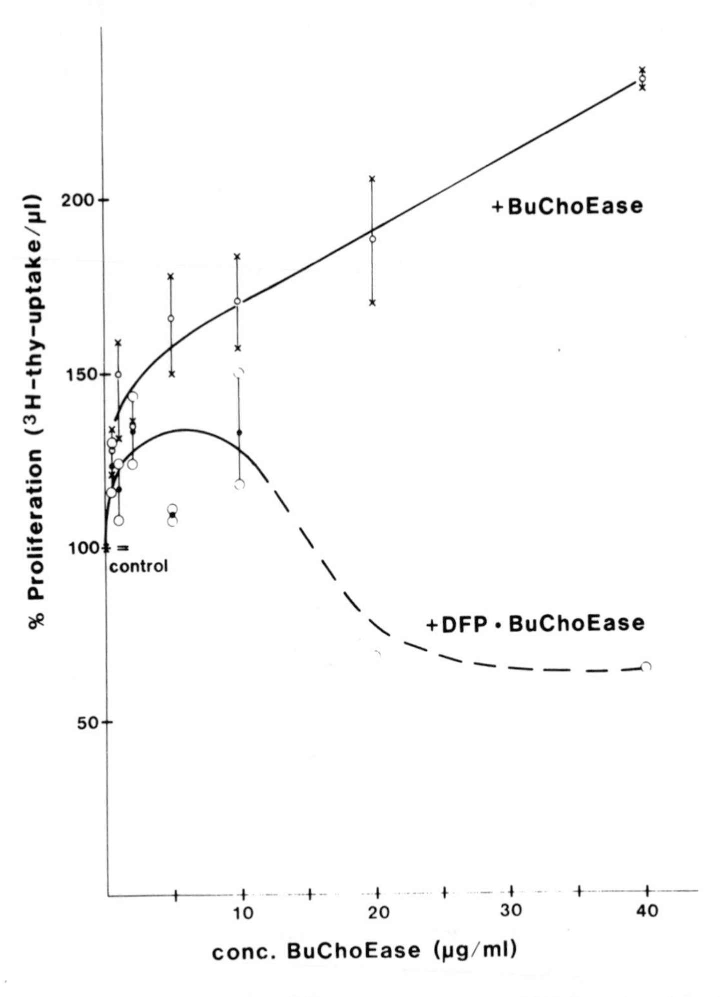

Fig. 2. BChE increases thymidine uptake into retinal single cells from an E6-embryo within a 2-day culture period. A DFP-inactivated, thoroughly dialyzed sample of BChE does not increase thymidine uptake (see text).

a) with the switch of cell proliferation into differentiation and b) with AChE expression. In the light of the close spatial association of the two enzymes, we therefore suggest that BChE acts by inducing neighboring cells to produce AChE and thereby become postmitotic. According to this interpretation, BChE acts as a differentiation switch.

III. AChE as a postmitotic marker to map patterns of primary brain development.

Using the AChE/G4 double staining technique (Layer et al. 1988b), the first biochemical and morphological differentiation of the young brain can be precisely studied. We have reconstructed developmental maps from AChE-stained serial sections of E1 -E4 chicken brains (Layer et al. 1988a; Weikert et al. 1990). The first two AChE-activation centers are detected almost simultaneously in the rhombencephalon and the diencephalon at stage 11 (Fig. 3; Layer & Alber, 1990). G4-positive neurites (stars) originate from AChE-activated areas (dots). Although the regulation of the AChE pattern in the brain is not understood, this result shows that AChE-producing cells are those which are going to establish the first long fiber tracts within the vertebrate brain. AChE as a sensitive differentiation marker has now been applied with similar results to most vertebrates (Moody and Stein 1988; Hanneman et al. 1988; Puelles et al. 1987).

IV. Cholinesterases at the origin, at the target and along pathways of spinal motor axons.

To investigate further the relationship of cholinesterase production with neurite growth for one of these long tract systems, the segmented spinal motor system of the trunk provides an excellent model, since all parts are segmentally and thus orderly organized; moreover, motor axons are restricted during their outgrowth to the anterior part of the corresponding sclerotome ("A-half"), while searching for their appropriate myotomal target within each somite (Keynes and Stern 1985). In the chick, the AChE/G4-double staining technique labels the entire

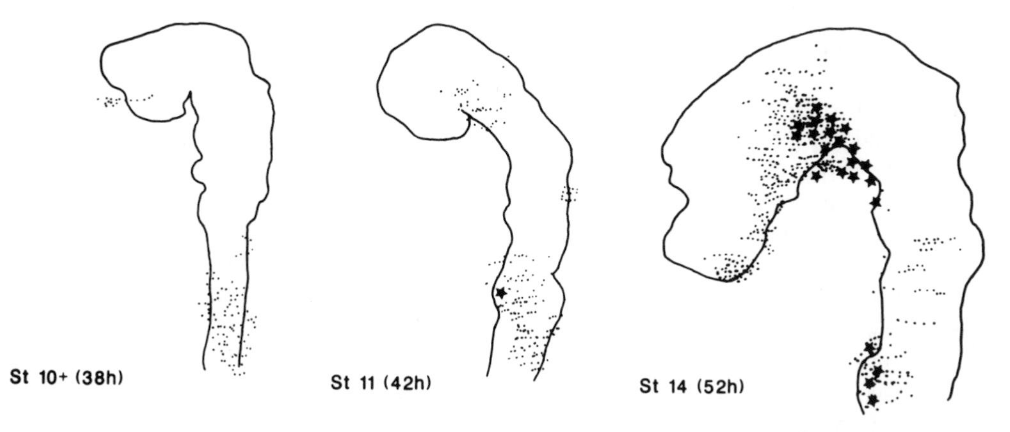

Fig. 3. Postmitotic AChE expression (dots) is polycentric and precedes the formation of major fiber tracts (stars) in the young chicken brain. Three early stages are shown (see text).

segmental motor unit including AChE-positive motoneuronal (MN) and myotomal (Myo) cell bodies and G4-positive motor axons (MAx) (Fig.4; Layer et al. 1988b). AChE is expressed in a nonsegmented rostro-caudal wave in motoneuronal cell bodies of the ventral horn and almost simultaneously in the corresponding dermomyotomes (Derm, Myo; see Fig.4). There, AChE is detectable in conjunction with BChE first in a rostromedial sector, thus representing the very first sign of muscle differentiation. Motor axons begin to grow only after the A-halves have elevated levels of BChE; and also they do so only after their origins and their target cells have been activated by AChE.

Thus, the entire process of early myotome differentiation, of motor axon outgrowth and of the first establishment of target contacts takes place within the A-half somite. Noticeably, both cholinesterases precede these processes. Thereby the origin and the target of a neuronal system undergo similar steps of differentiation, before they become interconnected by axons (see also Layer, 1990b).

V. A role for cholinesterases in the modulation of axon growth ?

A number of molecules are localized asymmetrically in one or the other sclerotome half (Chiquet 1989; Mackie et al. 1988; Rickmann et al. 1985; Hoffman et al. 1988), BChE is one of the earliest molecules expressed in the anterior sclerotome, and therefore it is a good candidate for regulating motor axon growth. By using *in-vitro* cultures, we have started to test whether cholinesterases can modulate neurite navigation patterns. In dissociated tectal cells kept in culture for two days, an extensive fiber growth is observed (Fig. 5, control), the amount and the morphology of which is sensitive to the application of anticholinesterases (Fig. 5). We have determined the amount of growing neurites quantitatively by an ELISA, using the G4-antibody as antigen. Both, BW 284C51 inhibiting AChE and Ethopropazine inhibiting BChE decrease fiber growth (not shown).

These results fit well with recent experiments that have shown that acetylcholine is secreted by some growing axons and that it can inhibit outgrowth of neurites (review Lipton and Kater 1989). If ACh would be released from growth cones, and if it also would inhibit neurite extension, then cholinesterases would attract and direct neurites by creating ACh-free spaces (see also Layer, 1990b).

VI. Conclusion: New roles of cholinesterases and cholinergic disorders?

In a number of neurological disorders levels of cholinesterases are significantly changed from normal. However, the causal involvement of these enzymes in the pathophysiology of these disorders has not been demonstrated conclusively. In most of these cases, cholinergic neurotransmission may not be distorted (for review see Rakonczay, 1988). The possible regulation of neurogenetic processes by cholinesterases as discussed in here could provide a new basis for understanding these defects. In particular, the regulation of the BChE gene expression needs further attention.

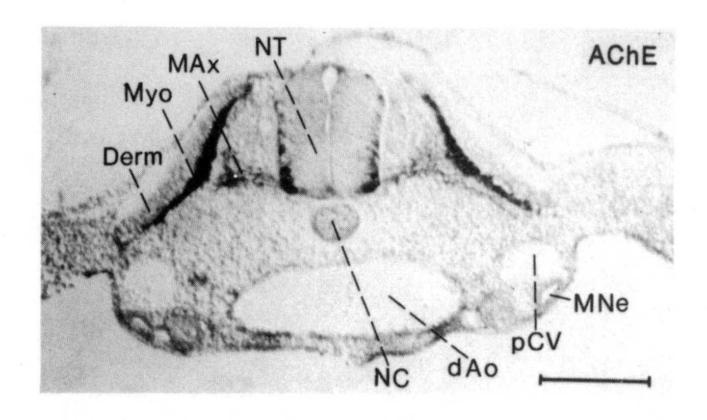

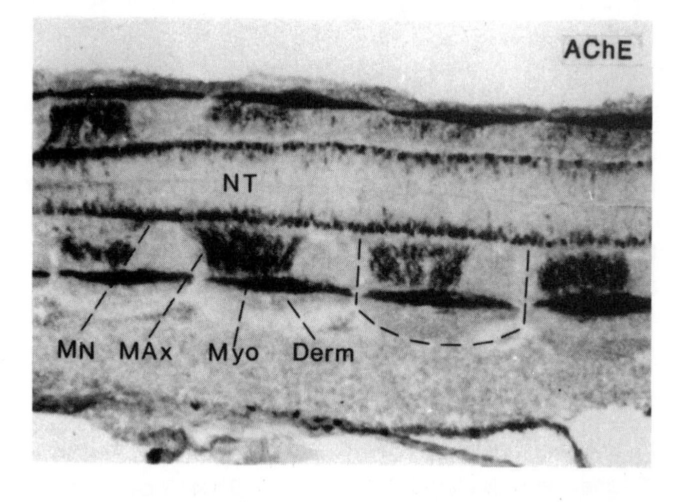

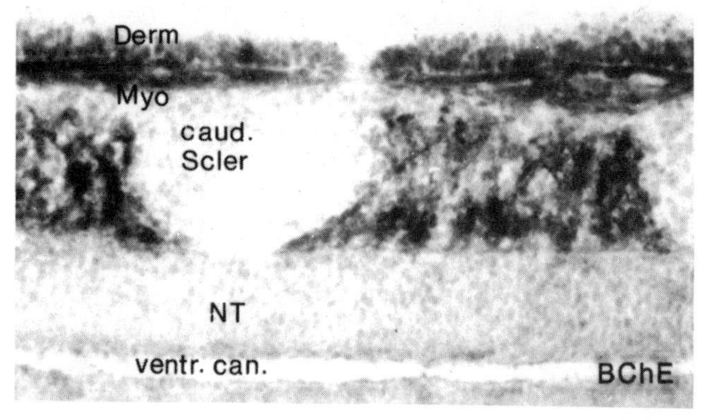

Fig. 4. Cholinesterases during development of the segmented spinal motor axons (MAx). A cross section of a HH19 embryo (upper) and parallel longitudinal frozen sections of a stage HH18 embryo are either stained by AChE plus G4-antibody (upper, middle), or BChE plus G4-antibody (lower). Further description see text. Bar = 400 μm (upper), = 100 μm (middle), = 200 μm (lower). Reproduced with permission from reference (Layer, 1990a).

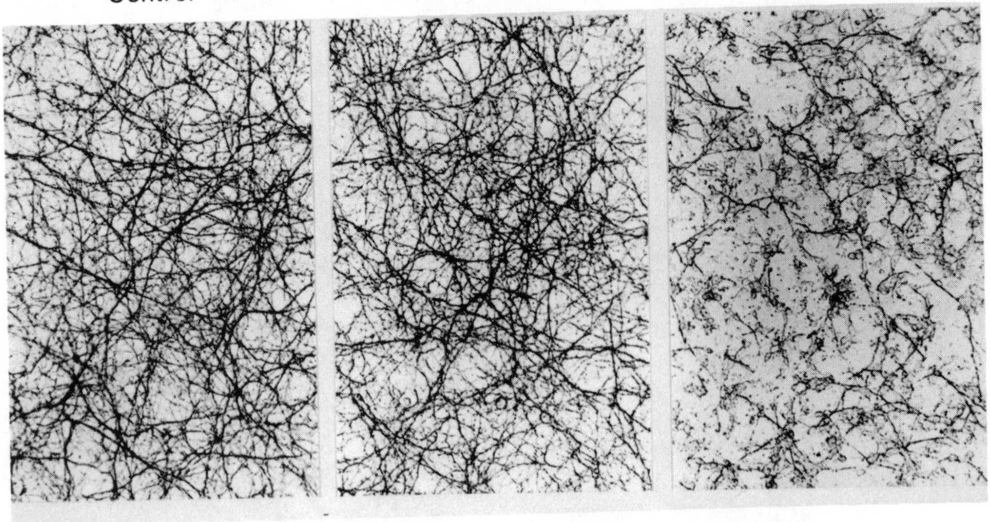

Control BW 284C51 Ethopropazine

Fig. 5. Anticholinesterases modulate neurite growth. 5 x 10^{-5} M BW 284C51 induces slight effects on the morphology and a significant decrease of G4-antigen expression (not shown). 10^{-5}M Ethopropazine changes both parameters significantly. Bar = 50 μm.

Acknowledgements

I wish to thank my colleagues: R. Alber, C. Ebert, G. Fischer von Mollard, R. Girgert, S. Kaulich, S. Kotz, L. Liu, P. Mansky, F.G. Rathjen, S. Rommel, O. Sporns, S. Treskatis, G. Vollmer, T. Weikert, E. Willbold. Secretarial assistance by C. Hug is acknowledged. This work was supported by the "Deutsche Forschungsgemeinschaft (DFG La 379/3-2; La 379/4-2)".

References

Chiquet, M. *Dev. Neurosci.,* **1989**, *11,* 266-275.

Hanneman, E., Trevarrow, B., Metcalfe, W.K., Kimmel, C.B., Westerfield, M. *Development,* **1988**, *103*, 49-58.

Hoffman, S., Crossin, K.L., Edelman, G.M., *J. Cell Biol.,* **1988**, *106*, 519-532.

Karnovsky, M.J., Roots, L.J., *J Histochem Cytochem*, **1964**, *12*, 219-221.

Keynes, R.J., Stern, C.D., *Trends Neurosci*, **1985**, *8*, 220-223.

Layer, P.G., *Proc Natl Acad Sci (USA)*, **1983**, *80*, 6413-6417.

Layer, P.G., *Cell & Mol Neurobiol*, **1990a**, in press.

Layer, P.G., *Bioassays,* **1990b**, submitted.

Layer, P.G., Alber, R., *Development*, **1990**, in press.

Layer, P.G., Sporns, O., *Proc Natl Acad Sci (USA),* **1987**, 84, 284-288.

Layer, P.G., Alber, R., Rathjen, F.G., *Development*, **1988b**, *102*, 387-396.

Layer, P.G., Alber, R., Sporns, O. *J. Neurochem*, **1987**, *49*, 175-182.

Layer, P.G., Rommel, S., Bülthoff, H., Hengstenberg, R., *Cell Tiss. Res.*, **1988a**, *251*, 587-595.

Lipton, S.A., Kater, S.B., *Trends Neurosci*, **1989**, *7*, 265-270.

Mackie, E.J., Tucker, R.P., Halfter, W., Chiquet-Ehrismann, R., Epperlein, H.H., *Development,* **1988**, *102*, 237-250.

Massoulié, J., Bon, S., *Ann. Rev.*

Neurosci., **1982**, *5*, 57-106.
Moody, S.A., Stein, D.B., *Dev. Brain Res.,* **1988**, *39*, 225-232.
Puelles, L., Amat, J.A., Martinez-de-la-Torre, M., *J. Comp. Neurol.,* **1987**, *266*, 247-268.
Rakonczay, Z., *Prog. Neurobiol.*, **1988**, *31*, 311-330.
Rathjen, F.G., Wolff, J.M., Frank, R., Bonhoeffer, F., Rutishauser, U., *J. Cell Biol.*, **1987**, *104*, 343-353.
Rickmann, M., Fawcett, J.W., Keynes, R.J., *J. Embryol. Exp. Morphol.,* **1985**, *90*, 437-455.
Silver, A., *North-Holland, Amsterdam*, **1974**.
Weikert, T., Rathjen, F.G., Layer, P.G., *J. Neurobiol.*, **1990**, *21*, 482-498.

Transiently Expressed Acetylcholinesterase Activity in Developing Thalamocortical Projection Neurons

Richard T. Robertson, Department of Anatomy and Neurobiology, College of Medicine, University of California, Irvine, California 92717

ABSTRACT

Acetylcholinesterase (AChE) activity is expressed transiently as a dense plexus of fibers in layer IV of primary sensory areas of cerebral cortex during the first two postnatal weeks of life. The distribution of transient AChE activity corresponds to the distribution of primary sensory thalamocortical axon terminals. Sensory nuclei of the dorsal thalamus also display prominent transient AChE staining. Lesions of the sensory nuclei of thalamus result in loss of the pattern of AChE in corresponding areas of cerebral cortex. These data indicate that transiently expressed AChE activity is characteristic of primary sensory thalamocortical neurons. This AChE activity is expressed during the time when thalamocortical axons are growing into layer IV and forming connections with cortical neurons.

Acetylcholinesterase (AChE) has been the subject of a great deal of investigation during recent years primarily because of its important role as the catabolic enzyme for acetylcholine in cholinergic neuro-transmission. However, because AChE can be found in systems that are believed to be neither cholinergic nor cholinoceptive, the hypothesis has been presented that AChE may have functions other than the hydrolysis of acetylcholine (Chubb et al., 1980; Greenfield, 1984).

One of the more intriguing possibilities is that AChE may have a functional role in neural development. Suggestions that AChE may have a morphogenic role have been made for several decades, but conclusive evidence remains elusive. Requirements for establishing a functional role for AChE in neural development would include morphological data indicating the presence of AChE in an identified neural system at a particular time of its development. Such evidence is presented in this chapter.

Transient AChE Activity in Primary Sensory Areas of Cerebral Cortex.

The photomicrographs presented in figure 1 illustrate patterns of AChE histochemical activity (Tago et al., 1984) in cerebral cortex of developing rats. It is clear that AChE activity is most intense in the primary sensory areas of cerebral cortex, including the primary auditory, somatosensory, and visual regions. Work from this and other laboratories has indicated that the transiently expressed AChE in each of these cortical regions shows similar laminar patterns and similar, but not identical, temporal characteristics (Kristt, 1979; Robertson, 1987).

Figure 2 provides a closer look at the AChE activity in the primary auditory area of temporal cortex. Most noticable is the intense band of AChE staining in superficial layers of cortical area 41, primary auditory cortex. The intense AChE activity is found in cortical layer IV and the deepest portion of cortical layer III. This laminar pattern of AChE activity is characteristic of all three primary sensory cortical fields and corresponds to the laminar pattern of primary sensory thalamocortical projections. In addition to the transient pattern of AChE in superficial layers of cortex, a number of AChE-positive neuronal cell bodies and a population of rather crisply defined AChE-positive axons are seen throughout the cortical layers.

Interestingly, the band of AChE activity in cortical layers III and IV is transient. The dense band of AChE is first detectable early in the first postnatal week of life, appears to reach peak intensity between postnatal day 8 and 10, and then declines to adult levels by the end of the third postnatal week. Certainly all of cerebral cortex, including these primary sensory areas, show AChE activity in adults (Lysakowski et al., 1986), but the dense band of AChE activity is not seen in animals older than three weeks.

Transient AChE Activity in Primary Sensory Nuclei of Thalamus.

The regional distributions of AChE staining in primary sensory cortical areas are reminiscent of primary thalamocortical terminal fields. This correspondence between the location of transiently expressed AChE and the primary sensory thalamocortical terminal fields indicates that AChE may be expressed transiently within

0–8412–2008–5/91/0358$06.00/0

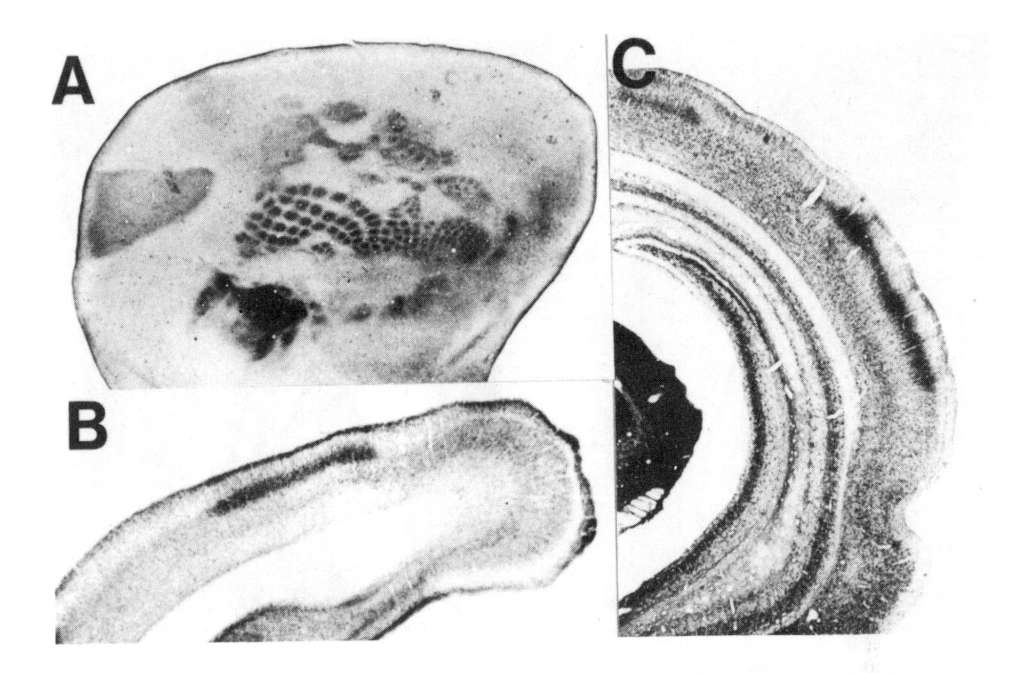

Fig. 1. AChE reaction product in infant rat cortex. **A**: Tangential section through layer IV of the right hemisphere. Transverse sections illustrate visual (**B**) and auditory (**C**) staining.

these thalamocortical neurons. If so, AChE should be present in neuronal somata in the primary sensory nuclei of the dorsal thalamus that send projections to sensory cortex.

Photomicrographs in figure 3 illustrate the pattern of AChE staining in dorsal thalamus of developing and adult rats. Note the intense AChE activity in the infant medial geniculate body (which sends projections to auditory cortex). Figure 3 further illustrates that AChE activity in these thalamic nuclei is transient, and is not found in adults. Activity is first detectable at or shortly after birth, reaches peak intensity around the end of the first postnatal week, and declines to adult levels (Lysakowski et al., 1986) by the third postnatal week.

Electron Microscopic Localization of Transiently Expressed AChE.

Our understanding of the function of transiently expressed AChE activity must be based on knowledge of the cellular compartments in which the AChE is located. Examination of fine structure of neuron cell bodies in the sensory nuclei of thalamus reveal that virtually all thalamocortical neurons of infant rats display prominent levels of AChE activity associated with granular endoplasmic reticulum. An example is presented in figure 4A. In animals older than 3 weeks of age, these thalamic neurons display essentially no AChE activity in their perikaryal cytoplasm, even though surrounding neuropil typically shows abundant AChE activity. Electron microscopic analysis of cerebral cortex of infant rats reveals that AChE activity is associated with axon terminals and with unidentified membranous structures in the neuropil (fig. 4B).

Sources of AChE Activity in Sensory Regions of Cerebral Cortex.

If the pattern of AChE activity in sensory cortex is associated with thalamocortical axon terminals, then this pattern of AChE should be affected by lesions that interfere with the thalamocortical projections. Figure 5 presents results from cases in which small electrolytic lesions were placed in subcortical regions. It is clear that placement of a lesion in the dorsal thalamus results in a marked loss of AChE activity in corresponding regions of cerebral cortex (fig. 5B). Similarly, neonatal placement of lesions that remove the major afferent projections to dorsal thalamus result in decrease in AChE staining in the thalamus and in the thalamocortical terminal fields in sensory cortex (fig. 5C).

Contrasting results are obtained following placement of lesions in the basal forebrain region, the major source of AChE-positive cholinergic afferents to cortex in adult animals.

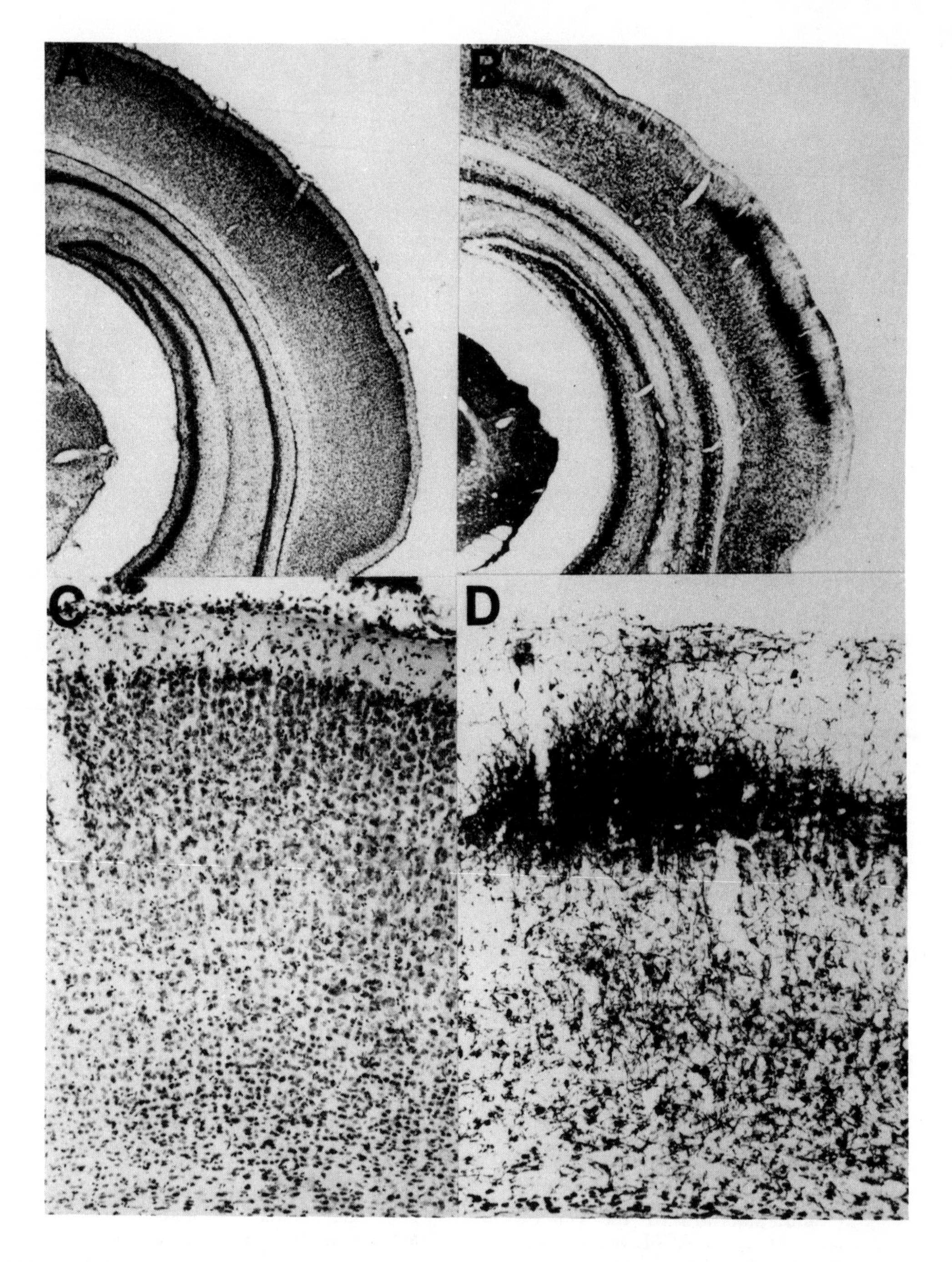

Fig. 2. AChE histochemical staining in cortical layer IV and deep III in infant rat. **A**: Nissl stained section. **B**: AChE stained section adjacent to 'A'. **C**: Nissl section showing cortical layers. **D**: AChE stained section adjacent to 'C'.

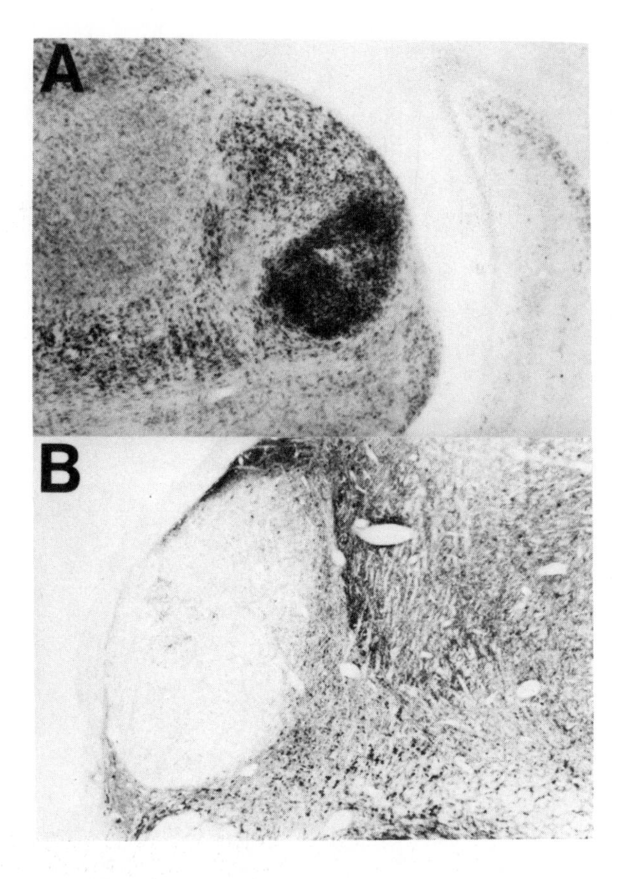

Fig. 3. AChE staining of dorsal thalamus of infant (**A**) and adult (**B**) rat medial geniculate bodies.

In these infants, basal forebrain lesions had minimal effect on AChE staining in cortical layers IV and III, but these lesions did eliminate the sparser AChE-positive axons normally seen throughout the cortex.

Is The Transiently Expressed AChE Associated with Cholinergic Function?

It seemed prudent to determine whether the transiently expressed AChE activity might be associated with a (possibly transient) cholinergic system. We believe this not to be the case, for two principle reasons. First, we have studied the development of choline acetyltransferase (ChAT) immuno-reactivity in cerebral cortex and thalamus, and found no regional or temporal pattern of ChAT that corresponds to the transient patterns of AChE. Second, manipulations that strongly affect AChE activity have no detectable impact on ChAT activity (Robertson et al., 1988), suggesting that the two enzymes are not found associated in the same neural system.

Molecular Forms of Transient AChE.

We have studied the molecular forms of AChE activity in cerebral cortex and thalamus of developing rats, to determine whether the transiently expressed AChE might be a novel or unique form of the enzyme. As illustrated in figure 6, cortical tissue from animals sacrificed before, during, and after the peak of transient AChE activity, all show essentially the same sucrose gradient profiles of AChE.

Interpretation.

The histochemical data presented here indicate that developing primary sensory thalamocortical neurons express AChE activity transiently during early postnatal development. This transiently expressed AChE is produced by thalamocortical projection neurons and transported via their thalamocortical axons to terminal fields in cortex. The timing of transiently expressed AChE corresponds closely to the time

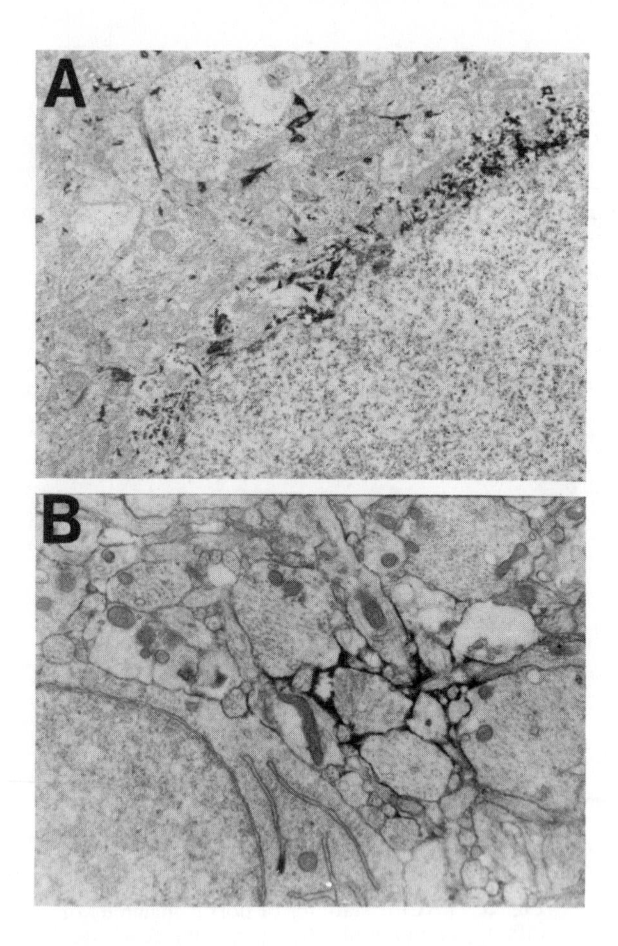

Fig. 4. Electron micrographs showing AChE staining of cytoplasm of a thalamocortical neuron (**A**) and of an axon terminal and neuropil (**B**) in cortex.

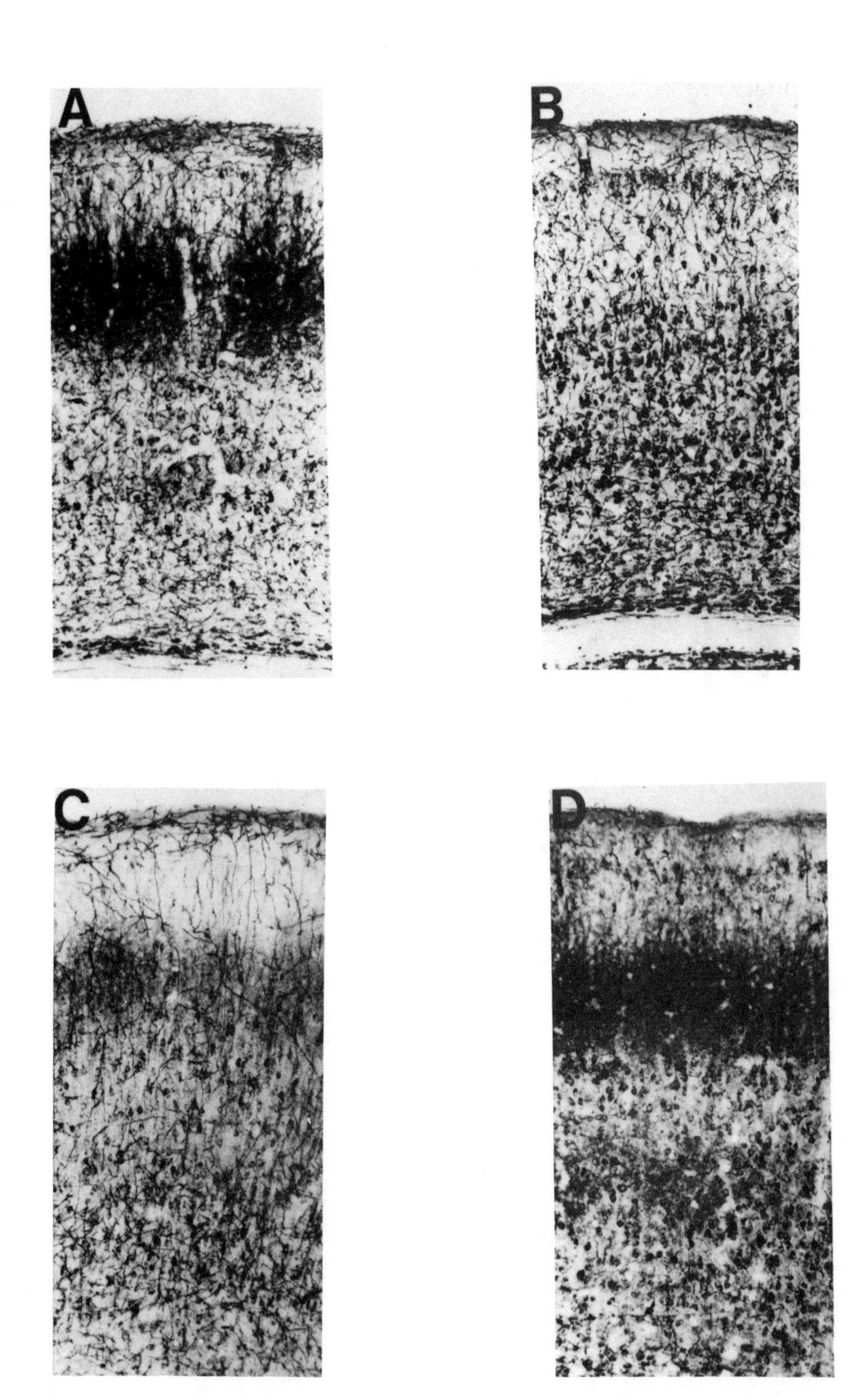

Fig. 5. Effects of subcortical lesions on patterns of AChE staining in infant temporal cortex. **A:** Normal AChE staining. **B**: Auditory cortex with lesion of medial geniculate body. **C**: Auditory cortex with a neonatal lesion of the inferior colliculus. **D**: Auditory cortex with lesion of the medial globus pallidus.

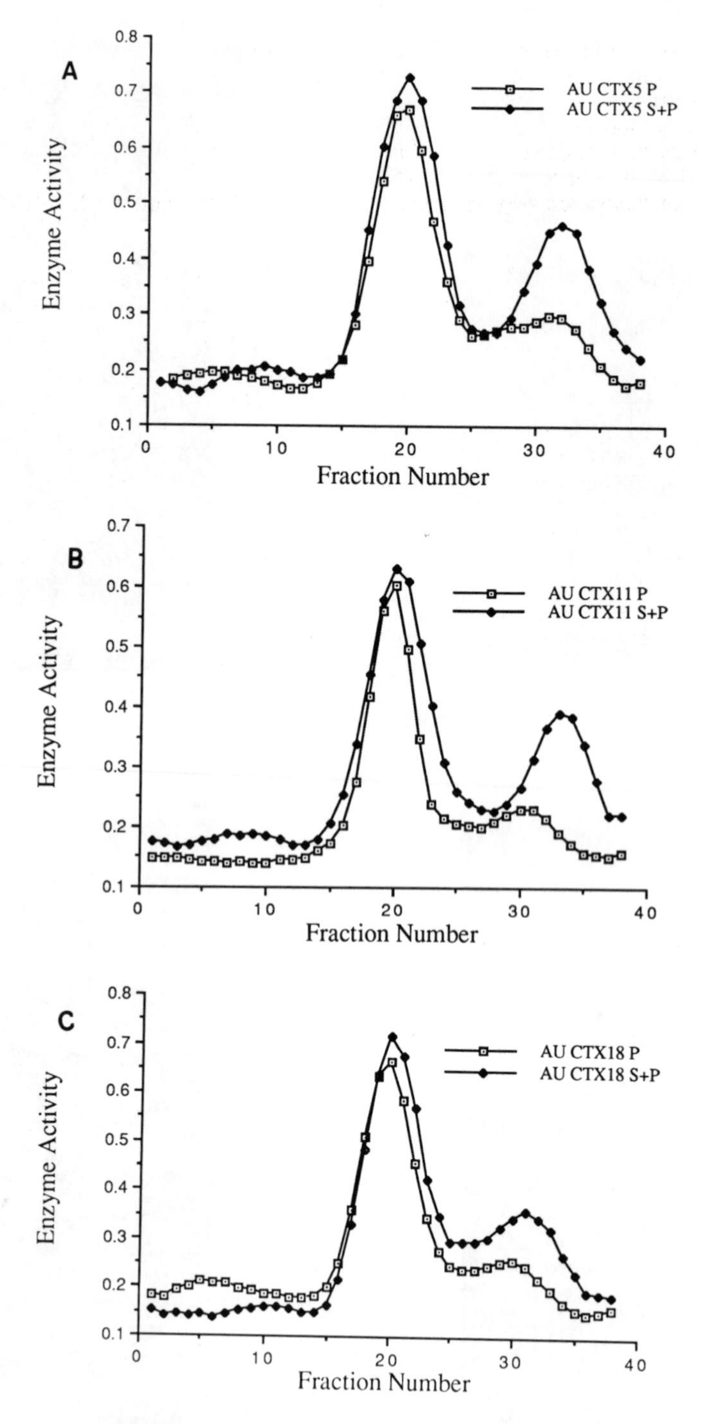

Fig. 6. Sucrose gradients of AChE from auditory cortex in 5 day (**A**), 11 day (**B**), and mature (**C**) rats.

of ingrowth and proliferation of primary thalamocortical axons within sensory cortex (Lund and Mustari, 1977; Wise and Jones, 1978). This close temporal association of AChE expression and growth of thalamocortical axons implies that the AChE may have a functional role in development of thalamocortical connectivity.

Acknowledgements

Supported in part by NIH grant NS 25674 and NSF grant BNS 87-08515. I thank K.A. Gallardo, G.H. Kageyama and J. Yu for their collaborations.

References

Chubb, I. W., A.J. Hodgson, and G.W. White, *Neurosci.*, **1980**, *5* , 2065-2072.

Greenfield, S., *Trends in Neurosci.*, **1984**, *8*, 364-368.

Kristt, D. A., *J. Comp. Neurol.*, **1979**, *186*, 1-16.

Lund, R.D. and M.J. Mustari, *J. Comp. Neurol.*, **1977**, *173*, 289-306.

Lysakowski, A., B.H. Wainer, D.B. Rye, G. Bruce, and L.B. Hersh, *Neurosci. Lett.*, **1986**, *64*, 102-108.

Robertson, R. T., *Neurosci. Lett.*, **1987**, *75*, 259-264.

Robertson, R.T., C.F. Höhmann, J.L. Bruce, and J.T. Coyle, *Dev. Brain Res.*, **1988**, *39*, 298-302.

Tago, H., H. Kimura and T. Maeda, *J. Histochem. Cytochem.*, **1986**, *34*, 1431-1438.

Wise, S.P. and E.G. Jones, *J. Comp. Neurol.*, **1978**, *178*, 187-208.

A Non-cholinergic Function for AChE in the Substantia Nigra

Susan A Greenfield, University Department of Pharmacology, Oxford, OX1 3QT, U.K.

Background

Within the substantia nigra, there is very little evidence for significant cholinergic transmission and it appears that the large amounts of AChE found there may have a novel, non-cholinergic function (Greenfield, 1984). This idea is particularly attractive in the light of the observation that AChE can exist not only in a membrane bound form, but as a soluble protein (Chubb and Smith, 1975). Indeed, within the substantia nigra ultrastructural clues have strongly suggested that AChE is secreted specifically from dopamine (DA)-containing nigrostriatal neurons, (Henderson and Greenfield 1984): first, it is claimed that the distinguishing morphological feature of nigrostriatal cells is the abundance of rough endoplasmic reticulum (Domesick et al.,1983), which would enable the rapid sythesis and turnover of protein, such as soluble AChE. Secondly, AChE is associated with the Golgi apparatus of nigrostriatal cells specifically: since soluble AChE must pass through the Golgi apparatus prior to 'externalisation' (Rotundo, 1984), the presence of the protein in the Golgi would suggest that it was destined for release. This idea is validated by the strong correlation, within individual cells, of an AChE-reactive Golgi apparatus, and AChE- reactive external plasma membrane (Henderson and Greenfield, 1984). A third clue to the functioning of nigrostriatal cells comes from the observation that AChE (Henderson and Greenfield, 1984) and indeed DA (Mercer et al.,1978) are stored in the smooth endoplasmic reticulum (SER) of the dendrites. This location and manner of storage suggests a non-quantal, and sustained process of release unlike that at the classic axonal synapse (see Wassef et al.,1981). Indeed these clues have proved valid: lesions with the selective neurotoxin 6-OHDA caused a marked decline in secretion of AChE in the substantia nigra (Greenfield et al., 1983a). However there are no dopaminergic axon terminals nor axon collaterals in this brain region, and hence we can only postulate that, as for DA itself (Nieoullon et al., 1977), AChE is secreted from nigrostriatal cell dendrites (see Greenfield 1985). Within the last ten years, it has been shown that dendritic release of AChE in the substantia nigra can be evoked by a diverse range of transmitters and related substances (Table 1). However, this evoked release is not necessarily linked to the corresponding excitability of the nigrostriatal cells from which it occurs (Greenfield 1985), nor indeed to the release of DA.

Release of AChE

Since dendritic release of AChE is not caused by a direct increase in the discharge rate of the neuron, it was clearly important to identify the ionic basis of evoked release of the protein. Previous intracellular recordings from nigrostriatal neurons <u>in vitro</u> had already provided a clue: it was found that nigrostriatal cells could generate a TTX-resistant, Ca-dependent voltage transient at sites remote from the soma, ie the dendrite (Llinas et al., 1984). This conductance ('High Threshold', HTSgCa) was only evoked by a very high threshold of stimulation and due to its apparent similarity in location and ionic requirements was a good candidate for the basis for dendritic release.

To test this hypothesis the following strategy was adopted: identify a transmitter ('X') in direct contact with a DA-containing dendrite; verify that 'X' can evoke release of AChE; investigate whether 'X' can enhance the HTSgCa. In brief, we have found that 5-HT immunoreactive boutons make direct synaptic contact with the dendrites of nigrostriatal cells (Nedergaard et al., 1988), that 5-HT evokes release of AChE in the

0–8412–2008–5/91/0366$06.00/0

Table 1

Evoked Release of AChE in the Substantia Nigra

TREATMENT	SPECIES	REF
Veratridine	Rat (*in vitro*)	Cuello et al., 1981
Potassium ions	Rat (*in vitro*)	Cuello et al., 1981
	Cat	Greenfield et al., 1980
	Rabbit	Greenfield et al., 1983
	Guinea pig	Taylor et al., 1988
Haloperidol	Rat	Weston & Greenfield, 1984
Amphetamine	Rabbit	Greenfield & Shaw, 1982
	Guinea pig	Taylor et al., 1988
Gamma-hydroxyburytrate	Rat	Weston & Greenfield, 1986
5-HT	Guinea pig	Taylor et al., 1988
Peripheral ES*	Guinea pig	Taylor et al., 1989c
Central ES	Guinea pig	Taylor et al., 1989b
Drinking	Guinea pig	Taylor et al., 1990

*ES Electrical Stimulation

substantia nigra (Taylor et al., 1988) and indeed that 5-HT enhances the HTSgCa (Nedergaard et al.,1988). Hence it seems that the HTSgCa does indeed mediate dendritic release of AChE; this conclusion has been substantiated by repeating the above procedure with diverse other transmitter related substances (Nedergaard et al., 1989). In conclusion, dendritic release of AChE can occur, independent of cell discharge, via calcium entry at dendritic sites electrotonically remote from the cell body upon activation of the synaptic contacts in that region: the corollary of this conclusion is that dendritic release occurs, not as a function of the excitability of the cell of origin, but of the inputs to it.

The features of dendritic release of DA and AChE are strikingly similar and are shown in Table 2. The general picture that emerges is one not completely comparable with familiar events at the classic axonal synapse. Rather, it seems that the sustained and less specific secretion of the transmitter and protein might represent a more 'modulatory' phenomenon. However, until very recently, studies on dendritic secretion of AChE were hampered by the poor time resolution (ten minutes) of the conventional assay/push-pull cannulae perfusion. Although it was possible to study the pharmacology of AChE release using this procedure, attempts at exploring the functional aspects of the phenomenon required a time scale commensurate with

Table 2

Similarities Between Dopamine and Acetylcholinesterase in the Substantia Nigra

(i)	Co-localized in nigrostriatal neurons (Butcher et al., 1975).
(ii)	Not stored in vesicles (Wassef et al., 1981), but SER of dendrites (Mercer et al., 1978; Liesli et al., 1980; Henderson & Greenfield, 1984).
(iii)	Accumulate in parallel during development (Butcher et al., 1975).
(iv)	Spontaneous and evoked release from somata/dendrites of nigrostriatal cells (Greenfield, 1985).
(v)	Release not related to firing frequency (Greenfield, 1985).
(vi)	Release not blocked by TTX or gamma-butyrylactone (Weston & Greenfield, 1986).
(vii)	Release potassium evoked, calcium-dependent (Greenfield et al., 1983b).
(viii)	Release evoked by amphetamine (Greenfield & Shaw, 1982).
(ix)	Slow diffusion rates through the extracellular space of AChE: '1-2mm/hr (Kreutzberg & Kaizer, 1974); DA: $100\mu/10$ sec 1mm/17 mins (see Greenfield, 1985).

behavioural events, and hence were futile.

Nonetheless, within the last few years we have adapted a chemiluminescent assay for acetylcholinesterase (Israel and Lesbatts, 1981) for monitoring the secretion of AChE within the substantia nigra 'on line' both in vitro (Llinas and Greenfield, 1987) and in vivo (Taylor et al., 1989a, 1989b). Moreover, it has proved possible to use this system in the freely moving animal: secretion of AChE in the substantia nigra appears to be greater and to occur in a pulsatile manner when the animal is not anaesthetized (Taylor et al., 1990). We can thus combine the conclusions from the electrophysiological and behavioural studies to form the syllogism that at least some inputs to the substantia nigra relay signals relating to conscious movement.

Action of AChE

Once released, it appears that both exogenous (Lacey et al.,1987) and endogenous DA (Nedergaard et al.,1988; Kapoor et al.,1989;) hyperpolarize the nigrostriatal cell membrane. A similar effect is seen following application of AChE where the catalytic site is actually inactivated (Greenfield et al.,1988; Greenfield et al.,1989). The significance of the hyperpolarisation, in both cases, is dependent on an interesting property of the nigrostriatal cell membrane. When these neurons are hyperpolarised, a conductance is de-inactivated; this conductance ('Low Threshold Calcium Conductance, LTSgCa') is mediated by the sustained entry of calcium ions, following activation, ie depolarization, from the hyperpolarized level (Llinas et al.,1984). The consequence of generation of the LTSgCa is to evoke a pattern of bursts of action potentials, which in turn leads to a non-linear increase in release of DA from the terminals in the striatum (Gonon, 1988).

In a physiological situation, it has been proposed that the LTSgCa could be generated by activation from synaptic inputs from a hyperpolarized level induced by dendritically released DA and/or AChE (Greenfield, 1985). The corollary of this hypothesis would be that a function of dendritically released substances would be to enhance the sensitivity of nigrostriatal cells to synaptic input. This idea has been shown to be valid in that, independent of hydrolysis of acetylcholine, AChE can enhance the firing pattern of nigrostriatal cells following orthodromic stimulation from the striatum (Last and Greenfield, 1987) and indeed causes changes in motor behaviour indicative of enhanced excitibility in the nigrostriatal pathway (Greenfield et al.,1984; Weston and Greenfield, 1985).

Conclusions

It appears then that a soluble, secreted form of AChE may act in a novel non-cholinergic capacity, perhaps in conjunction with DA, to modulate the processing of signals in the substantia nigra, and hence to play a critical role in the control of movement.

If it is the case that dendritic release of AChE represents a novel mechanism of neuromodulation in the substantia nigra, then it seems not unreasonable that similar phenomena might exist incorporating either other secretory proteins and/or be generated in other brain regions. Already diverse lines of evidence suggest that these possibilities actually do occur.

Release of AChE has been reported in the striatum (Greenfield et al.,1980; Weston and Greenfield, 1986), the hippocampus (Appleyard and Smith, 1987), the cerebellum (Appleyard et al.,1988) and perhaps the hypothalamus (Romero and Smith, 1980; Greenfield and Smith, 1979). Furthermore, despite little evidence of cholinergic transmission, AChE is co-stored with 5-HT and noradrenaline (NA) in the Raphe nucleus (Butcher and Woolf, 1982) and locus coeruleus (Albanese and Butcher, 1979), respectively. Finally, there is indirect evidence that AChE could have a modulatory, non-cholinergic action on neurons in the locus coeruleus (Greenfield et al., 1986), and a recent direct demonstration of a non-classical action of AChE on cerebellar Purkinje cells (Appleyard and Jahnsen, 1989).

Regarding secretion of proteins from CNS neurons, we already know that aminopeptidase (Greenfield and Shaw, 1982), and unidentified proteins (Greenfield et al.,1983) are released in a physiological fashion from the striatum. In addition, dopamine-beta-hydroxylase (DBH) is secreted from the brain into cerebrospinal fluid, (De Potter et al., 1980) in a fashion analagous to that seen for AChE (Greenfield and Smith, 1979).

In summary, it seems that secretion of AChE in the substantia nigra is not a unique phenomenon, either in terms of brain area or substance. The substantia nigra, however, might be regarded as an example of where a confluence of features, including dendritic release and action of AChE, can yield a particular neuronal context for specific behaviour or function.

References

Albanese, A.; Butcher, L.L. *Neurosci. Lett.* **1979** *14*, 101-104.

Appleyard, M.E.; Smith, A.D. *Neurochem. Int.* **1987**, *11*, 397-406.

Appleyard, M.E.; Vercher, J.L.; Greenfield, S.A. *Neuroscience* **1988**, *25*, 133-138.

Appleyard, M.E.; Jahnsen, H. *Eur. J. Neurosci.* **1989**, In Press.

Butcher, L.L.; Talbot, K.; Bilezikjian *J. Neural Trans.* **1975** *37*, 127-153.

Butcher, L.L.; Woolf, N.J. *Prog. Brain Res.* **1982**, *55*, 3-40.

Chubb, I.W.; Smith, A.D. *Proc. Roy. Soc.* **1975**, *191*, 245-261.

De Potter, W.P.; De Potter, R.W.; De Smet, F.H.; De Schaepdryver, A.C. *Neuroscience* **1980**, 1969-1977.

Domesick, V.B.; Stinus, L. *Neuroscience* **1983**, *8*, 743-765.

Gonon, G.G. *Neuroscience* **1988**, *24*, 19-28.

Greenfield, S.A.; Smith, A.D. *Brain Res.* **1979**, *177*, 445-459.

Greenfield, S.A.; Shaw, S.G. *Neuroscience* **1982** *7*, 2883-2893.

Greenfield, S.A. *Trends Neurosci.* **1984**, *7*, 364-368.
Greenfield, S.A.; Chubb, I.W.; Grunewald, R.A.; Henderson, Z.; May, J.; Portnoy, S.; Weston, J; Wright, M.C. *Exp. Brain Res.* **1984** , *54*, 513-520.
Greenfield, S.A. *Neurochem. Int.* **1985**, *7*, 887-901.
Greenfield, S.A.; Grunewald, R.A.; Foley, P.; Shaw, S.G. *J. Comp. Neurol.* **1983a**, *214*, 87-92.
Greenfieldk S.A.; Cheramy, A.; Leviel, V.; Glowinski, J. *Nature* **1980**, 355-357.
Greenfield, S.A.; Cheramy, A.; Glowinski, J. *J. Neurochem.* **1983b** *40*, 1045-1057.

Greenfield, S.A.; Jack, J.J.B.; Last, A.T.J.; French, M. *Exp. Brain Res.* **1988**, *70*, 441-444.
Greenfield, S.A.; Nedergaard, S.; Webb, C.; French, M. *Neuroscience*, **1989**, *29*, 21-25.
Henderson, Z.; Greenfield, S.A. *J. Comp. Neurol.* **1984**, *230*, 278-286.
Israel, M.; Lesbats, B. *Neurochem. Int.* **1981**, *3*, 81-90.
Kapoor, R.; Webb, C.; Greenfield, S.A. *Exp. Brain Res.* **1989**, *74*, 653-657.
Lacey, M.; Mercuri, N.B.; North, R.A. *J. Physiol.* **1989**, *74*, 397-416.
Last, A.T.J.; Greenfield, S.A. *Exp. Brain Res.* **1987**, *67*, 445-448.
Liesli, P.,; Panula, P.; Rechardt, L. *Histochemistry* **1980** *70*, 7-18.
Llinas, R.; Greenfield, S.A.; Jahnsen, H.J. *Brain Res.* **1984**, *294*, 127-132.
Weston, J.; Greenfield, S.A. *Behav. Brain Res.* **1985**, *18*, 71-74.
Weston, J.; Greenfield, S.A. *Neuroscience* **1986**, *17*, 1079-1088.
Llinas, R.R.; Greenfield, S.A. *Proc. Nat. Acad. Sci. USA* **1987**, *84*, 3047-3050.
Mercer, L.; del Fiacco, M.; Cuello, A.C. *Experientia* **1978**, *35*, 101-103.
Nedergaard, S.; Bolam, J.P.; Greenfield, S.A. *Nature*, **1988**, *333*, 174-177.
Nedergaard, S.; Webb, C.; Greenfield, S.A. *Acta Physiol. Scand.* **1989**, *135*, 67-68.
Nieoullon, A.; Dusticier, N. *J. Neural. Trans.* **1982**, *53*, 133-146.
Romero, E.; Smith, A.D. *J. Physiol.* **1980**, *301*, 52P.
Rotundo, R. In *Cholinesterases: Fundamental and Applied Aspects*; Brizin, M.; Kiauta, T.; Barnard, E.A., Ed., Walter de Gruyter: The Hague, 1984, 203-218.
Taylor, S.J.; Bartlett, M.J.; Greenfield, S.A. *Neuropharmacology* **1988**, *27*, 507-514.
Taylor, S.J.; Haggblad, J.; Greenfield, S.A. *Neurochem. Int.* **1989a**, *15*, 199-205.
Taylor, S.J.; Greenfield, S.A. *Brain Res.* **1989b**, *505*, 153-156.
Taylor, S.J.; Jones, S.A.; Haggblad, K.; Greenfield, S.A. *Neuroscience* In Press.
Wassef, M.; Berod, A.; Sotelo, C. *Neuroscience* **1981**, *6*, 2125-2139.

The Aryl Acylamidase and Peptidase Activities of Human Serum Butyrylcholinesterase

A.S. Balasubramanian, Neurochemistry Laboratory, Christian Medical College Hospital, Vellore 632 004, India

Human serum butyrylcholinesterase exhibits an amine sensitive aryl acylamidase and a peptidase activity. The cholinesterase and aryl acylamidase could be localized in a 20 kDa fragment and the peptidase in a 50 kDa fragment obtained by limited alpha chymotrypsin digestion of purified butyrylcholinesterase. These fragments were identified to be parts of the butyrylcholinesterase molecule by N-terminal sequencing.

While the importance of acetylcholinesterase (AChE) in cholinergic transmission has been well established, the biological role of butyrylcholinesterase (BChE) which closely resembles AChE in several of its properties and amino acid sequence (54% identity) is not yet clearly known (Chatonnet and Lockridge, 1989). Several lines of evidence have indicated that the cholinesterases may have more than one function (Greenfield, 1984). The existence of AChE in areas of the brain where there is no cholinergic transmission or in erythrocyte membrane where it has no known functions have been cited as supporting evidence for this.

AChE from Basal Ganglia, Erythrocytes and Electric Eel Exhibit a Serotonin Sensitive Aryl Acylamidase Activity

We have observed that AChE purified from three different sources exhibit a serotonin sensitive aryl acylamidase (AAA) activity (George and Balasubramanian, 1980). The AAA activity catalyses the conversion of o-nitroacetanilide to o-nitroaniline and acetate, cleaving an acyl amide bond. Several criteria indicated that AAA was identical with the AChE molecule. Prominent among them were the co-purification of both activities in several affinity chromatographic procedures, co-elution from Sephadex G-100 columns, similar response to potent inhibitors of AChE, identical extent of solubilization by phosphatidyl inositol specific phospholipase C and co-precipitation by a polyclonal antibody at different dilutions.

Human Serum BChE also Exhibits a Serotonin Sensitive AAA Activity

BChE is a tetrameric glycoprotein with a monomer Mr of approximately 90 kDa as evidenced by SDS-gel electrophoresis of the purified enzyme. Its complete amino acid sequence is known (Lockridge et al, 1987). Since BChE has 54% amino acid sequence homology to AChE we investigated whether BChE also exhibited a serotonin sensitive AAA activity. The purified BChE exhibited a similar serotonin sensitive AAA activity as observed with AChE. An additional interesting property of the AAA associatd with BChE was its several fold stimulation specifically by tyramine (George and Balasubramanian, 1981; Tsujita and Okuda, 1983). The significance of the influence of

0–8412–2008–5/91/0371$06.00/0

serotonin and tyramine on the AAA activity associated with BChE is not yet understood.

A Peptidase Activity Exhibited by Human Serum BChE

We identified a peptidase activity in purified human serum BChE (Boopathy and Balasubramanian, 1987). The peptidase could cleave amino acids from the carboxy terminus of several peptides including enkephalins. For routine assay a dipeptide Phe-Leu was used as substrate that was cleaved into its constituent amino acids. The peptidase activity behaved identical to AAA and BChE in all respects except in its response to the potent inhibitors of BChE and AAA. These potent inhibitors not only did not inhibit the peptidase activity but also showed a slight activating effect.

Identification of the Domains Responsible for the AAA, Peptidase and BChE Activities in Purified BChE.

The efficient and reproductible method of purification of BChE to homogeneity from human serum prompted us to identify the domains responsible for the AAA, peptidase and BChE activities by limited protease digestion of the BChE molecule. Alpha chymotrypsin proved to be the most useful proteolytic enzyme for this purpose. Limited protease digestion with this enzyme cleaved the BChE molecule into three major fragments of approximate Mr 20 kDa, 21 kDa and 50 kDa as evidenced by SDS-gel electrophoresis of a chymotryptic digest of BChE. Of these the 20 kDa fragment isolated by Sephadex G-75 gel chromatography exhibited both AAA and BChE activity. The 50 kDa fragment retained the peptidase activity. Investigation into the glycosylated nature of the fragments based on lectin-Sepharose chromatography revealed that the 20 kDa fragment was devoid of galactose whereas the 50 kDa fragment retained galactose residues (Rao and Balasubramanian, 1989). The inhibition characteristics of AAA and peptidase activities in the fragments by various cholinesterase inhibitors were similar to the parent enzyme from which they were derived. N-terminal sequence analysis showed that the 20 kDa fragment was formed by cleavage at the Tyr(146)-Arg(147) bond and the 50 kDa fragment was formed by cleavage at the Phe(290)-Gly(291) bond of the BChE molecule.

The Implications of AAA and Peptidase Activities in Cholinesterase.

The limited alpha chymotryptic digestion and the N-terminal sequence analysis of the enzymatically active fragments obtained from human serum BChE provides convincing evidence that AAA and peptidase activies are part of the BChE molecule (Fig.1). These studies also suggest that the catalytic site of the peptidase is different from those of AChE and AAA. We do not know of a physiological

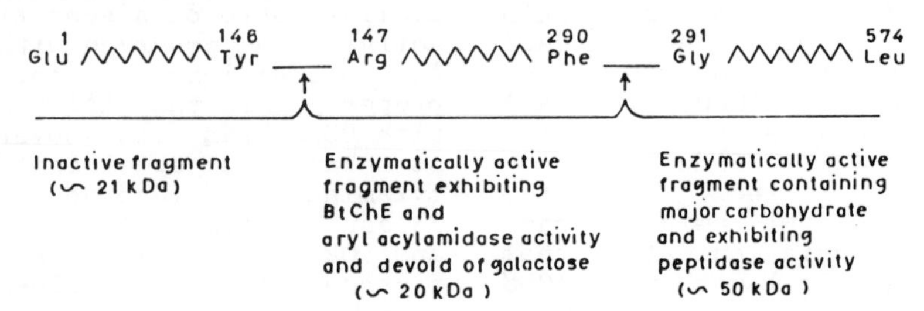

substrate for AAA so far but it is possible that it can act on an endogenous acylamide compound. The inhibition of AAA by serotonin and stimulation by tyramine provide an attractive hypothesis that the half-life of such an endogenous acylamide compound would depend upon these amines which influence AAA activity.

The peptidase activity found in BChE is relatively weak compared to many other proteases and peptidases present in the mammalian system. However the question that remains to be answered is whether tho co-existence of cholinesterase and peptidase has any special significance in terms of neuro-transmission. An unambiguous demonstration of peptidase activity in AChE which has several features in common with BChE would greatly facilitate research in this direction.

References

Boopathy, R., Balasubramanian, A.S., Eur. J. Biochem., 1987, 162, 191-197.

Chatonnet, A., Lockridge, O., Biochem. J., 1989, 260, 625-634.

George, S.T., Balasubramanian, A.S., Eur. J. Biochem., 1980, 111, 511-524.

George, S.T., Balasubramanian, A.S., Eur. J. Biochem., 1981, 121, 177-186.

Greenfield, S., Trend. Neurosci., 1984, 7, 364-368.

Lockridge, O., Bartels, C.F., Vaughan, T.A., Wong C.K., Norton, S.E., Johnson, L.L., J. Biol. Chem. 1987, 262, 549-557.

Rao, R.V., Balasubramanian, A.S., Eur. J. Biochem., 1989, 179, 639-644.

Tsujita, T., Okuda, H., Eur. J. Biochem., 1983, 133, 215-220.

Acetylcholinesterases: Proteases Regulating Cell Growth and Development?

*David H. Small, Department of Pathology, University of Melbourne, Parkville, Victoria 3052, Australia

The role of acetylcholinesterase in hydrolysing acetylcholine at cholinergic synapses is well established. However, the function of this enzyme in non-cholinergic tissues has been a mystery. Recent evidence suggests that some species of acetylcholinesterase could function as proteases which regulate cell division or neurite outgrowth.

The functions of acetylcholinesterase (AChE, EC 3.1.1.7) have remained elusive (Greenfield, 1984; Small, 1989) despite intensive research into the structure, activity and tissue localization of this enzyme (Massoulié and Bon, 1982). While AChE has been well studied for its ability to hydrolyse acetylcholine at cholinergic synapses, the distribution, developmental expression and existence of multiple molecular forms of AChE indicate that the enzyme may have non-cholinergic functions:

1. Not all AChE activity is localized with its presumed substrate acetylcholine. AChE is present in blood, where it is found in erythrocytes (Ott et al., 1982), platelets and megakaryocytes (Paulus et al, 1981). Even in the nervous system. not all AChE is localized with cholinergic neurons or their terminals (Eckenstein and Sofroniew, 1983).
2. AChE activity is expressed early in development, before the appearance of cholinergic structures (Layer et al., 1987; Massoulié and Bon, 1982).
3. Multiple forms of AChE differing in their molecular weights, solubilities and amino acid sequences have been identified (Massoulié and Bon, 1982). Some of this heterogeneity may relate to the mechanisms by which AChEs are anchored to membranes (Taylor et al., 1987). Alternatively, the presence of multiple forms may relate to whether AChE has cholinergic or non-cholinergic functions.

AChEs have Protease Activity

Studies in our laboratory have demonstrated that highly purified preparations of AChE from eel electroplax organ or foetal calf serum possess associated trypsin-like and metal ion-stimulated exopeptidase activities (Small et al., 1987; Small, 1988). While earlier studies suggested that the protease activity associated with AChE was weak and relatively non-specific (Chubb et al., 1980; 1983), more recent studies using polypeptide substrates demonstrate that this proteolytic activity is both rapid and specific (Ismael et al, 1986; Small et al., 1986).

Several lines of evidence indicate that these protease activities are intrinsic to AChE and not associated with contaminating enzymes:

1. Various purification procedures including affinity chromatography, high performance size-exclusion and ion-exchange chromatography, polyacrylamide gel electrophoresis (Chubb et al., 1980; 1983; Ismael et al., 1986; Small et al., 1986; 1987) and immunoprecipitation using a monoclonal antibody (Elec 106) to AChE (Table 1) fail to separate the protease activity from purified AChE.

While a recent report has demonstrated that the protease activity of a crude preparation of eel AChE can be partially separated from the esterase activity (Checler and Vincent, 1989), the differences between crude and purified AChE may reflect the fact that the esterase form is a zymogen of a protease (discussed later).

2. While AChEs are purified from different sources (serum, eel electroplax organ) using different procedures, their protease activities are quite similar. It is unlikely these preparations would contain identical contaminants.
3. Some of those AChEs for which the complete amino acid sequence is known

0–8412–2008–5/91/0374$06.00/0

Table 1. Immunoprecipitation of eel AChE using a monoclonal Ab (Elec 106) bound to protein A-Sepharose

Treatment	Enzyme activity	
	Protease (ug/h/ml)	Esterase (U/ml)
Control (Pr.A-Seph)	3.2 ± 1.2	36.5 ± 2.3
Elec106-Pr.A-Sepharose	not detected	0.6 ± 0.0

Table shows enzyme activity in supernatant after immunoprecipitation. Trypsin-like activity was measured with a synthetic peptide substrate (Small et al., 1987). Values are means ± S.D. (n = 3). (Elec 106 monoclonal Ab was a kind gift from Dr. J. Grassi.)

share sequence similarity to trypsin-like endopeptidases and serine carboxypeptidases (Small, 1990). While the forms of AChE which are cloned and sequenced may not be proteases, the existence of amino acid sequence similarity indicates that AChEs share a common evolutionary relationship with these 2 families of proteases.

The C-terminus of *Drosophila* AChE, for example, possesses 40% amino acid sequence similarity to the active site region of pancreatic trypsin (Hall and Spierer, 1986). This region of *Drosophila* AChE includes a serine residue (position 584) which is found in a position corresponding to the position of the active serine of trypsin (position 176 in trypsin) (Small, 1989). The serine residue in *Drosophila* AChE is distinct from the serine residue at the esteratic site (position 241). Interestingly, the start of the amino acid sequence similarity to trypsin coincides with the position of an exon-exon boundary. This suggests that alternative splicing of mRNA could be a means by which protease forms of AChE are produced.

AChEs also share amino acid sequence similarity to serine carboxypeptidases such as carboxypeptidase I including the active site regions of both enzymes (Small, 1990). While the degree of similarity (20% over 79 residues) is less than with trypsin, the existence of sequence similarity around the active esteratic site does suggest that the carboxypeptidase activity of AChEs may reside in close proximity to the esteratic site. Like eel AChE, serine carboxypeptidases have been reported to possess esterase and amidase activity.

Eel AChE is a Protease Zymogen

When purified eel AChE is incubated by itself in solution, it undergoes autolysis to generate a number of lower molecular weight polypeptides (Small and Simpson, 1988). Incubation also results in an increase in the level of trypsin-like activity, which correlates with the appearance of lower molecular weight polypeptides.

We have isolated a 25 kDa polypeptide which is closely associated with the trypsin-like activity (Small, submitted). This peptide possesses an N-terminal amino acid sequence similar to that of pancreatic trypsin (ile-val-gly-gly-tyr). The 25 kDa polypeptide is not a contaminant of AChE as it is tightly bound to the affinity-purified enzyme. SBTI- and benzamidine-Agarose fail to bind the 25 kDa polypeptide, indicting that it is not identical to pancreatic trypsin. Rabbit polyclonal antisera raised against this polypeptide recognize the 75 kDa catalytic subunit of eel AChE on "Western" blots. This demonstrates that the 25 kDa polypeptide is a proteolytic fragment of the 75 kDa catalytic subunit.

A model of eel AChE describing the relationship between the enzyme's esterase and protease activities is presented in Fig. 1. In this model, AChE contains 4 identical subunits of 75 kDa. Tetrameric AChE contains esterase activity, but no protease activity. Proteolysis of one of the subunits results in the formation of polypeptides of 50 and 25 kDa which remain tightly bound to the other 75 kDa subunits. Carboxypeptidase activity is associated with the N-terminal fragment, while trypsin-like activity is associated with the C-terminal fragment. Further proteolysis of AChE results in dissociation of the subunits and complete loss of esterase activity.

This model explains 2 features of the protease activity which have not previously been understood:

1. Highly purified AChE possesses only a weak protease activity, the total amount of which is variable between different preparations. We have shown that the total number of trypsin-like active sites is only 20% of the total number of DFP-reactive sites. As shown in Fig. 1, affinity purified AChE consists of a mixture of both intact and proteolytically cleaved subunits. Thus the total number of 25 kDa subunits is less than the total number of DFP-reactive sites.

2. It is possible to partially separate the protease and esterase activities from each other in a crude preparation of AChE (Sigma Type III)(Checler and Vincent, 1989), but not in a purified preparation. This difference can now be explained by the fact that crude Sigma preparation contains a number of lower molecular weight fragments of AChE with protease activity, which are easily separated from the higher molecular weight esterase forms. In contrast, in "affinity-purified" AChE, proteolytically cleaved fragments of the 75 kDa monomer remain tightly bound to the other subunits (Fig. 1).

The model presented here is consistent with models described for other proteases (Bond and Butler, 1987). For example, the existence of multi-catalytic proteases is now well established (Bond and Butler, 1987). Therefore, association of 2 protease activities with the same enzyme is not uncommon. Furthermore, for most high molecular weight trypsin-like serine endoproteases, the trypsin-like activity is generally associated with a C-terminal polypeptide which must be cleaved from the zymogen, but which remains associated with the other subunits through ionic or disulphide bonding (Teller et al., 1976). Several trypsin-like proteases (e.g., gamma-nerve growth factor) and serine carboxypeptidases (carboxypeptidase Y) also have esterase activity.

Evolution of AChEs

It is unlikely that a single gene product could encode proteins which function both as esterases (to hydrolyse acetylcholine) and as a proteases (to hydrolyse peptides or proteins). AChEs with protease activity probably represent an entirely separate family of enzymes which share a common evolutionary ancestry with the true AChEs. As discussed previously (Small, 1989), a single point mutation, or deletion, at a region corresponding to a zymogen cleavage site could have resulted in the formation of AChEs with esterase activity, but no protease activity. Thus, with the development of neuronal systems, protease forms may have been sequestered by neurons for the purpose of hydrolysing acetylcholine (Fig. 2).

It would seem likely that protease forms of AChE would be associated with a gene that is distinct from the gene for other AChEs. With the advent of

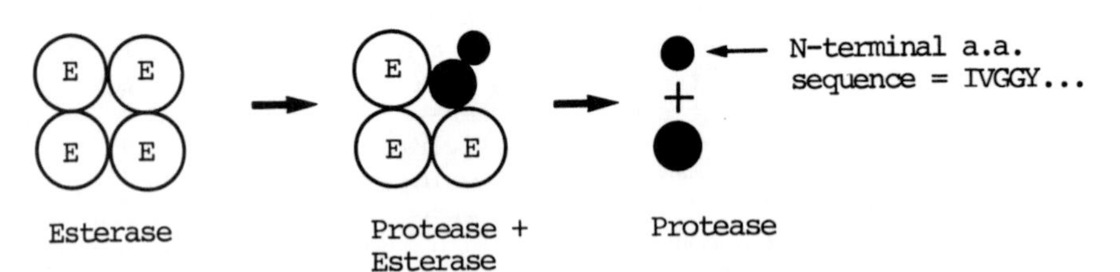

Fig. 1. Model of eel AChE showing the activation of protease activity by proteolysis of the intact 75 kDa catalytic subunit. AChE is a tetramer consisting of four esteratic (E) subunits. Proteolysis of one or more subunits activates the enzyme's protease activity. The activated enzyme copurifies with the esterase upon affinity chromatography. Further proteolysis results in the complete breakdown of the quaternary structure with loss of esterase activity and further activation of protease activity. Carboxypeptidase activity is associated with 40-50 kDa N-terminal fragments and trypsin-like activity with a 25 kDa C-terminal fragment.

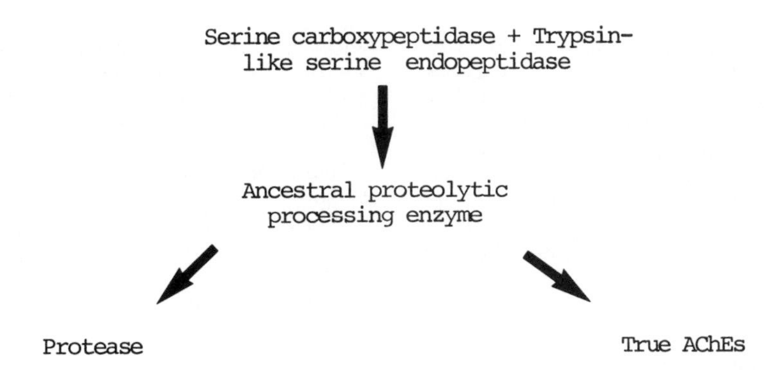

Fig. 2. Possible evolution of diverse forms of vertebrate and invertebrate AChE. The fusion of genes encoding a serine carboxypeptidase with a trypsin-like serine endopeptidase may have resulted in the formation of a multicatalytic proteolytic processing enzyme with esterase activity. With the development of neuronal systems, the enzyme may have been sequestered by neurons for the hydrolysis of acetylcholine. Protease forms may be found today in both neuronal and non-neuronal tissues (e.g., megakaryocytes) and may be involved in the regulation of cell growth. "True" AChEs are found solely in association with cholinergic or cholinoceptive cells (e.g. at neuromuscular junction) and are responsible for the hydrolysis of acetylcholine.

cDNA probes which recognize mammalian and non-mammalian AChE genes (Lapidot-Lifson et al., 1989), this question may soon be answered. There is little evidence, as yet, to indicate that there may be more than one AChE gene per species. However, existing probes may not hybridize well with DNA from genes which diverged early in evolution.

A Role for AChEs in Cell Growth

There is increasing evidence implicating serine proteases in mechanisms regulating cell growth, maturation and neurite outgrowth (Knauer and Cunningham, 1984; Monard, 1988). For example, a class of protease inhibitors known as protease nexins promote neurite outgrowth from neuroblastoma and pheochromocytoma cells (Monard, 1988). Protease nexin II, which is similar, if not identical, to the amyloid precursor protein (Van Nostrand et al., 1989) has been implicated in neuritic plaque formation associated with Alzheimer's disease. Other protease inhibitors such as soybean trypsin inhibitor (Hawkins and Seeds, 1986) also induce neurite outgrowth. Cell surface trypsin-like enzymes are now thought to be associated with transformation by oncogenic viruses, and with migration of neurons in the granular layer of the cerebellum (Krystosek and Seeds, 1981).

A number of studies implicate AChEs in the regulation of cell growth (reviewed by Massoulié and Bon, 1982). AChE activity is rich in foetal tissues as well as in various primary tumours (Razon et al., 1984) and in the sera of patients with carcinomas (Zakut et al., 1988). Developmental studies demonstrate that the appearance of AChE in the brain correlates well with neuronal proliferation (Layer et al., 1987). In rat PC12 cells, the induction of neurite outgrowth by nerve growth factor is associated with an increase in intracellular levels of AChE (Rieger et al., 1980). The localization of AChE in sympathetic neurons in association with growth cones and in regions in which lamellipodia and filopodia attach to the substratum (Rotundo and Carbonetto, 1987) is also consistent with such a role.

Recently, AChE genes have been found to be amplified in a number of patients with abnormal megakaryocyte proliferation and leukemias (Lapidot-Lifson et al., 1989). This amplification may be of importance, as a number of studies have shown that AChE may be involved in the process by which megakaryocyte forming units arrest mitotic proliferation and begin the process of forming mature platelets (Paulus et al., 1981). For example, the AChE inhibitor neostigmine stimulates megakaryocyte proliferation. Thus, the possibility that AChEs are proteases which are secreted into the extracellular environment to regulate cell growth certainly needs to be tested.

Acknowledgements

The author gratefully acknowledges the support and advice of Professor Ian W. Chubb, Dr. Bruce G. Livett and Professor Colin Masters on various aspects of this work.

References

Bond, J.S.; Butler, P.E. *Ann. Rev. Biochem.* **1987**, *56*, 333-364.

Checler, F.; Vincent, J.-P. *J. Neurochem.* **1989**, *53*, 924-928.

Chubb, I.W.; Hodgson, A.J.; White, G.H. *Neuroscience* **1980**, *5*, 2065-2072.

Chubb, I.W.; Ranieri, E.; White, G.H.; Hodgson, A.J. *Neuroscience* **1983**, *10*, 1369-1377.

Eckenstein, F.; Sofroniew, M.V. *J. Neurosci.* **1983**, *3*, 2286-2291.

Greenfield, S. *Trends Neurosci.* **1984**, *7*, 364-368.

Hall, L.M.C.; Spierer, P. *EMBO J.* **1986**, *5*, 2949-2954.

Hawkins, R.L.; Seeds, N.W. *Brain Res.* **1986**, *398*, 63-70.

Ismael, Z.; Millar, T.J.; Small, D.H.; Chubb, I.W. *Brain Res.* **1986**, *376*, 230-238.

Knauer, D.J.; Cunningham, D.D. *Trends Biochem. Sci.* **1984**, *9*, 231-233.

Krystosek, A.; Seeds, N.W. *Proc. Natl. Acad. Sci. USA* **1981**, *78*, 7810-7814.

Lapidot-Lifson, Y.; Prody, C.A.; Ginzberg, D.; Meytes, D.; Zakut, H.; Soreq, H. *Proc. Natl. Acad. Sci. USA* **1989**, *86*, 4715-4719.

Layer, P.G.; Alber, R.; Sporns, O. *J. Neurochem.* **1987**, *49*, 175-182.

Massoulié, J.; Bon, S. *Ann. Rev. Neurosci.* **1982**, *5*, 57-106.

Monard, D. *Trends Neurosci.* **1988**, *11*, 541-544.

Ott, P.; Lustig, A.; Brodbeck, U.; Rosenbusch, J.P. *FEBS Lett.* **1982**, *138*, 187-189.

Paulus, J.-M.; Maigne, J.; Keyhani, E. *Blood* **1981**, *58*, 1100-1106.

Razon, N.; Soreq, H.; Roth, E.; Bartal, A.D.; Silman, I. *Exp. Neurol.* **1984**, *84*, 681-695.

Rieger, F.; Shelanski, M.L.; Greene, L.A. *Dev. Biol.* **1980**, *76*, 238-243.

Small, D.H. *Neurosci. Lett.* **1988**, *95*, 307-312.

Small, D.H. *Neuroscience* **1989**, *29*, 241-249.

Small, D.H. *Trends Biochem. Sci.* **1990**, in press.

Small, D.H.; Ismael, Z.; Chubb, I.W. *Neuroscience* **1986**, *19*, 289-295.

Small, D.H.; Ismael, Z.; Chubb, I.W. *Neuroscience* **1987**, *21*, 991-996.

Small, D.H.; Simpson, R.J. *Neurosci. Lett.* **1988**, *89*, 223-228.

Taylor, P.; Schumacher, M.; MacPhee-Quigley, K.; Friedman, T.; Taylor, S. *Trends Neurosci.* **1987**, *10*, 93-95.

Teller, D.C.; Titani, K.; de Haen, C. In *Handbook of Biochemistry and Molecular Biology -- Proteins*; Fasman, G.D., Ed.; CRC Press: Cleveland, OH, 1976, Vol. 3; pp. 340-355.

Van Nostrand, W.E.; Wagner, S.L.; Suzuki, M.; Choi, B.H.; Farrow, J.S.; Geddes, J.W.; Cotman, C.W.; Cunningham, D.D. *Nature* **1989**, *341*, 546-549.

Zakut, H.; Even, L.; Birkenfeld, S.; Malinger, G.; Zisling, R.; Soreq, H. *Cancer* **1988**, *61*, 727-739.

Non-cholinergic Effects of Acetylcholinesterase upon Neuronal Membrane Properties and Synaptic Transmission in the Mammalian Cerebellar Cortex

***Margaret E. Appleyard** and **Henrik Jahnsen**, Institute of Neurophysiology, Panum Institute, Blegdamsvej 3c, DK 2200, Copenhagen.

The distribution of acetylcholin-esterase (AChE) in the brain is not restricted to the cholinergic system, suggesting that it may have functions other than the hydrolysis of acetylcholine. In the cerebellar cortex there is scant evidence for cholinergic transmission, but levels of AChE are high, and the protein can be released in a Ca^{2+}-dependent manner by stimulation of climbing fibres (Appleyard et al., 1988). In order to measure possible effects of AChE on cerebellar neurones slices of guinea-pig cerebella were cut and maintained using standard techniques. Intracellular recordings from Purkinje cells revealed three different actions of AChE when added to the superfusing medium (10-20U/ml). 1: The threshold for firing of Ca^{2+}-spikes in response to depolarizing intracellular current injection was increased; 2: The late part of the synaptic response to climbing fibre stimulation was enhanced so that Na^{+} action potentials were often generated for tens of milliseconds after stimulation and 3: Responses to the excitatory amino acids glutamate and aspartate were increased in both amplitude and duration. All the effects persisted when the catalytic site of AChE was blocked by Soman.

There was, however, no consistent effect of AChE upon either the membrane potential or the input resistance of Purkinje cells, even though AChE has previously been shown to hyperpolarise neurones in the substantia nigra (Greenfield et al., 1988).

These results are the first demonstration of a direct modulatory action of AChE upon synaptic responses. Such actions of AChE could occur in other brain regions with important consequences for the activity of target cells.

References

Appleyard, M.E., Vercher, J.L., Greenfield, S.A., *Neuroscience* **1988**, *25*, 133-138.

Greenfield, S.A., Jack, J.J.B., Last, A.T., French, M., *Expl Brain Res.* **1988**, *70*, 441-444.

*FAX: (0)71 433 1921 (England)
Present address: Dept. of Physiology, Royal Free Hospital School of Medicine, Rowland Hill Street, London NW3 2PF.

AChE Release from the Substantia Nigra is Associated with Certain Types of Movement: Direct Evidence "On-line"

S.A. Jones and S.A.Greenfield, University Department of Pharmacology, South Parks Road, Oxford, OX1 3QT, U.K.

It is now well established that a soluble form of acetylcholinesterase (AChE) is released from the substantia nigra,in a non cholinergic capacity. Although the characteristics of this phemomenon suggest that it may underlie a modulatory mechanism (Greenfield 1985) little is known of the actual brain functions with which it might be associated. Recently, however, we have developed a system whereby secretion of AChE from central neurones can be monitored on a time scale commensurate with actual behaviour i.e. "on-line" (Taylor et al. 1989). Using this novel approach it has already been shown that release of AChE in the substantia nigra is greatly enhanced in the awake and freely moving guinea pig (Taylor et al. 1990). The aim of this study was to attempt to identify specific types of movement that might be associated with increases in release of the protein.

Results and Conclusions

It was found that this release occurs in a pulsatile fashion, composed of pulses or peaks. Analysis of the characteristics of these peaks shows that their frequency were highly significantly different under the two experimental conditions. Despite this effect of behavioural paradigm the absolute amount of the protein within a peak was similar in both conditions. During repeated bouts of locomotor activity there was a decrease in the likelihood of peaks occuring. Indeed even within a single bout of movement the peak of released AChE declined. Therefore AChE release from the substantia nigra does not simply and continuously reflect ongoing movements.Rather this phenomenon might be implicated in a more active "cognitive" role in motor control.

References

Taylor, S.J.; Haggblad, J.; Greenfield, S.A. *Neurochem. Int.* **1989**, *15*, 199-205.

Taylor S.J.; Jones, S.A.; Haggblad, J.; Greenfield, S.A. *Neuroscience* **1990**, In Press.

0–8412–2008–5/91/0380$06.00/0

Distinct Dose-Dependency of the Non-Cholinergic Effects of AChE on Circling Behaviour

Denyse Lévesque and Susan A. Greenfield, University Dept. of Pharmacology, South Parks Rd., Oxford, OX1 3QT, U.K.

It has been demonstrated that exogenous acetylcholinesterase (AChE) can have two different actions on nigrostriatal cells, independent of cholinergic transmission. In response to striatal stimulation, the firing rate is increased by low local concentrations of AChE in the extracellular space (Last and Greenfield, 1987) but decreased by high concentrations (Greenfield et al., 1981). The aim of this study was to see whether this dose-dependency had any functional significance.

Method, Results and Discussion

The behavioural model used was that of rotation, where circling occurs due to an imbalance of available dopamine in the two nigrostriatal pathways (Greenfield et al., 1984). Rats were chronically implanted with guide cannulae for unilateral microinfusion of AChE directly into one substantia nigra.

Following application of a "low" concentration of AChE (Sigma, 100 mU.), intraperitoneal amphetamine evoked circling behaviour implying a net excitation of the nigrostriatal pathway. This observation confirmed a previous behavioural study (Greenfield et al., 1984) and was indeed consistent with electrophysiological findings (see Introduction). On the other hand, "high" doses of AChE (1 000 mU.) resulted in the opposite direction of circling, suggestive of a net inhibitory action. Again, this result is consistent with electrophysiological data (see Introduction). Histological examination revealed that the effects were not due to differential cannulae placements in the substantia nigra but were truly attributable to differences in doses.

It thus appears that the duality in responses to different doses of AChE at the level of single cells, can ultimately be reflected in motor behaviour. The actual ionic mechanisms underlying these two effects can be explained in the light of the properties of specific conductances in nigrostriatal cells.

Conclusion

Our results reveal that: (1) high and low doses of AChE have differential effects on motor behaviour irrespective of the site of the injection within the substantia nigra and (2) these differential effects correlate with dose-dependent electrophysiological actions of AChE on nigrostriatal neurones. The fact that dose-dependent effects of AChE can ultimately be reflected in behaviour may be of relevance to our understanding of the physiology, or perhaps even the pathology, of the nigrostriatal pathway.

References

Greenfield, S.A.; Stein, J.F.; Hodgson, A.J.; Chubb, I.W. *Neuroscience*, **1981**, 6, pp 2287-2295.

Greenfield, S.A.; Chubb, I.W.; Grunewald, R.A.; Henderson, Z.; May, J.; Portnoy, S.; Weston, J.; Wright, M.C. *Exp. Brain Res.*, **1984**, 54, pp 513-520.

Last, A.T.J.; Greenfield, S.A. *Exp. Brain Res.*, **1987**, 67, pp 445-448.

Llinas, R.; Greenfield, S.A.; Jahsen, H.J. *Brain Res.*, **1984**, 294, pp 127-132.

0–8412–2008–5/91/0381$06.00/0

Characterization of a Non-cholinergic Action of AChE on the Membrane Properties of Pars Compacta Neurones

C. Webb and S.A. Greenfield, University Department of Pharmacology, South Parks Road, Oxford OX1 3QT, U.K.

AChE hyperpolarizes the membrane of pars compacta neurons *in vitro*, independent of cholinergic transmission. Yet although this effect has been demonstrated by two distinct methods, i.e. bolus application (Greenfield et al.,1988) and pressure ejection (Greenfield et al.,1989) of the protein, neither have proved sufficiently quantitative to study this novel phenomenon further. In these experiments, we have applied exogenous AChE (Sigma Electic Eel Type V-S) to the perfusate reservoir such that slices of guinea pig midbrain containing substantia nigra are in constant and sustained contact with a known concentration of AChE (125 U/ml) for fifteen to thirty minute periods.

Results and Conclusions

When the protein is applied repeatedly on the same cell, there is a hyperpolarization of similar amplitude following each treatment i.e. no tachyphylaxis. Furthermore AChE-induced hyperpolariz- ation is resistant to blockade of potassium channels by tetraethyl- ammonium chloride (TEA). Since dopamine receptor activation displays both tachyphylaxis and TEA sensitivity (see Nedergaard et al.,1988), it would appear that the effects of AChE do not involve this particular target. In addition, an enzymatic action of any type can be excluded since application of AChE solution, pre-boiled for up to 30 minutes, still evoked a marked hyperpolarization. Artefactual or non-specific effects seem unlikely to account for the AChE-induced hyperpolarization since a sub population of nigral neurons were unaffected by AChE. It is concluded that, independent of hydrolysis of acetylcholine, extracellular AChE can have a selective hyperpolarizing action on the membrane properties of dopaminergic neurons via a TEA-independent potassium channel and independent of the dopamine receptor itself.

References

Greenfield S.A.; Jack, J.J.B.; Last, A.T.J.; French, M. *Exp. Brain Res.* **1988**, *70*, 441-444.

Greenfield, S.A.; Nedergaard, S.; Webb, C.; French, M. *Neuroscience* **1989**, *29*, 21-25.

Nedergaard, S.; Hopkins, C.; Greenfield, S.A. *Exptl. Brain Res.* **1988**, *69*, 444-448.

Embryonic Cholinesterase and Morphogenesis

Ulrich Drews, Frauke Lohmann, Maria Lammerding-Köppel, Wolfgang Mengis, and Günter Oettling

Institute of Anatomy, University of Tübingen, D-7400 Tübingen, Federal Republic of Germany

Several years ago we described single cholinesterase (ChE) - active droplet cells during gastrulation of the chick embryo. ChE activity in embryonic cells turned out to be a general phenomenon. In all organs studied in various species, expression of cholinesterase was correlated with morphogenetic movements. The activity disappeared when the respective organ structure had formed (Drews 1975).

Ultrastructural localization of ChE

In the chick limb bud, ChE activity is found in the apical ectodermal ridge, the subridge mesenchyme and in the chondrogenic core. It is localized in the endoplasmic reticulum and the perinuclear cisterna (Vanittanakom et al. 1985).

Choline acetyltransferase (ChAT)

In order to demonstrate the presence of acetylcholine we measured the acetylcholine synthesizing enzyme in the limb bud during the phase of ChE activity.

Muscarinic receptors

By binding studies and autoradiography with ^{3}H-QNB we demonstrated a specific muscarinic receptor in the chick limb bud and in the whole embryo.

Ca^{2+} Mobilization

On stimulation with muscarinic compounds the embryonic cells respond with intracellular Ca^{2+} mobilization, which we measured by monitoring chlorotetracycline or Fura-2-AM fluorescence in cell suspensions. The reaction was used for pharmacological characterization of the muscarinic receptor and for compilation of doese response curves (Oettling et al. 1985, 1988).

Muscarinic induced contraction wave

Chick blastoderms were explanted and cellular movements were recorded by time lapse video filming. Perfusion with acetylcholine induced a contraction wave in the blastoderm which started at the point of entrance of the drug and proceeded within 8-10 min. through the whole blastoderm. Perfusion with the muscarinic antagonist pilocarpine in turn induced a wave of relaxation (Drews and Mengis 1990).

Conclusion

Embryonic cholinesterase is part of a muscarinic cholinergic system expressed during phases of morphogenesis. We assume that autocrine secretion of the transmitter acetylcholine enables the cells to respond to embryonic inducers or chemotactic agents with pulses of cellular movements. Stimulation of the receptor triggers intracellular Ca^{2+} release which in turn leads to contraction of actin and myosin filaments.

Drews, U., Progr. in Histochem. and Cytochem. **1975**, Vol. 7 No. 3.

Drews, U., Mengis, W., Submitted.

Oettling, G., Schmidt, H., Drews, U., J. Cell. Biol. **1985**, 100, 1073-1081.

Oettling, G., Schmidt, H., Show-Klett, A., Drews, U., Cell Diff., **1988**, 23, 77-86.

Vanittanakom, P. Drews, U., Anat. Embryol., **1985**, 172, 183-194.

0–8412–2008–5/91/0383$06.00/0

Cholinergic Properties of the Gastrulating Chick Embryo

Tiit Laasberg, Laboratory of Molecular Genetics, Institute of Chemical Physics and Biophysics, Akadeemia tee 23, Tallinn 200026, Estonia

The gastrulating chick embryo has cholinergic characteristics whose presence is indicated by acetylcholinesterase (AChE, EC 3.1.1.7) and choline acetyltransferase (CAT, EC 2.3.1.6) activity, and by the existence of functionally active muscarinic receptors which are coupled with phosphatidylinositol turnover.

A sharp increase in AChE activity occured in the gastrulating embryo. The highest activity was associated with hypoblast cells. By sucrose density gradient centrifugation three molecular forms of AChE with sedimentation coefficients 4.7 S, 6.8 S and 10.9 S were determined. During the gastrulation there was no significant change in CAT activity. No reliable differencies between the germ layers were observed either.

It may be that during the gastrulation of chick embryo the metabolism of ACh is mainly regulated by AChE. In the epiblast the activity of AChE was about 200-fold, in the primitive streak 600-fold and in the hypoblast about 2600-fold higher than the activity of CAT.

The intracellular concentration of acetylcholine (ACh) in gastrulating chick embryo cells was 10^{-6}-10^{-5} M. ACh, via muscarinic receptors increased the level of inositol phosphates, raised the intracellular level of cGMP 2-fold and $[Ca^{2+}]_i$ from the basal level of 120 nM to the 200 nM (measured with indo-1).

Calculating the amount of bound QNB with respect to the cell number we found that the number of muscarinic receptors is 10.000-20.000 receptors per gastrulating chick embryo cells.

The establishment of dose-response curve by the measurement of $[Ca^{2+}]_i$ mobilization in dissociated embryo cells at stages 3 and 4-5 by Hamburger and Hamilton (1951) gave the ED_{50} value for ACh $4(\pm 0.5) * 10^{-6}$ M.

Thus, one role of ACh in the early embryogenesis is to regulate the calcium homeostasis within the cell. During the gastrulation the chick embryo cells are actively migrating through the primitive streak to form germ layers. It may be supposed that ACh is also involved in this process, influencing through Ca^{2+} the contractile elements of cells.

References

Hamburger, V., Hamilton, H.L. J. Morphol. **1951**, 88, 49-92.

Cholinesterase Activity and Neurotransmitter Systems in Amphibian Gametogenesis

C. Falugi, G. Faraldi, G. Tagliafierro, Istituto di Anatomia Comparata dell'Universita' di Genova, Viale Benedetto XV, 5 - 16132 Genova (Italy).

Cell interactions are known to be implicated in development, although the nature and mechanisms of such interactions are still obscure. In embryonic cells, a cholinergic system, transiently expressed, was hypothesized to regulate intercellular communications (Drews, 1975; Buznikov, 1987). As acetylcholine (ACh) receptors are known to be present in the amphibian oocyte membrane (Kusano et al., 1982), and in the sperm tail of vertebrates and invertebrates, a study was made using histochemical and immunohistochemical methods to detect the presence and localization of cholinesterase and acetylcholinesterase activity (ChE), choline-acetyl transferase (ChAT), ACh and bungarotoxin-binding sites (BuTx), in male and female gonads of the amphibian *Triturus cristatus carnifex,,* to give a contribution to the knowledge of cholinesterase pre-nervous functions.

ChE activity was detected by a direct thiocholine method, while catecholamines by the FIF method; indirect immunofluorescence methods were carried out by the use of antibodies anti-ChE and anti-ChAT (Seralab, U.K.), anti-ACh (Biosys, F), and anti-5 HT (Immunonuclear, USA); fluorescent BuTx was obtained by Sigma (USA).

The cholinergic system, including BuTx-binding sites, was found to be present mainly during sperm maturation and in the sperm lineage cells; cytoplasmic location of the enzymes was found in mature oocytes and in follicular cells around them. 5 HT was present in male gonads at every stage of spermatogenesis, but was not found in maturing and mature sperms.

Aldehyde induced fluorescence was present in both male and female gonads, suggesting the presence of NA or DA. The presence of Ca++ was also checked in gonads.

Our results suggest the implication of cholinesterase activity in the regulation of gamete differentiation in amphibians, with a cholinergic-like role, possibly by modulating intracellular dynamics at the different developmental events.

References

Buznikov, G.A., *Sov. Sci. Rev. F. Physiol. Gen. Biol.*,**1987**, *1*, 137.

Drews, U., *Prg. Histochem. Cytochem.*,**1975**, *7(3)*, 1.

Kusano, K. *et al., J. Physiol.*, **1982**, *328*, 143.

This work was supported by a MPI grant (60%, 40%).

0–8412–2008–5/91/0385$06.00/0

Distribution and Role of Non-specific Cholinesterase Molecules in Growing Peripheral Nerves, in Schwann Cell Progenitors and Mesenchymal Cells in Contact with Growth Cones

P. Haninec, Department of Neurosurgery, Charles University, Prague 6, 169 02, Czechoslovakia
P. Dubový, Department of Anatomy, Masaryk's University, Brno, Komenského nám. 2, Czechoslovakia

The quail-hick and chick-quail chimeras were constructed by grafting isotopically the limb bud of quail embryos into a chick and vice versa prior to the entry of nerve fibres and before the immigration of neural crest cells into the limb. The chimeras were sacrificed at the time of main nerve trunks formation. In all the chimeric embryos the growth cones were in contact with mesenchymal cells of the limb bud origin as well as with Schwann cells progenitors (Fig.1,2). On the basis of this observations one can conclude that these mesenchymal cells may be a source of the information for pattern formation of peripheral nerves.

The reaction product indicating nCHE activity was found in the perinuclear envelope, rough endoplasmic reticulum and on the outer surface of the plasma membrane of cells (Schwann cells progenitors and mesenchymal cells) in close contact with growth cones (Fig.3). The nCHE reaction product was deposited also on the axolemma of nerve fibres and

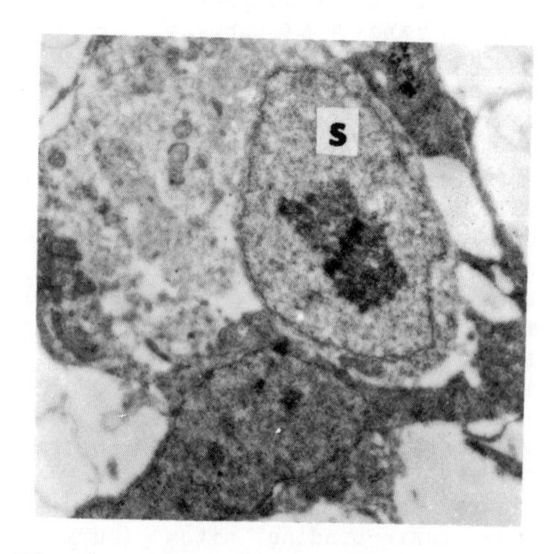

Fig.2. The chick-quail chimera, Schwann /quail/ cell (s) is in contact with growth conus

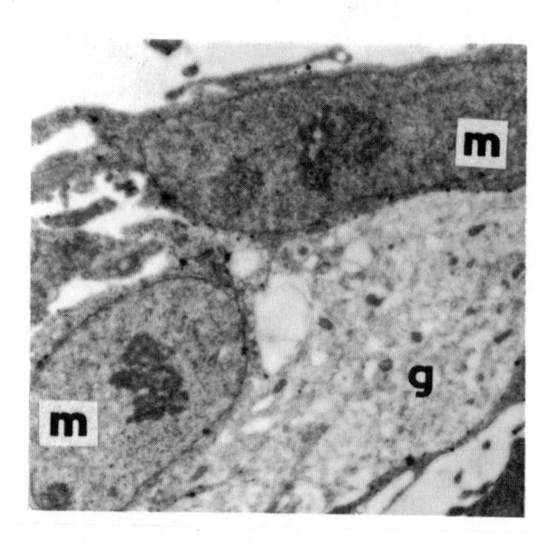

Fig.1. The quail-chick chimera, the growth conus (g) is in contact with mesenchymal /quail/ cells (m)

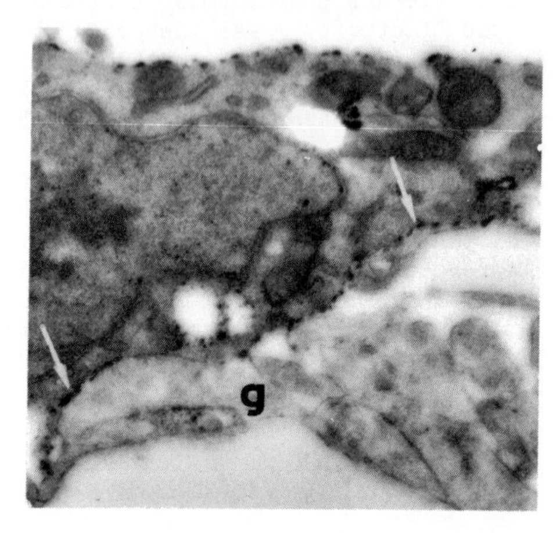

Fig.3. The non-specific cholinesterase is distributed on the plasma membrane of the cell in contact with growth conus (g) and on the axolemma (arrows)

0–8412–2008–5/91/0386$06.00/0

their growth cones.

Several possible mechanism of nCHE action on growing axons can be presumed: i. nCHE molecules may be involved in the modulation of intracellular Ca2+ level in growth cones and may thus influence growth cones motility. ii. The growing axons could be influenced directly via the metabolic product of nCHE activity, e.g. butyrate. Butyrate causes changes in the distribution of cytoskeletal proteins. iii. The localization of nCHE molecules on the plasma membrane of cells in contact with growing axons as well as structural similarities of nCHE and NCAM support the concept that nCHE molecules are involved in adhesive interactions.

Acetylcholinesterase Localization in Chick Ovarian Cortex during Morphogenesis

L. Mastrolia, Faculty of Sciences, University "La Tuscia", Viterbo, Italy
V.P. Gallo & A. Civinini, Department of Animal and Human Biology, University "La Sapienza", Roma, Italy

The acetylcholinesterase (AChE) was evidenced at the ultrastructural level in the germinal epithelium and cortex in the left ovary of chick embryos at 7 and 12.5 days of incubation.

The enzyme was localized in some somatic cells and in some gonocytes. The positive somatic cells in the earlier stage are spread throughout the germinal epithelium and later they are present only in the deepest zone of the cortex. The precipitates are found in the perinuclear cisterna, in the endoplasmic reticulum and on the plasma membrane. The positive gonocytes, few in the earlier stage and more numerous later, show the reaction products in the perinuclear cisterna and in the endoplasmic reticulum, this latter often tightly connected to mitochondria.

The AChE activity is not present in the subsequent considered stages, i.e. at 14 and 19 days of incubation.

The presence of the enzyme in the somatic cells may be correlated to an embryonic cholinergic system stimulated by muscarinic receptors during morphogenesis (Oettling et al., 1985; Drews et al., 1986). This system could regulate cellular movements during morphogenesis (Drews, 1975; Oettling et al., 1988). In the chick ovary this system could be implicated in regulating the various phases of cells migration during the morphogenesis of the cortex.

In the germ cells the positivity may be connected to some migratory capacity of these cells. The close relation between positive cisternae and mitochondria could also suggest a role of the enzymatic activity in the proliferation and/or in functional changes of these organelles.

References

Drews,U., Progr.Histochem.Cytochem.7, **1975,** 1-49.

Drews,U.,Schmidt,H.,Oettling,G.,Vanittanakom,P., Acta Histochem.Suppl.32, **1986,** 133-137.

Oettling,G.,Schmidt,H.,Drews,U.,J.Cell Biol.100, **1985,** 1073-1081.

Oettling,G.,Schmidt,H.,Show-Klett,A., Drews,U., Cell Differ.23,**1988,** 77-86.

Development of Cholinesterase (ChE) Activity in the Ontogenesis of Mytilus (Mollusca, Bivalvia)

Margherita Raineri, Maxim Ospovat, Institute of Comparative Anatomy, University of Genova, 16132 Genova, Italy.

As a histochemical marker ChE is use ful to investigate the morpho-functional differentiation of nervous and muscular cells, particularly, in small organisms, to get pictures of the whole nervous sys tem (Falugi, 1988; Raineri, 1989).

Moreover, ChE takes part in the regulation of cell activities and interactions, playing a role in other developmental processes (Drews,1975).

In the mussel Mytilus galloprovincialis a weak enzyme activity with both acetylcholinesterase (AChE) and pseudocholinesterase (BuChE) properties is detected in the egg at fertilization and the blastomeres during segmentation. In the conchostoma larva increasing AChE staining is localized in two superficial cells close to each oth er, bilaterally symmetrical in the postero-ventral area, which develop cilialike protrusions. Some inner, most prob ably mesodermal cells, show plasma membrane-associated ChE activity. In the early trochophora other ChE-active cells which can be called neuroblasts differentiate postero-laterally close to the earliest two ones. These develop into a middle (sensory?) structure with the strongest enzyme staining that can bulge out, while on each side the nerve cells increase in number and form two bilateral clusters (viscero-parietal ganglion). Under the presumptive velum a cell membrane ChE staining is detected; here,as the velum develops, two rows of ChE-active cells and fibres, most of them with the appearance of bipolar neurons, differentiate. They run from the viscero-parietal ganglion to a smaller cluster of nerve cells (pleural ganglion) which develops anteriorly on each side of the veliger larva and includes ChE-active neurons. Other histo chemically stained cells appear in the region of the presumptive foot in the late veliger. This pattern already corresponds to the adult nervous system.

References

Drews, U., *Prg. Histoche. Cytochem.*, **1975**, 7, 1-52.

Falugi, C., *Acta Embryol. Morph.Exp.*, **1988**, 9, 133-156.

Raineri,M. In *Cell and Molecular Biology of Artemia Development;* A.H. Warner,T.H. MacRae, J.C. Bagshaw, Eds.;ASI-NATO;Plenum Press,N.Y., U.S.A.; pp.131-156.

0–8412–2008–5/91/0389$06.00/0

Cholesterol Ester Hydrolyzing Activity by Pseudo Choline Esterase (PchE) and Its Role in Lipoprotein Metabolism

K. Shirai, H. Inadera, H. Kurosawa, Y. Saito, S. Yoshida, Second Department of Internal Medicine, Chiba University School of Medicine, Chiba 280, Japan

Possible role of PchE as cholesterol esterase was evaluated. Purified human serum PchE could hydrolyze cholesterol oleate emulsified with lysophosphatidylcholine. Purified PchE hydrolyzed cholesterol oleate much more in β-VLDL and IDL than those in LDL or HDL. When [^{3}H]cholesterol linoleate-β-VLDL was incubated in LCAT deficient patient's plasma, free [^{3}H]cholesterol was increased in HDL fraction. The ratio of [^{3}H]cholesterol to [^{3}H]cholesterol ester in HDL increased, indicating that released [^{3}H]cholesterol from β-VLDL was transferred to HDL.

PchE is essentially known to hydrolyze short chain fatty acid esters such as tributyrin, but also hydrolyzed triolein in the presence of serum factor (Shirai, et al., 1985) and phospholipids (Shirai, et al., 1988). PchE is associated with β-VLDL or IDL in type III patient. But the role of PchE is not fully understood. We try to elucidate cholesterol ester hydrolyzing activity in PchE.

Above results suggested that PchE may hydrolyze cholesterol ester in IDL or β-VLDL, and released cholesterol is transferred to HDL, so as to promote the clearance of IDL. Midband observed in disc electrophoresis is mainly representative of IDL. In type II hyperlipidemias, midband was more frequently observed in low PchE group that in high PchE group. This result is also consistent with the idea that PchE may work on IDL cholesterol clearance.

References

Shirai, K., Saito, Y., Yoshida, S., *J. Biol. Chem.*, **1985**, 5225-5227.

Shirai, K., Saito, Y., Yoshida, S., *Biochim. Biophysis. Acta*, **1988**, *962*, 377-383.

Comparative Studies of Acetylcholinesterases from Plant Tissues and from Human Erythrocytes

O.V. Antonova, V.V. Kuusk, R.S. Agabekyan, I.V. Morozova, V.O. Chistyakov, R.G. Gafurov, Institute of Physiologically Active Substances, Academy of Science of USSR, Chernogolovka, Moscow Region, 142432, USSR

It was found that derivatives of N,N-dialkylpiperidine did not inactivate acetylcholinesterase from pea roots (AChE-p1), but they were reversible inhibitors for acetylcholinesterase from human erythrocytes. By means of AChE-p1 active site modifications it was shown that both carboxylate anion and tryptophan are essential for enzymatic hydrolysis of acetylthiocholine, while enzymatic hydrolysis of indophenylacetate and butylthioacetate was not sensitive to modifications of carboxylate anion.

A series of quaternary ammonium compounds such as N-(2-benz-oxyethyl)-triethylammonium iodide (I), N-(2-ethoxyethyethyl)trimethyl-ammonium chloride (II), N-benzyl-(N-2-benzoxy-ethyl)dimethylammonium chloride (III) and some derivatives of N,N-dialkylpiperidine (IV) have been synthesized and the kinetics of their interaction with AChE-p1 which was purified as described in (Solyakov et al., 1989) and AChE-er has been studied. It was found that I,II and III inactivate both enzymes reversibly with inhibition constant K_i equaled to $4.3x10^{-3}$ M, $3.7x10^{-2}$ M and $5.3x10^{-3}$ M for AChE-p1 and $1.9x10^{-4}$ M and $6.5x10^{-5}$ M for AChE-er, respectively. The IV did not inactivate AChE-p1, but they are reversible inhibitors of noncompetitive type for AChE-er (K_1 ranged from $1.5x10^{-4}$ M to $4.6x10^{-5}$ M). With the aim of understanding these data and the results of our preceding investigations (Solyakov et al., 1989), two kinds of active site modifications were made: of carboxylate anion and of tryptophan residue by N,N-dimethyl-2-phenylaziridinium ion (Purdie, J.E., 1969) and by N-bromsuccinimide (Privat, J-P. et al., 1976), respectively. It was shown that enzymatic hydrolysis of acetylthiocholine (ATCh) is affected by modification of carboxylate anion of AChE-p1, whereas hydrolysis of indophenyacetate (IPhA) and butylthioacetate (BuTA) is not sensitive to this modification (see Table 1).

Table 1. Michaelis Constants of Substrate Hydrolysis by Modified and Unmodified AChE-P1

Substrate	K_m (M) Modified	K_m (M) Unmodif.
ATCh	$8.3x10^{-4}$	$7.2x10^{-5}$
IPhA	$1.1x10^{-4}$	$1.0x10^{-4}$
BuTA	$1.3x10^{-3}$	$1.1x10^{-3}$

This data allow us to propose the presence of anionic center of active site of AChE-p1 (analogous to the one of AChE-er. The fact that modification of tryptophan residue inhibit enzymatic hydrolysis of ATCh allow us to conclude that this moity play an essential role in catalytic hydrolysis of specific substrates by AChE-p1.

References

Privat, J-P., Lotan, R. et al., Eur. J. Biochemistry, **1976**, 68, 563-572

Purdie, J.E., Biochim. et Biophys. acta, **1969**, 185, 122-133

Solyakov, L.S., Sablin, S.O., Kuusk, V.V., Agabekyan, R.S., Biochemistry USSR, **1989**, 54, No. 1, part 2, 67-72

0–8412–2008–5/91/0391$06.00/0

Possible Light-Sensitive Role of Acetylcholine (ACh) in the Regulation of a Lichen Growth

Margherita Raineri, Institute of Comparative Anatomy, 16132 Genova, Italy
Paolo Modenesi, Institute of Botany "Hanbury", 16136 Genova, Italy

In animal development ACh acts as a chemical messenger in the regulation of cell division, migration, differentiation and inductive signalling (Drews, 1975; Falugi & Raineri, 1985). A comparable morphogenetic role is documented in plants and even lichens (alga-fungus symbiosis).In the latter, a well-defined pattern of cholinesterase (ChE) activity in the alga and the fungus (Raineri & Modenesi, 1986), coincident with a pattern of higher Ca^{++} concentration (Raineri & Modenesi, 1988),is associated with the development of propagative structures, the soredia. This involves inductive-like interactions between the symbionts, which determine an oriented growth of medullary hyphae towards more actively dividing algae, and eventually the formation of a hyphal sheath around them.Likewise in animal cells (Oettling et al., 1985), the activation of an ACh receptor in the lichen may trigger a a free Ca^{++} rise, with a role in morphogenetic processes. A light sensitivity of the cholinergic mechanism is reported in animals, but mostly in plants, where ACh mimics some effects of the activated photopigment phytochrome (Miura & Shih, 1984).

We have recorded the effects of exogenous ACh, atropine, d-tubocurarine and inhibitors of acetylcholinesterase (AChE) and pseudocholinesterase (BuChE) in the lichen Parmelia caperata (L.)Ach. on thalli growing either in shaded or sunny habitat for a period of 3 months.

No effect was observed in shade, but in sunny habitat ACh-treated thalli became brownish at the growing edges of the lobes. Here light microscopy showed increased size of the algal chloroplast followed by enhanced, but atypical aplanosporogenesis which led to a thicker algal layer with a dramatic increase in number of the algae: $7.21 \times 10^6/cm^2$ vs. $2.76 \times 10^6/cm^2$ of the controls. Other structural changes, as those related to the differentiation of soredia,were not evident in the ACh-treated thalli, but peripherally to the areas of their greatest concentration the algae degenerated and died. In the same thalli histochemically detected ChE activity was much stronger than in controls in the areas of higher algal proliferation, but weaker or absent in those of algal degeneration.Substrates hydrolyzed preferentially by AChE gave more evident, eserine-sensitive enzyme reaction in the algae, mostly associated with cell membranes and the chloroplast. With substrates for BuChE a stronger enzyme staining scarcely inhibited by eserine was localized in fungal hyphae of the algal layer in close contact with the algae.

The results support a light-sensitive cholinergic regulation of the lichen morphogenesis, particularly of the algal division which also is the earliest visible sign of differentiation of soredia, and the presence of different cholinesterases in the symbionts (Raineri & Modenesi,1986).

References

Drews, U., Prg. Histochem. Cytochem.,1975, 7, 1-52.
Falugi, C., Raineri, M., J. Embryol. Exp., 1985,86, 89-108.
Miura, G.A., Shih, T-M., Physiol. Plant., 1984, 61, 417-421.
Oettling, G., Schmidt, H., Drews, U., J. Cell Biol., 1985, 100, 1073-1081.
Raineri, M., Modenesi, P., Histochem.J., 1986, 18, 647-657.
Raineri, M., Modenesi, P., Histochem. J., 1988, 20, 81-87.

*This research was supported by a fund (60%) of M.P.I.

0-8412-2008-5/91/0392$06.00/0

APPENDIX

Alignment of Amino Acid Sequences of Acetylcholinesterases and Butyrylcholinesterases

Mary K. Gentry and Bhupendra P. Doctor, Division of Biochemistry, Walter Reed Army Institute of Research, Washington, DC 20307–5100 USA

Sources of the sequences are as follows: fetal bovine serum acetylcholinesterase (FBS AChE), amino acid sequence, Doctor et al., 1990; human acetycholinesterase, from cDNA sequence, Soreq et al., 1990, GenBank accession number M55040; mouse acetylcholinesterase and mouse butyrylcholinesterase (BuChE), both deduced from cDNA sequences, Rachinsky et al., 1990; *Torpedo californica* acetylcholinesterase, from cDNA sequence, Schumacher et al., 1986, GenBank accession number X03439; *Torpedo marmorata* acetylcholinesterase, from cDNA sequence, Sikorav et al., 1987, GenBank accession number X05497; *Drosophila melanogaster* acetylcholinesterase, from cDNA sequence, Hall and Spierer, 1986, GenBank accession number X05893; human serum butyrylcholinesterase (BuChE), amino acid sequence, Lockridge et al., 1987; rabbit butyrylcholinesterase, from cDNA sequence, Jbilo and Chatonnet, 1990; rat liver carboxylesterase (CE), from cDNA sequence, Long et al., 1988, GenBank accession number M20629; rabbit liver carboxylesterase, Korza and Ozols, 1988; rat pancreas lysophospholipase (LPL), from cDNA sequence, Han et al., 1987, GenBank accession number M15893; *Drosophila* esterase–6 (E–6), from cDNA sequence, Oakshott et al., 1987, GenBank accession number M15961; *Dictyostelium discoideum* D2 gene from cDNA sequence, Rubino et al., 1989, GenBank accession number M15966; bovine thyroglobulin, from cDNA sequence, Mercken et al., 1985, GenBank accession number X02815; Glutactin, from cDNA sequence, Olson et al., 1990.

Acetylcholinesterases and butyrylcholinesterases were aligned using the multiple sequence alignment program MACAW (Schuler et al., 1990). The sequence of cholinesterase from fetal human tissues (Prody et al., 1987) has not been included since it is identical with the sequence for human serum butyrylcholinesterase. Sequences of the seven additional enzymes and proteins are presented for comparison but have not been aligned.

References

Doctor, B.P., Chapman, T.C., Christner, C.E., Deal, C.D., De La Hoz, D.M., Gentry, M.K., Ogert, R.A., Rush, R.S., Smyth, K.K., Wolfe, A.D., *FEBS Lett.*, **1990**, *266*, 123–127.

Hall, L.M.C., Spierer, P., *EMBO J.*, **1986**, *5*, 2949–2954.

Han, J.H., Stratowa, C., Rutter, W.J., *Biochemistry*, **1987**, *26*, 1617–1625.

Jbilo, O., Chatonnet, A., *Nucleic Acids Res.*, **1990**, *18*, 3990.

Korza, G., Ozols, J., *J. Biol. Chem.*, **1988**, *263*, 3486–3495.

Lockridge, O., Bartels, C.F., Vaughan, T.A., Wong, C.K., Norton, S.E., Johnson, L.L., *J. Biol. Chem.*, **1987**, *262*, 549–557.

Long, R.M., Satoh, H., Martin, B.M., Kimura, S., Gonzalez, F.J., Pohl, L.R., *Biochem. Biophys. Res. Commun.*, **1988**, *156*, 866–873.

Mercken, L., Simons, M.–J., Swillens, S., Massaer, M., Vassart, G., *Nature*, **1985**, *316*, 647–651.

Oakeshott, J.G., Collet, C., Phillis, R.W., Nielsen, K.M., Russell, R.J., Chambers, G.K., Ross, V., Richmond, R.C., *Proc. Natl. Acad. Sci. USA*, **1987**, *84*, 3359–3363.

Olson, P.F., Fessler, L.I., Nelson, R.E., Sterne, R.E., Campbell, A.G., Fessler, J.H., *EMBO J.*,**1990**, *9*, 1219–1227.

Prody, C.A., Zevin–Sonkin, D., Gnatt, A., Goldberg, O., Soreq, H., *Proc. Natl. Acad. Sci. USA*, **1987**, *84*, 3555–3559.

Rachinsky, T.L., Camp, S., Li, Y., Ekström, T.J., Newton, M., Taylor, P, *Neuron*, **1990**, *5*, 317–327.

Rubino, S., Mann, S.K.O., Hori, R.T., Pinko, C., Firtel, R.A., *Developmental Biol.*, **1989**, *131*, 27–36.

Schuler, G.D., Altschul, S.F., Lipman, D.J. In *Proteins: Structure, Function, and Genetics*; Alan R. Liss, Inc., New York, NY, **1990**. In press.

Schumacher, M., Camp, S., Maulet, Y., Newton, M., MacPhee–Quigley, K., Taylor, S.S., Friedmann, T., Taylor, P., *Nature*, **1986**, *319*, 407–409.

Sikorav, J.–L., Krejci, E., Massoulié, J., *EMBO J.*, **1987**, *6*, 1865–1873.

Soreq, H., Ben–Aziz, R., Prody, C.A., Seidman, S., Gnatt, A., Neville, L., Lieman–Hurwitz, J., Lev–Lehman, E., Ginzberg, D., Lapidot–Lifson, Y., Zakut, H., *Proc. Natl. Acad. Sci. USA*, **1990**. In press.

A

```
                                                                50
          FBS AChE   EGPEDPELLVMVRGGELRGLRLMAPRGPVSAFLGIPFAEPPVGPRRFLPP
        Human AChE   EGREDAELLVTVRGGRLRGIRLKTPGGPVSAFLGIPFAEPPMGPRRFLPP
        Mouse AChE   EGREDPQLLVRVRGGQLRGIRLKAPGGPVSAFLGIPFAEPPVGSRRFMPP
T. californica AChE     DDHSELLVNTKSGKVMGTRVPVLSSHISAFLGIPFAEPPVGNMRFRRP
  T. marmorata AChE     DDDSELLVNTKSGKVMGTRIPVLSSHISAFLGIPFAEPPVGNMRFRRP
    Drosophila AChE   VCGVIDRLVVQTSSGPVRGRSVTVQGREVHVYTGIPYAKPPVEDLRFRKP
 Human Serum BuChE       EDDIIIATKNGKVRGMNLTVFGGTVTAFLGIPYAQPPLGRLRFKKP
      Rabbit BuChE        EDVIITTKNGRIRGINLPVFGGTVTAFLGIPYAQPPLGRLRFKKP
       Mouse BuChE       EEDFIITTKTGRVRGLSMPVLGGTVTAFLGIPYAQPPLGSLRFKKP
                     __________________________________________________

      Rat Liver CE   HPSSPPVVDTTKGKVLGKYVSLEGFTQPVAVFLGVPFAKPPLGSLRFAPP
   Rabbit Liver CE   HPSAPPVVDTVKGKVLGKFVSLEGFAQPVAVFLGVPFAKPPLGSLRFAPP
           Rat LPL   AAACAAKLGALYTEGGFVEGVNKKLSLLGGDSVDIFKGIPFATAKTLENP
   Drosophila E-6    ASDTDDPLLVQLPQGKLRGRDNGSYYSYESIPYAEPPTGDLRFEAPEPYK
 Dictyostelium D2    MNKLVFILLLLLLINISFARKRSYIKNTDTSIVVTQFGAIKGIVEDTHRV
     Thyroglobulin   LLGRSQAIQVGTSWKPVDQFLGVPYAAPPLGEKRFRAPEHLNWTGSWEAT
         Glutactin   MKPLLLVLALCGAQVHAHSVGLRPDYNDYSDEDTRRDWLPEPLKPVPWQS
```

```
                                                               100
          FBS AChE   EPKRPWPGVLNATAFQSVCYQYVDTLYPGFEGTEMWNPNRELSEDCLYLN
        Human AChE   EPKQPWSGVVDATTFQSVCYQYVDTLYPGFEGTEMWNPNRELSEDCLYLN
        Mouse AChE   EPKRPWSGVLDATTFQNVCYQYVDTLYPGFEGTEMWNPNRELSEDCLYLN
T. californica AChE   EPKKPWSGVWNASTYPNNCQQYVDEQFPGFSGSEMWNPNREMSEDCLYLN
  T. marmorata AChE   EPKKPWSGVWNASTYPNNCQQYVDRQFPGFPGSEMWNPNREMSEDCLYLN
    Drosophila AChE   VPAEPWHGVLDATGLSATCVQERYEYFPGFSGEEIWNPNTNVSEDCLYIN
 Human Serum BuChE   QSLTKWSDIWNATKYANSCCQNIDQSFPGFHGSEMWNPNTDLSEDCLYLN
      Rabbit BuChE   QSLTKWSDIWNATKYANSCCQNIDQSFPGFHGSEMWNPNTDLSEDCLYLN
       Mouse BuChE   QPLNKWPDIHNATQYANSCYQNIDQAFPGFQGSEMWNPNTNLSEDCLYLN
                     __________________________________________________

      Rat Liver CE   EPAEPWSFVKNTTTYPPMCSQDGVVGKLLADMLSTGKESIPLEFSEDCLY
   Rabbit Liver CE   QPAESWSHVKNTTSYPPMCSSDAVSGHMLSELFTNRKENIPLKFSEDCLY
           Rat LPL   QRHPGWQGTLKATDFKKRCLQATITQDDTYGQEDCLYLNIWVPQGRKQVS
   Drosophila E-6    QKWSDIFDATKTPVACLQWDQFTPGANKLVGEEDCLTVSVYKPKNSKRNS
 Dictyostelium D2    FYGVPFAQPPVNQLRWENPIDLKPWENVRETLTQKSQCAQKCNLGPGVCS
     Thyroglobulin   KPRARCWQPGIRTPTPPGVSEDCLYLNVFVPQNMAPNASVLVFFHNAAEG
         Glutactin   ETRYAQPQEAVVQAPEVGQILGISGHKTIANRPVNAFLGIRYGTVGGGLA
```

B

```
                                                               150
          FBS AChE   VWTPYPRPSSPTPVLVWIYGGGFYSGASSLDVYDGRFLVQAEGTVLVSMN
        Human AChE   VWTPYPRPTSPTPVLVWIYGGGFYSGASSLDVYDGRFLVQAERTVLVSMN
        Mouse AChE   VWTPYPRPASPTPVLIWIYGGGFYSGAASLDVYDGRFLAQVEGAVLVSMN
T. californica AChE   IWVPSPRPKSTT-VMVWIYGGGFYSGSSTLDVYNGKYLAYTEEVVLVSLS
  T. marmorata AChE   IWVPSPRPKSAT-VMLWIYGGGFYSGSSTLDVYNGKYLAYTEEVVLVSLS
    Drosophila AChE   VWAPAKARLRHG-[1]ILIWIYGGGFMTGSATLDIYNADIMAAVGNVIVASFQ
 Human Serum BuChE   VWIPAPKPKNAT-VLIWIYGGGFQTGTSSLHVYDGKFLARVERVIVVSMN
      Rabbit BuChE   VWIPTPKPKNAT-VMIWIYGGGFQTGTSSLQVYDGKFLTRVERVIVVSMN
       Mouse BuChE   VWIPVPKPKNAT-VMVWIYGGGFQTGTSSLPVYDGKFLARVERVIVVSMN
                     __________________________________________________

      Rat Liver CE   LNIYSPADLTKNSRLPVMVWIHGGGLIIGGASPYSGLALSAHENVVVVTI
   Rabbit Liver CE   LNIYTPADLTKRGRLPVMVWIHGGGLMVGGASTYDGLALSAHENVVVVTI
           Rat LPL   HDLPVMVWIYGGAFLMGSGQGANFLKNYLYDGEEIATRGNVIVVTFNYRV
   Drosophila E-6    FPVVAHIHGGAFMFGAAWQNGHENVMREGKFILVKISYRLGPLGFVSTGD
 Dictyostelium D2    PMGTSEDSLYQDIFTPKDARPNSKYPVIVYIPGGAFSVGSGFLYLFMMLL
     Thyroglobulin   KGSGDRPAVDGSFLAAVGNLIVVTASYRTGIFGFLSSGSSELSGNWGLLD
         Glutactin   RFQAAQPIGYQGRVNATVQSPNCAQFPELDRLRLSESRGENVDDCLTLDI
```

Fig 1. Alignment of acetylcholinesterases and butyrylcholinesterases. The sequences of other similar enzymes and proteins are shown for comparison but are not aligned.

[1] A series of 33 amino acids has been omitted at this point from the *Drosophila* sequence in order to align it with the other sequences.

Continued on next page

```
                                                                                    200
           FBS AChE  YRVGAFGFLALPGSR------EAPGNVGLLDQRLALQSVQENVAAFGGDPTSVTLF
         Human AChE  YRVGAFGFLALPGSR------EAPGNVGLLDQRLALQWVQENVAAFGGDPTSVTLF
         Mouse AChE  YRVGTFGFLALPGSR------EAPGNVGLLDQRLALQWVQENIAAFGGDPMSVTLF
T. californica AChE  YRVGAFGFLALHGSQ------EAPGNVGLLDQRMALQWVHDNIQFFGGDPKTVTIF
  T. marmorata AChE  YRVGAFGFLALHGSQ------EAPGNMGLLDQRMALQWVHDNIQFFGGDPKTVTLF
    Drosophila AChE  YRVGAFGFLHLAPEMPSEFAEEAPGNVGLWDQALAIRWLKDNAHAFGGNPEWMTLF
 Human Serum BuChE  YRVGALGFLALPGNP------EAPGNMGLFDQQLALQWVQKNIAAFGGNPKSVTLF
      Rabbit BuChE  YRVGALGFLALPGNP------EAPGNMGLFDQQLALQWVQKNIAAFGGNPKSVTLF
       Mouse BuChE  YRVGALGFLAFPGNP------DAPGNMGLFDQQLALQWVQRNIAAFGGNPKSITIF

     Rat Liver CE  QYRLGFGGLFSTGDEHSRGNWAHLDQLAALRWVQDNIANFGGNPDSVTIF
  Rabbit Liver CE  QYRLGIGGFGFNIDELFLVAVNRWVQDNIANFGGDPGSVTIFGESAGGQS
           Rat LPL  GPLGFLSTGDANLPGNFGLRDQHMAIAWVKRNIAAFGGDPDNITIFGESA
    Drosophila E-6  RDLPGNYGLKDQRLALKWIKQNIASFGGEPQNVLLVGHSAGGASVHLQML
 Dictyostelium D2  NLLHSSVIVVNINYRLGVLGLMGTDLMHGNYGFLDQIKALEWVYNNIGFL
      Thyroglobulin  QVVALTWVQTHIQAFGGDPRRVTLAADRGGADIASIHLVTTRAANSRLFR
          Glutactin  YAPEGANQLPVLVFVHGEMLFDGGSEEAQPDYVLEKDVLLVSINYRLAPF
```

C

```
                                                                                250
           FBS AChE  GESAGAASVGMHLLSPPSRGLFHRAVLQSGAPNGPWATVGVGEARRRATL
         Human AChE  GESAGAASVGMHLLSPPSRGLFHRAVLQSGAPNGPWATVGMGEARRRATQ
         Mouse AChE  GESAGAASVGMHILSLPSRSLFHRAVLQSGTPNGPWATVSAGEARRRATL
T. californica AChE  GESAGGASVGMHILSPGSRDLFRRAILQSGSPNCPWASVSVAEGRRRAVE
  T. marmorata AChE  GESAGGASVGMHILSPGSRDLFRRAILQSGSPNCPWASVSVAEGRRRAVE
    Drosophila AChE  GESAGSSSVNAQLMSPVTRGLVKRGMMQSGTMNAPWSHMTSEKAVEIGKA
 Human Serum BuChE  GESAGAASVSLHLLSPGSHSLFTRAILQSGSFNAPWAVTSLYEARNRTLN
      Rabbit BuChE  GESAGAASVSLHLLSPRSHPLFTRAILQSGSSNAPWEVMSLHEARNRTLT
       Mouse BuChE  GESAGAASVSLHLLCPQSYPLFTRAILESGSSNAPWAVKHPEEARNRTLT

     Rat Liver CE  GESAGGVSVSALVLSPLAKNLFHRAISESGVVLTTNLDKKNTQAVAQMIA
  Rabbit Liver CE  VSILLLSPLTKNLFHRAISESGVALLSSLFRKNTKSLAEKIAIEAGCKTT
           Rat LPL  GGAIVSLQTLSPYNKGLIRRAISQSGVALSPWAIQENPLFWAKTIAKKVG
    Drosophila E-6  REDFGQLARAAFSFSGNALDPWVIQKGARGRAFELGRNVGCESAEDSTSL
 Dictyostelium D2  GGNKEMITIWGESAGAFSVSAHLTSTYSRQYFNAAISSSSPLTVGLKDKT
      Thyroglobulin  RAVLMGGSALSPAAVIRPERARQQAAALAKEVGCPSSSVQEMVSCLRQEP
          Glutactin  GFLSALTDELPGNVALSDLQLALEWLQRNVVHFGGNAGQVTLVGQAGGAT
```

```
                                                                                300
           FBS AChE  LARLVGCPPGGAGGNDTELVACLRARPAQDLVDHEWRVLPQEHVFRFSFV
         Human AChE  LAHLVGCPPGGTGGNDTELVACLRTRPAQVLVNHEWHVLPQESVFRFSFV
         Mouse AChE  LARLVGCPPGGAGGNDTELIACLRTRPAQDLVDHEWHVLPQESIFRFSFV
T. californica AChE  LGRNLNC----NLNSDEELIHCLREKKPQELIDVEWNVLPFDSIFRFSFV
  T. marmorata AChE  LGRNLNC----NLNSDEELIQCLREKKPQELIDVEWNVLPFDSIFRFSFV
    Drosophila AChE  LINDCNCNASMLKTNPAHVMSCMRSVDAKTISVQQWN--SYSGILSFPSA
 Human Serum BuChE  LAKLTGC----SRENETEIIKCLRNKDPQEILLNEAFVVPYGTPLSVNFG
      Rabbit BuChE  LAKFVGC----STENETEIIKCLRNKDAQEILLNEVFVVPFDSLLSVNFG
       Mouse BuChE  LAKFTGC----SKENEMEMIKCLRSKDPQEILRNERFVLPSDSILSINFG

     Rat Liver CE  TLSGCNNTSSAAMVQCLRQKTEAELLELTVKLDNTSMSTVIDGVVLPKTP
  Rabbit Liver CE  TSAVMVHCLRQKTEEELMEVTLKMKFMALDLVGDPKENTAFLTTVIDGVL
           Rat LPL  CPTEDTAKMAGCLKITDPRALTLAYRLPLKSQEYPIVHYLAFIPVVDGDF
    Drosophila E-6  KKCLKSKPASELVTAVRKFLIFSYVPFAPFSPVLEPSDAPDAIITQDPRD
 Dictyostelium D2  TARGNANRFATNVGCNIEDLTCLRGKSMDEILDGPRKIGLTFGYKILDAF
      Thyroglobulin  ARILNDAQTKLLAVSGPFHYWGPVVDGQYLRETPARVLQRAPRVKVDLLI
          Glutactin  LAHALSLSGRAGNLFQQLILQSGTALNPYLIDNQPLDTLSTFARLARCPP
```

Fig 1. Continued

D

	350
FBS AChE	PVVDGDFLSDTPEALINAGDFVGLQVLVGVVKDEGSYFLVYGAPG-FSKDN
Human AChE	PVVDGDFLSDTPEALINAGDFHGLQVLVGVVKDEGSYFLVYGAPG-FSKDN
Mouse AChE	PVVDGDFLSDTPEALINTGDFQDLQVLVGVVKDEGSYFLVYGVPG-FSKDN
T. californica AChE	PVIDGEFFPTSLESMLNSGNFKKTQILLGVNKDEGSFFLLYGAPG-FSKDS
T. marmorata AChE	PVIDGEFFPTSLESMLNAGNFKKTQILLGVNKDEGSFFLLYGAPG-FSKDS
Drosophila AChE	PTIDGAFLPADPMTLMKTADLKDYDILMGNVRDEGTYFLLYDFIDYFDKDD
Human Serum BuChE	PTVDGDFLTDMPDILLELGQFKKTQILVGVNKDEGTAFLVYGAPG-FSKDN
Rabbit BuChE	PTVDGDFLTDMPDTLLQLGQLKKTQILVGVNKDEGTAFLVYGAPG-FSKDN
Mouse BuChE	PTVDGDFLTDMPHTLLQLGKVKKAQILVGVNKDEGTAFLVYGAPG-FSKDN
Rat Liver CE	EEILTEKSFNTVPYIVGFNKQEFGWIIPTMMGNLLSEGRMNEKMASSFLK
Rabbit Liver CE	LPKAPAEIYEEKKYNMLPYMVGINQQEFGWIIPMQMLGYPLSEGKLDQKT
Rat LPL	IPDDPINLYDNAADIDYLAGINDMDGHLFATVDVPAIDKAKQDVTEEDFY
Drosophila E-6	VIKSGKFGQVPWAVSYVTEDGGYNAALLLKERKSGIVIDDLNERWLELAP
Dictyostelium D2	TIWSPVIDGDIIPMQTLTDSKGRSKHMMFQHYWKCKHEAIPFIYSFSKIV
Thyroglobulin	GSSQDDGLINRAKAVKQFEESQGRTSSKTAFYQALQNSLGGEAADAGVQA
Glutactin	PSINPSAQGLKPLYDCLARLPTSQLVAAFEQLLLQNEHLGLTQLGGFKLV

	400
FBS AChE	ESLISRAQFLAGVRVGVPQASDLAAEAVVLHYTDWLHPEDPARWREALSD
Human AChE	ESLISRAEFLAGVRVGVPQVSDLAAEAVVLHYTDWLHPEDPARLREALSD
Mouse AChE	ESLISRAQFLAGVRIGVPQASDLAAEAVVLHYTDWLHPEDPTHLRDAMSA
T. californica AChE	ESKISREDFMSGVKLSVPHANDLGLDAVTLQYTDWMDDNNGIKNRDGLDD
T. marmorata AChE	ESKISREDFMSGVKLSVPHANDLGLDAVTLQYTDWMDDNNGIKNRDGLDD
Drosophila AChE	ATALPRDKYLEIMNNIFGKATQAEREAIIFQYTSW-EGNPGYQNQQQIGR
Human serum BuChE	NSIITRKEFQEGLKIFFPGVSEFGKESILFHYTDWVDDQRPENYREALGD
Rabbit BuChE	TSIITRKEFQEGLKIFFPGVSEFGKESILFHYTDWVDEQRPENYREALDD
Mouse BuChE	DSLITRKEFQEGLNMYFPGVSRLGKEAVLFYYVDWLGEQSPEVYRDALDD
Rat Liver CE	RFSPNLNISESVIPAIIEKYLRGTDDPAKKKELLLDMFSDVFFGIPAVLM
Rabbit Liver CE	ATELLWKSYPIVNVSKELTPVATEKYLGGTDDPVKKKDLFLDMLADLLFG
Rat LPL	RLVSGHTVAKGLKGTQATFDIYTESWAQDPSQENMKKTVVAFETDILFLI
Drosophila E-6	YLLFYRDTKTKKDMDDYSRKIKQEYIGNQRFDIESYSELQRLFTDILFKN
Dictyostelium D2	VGIDYYRVLVAIVFPLNAMKILPLYPAAPRGQDSRPILSELITDYLFRCP
Thyroglobulin	AATWYYSLEHDSDDYASFSRALEQATRDYFIICPVIDMASHWARTVRGNV
Glutactin	VGDPLGFLPSHPASLATNSSLALPMIIGATKDASAFIVSRIYDQLARLQS

E

	450
FBS AChE	VVGDHNVVCPVAQLAGRLAAQGARVYAYIFEHRASTLSWPLWMGVPHGYE
Human AChE	VVGDHNVVCPVAQLAGRLAAQGARVYAYVFEHRASTLSWPLWMGVPHGYE
Mouse AChE	VVGDHNVVCPVAQLAGRLAAQGARVYAYIFEHRASTLTWPLWMGVPHGYE
T. californica AChE	IVGDHNVICPLMHFVNKYTKFGNGTYLYFFNHRASNLVWPEWMGVIHGYE
T. marmorata AChE	IVGNHNVICPLMHFVNKYTKFGNGTYLYFFNHRASNLVWPEWMGVIHGYE
Drosophila AChE	AVGDHFFTCPTNEYAQALAERGASVHYYYFTHRTSTSLWGEWMGVLHGDE
Human Serum BuChE	VVGDYNFICPALEFTKKFSEWGNNAFFYYFEHRSSKLPWPEWMGVMHGYE
Rabbit BuChE	VVGDYNFICPALEFTKKFSEWGNNAFFYYFEHRSSKLPWPEWMGVMHGYE
Mouse BuChE	VIGDYNIICPALEFTKKFAELENNAFFYFFEHRSSKLPWPEWMGVMHGYE
Rat Liver CE	SRSLRDAGAPTYMYEFQYRPSFVSDQRPQTVQGDHGDEIFSVFGTPFLKE
Rabbit Liver CE	VPSVNVARHHRDAGAPTYMYEYRYRPSFSSDMRPKTVIGDHGDEIFSVLG
Rat LPL	PTEMALAQHRAHAKSAKTYSYLFSHPSRMPIYPKWMGADHADDLQYVFGK
Drosophila E-6	STQESLDLHRKYGKSPAYAYVYDNPAEKGIAQVLANRTDYDFGTVHGDDY
Dictyostelium D2	DRYHTVTNAKKLSSPTYHYHYVHVKSTGHSLDACDDKVCHGTELSLFFNS
Thyroglobulin	FMYHAPESYSHSSLELLTDVLYAFGLPFYPAYEGQFTLEEKSLSLKIMQY
Glutactin	RNVSDYIDVVLRHTAPPSEHRLWKQWALREIFTPIQEQTASLQTVAPGLL

Fig 1. Continued

Continued on next page

```
                                                                                          500
               FBS AChE   IEFIFGLPLEPSLNYTIEERTFAQRLMRYWANFARTGDPNDPRAPKAPQW
             Human AChE   IEFIFGIPLDPSRNYTAEEKIFAQRLMRYWANFARTGDPNEPRDPKAPQW
             Mouse AChE   IEFIFGLPLDPSLNYTTEERIFAQRLMKYWTNFARTGDPNDPRDSKSPQW
     T. californica AChE   IEFVFGLPLVKELNYTAEEEALSRRIMHYWATFAKTGNPNEPHSQES-KW
       T. marmorata AChE   IEFVFGLPLVKELNYTAEEEALSRRIMHYWATFAKTGNPNEPHSQES-KW
         Drosophila AChE   IEYFFGQPLNNSLQYRPVERELGKRMLSAVIEFAKTGNPAQD----GEEW
     Human Serum BuChE   IEFVFGLPLERRDNYTKAEEILSRSIVKRWANFAKYGNPNETQNNST-SW
          Rabbit BuChE   IEFVFGLPLERRVNYTKAEEILSRSIMKRWANFAKYGNPNGTQNNST-RW
           Mouse BuChE   IEFVFGLPLGRRVNYTRAEEIFSRSIMKTWANFAKYGHPNGTQGNST-MW

          Rat Liver CE   GASEEETNLSKLVMKFWANFARNGNPNGEGLPHWPKYDQKEGYLQIGATT
       Rabbit Liver CE   APFLKEGATEEEIKLSKHVMKYWANFARNGNPNGEGLPQWPAYDYKEGYL
               Rat LPL   PFATPLGYRAQDRTVSKAMIAYWTNFAKSGDPNMGNSPVPTHWYPYTMEN
       Drosophila E-6   FLIFENFVRDVEMRPDEQIISRNFINMLADFASSDNGSLKYGECDFKDSV
     Dictyostelium D2   YELMGERLDNDEKELAIDINNYIVNLQLLINPNTGLDVPVQWRQVTCTQN
         Thyroglobulin   FSNFIRSGNPNYPHEFSRRAPEFAAPWPDFVPRDGAESYKELSVLLPNRQ
             Glutactin   ELSNYILYRAPYINSISQSYRSVPAYLYTFDYRGEHHRFGHLSNPLPFGV
```

F

```
                                                                                          550
               FBS AChE   PPYTAGAQQYVSLNLRPLGVPQASRAQA--CAFWNRFLPKLLNATDTLDEAE
             Human AChE   PPYTAGAQQYVSLDLRPLEVRRGLRAQA--CAFWNRFLPKLLSATDTLDEAE
             Mouse AChE   PPYTTAAQQYVSLNLKPLEVRRGLRAQT--CAFWNRFLPKLLSATDTLDEAE
     T. californica AChE   PLFTTKEQKFIDLNTEPMKVHQRLRVQM--CVFWNQFLPKLLNATETIDEAE
       T. marmorata AChE   PLFTTKEQKFIDLNTEPIKVHQRLRVQM--CVFWNQFLPKLLNATETIDEAE
         Drosophila AChE   PNFSKEDPVYYIFSTDDKIEKLARGPLAARCSFWNDYLPKVRSWAGTCDGDS
     Human Serum BuChE   PVFKSTEQKYLTLNTESTRIMTKLRAQQ--CRFWTSFFPKVLEMTGNIDEAE
          Rabbit BuChE   PVFKSTEQKYLTLNTESPRIYTKLRAQQ--CRFWTLFFPKVLEMTGNIDEAE
           Mouse BuChE   PVFTSTEQKYLTLNTEKSKIYSKLRAPQ--CQFWRLFFPKVLEMTGDIDETE

          Rat Liver CE   QQAQKLKGEEVAFWTELLAKNPPQTEHTEHT
       Rabbit Liver CE   QIGATTQAAQKLKDKEVAFWTELWAKEAARPRETEHIEL
               Rat LPL   GNYLDINKKITSTSMKEHLREKFLKFWAVTFEMLPTVVGDHTPPEDDSEA
       Drosophila E-6   GSEKFQLLAIYIDAARIGSMWNFRKLHE
     Dictyostelium D2   STLILETTIETKVSFTNDPKCNALDLTYYRNQVRPU
         Thyroglobulin   GLKKADCSFWSKYIQSLKASADETKDGPSADSEEEDQPAGSGLTEDLLGL
             Glutactin   DASLSDDSVYLFPYPPEASRLNPLDRSLSRALVTMWVNFATTGVPNPSSG
```

```
                                                        583
               FBS AChE   RQWKAEFHRWSSYMVHWKNQF-DHYSKQDRCSDL
             Human AChE   RQWKAEFHRWSSYMVHWKNQF-DHYSKQDRCSDL
             Mouse AChE   RQWKAEFHRWSSYMVHWKNQF-DHYSKQERCSDL
     T. californica AChE   RQWKTEFHRWSSYMMHWKNQF-DHYSRHESCAEL
       T. marmorata AChE   RQWKPEFHRWSSYMMHWKNQF-DQYSRHENCAEL
         Drosophila AChE   GSASISPRLQLLGIAALIYICAALRTKRVF
     Human Serum BuChE   WEWKAGFHRWNNYMMDWKNQFNDYTSKKESCVGL
          Rabbit BuChE   QEWKAGFHRWNNYMMAWKNHFNDYTSKKERCAGF
           Mouse BuChE   QEWKAGFHRWSNYMMDWQNQFNDYTSKKESCTAL

               Rat LPL   APVPPTDDSQGGPVPPTDDSQTTPVPPTDNSQA
         Thyroglobulin   PELASKTYSK
             Glutactin   VWPQATSEYGPFLRFTNNQQSPLELDPHFGEGIYLPNYRVIYKPTTNFSP
```

Fig 1. Continued

INDEX

Index

D

E

F

G

H

N

Production: Margaret J. Brown
Indexing: Deborah H. Steiner
Acquisition: Robin M. Giroux
Cover design: Amy Meyer Phifer

Printed and bound by Maple Press, York, PA

Author FAX Numbers*

Adler, M., 301–676–7045
Amitai, G., 972–8–401404
Appleyard, Margaret E., 0–71–433–1921
Arnaud, J., 61–31–97–52
Asianian, Dimitrina, 33–1–43–25–47–59
Atack, John R., 0279–440390
Babu, S. R., 217–782–0988
Bacou, Francis, 33–67–54–56–94
Bartels, Cynthia, 313–763–4450
Berman, Harvey Alan, 716–636–3850
Brimijoin, S., 507–284–9111
Brock, Axel, 45–86434930
Brock, V., 45–65930457
Brzin, M., 61–311–540
Bütikofer, P., 41–31–65–37–37
Busker, R. W., 31–15–843989
Chatonnet, Arnaud, 67–54–56–94
Davis, Robert E., 313–996–7178
De Bisschop, H. C. J. V., 32–2–7352421
de Jong, L. P. A., 31–15–843991
Dettbarn, Wolf-D., 615–343–7588
Doctor, Bhupendra P., 202–576–1304
Ehret-Sabatier, L., 33–88–66–01–90
Eichler, Jerry, 972–2–794010
Espinoza, Bertha, 972–8–466966
Falugi, C., 010–3538047
Fernandez, Hugo L., 816–861–1110
Friboulet, Alain, 33–44–20–39–10
Giacobini, Ezio, 217–782–0988
Gisiger, Victor, 514–343–2459
Grassi, J., 1–69–08–73–00
Greenfield, Susan A., 865–275161
Grubič, Z., 61–311–540
Heider, Harald, 41–31–65–37–37
Hodges-Savola, Cheryl A., 816–861–1110
Jäger, Karin, 41–31–65–37–37
Jensen, Frank Samsøe, 00945–35–37–66–45
Johnson, Carl D., 617–225–2741
Joron, Laurent, 33–44–20–39–10
Kambam, J. R., 615–343–7246
Kesvatera, T., 0142–52–95–79
Khaskiye, A., 33–40–29–32–51
Koelle, George B., 215–898–7729
Kounenis, Georgios, 30–31–206138
Laasberg, Tiit, 0142–529579
La Du, Bert N., 313–763–4450
Layer, Paul G., 7071–601–300
Lenz, D. E., 301–676–7045
Liao, Jian, 41–31–65–37–37
Lockridge, Oksana, 402–559–4238
Marty, Jean-Louis, 33–68–67–12–21
Masson, Patrick, 33–76–54–57–37
Massoulié, Jean, 33–1–43–29–81–72
Mastrolia, L., 761–225785
Maulet, Yves, 06221–56–68–09
Maxwell, D. M., 301–676–7045
Mazzanti, L., 071–5893667
Morelis, R. M., 33–72–44–28–34
Mutero, Annick, 33–93–67–88–25
Nio, Jacques, 33–64–93–52–66
Nordberg, A., 46–18–559718
Nordgren, Ingrid, 46–8–33–44–67
Olivieri-Sangiacomo, C., 39–6–3388066
Palumaa, Peep, 014–34–35440
Pezzementi, Leo, 33–1–43–29–81–72
Principato, Giovanni B., 075–585–2067
Quinn, Daniel M., 319–335–2951
Raba, R., 0142–529579
Radić, Zoran, 38–41–274–572
Raineri, Margherita, 010–3538047
Rakonczay, Zoltan, 36–62–26–444
Ramírez, Galo, 34–1–3974799
Randall, William R., 301–328–3991
Rasmussen Loft, A. G., 45–3195–2869
Reiner, Elsa, 38–41–274–572
Robertson, Richard T., 714–856–8549
Rosenberry, Terrone L., 216–368–3395
Rotundo, Richard L., 305–545–7166
Scarella, G., 39–6–49912351
Schwarz, R. D., 313–996–7178
Seto, Yasuo, 81–3–261–9986
Sharma, M., 415–476–9516
Silman, Israel, 972–8–466–966
Simeon, Vera, 38–41–274–572
Sine, J-P., 33–40–74–50–00
Skau, Kenneth A., 513–558–4372
Sket, Dušan, 61–311–540
Šketelj, J., 61–311–540
Škrinjarić-Špoljar, Mira, 38–41–274–572
Small, David H., 61–3–344–4004
Smith, Margaret E., 021–414–6924
Soreq, Hermona, 972–2–666804
Spinedi, A., 39–6–2493500
Sussman, Joel, 972–8–466–966
Svensson, Leif-Å., 46–46–166667
Taylor, Palmer, 619–534–6833
Testylier, Guy, 33–76–54–57–35
Toutant, Jean-Pierre, 33–67–63–28–02
Vallette, François-Marie, 33–1–43–29–81–72
Vidal, Cecilio J., 34–68–835418
Weber, Michel, 61–17–59–93
Weise, Christoph, 030–838–64–03
Yassine, Mohamed, 33–67–54–56–94
Zador, Erno, 041–330–5994

*These numbers were provided by the authors.